Specialist Periodical Reports Online

Application for Free Access to Electronic Chapters

Spectroscopic Properties of Inorganic and Organometallic Compounds

Volume 43

Customers purchasing a print volume are now entitled to free site-wide access to the electronic version of that title.
(For the definition of a site, please consult **www.rsc.org/subagree**)

To apply for free access, please complete and return this form.

Contact Name: ______________________________

Job Title: ______________________________

Organisation: ______________________________

Address: ______________________________

____

____

Town: ______________________________

Country: ______________________________

Post/Zip Code: ______________________________

E-mail: ______________________________

Telephone: ______________________________

Continued over

Please give IP addresses of the site overleaf in the format w.x.y.z.

Wildcards and ranges are allowed

e.g. for Class B 128.128.*.*

for Class C 192.192.192.*

for a range 123.456.1-99.*

Please put each entry on a new line:

____

____

____

____

(please use a continuation page if necessary)

I have read and agree to the terms of the Electronic Information Licence Agreement at **www.rsc.org/subagree**

Signed: ____________________ **Date:** _______________

☐ Please contact me about access for additional sites outside the site definition in the Electronic Information Licence Agreement

☐ Please arrange username/password access and not IP address control

Please return this form by non-electronic means to the Royal Society of Chemistry at the address below.

Photocopies or facsimiles are not acceptable.

Books Dept

Royal Society of Chemistry · Thomas Graham House

Science Park · Milton Road · Cambridge · CB4 0WF · UK

T +44(0)1223 420066 · **F** +44(0)1223 420247

E SPR@rsc.org

The Royal Society of Chemistry will store the information you supply on its electronic records in order that information about its activities, products and services may be sent to you by mail, telephone, email or fax.

If you DO NOT wish to receive information, please put a tick in the box ☐.

Spectroscopic Properties of Inorganic and Organometallic Compounds

Techniques, Materials and Applications

Volume 43

A Specialist Periodical Report

Spectroscopic Properties of Inorganic and Organometallic Compounds

Techniques, Materials and Applications
Volume 43

A review of the recent literature

Editors
J. Yarwood, *Sheffield Hallam University, UK*
R. Douthwaite, *University of York, UK*
S. B. Duckett, *University of York, UK*

Authors
Andrew M. Beale, *Utrech University, The Netherlands*
Natalia V. Belkova, *Russian Academy of Sciences, Russia*
Wim Bras, *Netherlands Organization for Scientific Research*
Hergen Breitzke, *Technical University Darmstadt, Germany*
Gerd Buntkowsky, *Technical University Darmstadt, Germany*
C. Casiraghi, *University of Manchester, UK*
Christophe Copéret, *ETH Zürich, Switzerland*
Keith B. Dillon, *Durham University, UK*
Trevor J. Dines, *University of Dundee, UK*
Sylvia M. Draper, *University of Dublin, Ireland*
Paul Dumas, *Synchrotron SOLEIL, France*
Lina M. Epstein, *Russian Academy of Sciences, Russia*
Oleg A. Filippov, *Russian Academy of Sciences, Russia*
David Gajan, *ETH Zürich, Switzerland*
R. Van Grieken, *University of Antwerp, Belgium*
Anna Grünberg, *Technical University Darmstadt, Germany*
Giuseppe Lapadula, *ETH Zürich, Switzerland*
Robert C. Maher, *Imperial College London, UK*
Gearóid M. Ó Máille, *University of Dublin, Ireland*
Michael C. Martin, *Lawrence Berkeley National Laboratory, USA*
J. H. Potgieter, *Manchester Metropolitan University, UK*
S. Potgieter-Vermaak, *Manchester Metropolitan University, UK*
Martin Schwarzwälder, *ETH Zürich, Switzerland*
Elena S. Shubina, *Russian Academy of Sciences, Russia*

RSC Publishing

If you buy this title on standing order, you will be given FREE access to the chapters online. Please contact sales@rsc.org with proof of purchase to arrange access to be set up.

Thank you.

ISBN: 978-1-84973-395-3
ISSN: 0584-8555
DOI: 10.1039/9781849734899

A catalogue record for this book is available from the British Library

Published by The Royal Society of Chemistry,
Thomas Graham House, Science Park, Milton Road,
Cambridge CB4 0WF, UK

Registered Charity Number 207890

For further information see our web site at www.rsc.org

Preface

DOI: 10.1039/9781849734899-FP005

In keeping with the changes introduced in 2009 to this volume of the series, the current edition encompasses predominantly critical reviews from a wide spectrum of inorganic materials science and spectroscopy. The reader should be able to find unfamiliar material described in an accessible style, providing the critical context and important recent work in the field.

In this volume the articles include techniques and applications ranging from 15th Century steel analysis to thermally sensitive small molecules that interact *via* some of the weakest hydrogen derived chemical bonds. Advances in spectroscopic methods are partially determined by developments in available instrumentation. For example, combining multiple techniques in a single experiment is potentially beneficial and some instructive examples are provided by time resolved X-ray techniques that take advantage of developments at synchrotron facilities. NMR, IR and Raman spectroscopies feature heavily in this issue particularly for catalytic applications and related technologies such as fuel cells. The characterisation and investigation of some new materials including graphene and the photophysics of molecular switches is also described as well as particles prevalent in the environment.

Collectively the articles represent a wide cross section of activity in the field, however there are clearly common themes to the spectroscopy applied to a diverse range of materials. We hope the reader will at least learn something new or find an alternative perspective to a familiar topic and perhaps extract a little inspiration.

The editors are of course indebted to all our authors for agreeing to contribute and submitting their articles in time for publication. The editing for the majority of the articles was uneventful and of course we take responsibility for any errors and apologise in advance. We also thank Dr Merlin Fox and Mrs Alice Toby-Brant of the RSC for sending us timely reminders and keeping us on track during the commissioning and editing processes.

Jack Yarwood
Richard Douthwaite
Simon Duckett

CONTENTS

Cover

The cover shows the molecular structure of a molybdenum dioxocomplex with thioester functionalised amido ligands superimposed on a spectral background. Adapted from an image created by Nadia C. Mösch-Zanetti and reproduced with permission.

IR spectroscopy of hydrides and its application to hydrogen bonding and proton transfer studies

Natalia V. Belkova,*† Lina M. Epstein, Oleg A. Filippov and Elena S. Shubina

DOI: 10.1039/9781849734899-00001

This review is devoted to the IR spectra of transition metal and main group element hydrides focusing on the ν_{MH} bands and their response to hydrogen bonding and proton transfer. The aim is to show what advantages can be provide by variable temperature IR spectroscopy in structure and reactivity studies of hydrides.

Introduction

Transition metal and main group element hydrides keep attracting research interest because of their importance in different chemical and biochemical processes. Their chemical behaviour, and transformations, are usually studied by NMR spectroscopy. IR is employed mostly for traditional compound characterisation upon synthesis, however it provides information that would not otherwise be readily obtainable.[1] Indeed ^{1}H NMR is very convenient for studying hydrides since hydride resonances appear in a strong field without overlap with other hydrogens and there are distinct signals of both classical and non-classical (η^2-H_2) complexes. Development and availability of variable temperature NMR set-ups gave another advantage to NMR spectroscopy and made it a widespread (almost routine) technique. In contrast, variable temperature IR measurements are still not widely used. However IR spectroscopy operates at a different time-scale (10^{-11}–10^{-14} s) and thus can be used to study processes, which are too fast on the NMR timescale (10^1–10^{-6} s).‡ This includes regular IR or FTIR spectroscopy not to mention recently developed time resolved IR techniques.[1] Among the processes conveniently studied by IR spectroscopy are complexation (formation of intermolecular adducts) and proton transfer. The latter is an important step in many catalytic or stoichiometric transformations of hydrides and is one of our research interests.

The aim of this report is to show what advantages can be provided by variable temperature IR spectroscopy and our approach to study the reactivity of transition metal and main group element hydrides. We focus on the ν_{MH} bands and their response to hydrogen bonding and proton transfer, but will also mention behaviour of other bands such as ν_{OH}, ν_{CO} *etc*. These

A. N. Nesmeyanov Institute of Organoelement Compounds, Russian Academy of Sciences, 28 Vavilov Street, 119991 Moscow, Russia; E-mail: nataliabelk@ineos.ac.ru

† The authors thank the Russian Foundation for Basic Research (projects 11-03-01210, 10-03-00960) and the Division of Chemistry and Material Science of RAS for financial support of their recent work included herein.

‡ Another circumstance, which often unfavorably distinguishes NMR from IR, is the natural demand of the method for higher concentrations and the use of deuterated solvents.

studies gain a lot from combination with NMR and quantum chemical and frequency calculations, to which we will refer as well.

M-H stretching vibrations

Transition metal complexes containing terminal hydrido ligands exhibit stretching (ν_{MH}) and deformational (δ_{MH}) vibrations at 2200–1600 and 800–600 cm^{-1}, respectively.[2] The δ_{MH} vibrations are not characteristic but mixed with other modes.[3,4] Their region is more difficult to access due to the overlap with vibrations of other ligands and solvents commonly used in inorganic/organometallic chemistry. Therefore δ_{MH} will not be discussed here.

In the case of di- and polyhydrides the first question is the assignment of the ν_{MH} bands. Calculations of simple homoleptic hydrides (MH_2 and $MH_2 \cdot H_2$ complexes (M = Ti, V, Cr) and comparison with the matrix isolation IR measurements have shown that symmetric (ν^s) and antisymmetric (ν^{as}) M-H stretching vibrational bands in IR spectra can appear near to each other and differ widely (from 2 to 10 times) in intensity, the intensity of the ν^s_{MH} band being lower than that of the ν^{as}_{MH} band.[5] This work has been expanded by L. Andrews and co-workers who a trapped variety of homoleptic metal hydrides obtained by the reaction of transition metal atoms (from group 3 to 12) with dihydrogen for matrix isolation IR spectroscopic studies.[6–10] However, these "naked" metal hydrides and dihydrogen complexes are fundamentally different from organometallic hydride complexes. Being unsaturated, they exhibit high reactivity, *e.g.* interact differently with molecular H_2. The spectroscopic and theoretical data on these compounds are discussed in a review.[11] They confirm the early observations (above) that very often only more intense ν^{as} bands are observed in the IR spectra of hydrides.

The analysis of IR and Raman spectra of bis-cyclopentadienyl metal hydrides Cp_2MH_n (Cp = η^5-C_5H_5; M = Re (n = 1); Mo, W (n = 2); Nb, Ta (n = 3))[3] having C_{2v} symmetry led to the conclusion that for Cp_2MH_2 only symmetric stretching vibrations appear in both spectra. The measurements for (η^5-C_5D_5)$_2MH_2$ derivatives have shown very little $\nu^s_{MH_2}$ band shift as a result of the Cp-ring deuteration evidencing the absence of vibrational coupling with other modes. However, the authors noticed unusually high bandwidth (in solution) and splitting (in matrices) of ν_{MH} bands for almost all the spectra measured. The proposed explanation for this observation was the steric interference with the Cp-rings during M-H vibrations that broadens MH stretching bands inhomogeneously in solution or splits them in matrices when Cp rotation is frozen.[3] Recently we reported for the first time on the observation of such conformational equilibrium for $[Cp^*MoH(PMe_3)_3]^{n+}$ (Cp* = η^5-C_5Me_5; n = 0, 1).[12] The two temperature and solvent dependant ν_{MH} bands in IR spectra (Fig. 1 gives an example of the neutral hydride) were assigned to the two rotamers originating from hindered rotation of the Cp*-ring. Despite the seemingly small difference in the geometry of the two rotamers (change of C-ring center-Mo-H dihedral angle by *ca.* 20°) the Mo-H bond lengths and the calculated ν_{MoH} frequencies are different (Fig. 2). A detailed look into the structure reveals

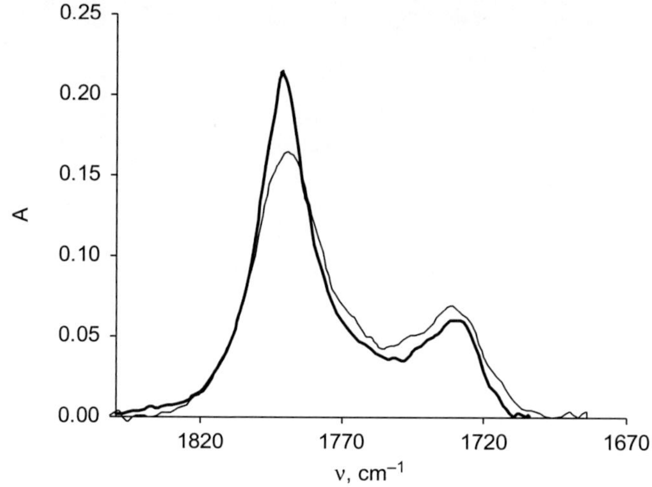

Fig. 1 Infrared spectra of $Cp^*Mo(PMe_3)_3H$ (0.08 M, 0.4 mm pathlength) at 200 K in hexane (thinner line) and THF (thicker line). Adapted with permission from ref. 13. Copyright © 2009 American Chemical Society.

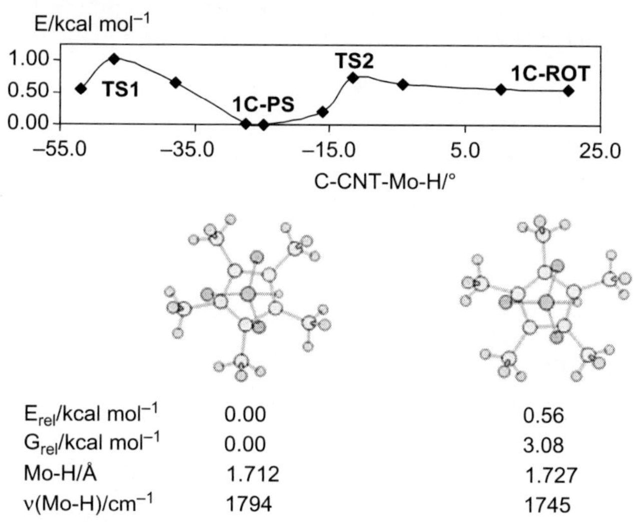

Fig. 2 Energy of $Cp^*Mo(PMe_3)_3H$ as a function of the C-CNT-Mo-H dihedral angle. Views of the optimized minima emphasizing the Cp* dihedral angle (the PMe_3 methyl groups have been removed for clarity) and their calculated Mo-H distances, Mo-H vibrational frequencies, and relative energies and free energies, are also shown. Adapted with permission from ref. 13. Copyright © 2009, American Chemical Society.

different lengths of Mo-H···HCH_2(ring) contacts, which are *ca.* 0.1 Å shorter for the lower energy rotamers.

IR and polarized Raman measurements performed for $Os(L)H_2(Mes)$ (Mes = 2,4,6-$Me_3C_6H_3$; L = CO, CH_3CN)[13] have shown that for $Os(CO)H_2(Mes)$ a strong Raman band at 2066 cm^{-1} belongs to the ν^{s}_{OsH2} mode and can't be observed in the IR spectrum. The corresponding ν^{as}_{OsH2} is hidden beyond the symmetric stretch but can be detected in the polarized Raman ($I_{\perp}$) spectra or in IR spectra at 2062 cm^{-1}. For

$Os(CH_3CN)H_2(Mes)$ these bands appear at lower frequencies (2026 and 2020 cm^{-1}, respectively) as a consequence of the lower electron accepting ability of the CH_3CN ligand.[13] The normal-coordinate analysis was performed under approximation of the ring ligand as a point mass[13] and then for the full $Os(CO)H_2(Mes)$ molecule.[4] Comparison of the results have shown that the $Os(CO)H_2$ moiety can be treated as having C_{2v} symmetry and giving rise to a polarized and depolarized pure Os-H stretching modes without vibrational coupling to other vibrations. There is only a small difference in force constants of the internal coordinates of $Os(CO)H_2$ obtained under point mass approximation and in the complete calculations.[4,13]

Similar considerations could be applied to *cis*-$[Cp^*OsH_2(dppe)]^+$ (dppe = 1,2-diphenylphosphinoethane). However two wide bands observed in the IR spectra at 2126 and 2177 cm^{-1} in CH_2Cl_2 at 200 K[14] are individual ν_{OsH} vibrations of each of the two hydride ligands as revealed by the frequency calculations (Table 1). The same is true for the *trans*-$[Cp^*M(H)_2(dppe)]^+$ complexes (M = Fe, Ru, Os). Experimental spectroscopic data for these compounds are given below (Table 1). Note, that hydride ligands in these complexes are equal on the NMR time scale exhibiting only one δ_{MH} resonance, but they give two well separated bands in IR spectra.[14–17] The MH_2P_2 moiety in these complexes has local C_{2v} symmetry, which is disrupted by the ethylene bridge and phenyl rings (Fig. 3).

Table 1 Calculated frequencies (ν_{MH}, in cm^{-1}), intensities (A in 10^4 L $mol^{-1}cm^{-2}$), force constants (F, mdyn/Å) and potential energy distribution (PED, %) of MH stretching vibrations in *trans*-$[Cp^*M(H)_2(dppe)]^+$ complexes.[a]

M	r_{MH}	ν^1_{MH}	F	A	PED[b]	r_{MH}	ν^2_{MH}	F	A	PED[b]
Fe	1.476	2062	2.48	0.20	90 ν_{MH1}	1.479	2022	2.38	0.40	89 ν_{MH2}
Ru	1.603	2069	2.52	0.55	80 ν_{MH1} + 15 ν_{MH2}	1.605	2034	2.43	0.31	75 ν_{MH2} + 15δ_{PP}
Os	1.628	2169	2.78	0.41	77 ν_{MH1} + 15 ν_{MH2}	1.631	2122	2.66	0.14	74 ν_{MH2} + 14δ_{PP}
	1.602	*2257*[c]	*3.01*[c]	*0.18*[c]	*97* ν_{MH1}[c]	*1.610*	*2192*[c]	*2.84*[c]	*0.12*[c]	*98* ν_{MH2}[c]

[a] DFT/B3PW91 calculations for real complexes. No scaling factors applied to the frequencies.
[b] PED calculations were performed for gas phase optimised complexes with VEDA 4 program[18] using equilibrium geometries and Cartesian force constants from Gaussian output files.
[c] *cis*-$[Cp^*Os(H)_2(dppe)]^+$.

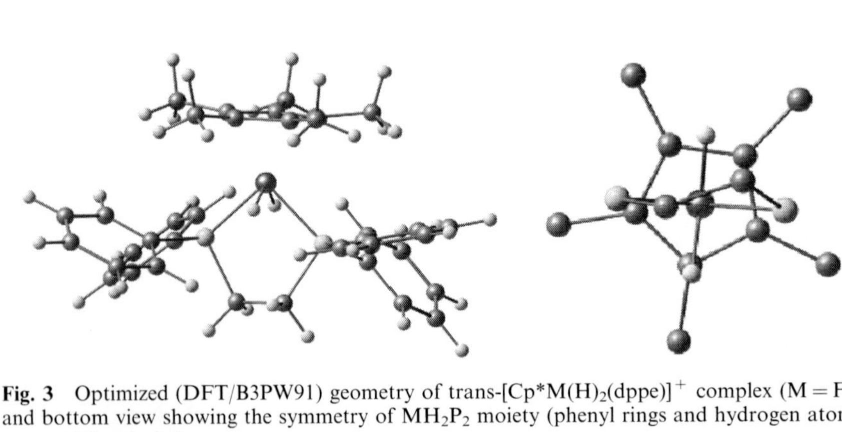

Fig. 3 Optimized (DFT/B3PW91) geometry of trans-$[Cp^*M(H)_2(dppe)]^+$ complex (M = Fe) and bottom view showing the symmetry of MH_2P_2 moiety (phenyl rings and hydrogen atoms removed for clarity).

Further influence is exerted by the Cp* ring; being closer to the metal in case of iron it adds more asymmetry and the ν_{MH} vibrations become individual stretches of two M-H bonds (Table 1). In the case of Ru and Os the ν_{MH} modes are less characteristic; the individual M-H stretches (*ca.* 80% PED) are mixed with in-phase stretches of the second hydride (ν^1_{MH} mode, Table 1) or with out-of-plane bending of the dppe ligand (δ_{PP} in ν^2_{MH} mode).

The octahedral complexes $(PP_3)MH_2$ (M = Fe, Ru, Os; PP_3 = κ^4-$P(CH_2CH_2PPh_2)_3$)[19,20] may be considered to have C_{2v} local symmetry if one does not take into account the ethylene backbones of the PP_3 ligand (Fig. 4). In this case only one ν_{MH} band would be expected in the IR spectra. However, the two hydride ligands in $(PP_3)MH_2$ located *trans* to the bridgehead phosphorus atom (axial ligand, H^{ax}) and to the terminal phosphorus atom (equatorial H^{eq}) exhibit two well separated ν_{MH} bands. These bands have similar polarisation in Raman spectra ($I_{\perp}/I_{\parallel} = 0.16$–$0.17$) and thus are individual symmetrical stretching vibrations (A_1) of each hydride ligand.[21] These two ν_{MH} bands have close intensity, the $\nu_{MH}{}^{eq}$ being slightly weaker. Note that $(PP_3)MH_2$ complexes are stereochemically non-rigid on the NMR time scale where the two hydride sites undergo a temperature-dependent scrambling process in solution. The slow exchange regime for the osmium complex was attained at 270 K, while the exchange process was frozen out only at 210 and 190 K for the ruthenium and iron complexes, respectively.[19,20,22,23]

The above examples illustrate how diverse may be the M-H vibrational modes in case of dihydrido complexes and that correct assignment of ν_{MH} bands may necessitate combination of IR and polarized Raman measurements or vibrational frequency calculations. One definitely has to keep in mind the possibility of other ligands influence which could diversify hydride ligands, affect band widths, their intensity and even lead to band splitting. However in all the cases discussed above the M-H stretching vibrations remain independent of vibrations of other ligands.

A rather unique example of complexes where M-H stretches are coupled with vibrations of a *trans*-ligand is provided by $WH(CO)_2(NO)(PR_3)_2$ and *cis,trans*-$ReH_2(CO)(NO)(PR_3)_2$ (R = Me, Et, ^{i}Pr, ^{i}PrO, MeO).[25–28] In the

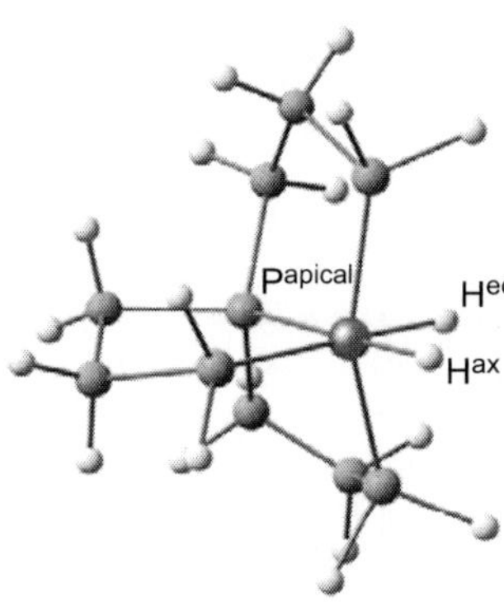

Fig. 4 Optimized geometry of model [κ^4-$P(CH_2CH_2PH_2)_3$]MH_2 dihydride (M = Fe, DFT/B3PW91).

Table 2 Measured[24] and calculated (B3LYP) vibrational frequencies (in cm^{-1}), calculated intensities (A in $10^4\,L\,mol^{-1}cm^{-2}$), and potential energy distribution (PED, %) of $WH(CO)_2(NO)(PMe_3)_2$.[a]

	IR	calc	A	PED
ν^s_{CO}		1955	0.10	49 ν_{CO1} + 44 ν_{CO2}
ν^{as}_{CO}	1914	1878	17.37	51 ν_{CO2} − 47 ν_{CO1}
ν_{WH}	1654	1686	5.92	56 ν_{WH} − 35 ν_{NO}
ν_{NO}	1568	1596	2.13	51 ν_{NO} + 42 ν_{WH}

[a] Negative sign means antisymmetric stretch.

tungsten complexes the ν_{MH} intensity is enhanced due to the resonance with the ν_{NO} vibration as confirmed by the measurements of deuterated analogues and by the frequency calculations (Table 2). In $ReH_2(CO)(NO)(PR_3)_2$ the difference in the electronic properties of carbonyl and nitrosyl ligands makes hydride ligands quite different. These dihydrides possess two well separated ν_{ReH} stretching vibrational modes which do not couple with each other.[29] Instead the ν_{ReH} vibrations couple with the ν_{XO} stretching mode (X = N, C) of the ligand in the *trans* position and significantly differ in their IR intensities, so that only the ν_{ReH} band of the H-ligand *trans* to the nitrosyl ligand can be observed experimentally.[27,29]

The ν_{MH} band assignment could be even more complicated for complexes with higher number of hydride ligands. In half-sandwich hydride complexes $Cp^*MH_3(dppe)$ (M = Mo, W) the three hydride ligands are asymmetrically disposed, one on one side and two on the opposite side of the plane that contains the two P donors and the Cp* ring centroid (Fig. 5). Atoms H1 and H2 are situated rather symmetrically across the ideal plane that contains the metal atom, H3 and the ring centroid. The three hydride ligands are equivalent in the NMR spectra: only one hydride signal is observed at all temperatures.[30–32] However they are not equivalent in terms of the M-H distances (Fig. 5). A decomposition of ν_{MH} bands in the IR spectra gives three bands at 1908, 1831, 1785 cm^{-1} for $Cp^*MoH_3(dppe)$ and 1964, 1882, 1846 cm^{-1} for $Cp^*WH_3(dppe)$ in THF.[30,31] The computational analysis of the M-H normal modes (Fig. 6) has shown that each normal mode is essentially a pure stretching vibration of the corresponding hydride ligand. The most characteristic is the highest-energy ν_1 vibration of the H1 ligand, which shows the shortest (and therefore strongest) bond.[30]

The solid-state IR spectrum of Cp^*OsH_5 features three Os-H stretching bands: 2214 (weak), 2083 (strong), and 2065 (shoulder) cm^{-1}, whereas the Raman spectrum contains $\nu_{Os\text{-}H}$ bands at 2214 (polarized) and 2096 cm^{-1}.[38] The IR spectrum resembles that calculated for the observed pseudo-octahedral structure of pseudo-C_{4v} symmetry (under point mass approximation for the Cp* ring), but is rather different from that predicted[39] for a five-legged piano stool geometry of C_{5v} symmetry.

Main group element hydrides present an important group of polyhydrido species and often have high symmetry. Of particular interest for us are boron hydrides and their heavier analogues. The most simple group 13 tetrahydrido anions EH_4^- (E = B, Al, Ga) have tetrahedral geometry of T_d symmetry and possess two E-H stretching vibration modes, of which only

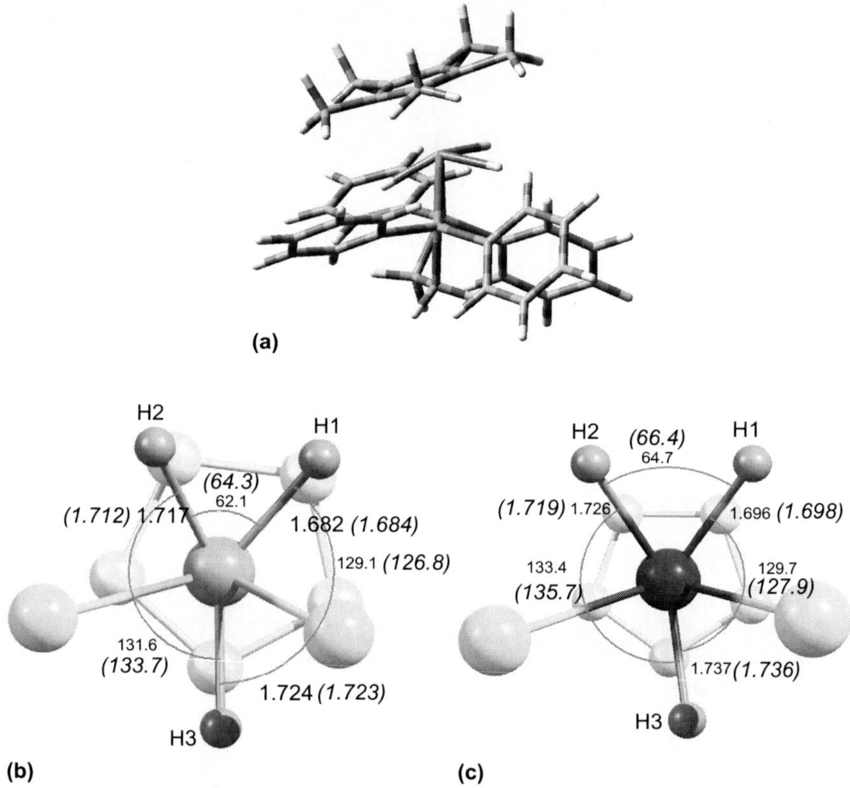

Fig. 5 Optimized geometry of $Cp^*W(dppe)H_3$ (a) and main geometrical parameters (distances in Å, angles in degrees) for the MoH_3 (b) and WH_3 (c) moieties. The values in parentheses correspond to those optimized for the full system. Adapted with permission from refs. 30, 31. Copyright © 2008 WILEY-VCH Verlag GmbH & Co. KGaA, Weinheim.

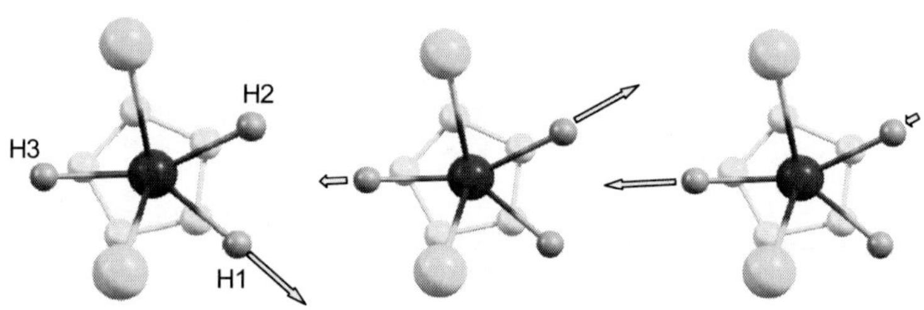

	ν_1 = 1948	ν_2 = 1864	ν_3 = 1812
A	0.27	0.48	0.67
W-H1	***84.4***	4.8	6.4
W-H2	3.5	***71.8***	–18.6
W-H3	9.1	19.4	***67.2***

Fig. 6 The W–H normal modes of the DFT optimized $CpW(dhpe)H_3$ with their frequency, intensity (A, 10^4 L $mol^{-1}cm^{-2}$), and potential energy distribution (in %, major components are in bold italic). The Cp- and P-bonded H atoms and ethylene backbone have been omitted for clarity. Negative sign means antisymmetric stretch. Adapted with permission from ref. 30. Copyright © 2008 WILEY-VCH Verlag GmbH & Co. KGaA, Weinheim.

one is IR active.[2] Coordination to a transition metal,[40] hydrogen bonding (see below) or formal replacement of at least one of the hydrogen atoms to form EH_3X^- compounds[41] lowers the symmetry to C_{3v} or lower. Thus all vibrational modes become IR active and spectra become more complicated. For example, the ν_{EH} bands of BH_4^- and GaH_4^- studied in solution (CH_2Cl_2 and THF) at 180–290 K have complex shape with the main band observed at 2200 and 1700 cm^{-1}, respectively.[42–44] In the case of BH_4^- the main ν_{BH} band is accompanied by higher 2280 cm^{-1} and lower (2155 cm^{-1}) frequency bands assigned as the $\delta^{s}_{BH} + \delta^{as}_{BH}$ combination and the overtone ($2\delta^{as}_{BH}$).[45] According to the C_{3v} symmetry of BH_3CN^- there should be two ν_{BH} bands in the IR spectra belonging to A_1 and two degenerate E_1 modes. However, the band decomposition showed the presence of three overlapped ν_{BH} bands: 2318 (s), 2284 (m) and 2268 (m) cm^{-1}. This could be due to removing the E_1 mode degeneracy as the result of cation and solvent effects. In addition to these ν_{BH} bands two weak bands are observed at 2230 and 2215 cm^{-1} belonging to the overtones of BH deformation modes.[41]

It is well known that the position of the E-H stretching vibrational bands for EH_n "hydrides" of main group elements (group 1–2, 14–17) increases on going to the right along the period and decreases on descending the group.[2] Removal of bonding electrons or protonation decreases the ν_{EH} frequency.[2] When E is a group 13 element forming the family of EH_4^- hydrides, the ν_{EH} stretching frequencies and the corresponding force constants change in the order $\nu_{BH} > \nu_{GaH} > \nu_{AlH}$ (Table 3). The deviation of AlH_4^- is in line with a longer E-H bond length and a force constant slightly smaller than that of a GaH bond (Table 3). Interestingly, the ν_{EH} frequencies and the force constants of diatomic EH species decrease evenly down the group (Table 3). Note much shorter E-H bond length and substantially higher force constants in the case of boron hydrides (both BH_4^- and diatomic BH) indicating a much stronger bond.

Transition metals exhibit the same trend as main group element hydrides: the ν_{MH} stretching frequency increases along the transition row from group 3 to group 12 metals. However, in contrast to non-metal hydrides the ν_{MH} of transition metal hydrides increases on descending the group. This was first analysed by Perutz *et al.* for Cp_2MH_n complexes[3] and our analysis

Table 3 The frequencies of ν_{EH} vibrations (in cm^{-1}) measured and calculated for group 13 element hydrides, the corresponding force constants (F, in mdyne/Å) and calculated bond lengths ($r_{E\text{-}H}$, in Å).

		EH_4^-					EH		
	μ^a	$\nu_{E\text{-}H}$ (A_1) calc[33]	$\nu_{E\text{-}H}$ (T_2) calc[33]	F	$r_{E\text{-}H}$[33]	$\nu_{E\text{-}H}$ (T_2)^b	$\nu_{E\text{-}H}{}^c$	F^c	$\nu_{E\text{-}H}{}^b$
B	0.922	2263	2230	2.78	1.238	2212	2435	3.22	2259
Al	0.972	1732	1644	1.72	1.645	1609	1683	1.62	1591
Ga	0.994	1762	1685	1.82	1.623	1773	1605	1.51	1531
In							1475	1.28	

^{a} μ – reduced mass.
^{b} in Ar/H_2 matrix.[34–36]
^{c} data from ref. 37.

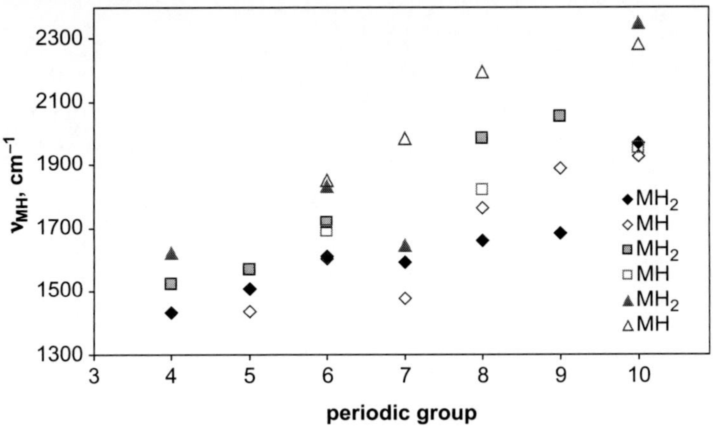

Fig. 7 Dependence of ν_{MH} frequency of MH and MH_2 hydrides on the metal position in the Periodic table for transition metals of 3rd (rhombi), 4th (squares) and 5th (triangles) row. Data from refs. 11, 46–51.

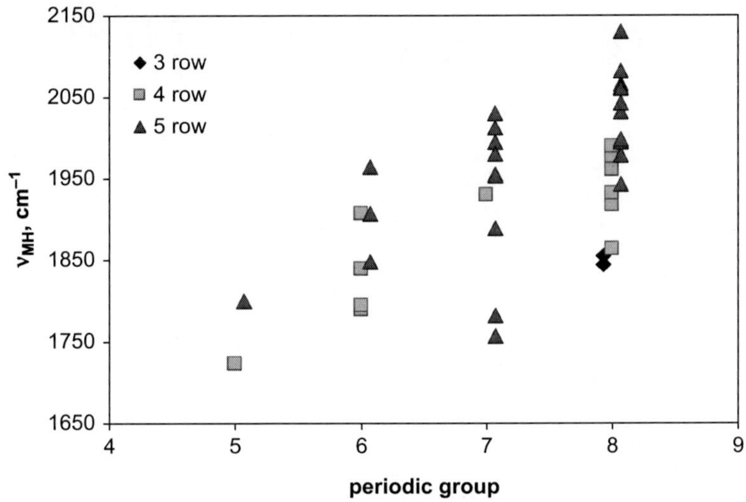

Fig. 8 Dependence of ν_{MH} frequency of transition metal hydrido complexes bearing various ligands on the metal position in the Periodic table. Data from refs.[3,12,21,30,52–57]

shows this increase does not depend on the complexes structure. In other words it can be traced for different hydrides, both "naked" hydrides and coordination hydrido complexes, see for example refs. 11, 52 (Fig. 7, Fig. 8). However, as can be appreciated from Fig. 9 the ν_{MH} increase with the metal atomic number is only valid for isostructural compounds featuring same ligand environment. At any rate the ν_{MH} increase indicates the increase of the M-H force constants because the reduced mass of M-H moiety, μ, the increase of which should diminish the frequencies, barely depends on M for transition metal hydrides. Since force constants correlate with the bond strength, their increase can be considered as the evidence for the increase of the M-H bond strength on going from 3d to 5d metal despite the bond lengthening (Table 4).

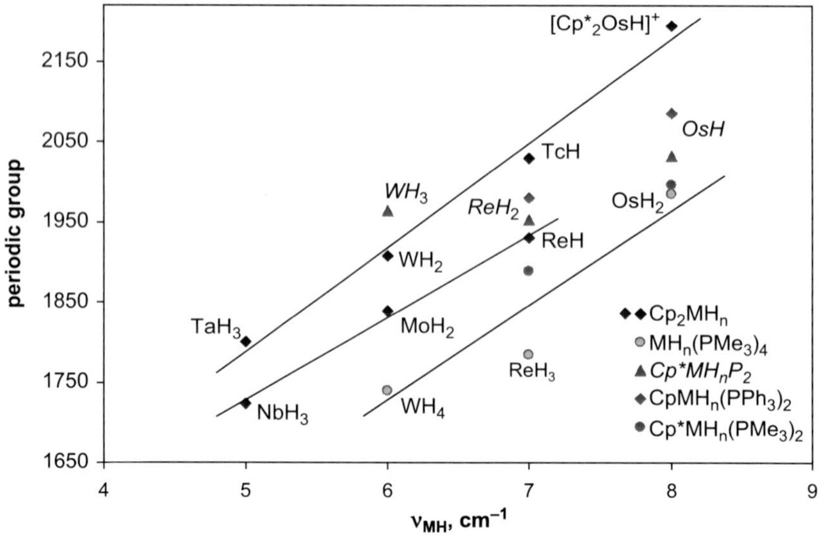

Fig. 9 Dependence of ν_{MH} frequency of sandwich and half-sandwich transition metal hydrido complexes on the metal position in the Periodic table. Data from refs. 3, 52, 59–61.

Table 4 ν_{MH} frequencies (in cm^{-1}) measured and calculated for Cp*MH(dppe), the corresponding force constants (F, in mdyne/Å) and calculated bond lengths($r_{M\text{-}H}$, in Å).[a]

	μ	ν_{MH} IR	F	ν_{MH} calc	F	$r_{M\text{-}H}$
Fe	0.990	1844	1.98	1920	2.15	1.494
Ru	0.998	1918	2.16	2007	2.37	1.605
Os	1.003	2033	2.44	2141	2.71	1.627

[a] data from refs.[14,16,17,58]

As can be seen from Fig. 8 and Fig. 9, the ranges of ν_{MH} stretches interfere for complexes of different metals. The change of ligands may produce a profound effect overwhelming the change of a core metal. Thus, variation of phosphine ligand or methylation of Cp ring changes ν_{MH} by *ca.* 100 cm^{-1} (Table 5). This shift is comparable to the difference in ν_{MH} frequencies measured for hydrides of different metals with the same ligands (compare, *e.g.*, Cp*OsH$(PMe_3)_2$ and Cp*ReH$_2$$(PMe_3)_2$, Table 5).

When an electron is removed from the metal-hydride σ-bonding valence orbital of CpRe(NO)(CO)H and Cp*Re(NO)(CO)H to yield $[Cp'Re(CO)(NO)H]^+$, the ν_{MH} frequencies shift by 100–200 cm^{-1} to lower wavenumbers, which indicates the bond lengthening (by 0.25 Å) and corresponds to a substantial weakening upon ionization.[62] This indicates also a predominantly localized rhenium-hydride bonding orbital similarly to main group element-hydride bonds. Moving from anionic to neutral or from neutral to cationic hydrides as the result of protonation increases the ν_{MH} frequency (see below) in contrast to the ν_{EH} decrease observed for example upon protonation of amines.

Table 5 The position of ν_{MH} bands (data in KBr; in cm^{-1}) for half-sandwich complexes of osmium and rhenium as a function of ligands.

complex	ν_{MH}	ref.	complex	ν_{MH}	ref.
$CpOsH(PPh_3)_2$	2085[a]	63	$CpReH_2(PPh_3)_2$	1980	61
$Cp^*OsH(PPh_3)_2$	1978	64			
$Cp^*OsH(PMe_3)_2$	1996	64	$Cp^*ReH_2(PMe_3)_2$	1889	60
$Cp^*OsH(PEt_3)_2$	2043	64	$Cp^*ReH_2(PMe_2Ph)_2$	1953	60
Cp*OsH(dppe)	2008	15			
Cp*OsH(dppm)	2060	64			
Cp*OsH(dmpm)	1995	64	CpRe(CO)(NO)H	2013	62
Cp*OsH(dmpe)	1999	64	Cp*Re(CO)(NO)H	1994	62

[a] 2060 cm^{-1} in nujol.

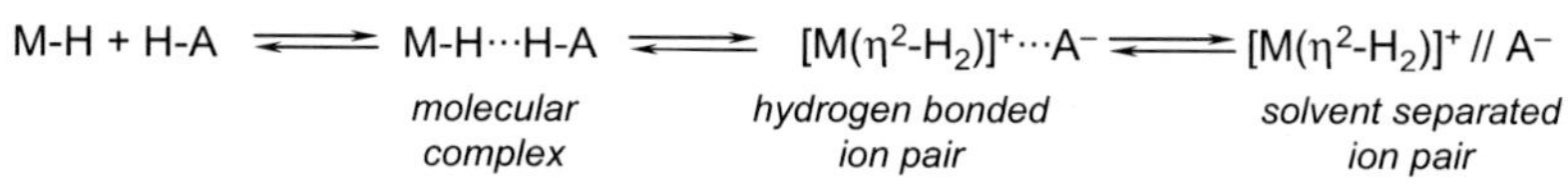

Scheme 1 General mechanism of proton transfer *via* dihydrogen bonding.

Thus, the M-H stretching vibrations are sensitive to the position of the core metal in the Periodic table, the ligand surrounding and protonation. Analysis of the data reveals that the trends are different for main group and trasition metal hydrides.

Hydrogen bonding and proton transfer

Proton transfer to transition metal and main group element hydrides is a complicated process.[65–67] It begins with formation of an unconventional hydrogen bond between a metal hydride (MH) and proton donor (HA), which nowadays is widely called a dihydrogen bond MH···HA (Scheme 1).[65] The next reaction step is proton transfer itself yielding non-classical di- or polyhydrides. In some cases classical polyhydrides can be formed without observation of η^2-H_2 intermediates.[66,68] In subsequent sections we discuss various aspects of spectroscopic studies of hydrogen bond formation and proton transfer paying particular attention to the use of variable temperature IR spectroscopy.

Hydrogen bond evidence and strength

Formation of a dihydrogen bond MH···HA, as any medium strength hydrogen bond, entails elongation of the proton donor H-A bond. Consequently, ν_{AH} vibrations shift to lower frequencies and a new, much wider and more intense band $\nu_{HA}{}^{bonded}$ appears with the band shift $\Delta\nu_{AH} = \nu_{AH}{}^{bonded} - \nu_{AH}{}^{free}$ ranging from -50 to $-350\,cm^{-1}$. These changes in the spectra (Fig. 10) are indicative of hydrogen bond formation and are observed for any base, organic or organometallic/inorganic.[69] The values of band shift ($\Delta\nu_{AH}$) and of intensity increase (ΔA_{AH}) depend on the hydrogen bond strength. Among various correlations, which allow estimating the enthalpy of the bonding, ΔH_{HB}, we have chosen correlation Eqs. 1–3 (ΔH_{HB} in kcal/mol). These correlations have been originally

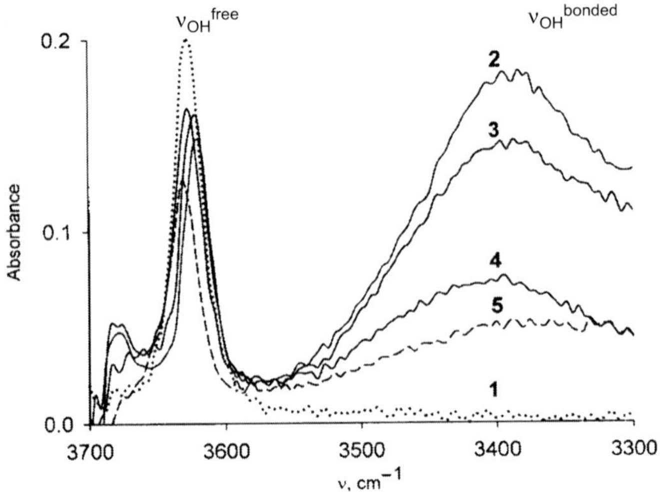

Fig. 10 IR spectra in the ν_{OH} absorption range of MeOH (0.015 mol L^{-1}, CH_2Cl_2) at 260 K (1); in the presence of $[Bu_4N]GaH_4$ (0.09 mol L^{-1}) at 200 K (2); 220 K (3); 260 K (4); 290 K (5). Reprinted with permission from ref. 42. Copyright © 2004 WILEY-VCH Verlag GmbH & Co. KGaA, Weinheim.

proposed for classical hydrogen bonds of organic acids and bases[70,71] and their applicability to dihydrogen bonding have been shown by us later.[24,43,72]

$$\Delta H_{HB} = -18\Delta\nu_{HA}/(\Delta\nu_{HA} + 720) \quad (1)$$

$$\Delta H_{HB} = -0.30\Delta\nu_{HA}{}^{1/2} \quad (2)$$

$$\Delta H_{HB} = -2.9\Delta A_{HA}{}^{1/2} = -2.9(A^{1/2}{}_{bonded} - A^{1/2}{}_{free}) \quad (3)$$

It is important to note that these data can be retrieved from measurements of IR spectra at room temperature in solution or even in the solid state[73] and need just a comparative analysis of the spectra of proton donor with and without base (Fig. 10). Application of Eqs. 1–2 does not require the knowledge of concentrations, but one should keep in mind that the base excess makes it easier to observe. Measurements at different temperatures provide not only the enthalpy (ΔH_{HB}) but entropy (ΔS_{HB}) of hydrogen bond formation as the parameters of the temperature dependence of hydrogen bond formation constant K_{HB}. The K_{HB} values (Eq. 4) can be obtained from the intensity drop of $\nu_{AH}{}^{free}$ band (Fig. 10). In this case the knowledge of proton donor and proton acceptor concentrations is vital. In the absence of proton donor self-association (relatively low HA concentrations) the hydrogen bonded complexes have 1:1 composition (one HA molecule interacts with one molecule of proton acceptor) and calculations are straightforward.

$$K_{HB} = \frac{[MH \cdots HA]}{[MH][HA]} \quad (4)$$

Most convenient for these studies appeared to be phenols and aliphatic fluorinated alcohols, but for more reactive main group element hydrides the

CH and NH proton donors like adamantyl acetylene, azoles, anilines were used.[42,74] The latter are potentially bidentate proton donors, thus formation of a chelated structure where two hydride ligands interact with both protons of the NH_2 group is possible. The NH_2 group of aniline and its substituted congeners has C_{2v} local symmetry with two identical N–H bonds, which gives a lower frequency symmetric ν^{s}_{NH2} and a higher frequency asymmetric ν^{as}_{NH2} stretching vibrations. Upon formation of a single hydrogen bond (involving only one NH proton) the local symmetry is lowered and the two N–H bonds are decoupled, which would place the stretching of free N–H at 10–20 cm^{-1} higher relative to the median of ν^{s}_{NH2} and ν^{as}_{NH2} as a rather narrow band. Such changes were observed for the interaction of *p*-nitroaniline with GaH_4^- in CH_2Cl_2 at 200 K[42] or aniline with BH_3NMe_3 in supersonic jets.[75] The second ν_{NH} band observed under these conditions is wider and substantially shifted to lower frequencies, and is a typical band of a NH group involved in a hydrogen bond. In the case of bidentate coordination the formation of a quite symmetric cyclic species is expected (Fig. 11).[74,75] Formation of such cyclic dihydrogen bonded structures was experimentally detected so far for only one system: trimethylamine alane $AlH_3(NMe_3)$ interacting with 2,3,5,6-tetrachloroaniline in hexane at low temperature. The new ν_{NH} bands observed in the IR spectra under these conditions are narrow and exhibit rather small shifts relative to free tetrachloroaniline ($\Delta\nu^{as}_{NH2} = -29$ cm^{-1} and $\Delta\nu^{s}_{NH2} = -17$ cm^{-1}) thus keeping the symmetry of the starting NH_2 moiety.[74]

The association and dissociation of the molecules forming hydrogen bonded complexes (exchange) is fast on the NMR time scale. This results in averaging of all NMR parameters in the temperature range available for common organic solvents. NMR spectra measured under the identical conditions show the signal of HA proton shifted to lower field by 2–4 ppm due to hydrogen bonding.[65,69,72]

The above-described changes in the IR and NMR spectra are diagnostic of hydrogen bond formation but provide no information about the site of coordination. Thus spectral criteria have been developed to distinguish between hydrogen bonded complexes of different types based on the changes of the ligand stretching vibrations (CO, NO, M-Hal, M-H) in IR spectra and corresponding signals in NMR spectra.

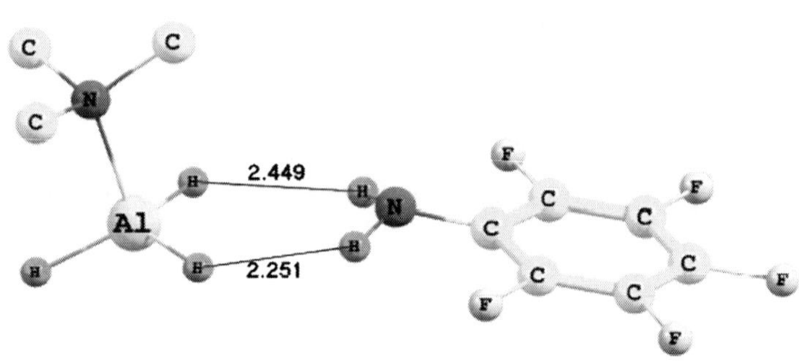

Fig. 11 Optimized geometry of dihydrogen bonded complex of Me_3NAlH_3 with perfluoroaniline featuring bidentate coordination (hydrogen atoms of NMe_3 are omitted for clarity). Reprinted with permission from ref. 74. Copyright © 2003 American Chemical Society.

Spectral criteria of dihydrogen bonding

Combining variable temperature IR and NMR spectroscopic studies of hydrogen bonding between several hydrides (transition metal and main group element) and proton donors with theoretical calculations allowed hydrogen bonds to different proton accepting sites to be distinguished: to sp lone pairs of ligand functional groups, to d-lone pair of a core transition metal atom or to the partly negative charge of a hydride ligand, and to establish the spectral criteria of their formation (Table 6).

The experimental ***IR evidence*** is the following:

a) When the ligand serves as the proton accepting site (ML···HA hydrogen bond) the corresponding stretching vibration band shifts to lower frequencies. This is applicable to *e.g.* carbonyl,[84] nitrosyl[28,73] and chloride[73,85] ligands, but CN ligand is the exception. High frequency shift of ν_{CN} vibration was shown for simple anions[86,87] and the coordination compound $[BH_3CN]^-$.[41]

b) When the metal atom is the proton accepting site (M···HA) the new high-frequency bands (or shoulders) appear for stretching vibrations of all the ligands including ν_{MH}.

c) The appearance of a new low-frequency $\nu_{M\text{-}H}$ band (or shoulder) in addition to new high-frequency bands (or shoulders) for the stretching vibrations of other ligands indicate that the hydride ligand is the proton accepting site, *i.e.* a dihydrogen bond MH···HA is formed.

For hydrides with very weak or not observable ν_{MH} bands especially important becomes the use of NMR techniques. The main ***^{1}H NMR evidence*** of dihydrogen bonding are the up-field shift of the hydride resonance by 0.1–0.8 ppm and the decrease of its longitudinal relaxation time (T_{1min}) 1.5–3 times (Table 6).[82] When the hydride ligand is not involved in the hydrogen bond, its resonance shifts to lower field.[21,80]

Attachment of an electron acceptor to the oxygen end of the CO ligand is known to shift the corresponding ^{13}C resonance downfield.[88] Our ^{13}C{^{1}H} NMR study[83] revealed that formation of a hydrogen bond to the carbonyl ligand of $Cp^*MoH(CO)(PMe_3)_2$ shifts the characteristic triplet of the CO resonance (δ 253.1, $^2J_{C\text{-}P} = 29$ Hz) downfield by 6.1 ppm in the presence of 2 equiv. of $(CF_3)_3COH$ at 200 K in toluene-d_8. Note, that the $\Delta\delta_{CO}$ values exceed considerably the ^{1}H resonance shifts for hydride ligands.

Table 6 The IR and NMR spectral changes[a] evidencing formation of hydrogen bonds of different types involving transition metal complexes (L = CO, NO, Cl, N_2).[24, 76–82 *etc.*]

	IR		^{1}H NMR		^{13}C	^{31}P
	$\Delta\nu_{MH}$	$\Delta\nu_L$	$\Delta\delta_{H(M)}$	T_{1min}(MH)/ T_{1min}(MH···H)	$\Delta\delta_{CO}$ (ppm)	$\Delta\delta_P$ (ppm)
ML···HA	+10 ÷ +20	−20 ÷ −60	0.05 ÷ 0.5[80]	no change	2 ÷ 7[83]	−0.4 ÷ −0.9
M···HA	+10 ÷ +20	+10 ÷ +25				
MH···HA	−15 ÷ −20	+10 ÷ +20	−0.2 ÷ −0.8	1.5 ÷ 2.0		−1 ÷ −3

[a] difference in the position (ΔX) of the stretching vibration in IR spectra or NMR resonance in hydrogen bonded complex and in free molecule, ΔX = X(complex) − X(free).

These IR and NMR changes are better observed in the presence of excess proton donor and at lower temperatures, both shifting the hydrogen bond formation equilibria to the right.[89] Lower polarity solvents favor hydrogen bond formation as well. Unfortunately, their use is not always possible due to low solubility of compounds. Some solvents, like dichloromethane, increase their polarity at low temperatures, but that potentially unfavorable change seems to be well compensated by the increase of formation constant at low temperatures.[89] The shifts of NMR signals depend on the proton donor amount and strength, the temperature and the solvent. The same is true for the decrease of T_{1min} of hydride resonance upon a dihydrogen bond formation. The reason for that is averaging of NMR parameters of free and hydrogen bonded compounds due to the fast exchange, the impact of each species being determined by their ratio and, thus, by the value of hydrogen bond formation constant.

In the IR spectra the band shifts for hydrogen bonded complexes in a given solvent depend on the interaction strength only, increasing with proton donating ability of HA. For example, the fluorinated alcohols induce the development of a low-frequency shift of the ν_{FeH} band of Cp*(dppe)FeH (1844 cm^{-1}) to 1836 cm^{-1} (in dihydrogen bonded complex with CF_3CH_2OH) and further down to 1828 cm^{-1} (in dihydrogen-bonded complex with $(CF_3)_3COH$).[17] This effect is observed even for stretching vibrational bands of ligands not involved in hydrogen bond formation. Thus, for dihydrogen bonded complexes of $WH(CO)_2(NO)(PEt_3)_2$ $\Delta\nu_{CO} = 11$ (PhOH) < 14 ($(CF_3)_2CHOH$) < 17 cm^{-1} ($(CF_3)_3COH$);[24] for $CpRuH(CO)(PCy_3)$ $\Delta\nu_{CO} = 11$ ($(CF_3)_2CHOH$) < 15 ($(CF_3)_3COH$) < 20 cm^{-1} (CF_3COOH).[84]

When vibrations of free and hydrogen bonded species are overlapped (most often due to rather high band width) the position of the maximum for the resulting envelope band depends on the equilibrium shift (like in NMR spectroscopy) as was observed *e.g.* for (triphos)Re$(CO)_2$H···HA[90] or (triphos)Ru(CO)H_2···HA.[91] Otherwise, the proton donor concentration and temperature effect only the intensity ratio of ν^{free} and ν^{bonded} bands allowing calculation of the equilibrium concentrations from the decrease of the ν^{free} band intensity for every temperature and then estimation of the hydrogen bond formation constants K_{HB}.[24,92]

In some complexes the presence of several ligands with potential proton accepting properties opens a possibility for their competition in hydrogen bonding and formation of more complicated adducts. In this situation ν_{CO} and ν_{NO} bands are sensitive to the hydrogen bonded adduct composition and provide valuable information about its structure. Thus, studying the interaction of $CpRuH(CO)(PCy_3)$ with $(CF_3)_3COH$ at different alcohol concentrations surprisingly we observed three ν_{CO} bands of hydrogen bonded complexes (Fig. 12) shifted to lower and higher frequencies with respect to the ν_{CO}(**1**) band of the free hydride. The high-frequency shifted band belongs to dihydrogen bonded complex (**2a**), whereas low-frequency shift ($-48\ cm^{-1}$) is typical for hydrogen bonding to a carbonyl group ν_{CO}(**2b**) (Scheme 2). At 6-fold alcohol excess (and higher) another band arises at 1892 cm^{-1} (ν_{CO}(**2c**), Fig. 12), which belongs to the hydrogen bonded complex with two $(CF_3)_3COH$ molecules (Scheme 2).

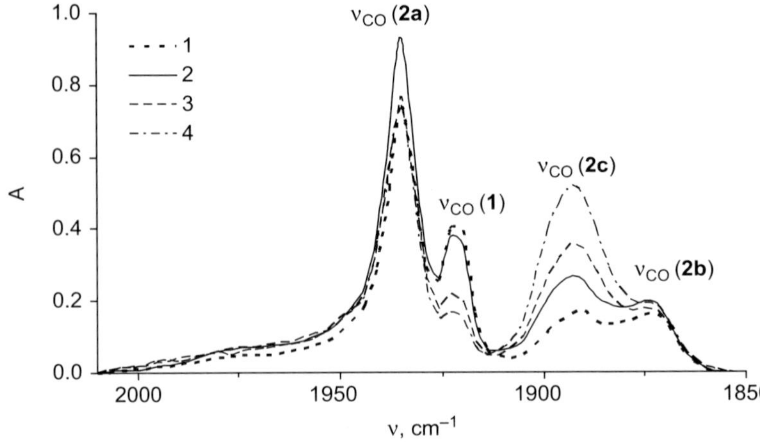

Fig. 12 IR spectra in the ν_{CO} range of CpRuH(CO)(PCy_3) (1) (0.004 mol L^{-1}) in the presence of excess $(CF_3)_3COH$ at 220 K in hexane; ratio 1/$(CF_3)_3COH$ = 1:6 (1), 1:8 (2), 1:12 (3), 1:16 (4). Reprinted with permission from ref. 84. Copyright © 2003 American Chemical Society.

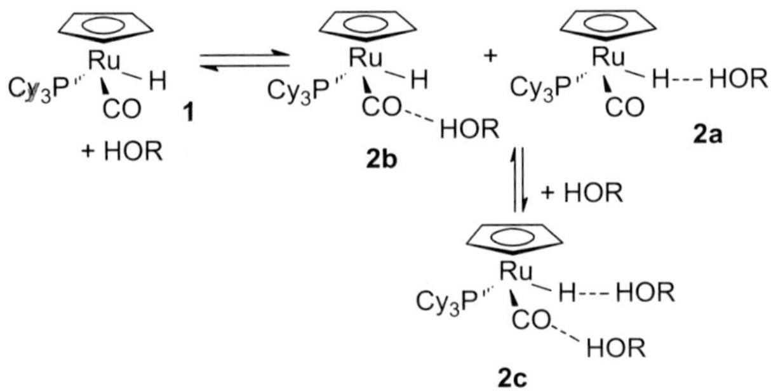

Scheme 2 Hydrogen bonded complexes formed at the first stage of CpRuH(CO)(PCy3) reaction with alcohols.

Table 7 IR data in the range of ν_{CO} and ν_{NO} vibrations (in cm^{-1}) for ReHX(CO)(NO)(PR_3) hydrides (**3**) and their hydrogen bonded complexes **4** and **5** (of 1:1 and 2:1 composition, respectively).[a] Adapted with permission from ref. 73. Copyright © 2000, Elsevier.

	ν_{CO}(**3**)	ν_{CO}(**4**)	ν_{CO}(**5**)	ν_{NO}(**3**)	ν_{NO}(**4**)/ν_{NO}(**4'**)[b,c]	ν_{NO}(**5**)[d]
3a	1962	1973	1986	1652	1671 (+19)/1597 (−55)	1686 (+15)/1622 (+25)
3b	1957	1968	1980	1650	1666 (+16)/1592(−58)	1613 (+21)
3c	1952	1963	1975	1645	1592 (−53)	1625 (+33)
3d	1966	1977	1987	1673	1689 (+16)	1710 (+21)

[a] X = H (**a–c**), Cl (**d**); R = Me (**a**), Et (**b**), iPr (**c**); in hexane at 200 K.
[b] $\Delta\nu_{NO}$ in parenthesis.
[c] $\Delta\nu_{NO} = \nu_{NO}(\mathbf{4}) - \nu_{NO}(\mathbf{3})$.
[d] $\Delta\nu_{NO} = \nu_{NO}(\mathbf{5}) - \nu_{NO}(\mathbf{3})$.

A similar situation was observed for the ν_{NO} band of rhenium hydrides $ReH_2(CO)(NO)(PR_3)_2$ (**3**, Table 7).[73] These hydrides form hydrogen-bonded complexes of ReH···HA (**4**) and NO···HA (**4′**) types, and complexes with two alcohol molecules containing both hydrogen bonds (**5**).

The ν_{NO} bands of the latter appeared at *ca.* 20 cm^{-1} higher than the ν_{NO} band in the ReNO···HA complex (Table 7).[73]

The lower frequency shift of ν_{MH} bands observed experimentally as the evidence of dihydrogen bond formation was confirmed by numerous DFT calculations which explained this effect by the elongation of M-H bond caused by the interaction. The theoretical calculations have also shown that spectral criteria developed during the studies of monohydrides can be applied to polyhydrides as well. Obviously, formation of a hydrogen bond changes the overall symmetry of the molecule under investigation, but it does not substantially perturb the local symmetry of MH_n fragments, so it entails rather small change in the M-H vibrations symmetry. Thus, the complex Cp_2NbH_3, having C_{2v} symmetry exhibits 3 ν_{MH} stretching vibration bands ($2A_1 + B_1$) in the IR spectrum which are not well resolved in solution (even in methylcyclopentane) and give an asymmetric envelope band at 1724 cm^{-1} (Fig. 13).[3,93] Upon dihydrogen bond formation (a proton donor can interact with either central or lateral hydride ligand, Fig. 14) the new ν_{MH} bands are observed at both lower and higher frequency shift for all three stretching vibrations ν_{NbH}, but their forms are retained at higher frequency (Fig. 13). The analysis of the calculated frequencies and forms of normal vibrations for the NbH_3 moiety shows that hydrogen bond formation leads to the frequency shift for all three stretching vibrations ν_{NbH}, but their forms are retained (Fig. 15). However intensity changes of the ν_{NbH} bands are different for H^{lat}···HA and H^{cent}···HA bonding. In the case of the H^{cent}···HA adduct the intensity increase calculated for ν_1 and ν_3 is accompanied by the intensity decrease of the ν_2 mode, whereas the ν_1 and ν_3 bands shift to higher and lower frequencies upon dihydrogen bond formation.

These changes become more prononced with the increase of the acidity of the alcohol and are in agreement with the measured IR spectra. Interestingly, two separate hydride resonances of H^{lat} and H^{cent} are observed in the ^1H NMR spectra, which provide evidence for preferential formation of a H^{cent}···HA diydrogen bond.[93]

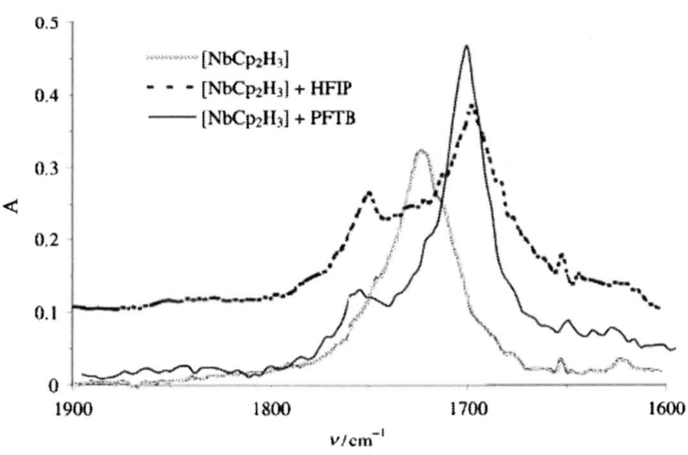

Fig. 13 IR spectra (ν_{MH} range) of Cp_2NbH_3 in the presence of excess $(CF_3)_2CHOH$ (HFIP) and $(CF_3)_3COH$ (PFTB). Methylcyclohexane, 200 K. Reprinted with permission from ref. 93 Copyright © 2004 WILLEY-VCH Verlag GmbH & Co. KGaA, Weinheim.

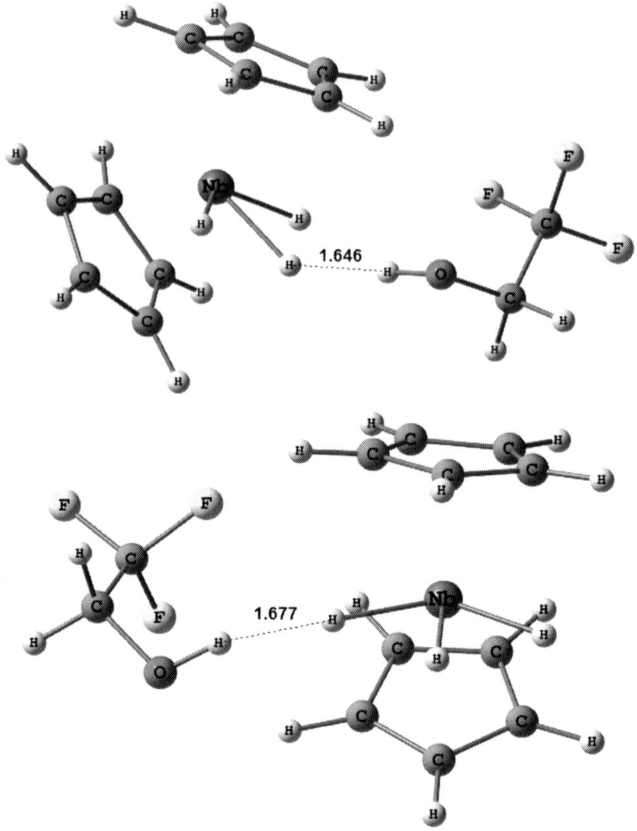

Fig. 14 DFT optimized structures of dihydrogen bonded complexes between Cp_2NbH_3 and CF_3CH_2OH featuring bonding to central (left) and lateral (right) hydride ligand.

The peculiarity of the Cp_2NbH_3 hydride is the absence of d-lone pair (d^0 configuration of the metal). Thus, it can not form M···HA hydrogen bonds and its MH···HA complexes were considered as an example of a "pure" dihydrogen bond. Hydrides of main group elements provide further evidence for hydride ligand ability to serve as hydrogen bond acceptor. Dihydrogen bond formation was proved on the basis of IR spectroscopic data, which conform to the corresponding spectral criteria, and by theoretical calculations.[33,41–43] For example the degeneracy of the T_2 mode of $\nu_{E\text{-}H}$ stretching vibrations of EH_4^- is naturally absent in their dihydrogen bonded complexes. Therefore, four frequencies describe these stretches in EH···HA complexes.[33] Three of these frequencies are shifted to higher values and were assigned to the non-bonded $\nu_{E\text{-}H}$ modes. The band shifted to the low-frequency range was assigned as $\nu_{E\text{-}H}^{bonded}$ (Fig. 16, Table 8). The $\Delta\nu_{GaH}$ values measured for dihydrogen bonded complexes of GaH_4^- with weak proton donors[42] are larger than the $\Delta\nu_{BH}$ values for the boron analogue[43,44] by about 10–30 cm^{-1}.

Measurements of the low temperature IR spectra of $Cp^*MH_3(dppe)$ ($M = Mo, W$) in the presence of excess fluorinated alcohols have shown that the low-frequency shift of the ν_{MH} band in the case of molybdenum and

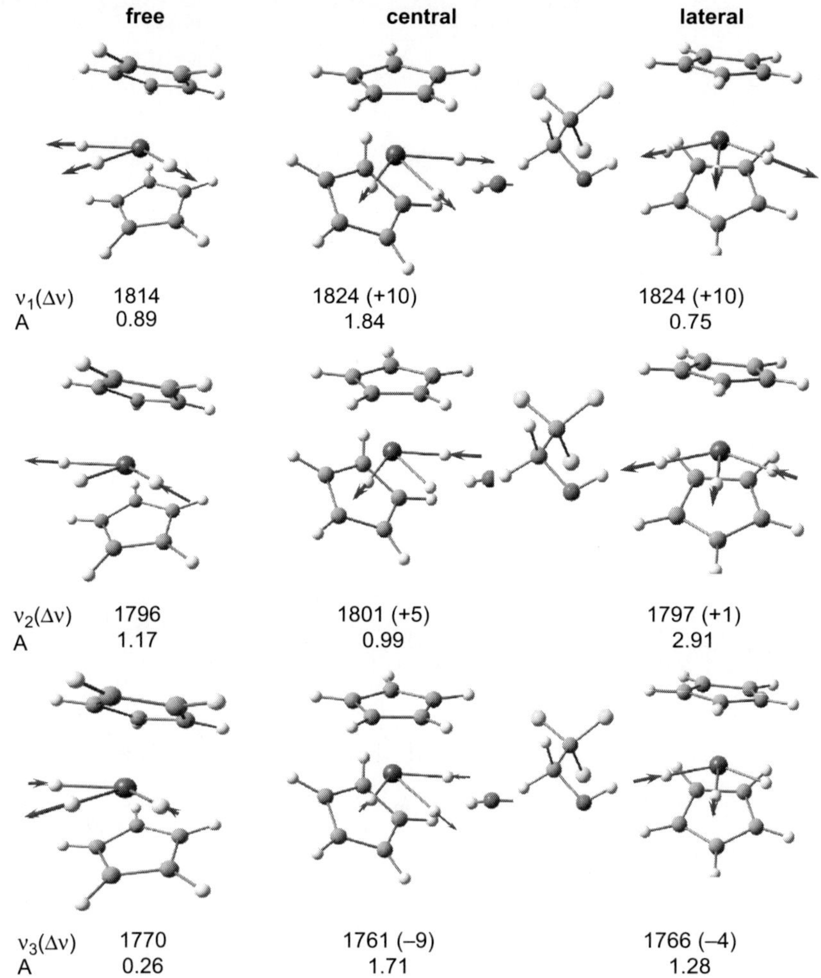

Fig. 15 Frequencies (ν, cm^{-1}), forms and intensities (A, 10^4 L mol cm^{-2}) of NbH stretching vibrations of Cp_2NbH_3 (left) and its dihydrogen bonded adducts with CF_3CH_2OH involving central (middle) or lateral (right) hydride ligands.

high-frequency ν_{MH} shift for tungsten.[30,31] Bonding to the metal is possible in this case because it has the d^2 configuration and was indeed initially suggested for tungsten on the basis of spectral data.[94] Dihydrogen bond formation was assumed for molybdenum trihydride. Analysis of the DFT optimised structures of hydrogen bonded complexes indicates the preferred structure[30] for both metals features bifurcate interactions with both M and H, rather than with either a metal lone pair or hydride ligand solely (Fig. 17). However, the analysis revealed that the metal-atom contribution increases on going from Mo to W.[30] Computed changes of ν_{MH} frequencies upon hydrogen bond formation are in agreement with those observed experimentally. These results support the applicability of spectral criteria of M-H···H-A and M···H-A bond formation to polyhydrides.

The bifurcate structure of the hydrogen bonded complex between $(NP_3)IrH_3$ ($NP_3 = N(CH_2CH_2PPh_2)_3$) and CF_3CH_2OH, similar to that of

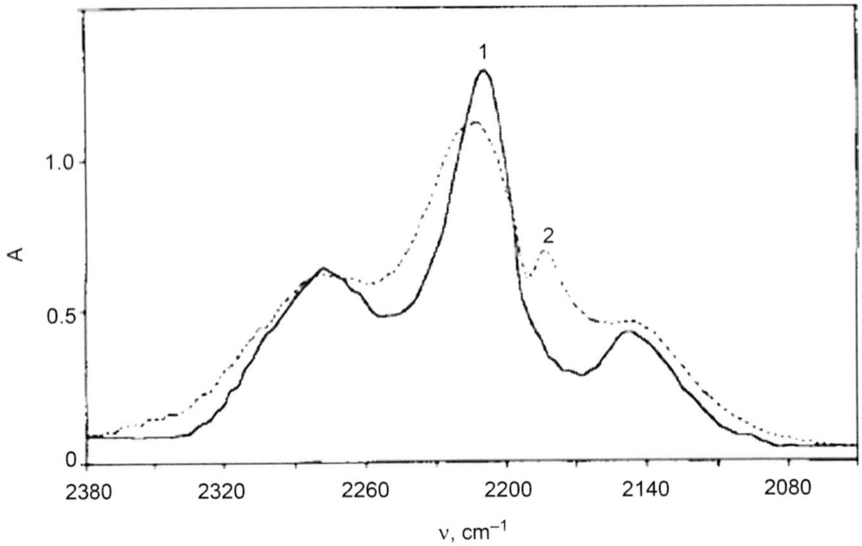

Fig. 16 IR spectra in the ν_{BH} range of (1) Bu_4NBH_4 and (2) Bu_4NBH_4 in the presence of 1 equiv $(CF_3)_2CHOH$. $c = 0.08$ mol/L, CH_2Cl_2. Reprinted with permission from ref. 43. Copyright © 1998 American Chemical Society.

Table 8 Calculated ν_{EH} band positions (in cm^{-1}), frequency shifts for dihydrogen bonded complex (in parentheses) with respect to free EH_4^- and potential energy distribution (PED, in %).

	BH_4^- [33]	Assignment (PED)	AlH_4^- [33]	Assignment (PED)	GaH_4^- [33]	Assignment (PED)
Free	2263	A_1	1732	$\nu AlH\ A_1$	1762	A_1
	2320	T_2	1644	$\nu AlH\ T_2$	1685	T_2
CH_3OH complex	2309	0.87 (νBH_1)	1768	νAlH_1 (0.30)	1791	νGaH_1 (0.35)
						νGaH_2 (0.32)
				νAlH_2 (0.28)		νGaH_3 (0.24)
				νAlH_3 (0.25)		
	2295	νBH_1 (0.12)	1695	νAlH_1 (0.62)	1732	νGaH_1 (0.55)
		νBH_2 (0.44)		νAlH_2 (0.34)		νGaH_2 (0.43)
		νBH_3 (0.43)				
	2287	νBH_2 (0.50)	1689	νAlH_2 (0.34)	1723	νGaH_2 (0.22)
		νBH_3 (0.51)		νAlH_3 (0.61)		νGaH_3 (0.70)
	2224	νBH_4 (0.95)	1657	νAlH_4 (0.76)	1666	νGaH_4 (0.70)

Cp^*MoH_3(dppe) with the same proton donor, was found by DFT calculations.[95] The calculated ν_{IrH} normal modes have different contributions from different Ir–H bond vibrations, and two out of three appear at lower frequencies in the dihydrogen bonded complex in comparison to those in free trihydride. These findings are in agreement with the experimental data: the ν_{IrH} intensity increase and the change of the band shape were observed in the IR spectra, whereas up-field shift and T_{1min} shortening were found for hydride resonance (δ_{IrH}) in the NMR spectra evidencing dihydrogen bond formation.[95]

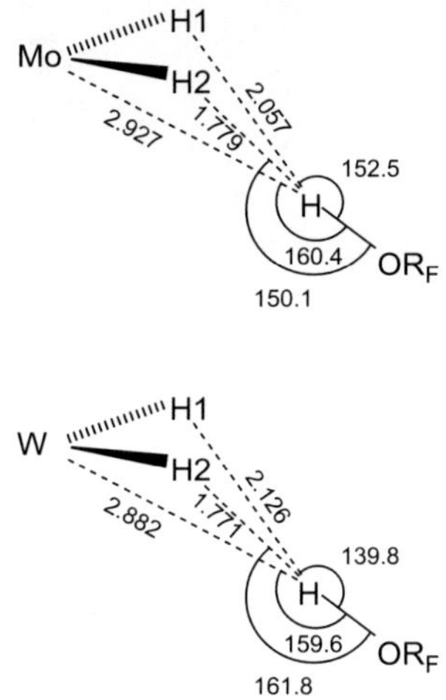

Fig. 17 Comparison of H-bond lengths and angles for the model complexes $CpM(dhpe)H_3 \cdot HOCH_2CF_3$ adducts. Adapted with permission from ref. 30. Copyright © 2008 WILEY-VCH Verlag GmbH & Co. KGaA, Weinheim.

Table 9 ν_{MH} band shifts in dihydrogen bonded complexes of $(PP_3)MH_2$ with CFH_2CH_2OH and of Cp*MH(dppe) with CF_3CH_2OH.[a]

	$(PP_3)MH_2 + CFH_2CH_2OH$ [a]			Cp*MH(dppe) + CF_3CH_2OH	
	$\Delta\nu_{MH}{}^{eq}$, cm^{-1}	$\Delta\nu_{MH}{}^{ax}$, cm^{-1}	E_j	$\Delta\nu_{MH}$, cm^{-1}	E_j
Fe	− 5	+ 8	1.12[21]	− 14[b]	1.34[17]
Ru	− 7	+ 13	1.33[21]	− 14[c]	1.39[16]
Os	− 12	+ 20	1.67[21]	intensity increase[c] − 5[d]	1.47[14]

[a] $\Delta\nu_{MH} = \nu_{MH}$(complex)-$\nu_{MH}$(free).
[b] in CH_2Cl_2 at 200 K.
[c] in n-hexane/CH_2Cl_2 (2:1 v/v) at 200 K.
[d] Indole in methylcyclopentane/CH_2Cl_2 (2:1 v/v) at 200 K.

As is mentioned above, the hydride ligand basicity increases on going from boron to gallium tetrahydride and the band shift, $\Delta\nu_{EH}$, caused by dihydrogen bond formation increases accordingly. A similar effect was observed for dihydrogen bonded complexes of the $(PP_3)MH_2$ hydride series (Table 9). In complexes with CFH_2CH_2OH the preferred hydrogen bonding site is the equatorial hydride ligand. This is evidenced by a low-frequency shift of $\nu_{MH}{}^{eq}$ and high-frequency shift of the non-bonded $\nu_{MH}{}^{ax}$ band.[21] Both $\Delta\nu_{MH}$ increase on descending the group as does the hydrides basicity (E_j).[65,96] A basicity increase on going from iron to osmium (although not that well pronounced) was also found for the Cp*MH(dppe) series. However, in this case the low-frequency shift of the ν_{MH} band was readily observed only for the iron congener. For ruthenium we had to find the conditions at which the low-frequency shift of ν_{RuH} is observable[16] and it turned to be the same as for the iron interaction with the same alcohol (Table 9). Under similar conditions only the increase of the ν_{OsH} band

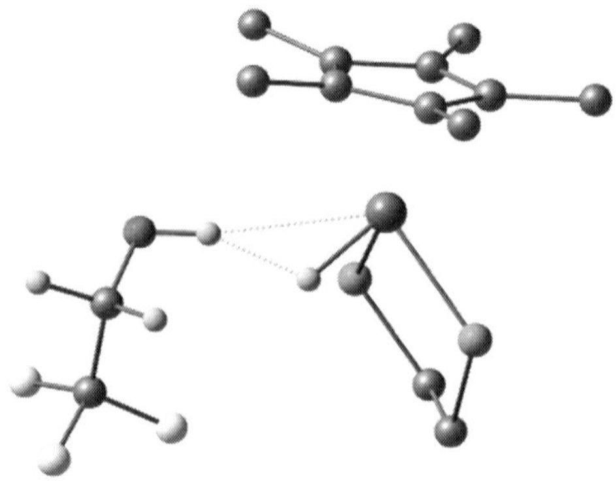

Fig. 18 DFT (B3PW91) optimised structure of dihydrogen bonded complex of Cp*MH(dppe) with CF_3CH_2OH (M = Os) showing simultaneous interaction of proton donor with hydride ligand and metal atom.

intensity was observed for Cp*OsH(dppe).[14] The extensive quantum-chemical analysis for dihydrogen bonded complexes of this series (Fig. 18)[14,97] have shown the substantial impact of hydrogen bonding to the metal atom that increases on descending the group. A M···HA hydrogen bond should cause M-H bond shortening and an high-frequency ν_{MH} shift.[14,58] Thus, working in the direction opposite to that of MH···HA bonding, the interaction with the metal diminishes the low-frequency ν_{MH} shift caused by dihydrogen bond formation. When the MH···HA and M···HA interactions are similarly strong like for Cp*(dppe)OsH···HA, only the ν_{MH} intensity increase is observed in the IR spectra. Note, that the hydride resonance shift in ^{1}H NMR spectra is also lower than expected for this system.[14]

Protonated species

Hydrogen bond formation is the first incipient step of proton transfer. Dihydrogen bonding is shown to precede the proton transfer to hydride ligand yielding η^2-H_2 complexes. When the contribution of the M···HA interaction in the dihydrogen bond is very strong like in the case of Cp*(dppe)OsH···HA or Cp*WH_3(dppe)···HA (see above) the classical cationic hydride is formed as the result of proton transfer. Vibrations of the M(η^2-H_2) moiety are discussed in details *e.g.* in refs. 98, 99. Unfortunately, very often IR spectra (measured in 3000–1500 cm^{-1} region) do not reveal any absorbance for $\nu_{H\text{-}H}$ and ν_{MH2} vibrational modes, since these bands are often hidden under strong CH vibrations of ligands or masked by the IR absorption bands of conventional hydrocarbon solvents.[100]

Cationic classical polyhydrides, $[L_mMH_n]^+$, form as the result of proton transfer to the core metal of neutral L_mMH_{n-1} hydrides or of cationic η^2-H_2 complex isomerisation. Their IR spectra feature ν_{MH} bands, which are shifted to higher frequencies by more than 60 cm^{-1} relative to their neutral precursors. The intensity of these bands is usually lower than that of the neutral analogues (L_mMH_n) or neutral precursor (L_mMH_{n-1}) but they still can be observed in the IR spectra. This difference was first analysed for Cp_2MH_n complexes[3] and traced later for many other hydrides.[52] For

Table 10 Computed parameters of M-H bonds and the IR spectroscopic data[a] for Cp*MH(dppe) hydrides and their protonation products, [Cp*MH_2(dppe)]$^+$. Reprinted from ref. 14.

	Cp*MH(dppe)					*trans*-[Cp*M(H)$_2$(dppe)]$^+$			
M	r_{MH}	ν_{MH} calc	F	ν_{MH} ($\Delta\nu_{1/2}$)	ε	ν^1_{MH} ($\Delta\nu_{1/2}$)	ε	ν^2_{MH} ($\Delta\nu_{1/2}$)	ε
Fe	1.494	1920	2.15	1844 (48)	12	1940 (40)	4		
Ru	1.605	2007	2.37	1918 (58)	34	2002 (56)	14	1968 (26)	8
Os	1.627	2141	2.71	2033 (56)	83	2102 (40)	41	2053 (24)	25
						2177 (85)[b]	*16*[b]	*2126 (55)*[b]	*29*[b]

[a] ν_{MH} in cm^{-1}; half-height full widths, $\Delta\nu_{1/2}$, in parenthesis; molar absorption coefficient, ε, in L mol^{-1} cm^{-1}; force constants, F, in mdyne/Å.
[b] *cis*-H_2.

example, low temperature protonation of Cp*MH(dppe) hydrides yields [Cp*M(η^2-H_2)(dppe)]$^+$ complexes for M = Fe, Ru[16,58] and classical *cis*-[Cp*M(H)$_2$(dppe)]$^+$ for M = Os.[15] For all three metal hydrides the ν_{MH} band of the starting hydride disappear upon protonation but only in the case of osmium new ν_{MH} bands appear in the IR spectra (Table 10). When these kinetic protonation products transform into thermodynamically more stable *trans*-[Cp*M(H)$_2$(dppe)]$^+$ the corresponding ν_{MH} bands appear in the IR spectra for all three metals (Table 10). Thus, such spectral changes can be used to monitor the protonation reaction.

We already discussed the symmetry of ν_{MH} vibrations in *trans*-[Cp*M(H)$_2$(dppe)]$^+$ complexes. Analysis of the data in Table 10 shows the decrease of the difference between ν^1_{MH} and ν^2_{MH} on ascending the group ($\Delta\nu = \nu^s_{MH2} - \nu^{as}_{MH2} = 51$ cm^{-1} for Os and 35 cm^{-1} for Ru) leading to band superposition for Fe. Another interesting feature is a decrease of the band shift upon protonation on descending the group. Taken as the difference between the starting hydride $\nu_{M\text{-}H}$ band and the higher intensity ν_{MH2} band of the *trans*-dihydride, the band shifts, Δν, are 100, 86 and 71 cm^{-1} for M = Fe, Ru and Os, respectively. A similar high-frequency shift of ν_{OsH} by 74 cm^{-1} was observed upon protonation of the structurally related complex OsH[η^5-$C_5H_4(CH_2)_2NMe_2$](P*i*Pr_3)$_2$ yielding the cationic dihydride.[55,57] The relative decrease of $\Delta\nu_{MH}$ on descending the group reflects the smaller change in the M-H force constants and bond length for the heavier metal (compare Table 1 and Table 10) upon a proton transfer and a change of the metal oxidation state (from d^6 to d^4 in this case).

Protonation causes a substantial change in the structure and properties of a metallohydride, and obviously vibrations of other ligands are sensitive to these changes. Thus, upon the formation of [(PP_3)MH(η^2-H_2)]$^+$ the two ν_{MH} bands of the starting dihydrides are replaced by weaker and broad high-frequency shifted bands, ν_{MH}^{eq}, due to the terminal hydride ligand in the corresponding protonated complex (1912 cm^{-1} for Ru and 2012 cm^{-1} for Os in CH_2Cl_2).[21] Carbonyl stretching vibrations turned to be sensitive not only to protonation but to the structure of the cationic hydride formed (Table 11). The ν_{CO} bands of neutral hydrides shift to higher frequency upon protonation but those of classical dihydrides appear at 25–100 cm^{-1}

Table 11 Change of ν_{CO} bands position (in cm^{-1}) in hydride complexes upon protonation and as the result of dihydrogen/dihydride dichotomy.

[MH]	ν_{CO}(MH)	[M(η^2-H_2)]	ν_{CO}(M(η^2-H_2))	ν_{CO}(M(H)$_2$)
Cp*OsH$(CO)_2$[101]	2001, 1941		2091, 2040[102]	2118, 2068[102]
Cp*MoH(CO)$(PMe_3)_2$[103]	1776^a (1751^b)		1852^a	1954^b
ReH_2(CO)(NO)$(PMe_3)_2$[73]	1962		2053	
		CpNb$(CO)_3$(η^2-H_2)[100]	2001, 1915, 1902	2053, 2006, 1966

^{a} in THF.
^{b} in CH_2Cl_2.

higher frequency than the ν_{CO} bands of the corresponding dihydrogen complexes. This is true even for neutral complexes, which exhibit dihydrogen-dihydride dichotomy (Table 11). Thus, IR spectra in liquid Xe show substantially higher frequencies (50–100 cm^{-1}) for the ν_{CO} bands of CpNb$(CO)_3(H)_2$ compared to the dihydrogen analog CpNb$(CO)_3$(η^2-H_2).[100] The difference is consistent with the higher formal oxidation state of dihydrides compared to dihydrogen complexes. For Cp*MoH(CO)$(PMe_3)_2$ we have shown also the correlation between the ν_{CO} frequency and position of the δ_{CO} chemical shifts in the ^{13}C NMR spectra for neutral and protonated forms.[103] So far such correlations were known for various carbonyl complexes,[104,105] the ^{13}C shifts and ν_{CO} frequencies being a function of the ligand substituent properties (like in 4-substituted pyridine ligands[106]).

Finally, we would like to mention that ν_{CO} bands can be sensitive not only to formation of hydrogen bond of molecular type, but to the strength of cation-anion interaction within the $[M(\eta^2\text{-}H_2)]^+A^-$ ion pair. For example, in the case of $[CpRu(CO)(PCy_3)(\eta^2\text{-}H_2)]^+OR^-$ obtained by protonation of the corresponding hydride the ν_{CO} bands appear at a different frequency depending on the HOR used; 2004 cm^{-1} for CF_3COOH, 1978 cm^{-1} for $(CF_3)_3COH$, 1972 cm^{-1} for $(CF_3)_2CHOH$.[84] These positions are lower than that of the BF_4^- salt, the band shift increasing with the anion basicity and, thus, with the $[M(\eta^2\text{-}H_2)]^+\cdots OR^-$ hydrogen bond strength.

Conclusions

Application of variable temperature IR spectroscopy can provide a valuable impact to the understanding of organometallic hydride structure and reactivity. It relies on the high localisation of the ν_{MH} normal modes and the convenience of the spectral range of MH stretching vibrations. IR spectra give very important insight into the dependence of the ν_{MH} frequency and, consequently, of the M-H bond strength on a metal atom position in the Periodic Table, its oxidation state and on ligand properties. The characteristic time of IR spectroscopy allows observing separately the bands of species which undergo fast exchange on the NMR time scale, *e.g.* different conformations of organometallic hydrides or equilibria between free and hydrogen bonded species. The studies of unconventional hydrogen bonds, especially dihydrogen bonds, are an example, where variable temperature

IR spectra provide the possibility to observe the separate bands of free and bonded proton donor groups, molecular and ionic complexes formed after proton transfer, and importantly to access the thermodynamic parameters of hydrogen bond and proton transfer equilibria. The combination of several spectroscopic techniques (IR, multinuclear NMR, UV-VIS, all at variable temperatures) with computational analysis is very useful in the band assignment, determination of structure and the thermodynamic parameters and, ultimately, of reaction mechanisms.

References

1 D. C. Grills, J. J. Turner and M. W. George, in *Comprehensive Coordination Chemistry II*, Pergamon, Oxford, 2003, vol. 2, pp. 91–101.
2 K. Nakamoto, *Infrared and Raman Spectra of Inorganic and Coordination Compounds*, 5th Edn, John Wiley & Sons Inc., New York, 1997.
3 R. B. Girling, P. Grebenik and R. N. Perutz, *Inorg. Chem.*, 1986, **25**, 31–36.
4 T. Polzer and W. Kiefer, *J. Organomet. Chem.*, 1994, **472**, 303–310.
5 B. Ma, C. L. Collins and H. F. Schaefer, *J. Am. Chem. Soc.*, 1996, **118**, 870–879.
6 L. Andrews, *Chem. Soc. Rev.*, 2004, **33**, 123–132.
7 X. Wang, L. Andrews, I. Infante and L. Gagliardi, *J. Am. Chem. Soc.*, 2008, **130**, 1972–1978.
8 X. Wang and L. Andrews, *Inorg. Chem.*, 2008, **47**, 8159–8166.
9 X. Wang and L. Andrews, *Organometallics*, 2008, **27**, 4273–4276.
10 X. Wang and L. Andrews, *J. Phys. Chem. A*, 2011, **115**, 14175–14183.
11 L. Andrews, *Chem. Soc. Rev.*, 2004, **33**, 123–132.
12 M. Baya, P. A. Dub, J. Houghton, J.-C. Daran, R. Poli, N. V. Belkova, E. S. Shubina, L. M. Epstein and A. Lledys, *Inorg. Chem.*, 2009, **48**, 209–220.
13 T. Polzer, A. Ellebracht, W. Kiefer, U. Wecker and H. Werner, *J. Organomet. Chem.*, 1992, **438**, 319–328.
14 P. A. Dub, N. V. Belkova, O. A. Filippov, G. A. Silantyev, J.-C. Daran, L. M. Epstein, R. Poli and E. S. Shubina, *Eur. J. Inorg. Chem.*, 2010, 1489–1500.
15 P. A. Dub, N. V. Belkova, K. A. Lyssenko, G. A. Silantyev, L. M. Epstein, E. S. Shubina, J.-C. Daran and R. Poli, *Organometallics*, 2008, **27**, 3307–3311.
16 N. V. Belkova, P. A. Dub, M. Baya and J. Houghton, *Inorg. Chim. Acta*, 2007, **360**, 149–162.
17 N. V. Belkova, P. O. Revin, L. M. Epstein, E. V. Vorontsov, V. I. Bakhmutov, E. S. Shubina, E. Collange and R. Poli, *J. Am. Chem. Soc.*, 2003, **125**, 11106–11115.
18 M. H. Jamroz, *Vibrational Energy Distribution Analysis VEDA*, 2004-2010, **4**.
19 C. Bianchini, P. J. Perez, M. Peruzzini, F. Zanobini and A. Vacca, *Inorg. Chem.*, 1991, **30**, 279–287.
20 C. Bianchini, K. Linn, D. Masi, M. Peruzzini, A. Polo, A. Vacca and F. Zanobini, *Inorg. Chem.*, 1993, **32**, 2366–2376.
21 E. I. Gutsul, N. V. Belkova, M. S. Sverdlov, L. M. Epstein, E. S. Shubina, V. I. Bakhmutov, T. N. Gribanova, R. M. Minyaev, C. Bianchini, M. Peruzzini and F. Zanobini, *Chem. Eur. J.*, 2003, **9**, 2219–2228.
22 C. Bianchini, F. Laschi, M. Peruzzini, F. M. Ottaviani, A. Vacca and P. Zanello, *Inorg. Chem.*, 1990, **29**, 3394–3402.
23 C. Bianchini, A. Meli, M. Peruzzini, P. Frediani, C. Bohanna, M. A. Esteruelas and L. A. Oro, *Organometallics*, 1992, **11**, 138–145.
24 E. S. Shubina, N. V. Belkova, A. N. Krylov, E. V. Vorontsov, L. M. Epstein, D. G. Gusev, M. Niedermann and H. Berke, *J. Am. Chem. Soc.*, 1996, **118**, 1105–1112.

25 A. A. H. van der Zeijden, C. Sontag, W. Bosch, V. Shklover, H. Berke, D. Nanz and W. von Philipsborn, *Helv. Chim. Acta*, 1991, **74**, 1194–1204.
26 H. Jacobsen, *Chem. Phys.*, 2008, **345**, 95–102.
27 H.-U. Hund, U. Ruppli and H. Berke, *Helv. Chim. Acta*, 1993, **76**, 963–975.
28 N. V. Belkova, E. S. Shubina, A. V. Ionidis, L. M. Epstein, H. Jacobsen, A. Messmer and H. Berke, *Inorg. Chem.*, 1997, **36**, 1522–1525.
29 H. Jacobsen, V. Jonas, D. Werner, A. Messmer, J. C. Panitz, H. Berke and W. Thiel, *Helv. Chim. Acta*, 1999, **82**, 297–307.
30 N. V. Belkova, M. Besora, M. Baya, P. A. Dub, L. M. Epstein, A. Lledós, R. Poli, P. O. Revin and E. Shubina, *Chem. Eur. J.*, 2008, **14**, 9921–9934.
31 N. V. Belkova, P. O. Revin, M. Besora, M. Baya, L. M. Epstein, A. Lledys, R. Poli, E. S. Shubina and E. V. Vorontsov, *Eur. J. Inorg. Chem.*, 2006, 2192–2209.
32 B. Pleune, D. Morales, R. Meunier-Prest, P. Richard, E. Collange, J. C. Fettinger and R. Poli, *J. Am. Chem. Soc.*, 1999, **121**, 2209–2225.
33 O. A. Filippov, A. M. Filin, V. N. Tsupreva, N. V. Belkova, A. Lledys, G. Ujaque, L. M. Epstein and E. S. Shubina, *Inorg. Chem.*, 2006, **45**, 3086–3096.
34 T. J. Tague and L. Andrews, *J. Am. Chem. Soc.*, 1994, **116**, 4970–4976.
35 X. Wang, L. Andrews, S. Tam, M. E. DeRose and M. E. Fajardo, *J. Am. Chem. Soc.*, 2003, **125**, 9218–9228.
36 X. F. Wang and L. Andrews, *J. Phys. Chem. A*, 2003, **107**, 11371–11379.
37 A. E. Shirk and D. F. Shriver, *J. Am. Chem. Soc.*, 1973, **95**, 5904–5912.
38 C. L. Gross and G. S. Girolami, *Organometallics*, 2007, **26**, 160–166.
39 C. A. Bayse, M. Couty and M. B. Hall, *J. Am. Chem. Soc.*, 1996, **118**, 8916–8919.
40 V. D. Makhaev, *Russ. Chem. Rev.*, 2000, **69**, 795–816.
41 O. A. Filippov, A. M. Filin, N. V. Belkova, V. N. Tsupreva, Y. V. Shmyrova, I. B. Sivaev, L. M. Epstein and E. S. Shubina, *J. Mol. Struct.*, 2006, **790**, 114–121.
42 N. V. Belkova, O. A. Filippov, A. M. Filin, L. N. Teplitskaya, Y. V. Shmyrova, V. V. Gavrilenko, L. M. Golubinskaya, V. I. Bregadze, L. M. Epstein and E. S. Shubina, *Eur. J. Inorg. Chem.*, 2004, 3453–3461.
43 L. M. Epstein, E. S. Shubina, E. V. Bakhmutova, L. N. Saitkulova, V. I. Bakhmutov, A. L. Chistyakov and I. V. Stankevich, *Inorg. Chem.*, 1998, **37**, 3013–3017.
44 E. S. Shubina, E. V. Bakhmutova, L. N. Saitkulova and L. M. Epstein, *Mend. Commun.*, 1997, **7**, 83–84.
45 S. Casper, J.H, *Spectrochim. Acta*, 1960, **16**, 1054–1059.
46 X. Wang and L. Andrews, *J. Phys. Chem. A*, 2002, **106**, 6720–6729.
47 X. Wang and L. Andrews, *J. Phys. Chem. A*, 2003, **107**, 570–578.
48 X. Wang and L. Andrews, *J. Phys. Chem. A*, 2003, **107**, 4081–4091.
49 X. Wang and L. Andrews, *J. Phys. Chem. A*, 2005, **109**, 9021–9027.
50 Z. L. Xiao, R. H. Hauge and J. L. Margrave, *J. Phys. Chem.*, 1991, **95**, 2696–2700.
51 S. Li, H. A. Weimer, R. J. Van Zee and W. Weltner, *J. Chem. Phys.*, 1997, **106**, 2583–2588.
52 K. Abdur-Rashid, T. P. Fong, B. Greaves, D. G. Gusev, J. G. Hinman, S. E. Landau, A. J. Lough and R. H. Morris, *J. Am. Chem. Soc.*, 2000, **122**, 9155–9171.
53 M. A. Esteruelas, A. V. Gomez, F. J. Lahoz, A. M. Lopez, E. Onate and L. A. Oro, *Organometallics*, 1996, **15**, 3423–3435.
54 E. Cayuela, F. A. Jalon, B. R. Manzano, G. Espino, W. Weissensteiner and K. Mereiter, *J. Am. Chem. Soc.*, 2004, **126**, 7049–7062.

55 M. A. Esteruelas, A. M. López, E. Oñate and E. Royo, *Organometallics*, 2004, **23**, 5633–5636.
56 R. Castro-Rodrigo, M. A. Esteruelas, A. M. Lopez, M. Olivan and E. Onate, *Organometallics*, 2007, **26**, 4498–4509.
57 M. A. Esteruelas, A. M. Lopez, E. Onate and E. Royo, *Inorg. Chem.*, 2005, **44**, 4094–4103.
58 N. V. Belkova, E. Collange, P. Dub, L. M. Epstein, D. A. Lemenovskii, A. Lledós, O. Maresca, F. Maseras, R. Poli, P. O. Revin, E. S. Shubina and E. V. Vorontsov, *Chem. Eur. J.*, 2005, **11**, 873–888.
59 W. C. Kwok, C. G. Howard, H. S. Rzepa, R. N. Sheppard, G. Wilkinson, A. M. R. Galas and M. B. Hursthouse, *Polyhedron*, 1982, **1**, 441–451.
60 W. A. Herrmann, H. G. Theiler, E. Herdtweck and P. Kiprof, *J. Organomet. Chem.*, 1989, **367**, 291–311.
61 D. Baudry, M. Ephritikhine and H. Felkin, *J. Organomet. Chem.*, 1982, **224**, 363–376.
62 D. L. Lichtenberger, N. E. Gruhn, A. Rai-Chaudhuri, S. K. Renshaw, J. A. Gladysz, H. Jiao, J. Seyler and A. Igau, *J. Am. Chem. Soc.*, 2002, **124**, 1417–1423.
63 T. Wilczewski, *J. Organomet. Chem.*, 1986, **317**, 307–325.
64 C. L. Gross and G. S. Girolami, *Organometallics*, 1996, **15**, 5359–5367.
65 N. V. Belkova, E. S. Shubina and L. M. Epstein, *Acc. Chem. Res.*, 2005, **38**, 624–631.
66 M. Besora, A. Lledos and F. Maseras, *Chem. Soc. Rev.*, 2009, **38**, 957–966.
67 R. Custelcean and J. E. Jackson, *Chem. Rev.*, 2001, **101**, 1963–1983.
68 N. V. Belkova, L. M. Epstein and E. S. Shubina, *ARKIVOC*, 2008, **iv**, 120–138.
69 G. A. Jeffrey, *An Introduction to Hydrogen Bonding*, Oxford University Press, New York, 1997.
70 A. V. Iogansen, *Spectrochim. Acta A*, 1999, **55**, 1585–1612.
71 A. V. Iogansen, *Theor. Experim. Khim.*, 1971, **7**, 312–317.
72 L. M. Epstein and E. S. Shubina, *Coord. Chem. Rev.*, 2002, **231**, 165–181.
73 N. V. Belkova, E. S. Shubina, E. I. Gutsul, L. M. Epstein, I. L. Eremenko and S. E. Nefedov, *J. Organomet. Chem.*, 2000, **610**, 58–70.
74 O. A. Filippov, V. N. Tsupreva, L. M. Golubinskaya, A. I. Krylova, V. I. Bregadze, A. Lledos, L. M. Epstein and E. S. Shubina, *Inorg. Chem.*, 2009, **48**, 3667–3678.
75 G. N. Patwari, T. Ebata and N. Mikami, *Chem. Phys.*, 2002, **283**, 193–207.
76 L. M. Epstein, E. S. Shubina and A. N. Krylov, *J. Mol. Struct.*, 1994, **322**, 345.
77 S. G. Kazarian, P. A. Hamley and M. Poliakoff, *J. Chem. Soc., Chem. Commun.*, 1992, 994–997.
78 P. A. Hamley, S. G. Kazarian and M. Poliakoff, *Organometallics*, 1994, **13**, 1767–1774.
79 E. S. Shubina, A. N. Krylov, N. V. Belkova, L. M. Epstein, A. P. Borisov and V. D. Mahaev, *J. Organomet. Chem.*, 1995, **493**, 275–277.
80 A. Messmer, H. Jacobsen and H. Berke, *Chem. Eur. J.*, 1999, **5**, 3341–3349.
81 S. Gruendemann, S. Ulrich, H.-H. Limbach, N. S. Golubev, G. S. Denisov, L. M. Epstein, S. Sabo-Etienne and B. Chaudret, *Inorg. Chem.*, 1999, **38**, 2550.
82 V. I. Bakhmutov, Dihydrogen bonds: Principles, Experiments and Applications, John Willey & Sons, Inc., Hoboken, New Jersey, 2008.
83 P. A. Dub, O. A. Filippov, N. V. Belkova, J.-C. Daran, L. M. Epstein, E. S. Shubina and R. Poli, *Dalton Trans.*, 2010, **39**, 2008–2015.
84 N. V. Belkova, M. Besora, L. M. Epstein, A. Lledós, F. Maseras and E. S. Shubina, *J. Am. Chem. Soc.*, 2003, **125**, 7715–7725.

85 D. V. Yandulov, K. G. Caulton, N. V. Belkova, E. S. Shubina, L. M. Epstein, D. V. Khoroshun, D. G. Musaev and K. Morokuma, *J. Am. Chem. Soc.*, 1998, **120**, 12553–12563.
86 L. M. Epstein, L. N. Saitkulova and E. S. Shubina, *J. Mol. Struct.*, 1992, **270**, 325–337.
87 P. W. Schultz, G. E. Leroi and A. I. Popov, *J. Am. Chem. Soc.*, 1996, **118**, 10617–10625.
88 C. P. Horwitz and D. F. Shriver, *Adv. Organometal. Chem.*, 1984, **23**, 219–305.
89 N. V. Belkova, L. M. Epstein and E. S. Shubina, *Eur. J. Inorg. Chem.*, 2010, 3555–3565.
90 N. V. Belkova, E. V. Bakhmutova, E. S. Shubina, C. Bianchini, M. Peruzzini, V. I. Bakhmutov and L. M. Epstein, *Eur. J. Inorg. Chem.*, 2000, 2163–2165.
91 V. I. Bakhmutov, E. V. Bakhmutova, N. V. Belkova, C. Bianchini, L. M. Epstein, D. Masi, M. Peruzzini, E. S. Shubina, E. V. Vorontsov and F. Zanobini, *Can. J. Chem.*, 2001, **79**, 479–489.
92 N. V. Belkova, A. V. Ionidis, L. M. Epstein, E. S. Shubina, S. Gruendemann, N. S. Golubev and H. H. Limbach, *Eur. J. Inorg. Chem.*, 2001, 1753–1761.
93 E. V. Bakhmutova, V. I. Bakhmutov, N. V. Belkova, M. Besora, L. M. Epstein, A. Lledós, G. I. Nikonov, E. S. Shubina, J. Tomas and E. V. Vorontsov, *Chem. Eur. J.*, 2004, **10**, 661–671.
94 J. Andrieu, N. V. Belkova, M. Besora, E. Collange, L. M. Epstein, A. Lledós, R. Poli, P. O. Revin, E. S. Shubina and E. V. Vorontsov, *Russ. Chem. Bull.*, 2003, **52**, 2679–2682.
95 A. Rossin, E. I. Gutsul, N. V. Belkova, L. M. Epstein, L. Gonsalvi, A. Lledos, K. A. Lyssenko, M. Peruzzini, E. S. Shubina and F. Zanobini, *Inorg. Chem.*, 2010, **49**, 4343–4354.
96 A. V. Iogansen, *Theor. Experim. Khim.*, 1971, **7**, 302–311.
97 O. A. Filippov, N. V. Belkova, L. M. Epstein, A. Lledós and E. S. Shubina, *ChemPhysChem*, 2012, *in press* DOI: 10.1002/cphc.201200097
98 B. R. Bender, G. J. Kubas, L. H. Jones, B. I. Swanson, J. Eckert, K. B. Capps and C. D. Hoff, *J. Am. Chem. Soc.*, 1997, **119**, 9179–9190.
99 G. J. Kubas, *Chem. Rev.*, 2007, **107**, 4152–4205.
100 M. W. George, M. T. Haward, P. A. Hamley, C. Hughes, F. P. A. Johnson, V. K. Popov and M. Poliakoff, *J. Am. Chem. Soc.*, 1993, **115**, 2286–2299.
101 J. K. Hoyano, C. J. May and W. A. G. Graham, *Inorg. Chem.*, 1982, **21**, 3095–3099.
102 R. M. Bullock, J.-S. Song and D. J. Szalda, *Organometallics*, 1996, **15**, 2504–2516.
103 P. A. Dub, N. V. Belkova, O. A. Filippov, J.-C. Daran, L. M. Epstein, A. Lledós, E. S. Shubina and R. Poli, *Chem. Eur. J.*, 2010, **16**, 189–201.
104 N. Legrand, A. Bondon, G. Simonneaux, C. Jung and E. Gill, *FEBS Letters*, 1995, **364**, 152–156.
105 W. T. Potter, J. H. Hazzard, M. G. Choc, M. P. Tucker and W. S. Caughey, *Biochemistry*, 1990, **29**, 6283–6295.
106 J. W. Box and G. M. Gray, *Inorg. Chem.*, 1987, **26**, 2774–2778.

Raman spectroscopy of graphene

C. Casiraghi[a,b]
DOI: 10.1039/9781849734899-00029

Graphene attracts enormous interest because of its unique properties. Giant intrinsic charge mobility at room temperature makes it a potential material for nano-electronics. Its optical and mechanical properties are ideal for micro- and nano-mechanical systems, thin-film transistors, transparent and conductive composites, electrodes and for photonics. This Chapter will show that Raman Spectroscopy is a very powerful tool for the investigation of graphene, being very sensitive to phonons, electronic states, defects and to the interaction between the fundamental excitations of graphene.

1 Introduction

Carbon can exist in different forms. Diamond and graphite are the most famous carbon-based materials. Diamond has a tetrahedral structure, where the carbon bond can be described by the sp^3 hybridization model. In contrast, graphite is a layered material, *i.e.* it is composed of many sheets piled one on the top of each other. Each sheet has a hexagonal honeycomb structure. The layers are held together by a weak interlayer Van der Waals force. The carbon bond in graphite can be described by the sp^2 hybridization model. This leads to different properties: diamond is transparent, it is an insulator and it is mechanically strong. Graphite is grey and most of its properties are highly anisotropic. For example, graphite can only conduct electricity within the plane of the layers.

Low-dimensional graphitic forms have been discovered in the last two decades such as fullerenes [1], nanotubes [2] and graphene [3], Fig. 1. They all show peculiarities in their electronic structure and outstanding mechanical and vibrational properties [4–9]. In particular, graphene is a flat monolayer of carbon atoms tightly packed into a two-dimensional (2D) honeycomb lattice, and it is a basic building block for graphitic materials of all other dimensionalities: it can be wrapped up into 0D-fullerenes, rolled into 1D-nanotubes or stacked into 3D-graphite.

Graphene is a single sheet of graphite, so one would expect these two materials to have similar properties. However, size and dimensionality strongly affect the properties of a material: graphene is almost transparent to light, it is extremely hard and its electrons flow through the plane like light goes through glass, *i.e.* they move as if they were massless through the hexagonal lattice, leading to giant intrinsic charge mobility [10]. This behaviour drastically changes with the number of layers, *i.e.* the unique properties of graphene are strictly related to its mono-atomic thickness [10]. Graphene shows a wide range of outstanding properties: it can be elastically stretched (up to 20%), it shrinks with increasing

[a] *School of Chemistry and Photon Science Institute, University of Manchester, UK.; E-mail: cinzia.casiraghi@manchester.ac.uk*
[b] *Fachbereich Physik, Freie Universität Berlin, D-1419, Germany.*

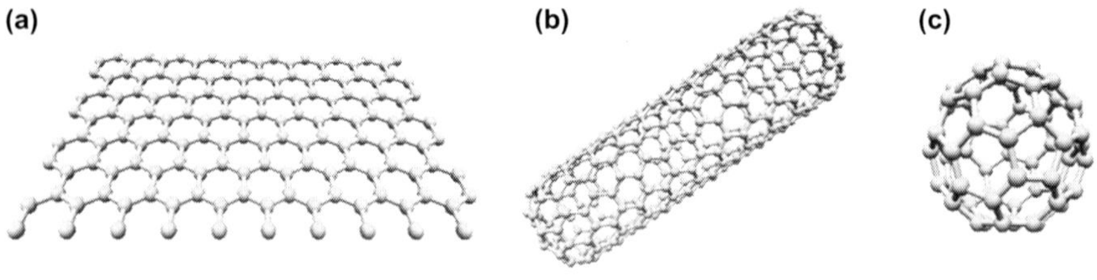

Fig. 1 Low dimensional graphitic materials: a) 2D graphene; b) 1D carbon nanotube and c) 0D fullerene.

temperature, it can be bent and cut in different shapes and it is impermeable to gases [11].

In term of applications, graphene has enormous potential for use in ultra-fast electronic transistors and flexible displays. Moreover, graphene can be used in composite materials and in electric batteries due to its large surface-to-volume ratio and high conductivity, for transparent membranes due to its atomic thickness, in micro-mechanical resonators due to its robustness and light weight, and chemical detectors due to it selective reactivity [11].

Raman Spectroscopy is an optical detection technique based on the inelastic scattering of light by a crystal. When light hits a crystal, most of the light is elastically scattered. However, a tiny fraction is able to see the crystal as composed of atoms, *i.e.* the light interacts with the atomic vibrations. This produces a change in the energy of the scattered light. By looking at the difference between incident and scattered light energy, one can derive information on the crystal's structure and chemistry. In the case of graphene, the Raman scattering process is strongly sensitive to the electronic structure because it involves real electronic states. Furthermore a strong electron-phonon coupling produces unique features in the phonon dispersion. This makes Raman spectroscopy a powerful tool for the investigation of graphene.

The present review aims at giving a general understanding of the Raman scattering process in graphene and at helping the reader to understand how to characterize graphene using Raman Spectroscopy. In Section 2, I will give a short overview of the structure and phonons of graphene; in Section 3, I will introduce the Raman scattering process in graphene, then I will show how to probe doping, defects, edges and strain by Raman Spectroscopy. In the last Section of this review I will shortly describe other spectroscopic techniques for graphene's characterization.

2 Graphene

2.1 Electronic structure

Graphene is a one-atom-thick planar layer of graphite, where the atoms are packed in a hexagonal ("honeycomb") crystal lattice. The carbon atoms are sp^2-bonded with a bond length of 1.42 Å. The crystal lattice has two atoms per unit cell, A and B, and it is rotationally symmetric for rotations of 120° around any lattice point. One can view the honeycomb lattice as a triangular Bravais lattice with a basis of two atoms per unit cell, Fig. 2a.

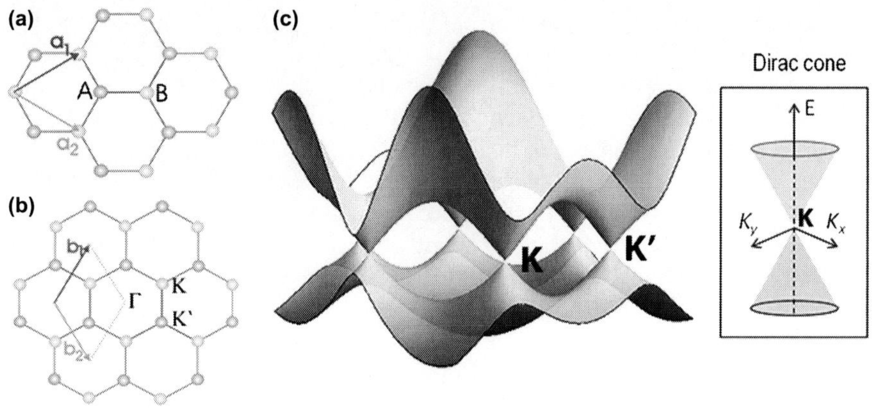

Fig. 2 (Colour online) a) Crystal lattice of graphene. This is composed by two sub-lattices defined by the atoms A and B; b) Brillouin zone of graphene, showing the high-symmetry points and the reciprocal vectors; c) Electronic structure of graphene calculated by tight-binding; close to **K** and **K**′ points the electronic structure is linear. This gives a "cone of carriers" at the corners of the 2D Brillouin zone (inset).

If we now move from real space into reciprocal space, the Brillouin zone associated with the crystal lattice is also hexagonal and it shows characteristic high-symmetry points: the centre is called Γ point, while two consecutive corners are denoted as **K** and **K**′ points, Fig. 2b.

Figure 2c shows the electronic structure of graphene described by a simple tight-binding Hamiltonian: the electronic wavefunctions from different atoms overlap. However, such an overlap between the $p_z(\pi)$ orbital and the p_x and p_y orbitals is zero by symmetry. Thus, the p_z electrons form the π band, and they can be treated independently from the other valence electrons. The two sub-lattices lead to the formation of two bands, π and π*, which intersect at the corners of the Brillouin zone. This yields the conical energy spectrum (Dirac cone, inset in Fig. 2c) near the points **K** and **K**′, which are called Dirac points. The bottom cone (equivalent to the HOMO molecular orbital) is fully occupied, while the top cone (equivalent to the LUMO molecular orbital) is empty. The Fermi level E_F is chosen as the zero-energy reference and lies at the Dirac point. Consequently, graphene is a special semimetal or zero-band-gap semicondutor, whose intrinsic Fermi surface is reduced to the six points at the corners of the two-dimensional Brillouin zone.

As a result, quasi-particles in graphene exhibit the linear dispersion relation: $E = \pm \hbar v_F |k|$, where v_F is the Fermi velocity ($\sim 10^6$ m/s), as if they were massless relativistic particles. Thus, graphene's quasiparticles behave differently from those in conventional metals and semiconductors, where the energy spectrum can be approximated by a parabolic dispersion relation. Electron transport in all known condensed-matter systems is described by the (non-relativistic) Schrödinger equation and relativistic effects are usually negligible. In contrast, the electrons of graphene are described by the (relativistic) Dirac equation, *i.e.* they mimic relativistic charged particles with zero rest mass and constant velocity [10].

Graphene is found to exhibit a pronounced ambipolar electric field effect, *i.e.* charge carriers can be tuned continuously between electrons to holes in concentrations as high as 10^{14} cm^{-2} [3]. From a chemical point of view, this means that graphene can participate in reactions as either reducing agent (electron donor) or an oxidizer (electron acceptor). The charge mobility μ is practically temperature independent and it can exceed 10000 cm^2/Vs for graphene deposited on an *oxidized silicon substrate* [10]. This means that charge carriers in these devices can travel submicron distances without scattering. The mobility of the charge carriers is a very important parameter because it is a measure of the electronic quality of a material: the higher the mobility, the higher the quality. However, the theoretically predicted mobility of graphene should be a few orders of magnitude higher than that one reported for devices on silicon [10]. This means that the transport properties of these devices are fully dominated by extrinsic scatterers such as impurities, ripples, defects and interaction with a rough substrate [12, 13]. Higher mobilities can be achieved by removing the substrate [14] or by depositing graphene on different substrates [15] (Section 3).

Because of the massless Dirac fermions, graphene shows a variety of unusual electronic properties, such as an anomalous integer Quantum Hall Effect, absence of Anderson localization, a "minimum conductivity", *etc.* These peculiarities in the electronic properties are described in dedicated reviews [10, 16].

2.2 Optical properties

Specially prepared graphene crystals, covering sub-millimeter apertures in a metal scaffold, have been used for optical studies [17]. A Xenon lamp was focused on the graphene membrane and the transmitted light intensity was measured by a spectrometer. The recorded signal was then compared with the one obtained by directing the light beam through either an empty space or, as a double check, another aperture of the same size but without graphene [17].

These measurements gives a graphene's opacity of $2.3 \pm 0.1\%$ and negligible reflectance ($<0.1\%$) in the visible range [17]. Thus, despite being one-atom thick, graphene is able to absorb a significant fraction of light and its transmittance is $\sim 97.7\%$. The opacity is practically independent on the excitation wavelength in the visible range [17]. The opacity is found to increase with the membranes thickness so that each graphene layer adds another 2.3%. This shows that few-layers graphene can be approximated as a stack of independent graphene planes, in agreement with Rayleigh scattering measurements (Section 4).

These unique optical properties are strongly related to the linear electronic structure of graphene: the theory suggests that for a material with a 2D Dirac spectrum with a conical dispersion relation, such as graphene, the optical conductivity G is constant and equal to $G_0 = e^2/4\hbar$, for a photon energy larger than temperature and E_F [18, 19]. This translates into constant transmittance (T), reflectance (R) and absorbance (A) [18]. In particular, $T = (1 + 2\pi G_0/c)^{-2} = (1 + 0.5\pi\alpha)^{-2} \sim 1\text{-}\pi\alpha$ for normal incident light, where c is the speed light and $\alpha = e^2/\hbar c$. From the optical measurements in the visible frequencies range, $\pi\alpha = 2.3\%$, so $G = (1.01 \pm 0.04)\ e^2/4\hbar$ in good agreement with the theory describing the optical properties of a system with ideal Dirac fermions [20].

2.3 Phonons

Raman Spectroscopy probes phonons, so to understand the Raman spectrum we have to look at the phonon dispersion of graphene. Since graphene contains two carbon atoms per unit cell (Fig. 2a), we have six phonon dispersions. Three of the six phonons are acoustic (A) and the other three are optic (O) phonon modes. One of the three modes is an out-of-plane (oT) phonon mode and the other two are in-plane modes, one longitudinal (L) and one transverse (iTO). If we now focus at the highest optical branches at **Γ**, we can see that LO and iTO phonons modes are degenerate. This phonon mode is Raman active and it corresponds to a E_{2g} mode. If we move away from the centre of the Brillouin zone, we can see that the degeneracy of the LO and iTO phonons disappears, but another degeneracy, involving the LA and LO branches, appears at the **K** point. This corresponds to a phonon with E' symmetry. Here the iTO branch is non degenerate and belongs to the A_1' irreducible representation.

Usually Raman Spectroscopy is able to probe only phonons close to **Γ**, but as I will show in Section 3, the Raman scattering process in graphene involves resonant and defect-activated processes, which allow probing also phonons very far from **Γ**.

Another unique characteristic of the Raman scattering process in graphene is given by the coupling of electrons and phonons, which cause an interesting effect known as a Kohn Anomaly [21]. The Kohn Anomaly is responsible for the two kinks observed in the phonon dispersion of the LO and TO phonon branches at the **Γ** and **K** points, marked by the dotted squares in Fig. 3. In general, atomic vibrations set up a potential due to the motion of the ions. The electrons move to screen this field. The ions will now interact with one another *via* this screened field. This produces a change in the phonon frequency of the atomic vibration. That is the origin of the Kohn anomaly effect [22]. In a metallic system, the screening produced by the electron can be very strong, *i.e.* it produces a large change of the phonon frequency. However, not all the phonons are affected: the screening depends on the shape of the Fermi surface because it occurs only for phonons with wave-vector $\boldsymbol{q} = 2\boldsymbol{k}_F$, where $\boldsymbol{k}_F$ is the wave-vector of the Fermi surface. In the case of (undoped) graphene, the Fermi surface lies in two points: at **K**

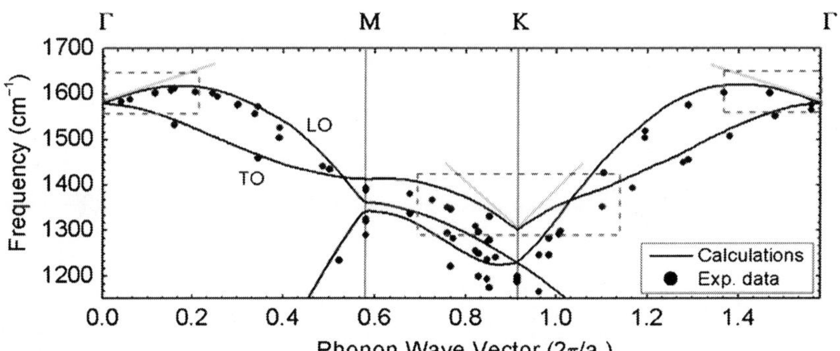

Fig. 3 (Colour online) Calculated and experimental phonon dispersion of the highest optical branches of graphene. Adapted from Ref. 25.

and at $\mathbf{K}'$. Thus, we do expect two Kohn anomalies in graphene. Since the two Dirac points are connected by the vector K, the Kohn anomaly occurs at $\mathbf{q} = \mathbf{\Gamma}$ and at $\mathbf{q} = \mathbf{K}$. Because of this, the TO branch is characterized by the highest electron-phonon coupling, which can be quantified by the slope of the TO band at **K** [21].

Due of the Kohn-anomalies, simple models based on force constants cannot reproduce the peculiar behaviour of the phonon branches at **Γ** and **K**. Only recent calculations have been able to correctly reproduce the phonon dispersion of graphene at **K** [23].

As we will show in the Section 3, the presence of the Kohn anomaly can be easily detected by Raman spectroscopy because the phonon affected by the Kohn-anomaly at **Γ** is Raman-active (called G peak, Section 3). Consequently, everything that affects the shape of the Fermi surface will also produce a change in the Raman spectrum and in particular on the G peak.

3 Raman spectroscopy

Optical detection relying on light scattering is especially attractive because it is fast, sensitive, and non-destructive. Light interaction with matter can be elastic or inelastic, and this corresponds to Rayleigh and Raman scattering, respectively. The Raman effect was first observed in molecules by Raman in 1930, but nowadays Raman spectroscopy is a standard technique for the analysis of crystals and semiconductors [24]. In the macroscopic description of the Raman scattering process in crystals, we consider an infinite crystalline medium at finite temperature. The atoms of the structure always undergo vibrations, whose normal modes are quantized by quasi-particles, called phonons. The interaction of light with the crystal induces emission and absorption of a phonon. This produces a change in the energy of the incident light and gives rise to the Raman effect. This description implies a direct interaction of the photon and phonon, which is rather weak in comparison to the electron-phonon interaction. Indeed, the microscopic description requires the use of electrons to describe the Raman scattering process. Within this framework, the Raman scattering process is described as a three-steps process:

i) An electron in its initial state is excited by an incoming photon of frequency ω_i to an excited state.

ii) The electron interacts with a phonon of frequency ω_0 due to electron-phonon coupling.

iii) The electron returns to its initial state by emitting a photon with frequency $\omega_s = \omega_i \pm \omega_0$. The radiation associated with the photons with frequency $\omega_i - \omega_0$ and $\omega_i + \omega_0$ are known as Stokes scattered and anti-Stokes scattered light, respectively. Since the Stokes component has very high intensity compared to the anti-Stokes, a Raman spectrum usually shows only the Stokes component.

Usually the electronic transitions of the three-steps process involve virtual states. When an electronic transition matches with real electronic states, then the Raman scattering process is *resonant*. The resonance can be single, double, triple and so on depending on how many resonant electronic transitions are involved in the Raman scattering process.

Raman scattering has recently emerged as a viable, non-destructive technique for the identification and characterization of graphene [5, 25]. This is because the Raman scattering process in graphene involves resonant and defect-activated processes. In addition, the Kohn-anomalies, introduced in Section 2, also produce strong effects in the Raman spectrum of graphene, which can be used for example to probe excess charge in graphene.

3.1 Background

All sp^2 bonded carbons show common features in their Raman spectra, the so-called G and D peaks, around 1580 and 1360 cm^{-1} respectively [26]. The G peak corresponds to the E_{2g} phonon at the $\mathbf{\Gamma}$ point. The D peak is due to the breathing modes of six-atom rings and requires a defect for its activation [27]. It comes from TO phonons around the **K** point [27, 28], it is active by double resonance [29], and is strongly dispersive with excitation energy due the Kohn Anomaly at **K** [21].

A Raman spectrum also shows overtones and combination modes. In the case of graphene a prominent feature is given by the second order of the D peak, historically named G′ since it is the second most prominent band always observed in the Raman spectrum of graphite and more recently renamed as 2D peak. Since this peak originates from a process where momentum conservation is satisfied by two phonons with opposite wave-vectors, it does not require any defect for its activation, and it is thus always visible in the Raman spectrum.

In general, the Raman scattering process involves the following steps:

i) a laser of frequency ω_{inc} induced excitation of an electron, *i.e.* an electron-hole couple is generated.

ii) electron-phonon scattering. The phonon has frequency ω_{ph} and momentum $\mathbf{q}\sim 0$ because of momentum conservation (*i.e.* Raman spectroscopy is able to probe only phonons close to the centre of the Brillouin zone);

iii) electron-hole recombination with emission of a photon with frequency ω_{sc}.

This simple description is valid only for the G peak, which is associated with a phonon at the centre of the Brillouin zone ($\mathbf{q}\sim 0$). In the case of the D and 2D peaks, the electron-phonon scattering involves an exchanged momentum $\mathbf{q}\sim\mathbf{k}$, so they need to be described with a more complex scattering mechanism.

Let us concentrate first on the G peak: because of the linear electronic dispersion of graphene, in principle the Raman scattering process of the G peak can be always resonant, *i.e.* the electronic transition can always match with real electronic states, no matter the incident excitation frequency. This is described in Fig. 4a in the case of incident resonance (scattered resonance is also possible). In general, a resonant process has a higher probability compared to a non-resonant process, so one would expect the G peak to be described only by resonant processes. However, it has been shown, both experimentally and theoretically, that quantum interference effects cancel the resonant contributions in the G peak Raman intensity. Consequently,

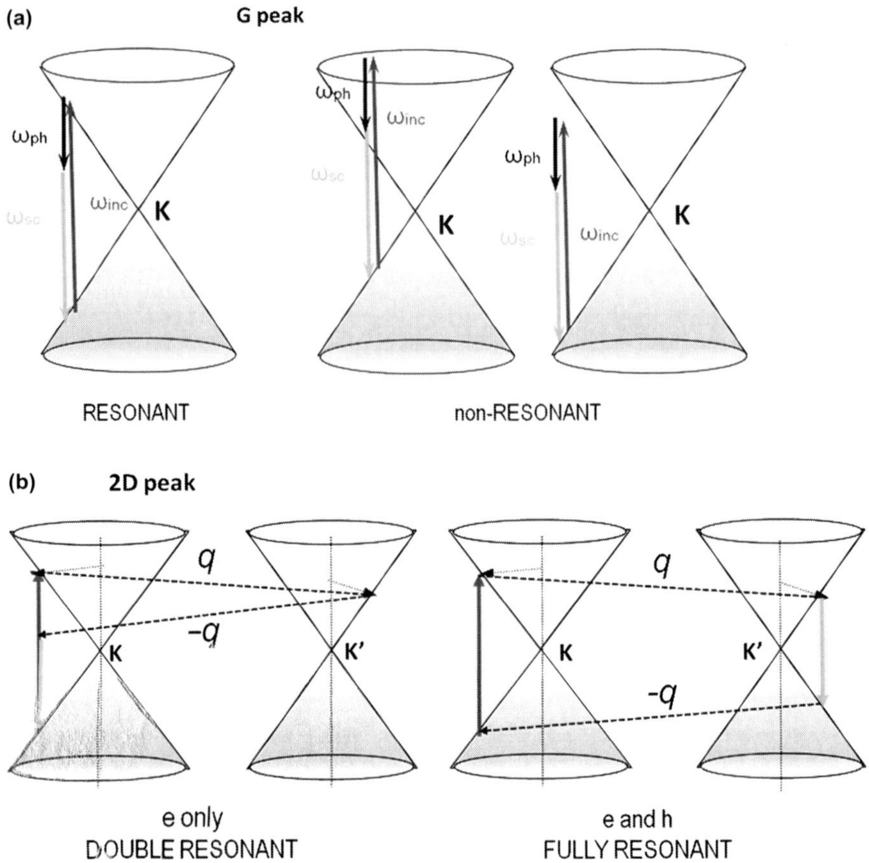

Fig. 4 Raman activation mechanism of the G peak (a) and 2D peak (b). The dotted line represents the phonon scattering with wave-vector $\mathbf{q} \sim \mathbf{k}$. Here, graphene is undoped: all the states are filled up to the Dirac point. The others are empty.

the Raman scattering process associated with the G peak activation is non-resonant [30–32].

In the case of the D and 2D peaks, the electron-phonon scattering involves an exchanged momentum $\mathbf{q} \sim \mathbf{k}$, so the electron-phonon scattering cannot be described by a vertical transition, as in the case of the G peak. The Raman scattering process is inter-valley, *i.e.* the process involves the electronic states of two inequivalent Dirac cones, as shown in Fig. 4b. It has been shown that in the case of the D and 2D peak the Raman scattering process is resonant: at least two steps of the Raman scattering process involve real electronic states [29]. In the case of the 2D peak, the presence of two-phonons with equal and opposite wavevectors allows momentum conservation, while a defect is necessary in the case of the D peak. However, there are two models one can use to describe the resonant activation mechanism: we can imagine that only electrons participate in the Raman scattering process, or we can assume that both electrons and holes are involved, as shown in Fig. 4b. The first model has been widely used in the past years for describing the D and 2D peaks [29]. The second model is based on a fully-resonant picture and it has been first described by Basko [33].

Recent calculations have shown that the resonant Raman scattering process is expected to involve both electron and holes [34].

In conclusion, the D and 2D peak are activated by an inter-valley resonant Raman scattering process. Consequently, these peaks are strongly sensitive to changes in the electronic structure, *i.e.* the phonon probed by Raman spectroscopy strongly depends on the electronic band structure. Thus, every change in the electronic structure will be reflected in a change of the D and 2D peaks. This effect allows identification of graphene.

3.2 Identification of graphene

Graphene has been theoretically studied for many years since it has been used as a model to describe the properties of graphite and carbon nanotubes. However, only recently it has been possible to produce ultra-thin graphitic films, including single-layers [35]. Graphene samples can be easily obtained by micro-mechanical cleavage of graphite, *e.g.* by slicing down a thick piece of graphite with adhesive tape to a single-layer of graphite [35]. Note that in the past few years there have been strong advances in the development of deposition techniques able to give large area and good quality graphene for mass production. Chemical Vapour Deposition is generating significant interest, since it allows uniform deposition of graphene on copper and nickel [36]. In addition, chemical methods to exfoliate graphite and stabilize individual sheets in solution are under investigation [37–40]. Most of the results presented in this review have been obtained from micro-mechanically exfoliated graphene.

Despite the wide use of micro-mechanical cleavage, the identification and counting of graphene layers is a major hurdle. Monolayers are a minority among accompanying thicker flakes. They cannot be seen in an optical microscope on most substrates. Currently, optically visible graphene layers are obtained by placing them on the top of oxidized silicon substrates (Section 4). However, the colour contrast of a single-layer is very weak and single-layer and bilayer can be easily confused since they have similar colour. A reliable identification technique is needed.

Atomic Force Microscopy (AFM) is viable but it has a very low throughput. Moreover, the different interaction forces between the AFM probe, graphene, and the silicon substrate lead to an apparent thickness of 0.5–1 nm even for a single layer [41], much bigger that what is expected from the interlayer graphite spacing (0.33 nm). Thus, in practice, it is only possible to distinguish between one and two layers by AFM if graphene films contain folds or wrinkles. High-resolution Transmission Electron Microscopy (TEM) is the most direct identification tool [42, 43]. However, it is destructive and very time-consuming, being viable only for fundamental studies. On the other side, Raman spectroscopy is a fast and non-destructive technique.

Figure 5a shows an optical micrograph of a flake composed of a different number of layers (N): a small mono-layer ($N=1$), a bilayer ($N=2$), a tri-layer ($N=3$) and a six-layers ($N=6$). Figure 5b shows the corresponding Raman spectra: no strong changes are observed for the G peak, while the 2D peak shape is strongly affected by the thickness of the sample. Figure 5c shows the 2D peak measured on single- and bi-layer graphene and graphite.

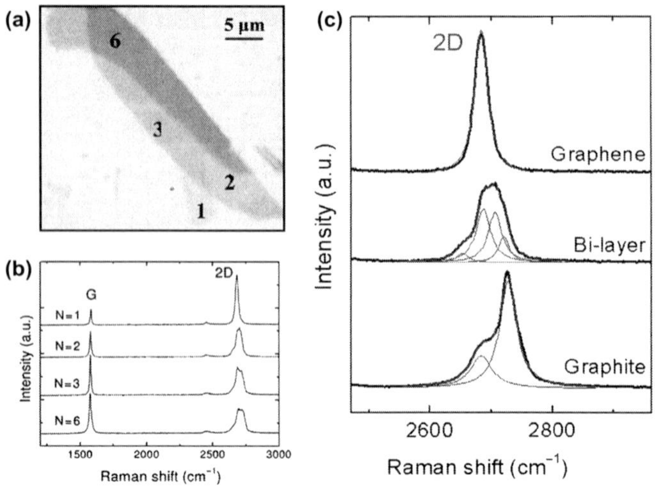

Fig. 5 (Colour online) a) Optical micrograph of a flake composed of different numbers of graphene layers (N); b) corresponding Raman spectrum; c) the 2D peak is a single and sharp peak for graphene, while it splits into 4 components for bilayer. In the case of few graphene layers and graphite the peak has at least two components. The 2D peak shape allows graphene identification. Adapted from Ref. 41.

Graphene shows a single and sharp 2D peak, with a typical Full Width at Half Maximum (FWHM) of $\sim 30\,\mathrm{cm}^{-1}$, while it splits into four components for the bilayer. In contrast, the 2D peak of graphite is composed of at least two components. Also the relative intensity between G and 2D peak is very different in graphene and graphite. Usually, the intensity of the 2D peak is $\sim$4–6 times higher than the intensity of the G peak in graphene [42]. However, identification of graphene based on the intensity ratio is questionable because this parameter is strongly affected by doping, as it will be shown in the next Section. Note that sometimes a small up-shift of the G peak is observed in graphene deposited on oxidized silicon substrate, compared to graphite, and this has been used for graphene identification [44]. As we will show in the next Section, this up-shift is caused by doping, so the G peak position is not a reliable parameter for graphene identification.

The changes in the shape of the 2D peak can be attributed to the resonant Raman scattering process: the resonance links the phonon dispersion with the electronic dispersion. Thus, every change in the electronic structure translates in a change of the phonon probed by Raman Spectroscopy [42]. When moving from single-layer to a bilayer, there is a strong change in the electronic structure: the linear dispersion of the electronic bands near the Fermi level is lost in bilayer as a consequence of the interactions between the layers. In a bilayer with AB stacking the unit cell consists of four carbon atoms and the interlayer interaction causes the π and π^* bands to divide into two hole and two electron parabolic bands. The incident laser light can couple only two pairs of the four bands. On the other side, the two phonons in the highest optical branch can couple all the electron bands among them. This gives rise to four possible optical transitions, involving four different phonons [42]. This produces the 2D four-peak structure observed in bilayer graphene. The changes in the electronic structure become smaller for increasing thickness, so it is impossible to "count" the number of layers by Raman Spectroscopy. The 2D band of a sample containing more than 7–8 layers is hardy distinguishable from that of bulk graphite.

In conclusion, the unique electronic structure of graphene is captured in its Raman spectrum. The shape of the 2D peak allows unambiguous, high-throughput, non-destructive identification of a single layer and bilayer.

3.3 Raman spectroscopy of doped graphene

An excess charge (doping) in the crystal lattice causes E_F to move away from the Dirac point: if electrons are introduced (n-doping) E_F shifts-up, while if holes are introduced (p-doping) E_F down-shifts. Since the charge concentration in graphene can be easily tuned by applying a gate voltage (V_g), electrical doping is one of the simplest methods to add an excess surface charge in graphene. The silicon substrate is used as back-gate and the Raman spectrum is recorded while tuning V_g [45, 46].

Raman spectroscopy can easily monitor doping in graphene. Changes in the Raman spectra are expected because the chemical-bond modification induced by the charge transfer can produce a variation of bond lengths and phonon frequencies [47]. Usually this variation is very small and difficult to measure. In the case of graphene, the Raman spectrum strongly changes with the charge concentration because of the presence of the Kohn-anomaly: doping moves E_F far from the Γ point. This affects the Kohn anomaly, which also moves away from Γ: the higher the doping level, the smaller the contribution of the electron-phonon coupling on the phonon associated with the G peak. In graphene the total response of doping is the sum of the charge transfer and the removal of the Kohn anomaly [47]. For a significant doping level, the effects of electron-phonon coupling can be neglected [47].

The Raman spectrum shows the following variations with doping:

i) The G peak position increases for increasing charge concentration and saturates for high doping [45, 46, 48]. This is due to the (non-adiabatic) removal of the Kohn anomaly at Γ [45, 47]. The G peak FWHM decreases for increasing doping and saturates when the electron-hole gap becomes higher than the phonon energy [45, 46, 48]. This is due to the blockage of the phonon decay into electron-hole pairs due to the Pauli exclusion principle [47]. Note that in order to reproduce the changes of the G peak with doping time-dependent perturbation theory needs to be used [45, 47]. This approach goes beyond the Born-Oppenheimer Approximation (BOA), which is not valid in the case of doped graphene: this approximation assumes the lighter electrons will adjust adiabatically to the motion of the heavier nuclei, remaining at any time in their instantaneous ground state [45]. In graphene the electrons do not have time to relax their momenta to reach the adiabatic ground state [45]. Note that BOA is well justified when the energy gap between ground and excited electronic states is larger than the energy scale of the nuclear motion, but graphene is gapless, so BOA is not appropriate to describe graphene [45].

ii) The 2D peak position increases for increasing p-doping, while it decreases for increasing n-doping, although this effect is well visible only at relatively high doping [48]. The 2D and G peaks show a different doping dependence because the effects of the non-adiabatic removal of the Kohn-anomaly are expected to be negligible on the 2D phonons, when measuring in the visible wavelength range [48].

iii) The ratio between the peaks' intensity (as integrated area), I(2D)/I(G), decreases for increasing doping at small charge concentration [48, 49]. In the framework of the fully Raman resonant process for the second-order 2D peak, the intensity dependence on the doping is due to the electron-electron scattering contribution, which increases for increasing charge concentration [49]. Two-phonon Raman spectroscopy is sensitive to the dynamics of the photo-excited electron-hole pair, in particular to the scattering processes it can undergo [49–51]. Thus, the 2D peak intensity is strongly affected by the rate of electron-electron and electron-defect interactions: any increase in the amount of doping and/or defects produces a decrease in the 2D intensity [49–51]. This effect has been used to derive the electron–phonon scattering rate in graphene [49, 50], which is a very important parameter because it determines the transport properties of graphene.

Note that these changes have been observed for relatively low doping ($E_F \ll$ half of the laser excitation energy). High doping levels ($E_F \sim 1eV$) can be obtained by using electro-chemical [31] and intercalation methods [52]. Strong changes are observed in the Raman intensities of highly doped graphene because, due to the shift of E_F from the Dirac point, some electronic transitions will be blocked by Pauli Exclusion Principle [31, 32].

In conclusion, by monitoring the G and 2D peak it is possible to quantify the doping level and distinguish between p- and n-doping in graphene.

3.4 Effect of the substrate

Strong variations in the Raman spectrum of several pristine graphene samples, deposited on oxidized silicon substrate, have been observed [53], as shown in Fig. 6. These variations match well with the changes observed in gated graphene with different doping level (Figs. 6b and 6d). Thus, the changes in the Raman spectrum have been attributed to unintentional doping: this can reach levels up to 10^{13} cm^{-2} under ambient conditions [53], it can change from sample to sample [53], and it is inhomogeneously distributed within the same sample [54].

Unintentional doping in graphene deposited on oxidized silicon substrate was first observed in gating experiments: the G peak position obtained at $V_g = 0$ did not correspond to the G peak position of an undoped graphene [45, 46]. This is expected since graphene needs to be processed in a device configuration, so residue from chemicals can easily dope graphene. Furthermore, doping can be produced by charge transfer from the gold contacts. Unintentional doping can strongly affect the transport properties of graphene [55, 56]: graphene devices prepared on oxidized silicon are highly disordered and exhibit charge mobility that is far inferior to the expected intrinsic properties of graphene.

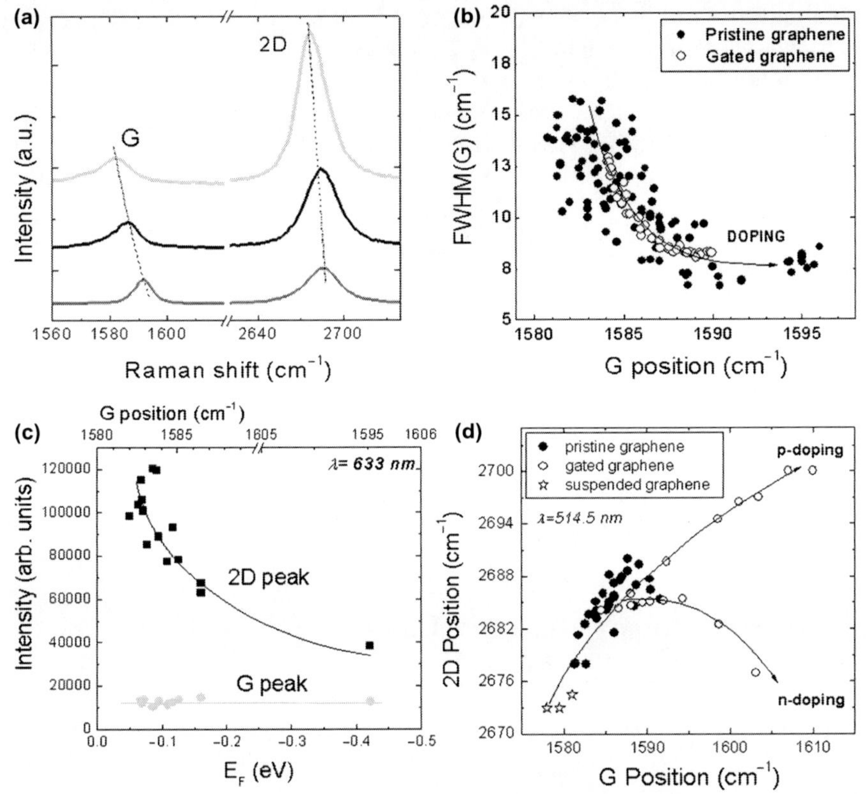

Fig. 6 a) Raman spectra of several pristine graphene samples deposited on the oxidized silicon; b) G peak position and FWHM measured on pristine graphene and gated graphene: the same behaviour is observed, indicating unintentional doping in pristine graphene; c) Raman intensity measured on different graphene samples. E_F is obtained from the G peak position; d) Correlation between G and 2D peak positions, which can be used to distinguish n-doping from p-doping. The data for suspended graphene are taken from Ref. 59.

Figure 6 shows that also pristine unprocessed graphene can exhibit significant unintentional doping. This can be attributed to adsorbed water on the surface, trapped water between silicon and graphene or to charged impurities in the substrate.

Oxidized silicon is the most commonly used substrate because it provides enhanced optical visibility of graphene (Section 4) and can be used for transport measurements. However, mechanically exfoliated graphene has also been deposited on other substrates such as quartz, silicon, glass, sapphire, NiFe, and GaAs [57, 58]: no strong effects on the Raman spectrum have been observed by using these substrates.

Raman mapping performed on suspended graphene over a trench has shown that by removing the substrate it is possible to produce graphene samples that are essentially undoped, have an high degree of spatial homogeneity and very little disorder [59, 60]. The same effect has been observed for bilayer, although these samples exhibit less pronounced differences between free-standing and supported areas [59]. Thus, it is believed that the doping observed in the pristine samples is mainly generated by the interaction of graphene with the substrate rather than an intrinsic property of exfoliated graphene held in ambient conditions [59]. In any case graphene is lipophilic, so hydrocarbon contamination cannot be excluded [17]. Transport measurements have shown that suspended graphene devices, after current annealing, show very high charge mobility: low-temperature mobility larger than 1,000,000 $cm^2 V^{-1} s^{-1}$ for carrier densities $\sim 10^{10}$ cm^{-2} have been reported [14, 61–63]. However, free standing graphene devices are very delicate and they cannot be implemented into commercial devices. Thus, graphene needs to be placed on a carefully selected substrate. Recent works have shown that by depositing graphene on top of atomically flat substrates such as hexagonal-Boron Nitride (h-BN) the mobility of the device can be higher than 200,000 cm^2 V^{-1} s^{-1} at room temperature [15]. Hexagonal-Boron Nitride is very similar to graphene: it is a layered material, with hexagonal structure and approximately the same lattice parameter as graphene. The use of h-BN as substrate allows a significant improvement of the electronic quality of graphene-based devices because it has an atomically smooth surface that is relatively free of dangling bonds and charge traps [15].

3.5 Raman spectroscopy of defected graphene

Raman Spectroscopy is able to probe disorder through defect-activated features, such as the D and D′ peaks, Fig. 7. The D peak is activated by an inter-valley double resonant scattering mechanism, as previously described. However, the double resonance can also happen as intra-valley process, *i.e.* connecting two electronic states belonging to the same cone around **K** (or **K**′). This process gives to the so-called D′ peak, which is visible at $\sim$1620 cm^{-1} in defected graphite. The second order of the D′ peak is also visible at $\sim$3250 cm^{-1} and it is named 2D′ peak. High quality graphene shows the G, 2D and 2D′ peaks, but not D and D′ peaks.

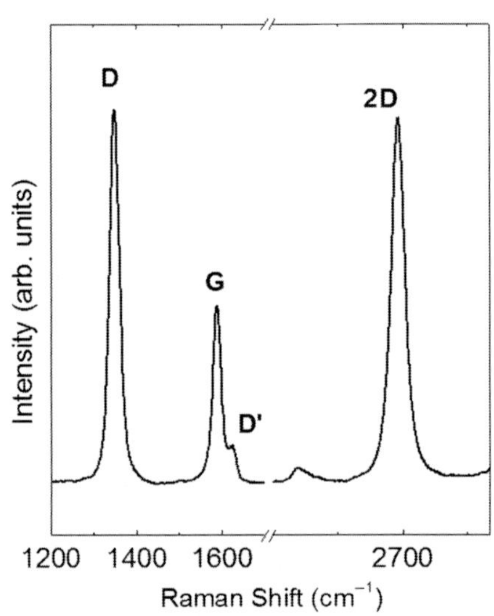

Fig. 7 Raman spectrum of defected graphene showing defect-activated features such as D and D′ peaks.

Note that all the positions of the peaks activated by inter- and intra-valley resonance strongly depend on the excitation energy. Since the electronic dispersion is linear, the higher the incident energy, the higher the phonon-wavevector and, because of the Kohn-anomaly, the higher the phonon energy (Section 2). The change of a peak position with the excitation energy is called dispersion. The D and 2D peak have a typical dispersion of $\sim 50\ cm^{-1}/ev$ and $\sim 100\,cm^{-1}/eV$, respectively [64, 65]. The D′ and 2D′ peak have smaller dispersion because the corresponding phonon does not lie at Γ, *i.e.* it is far from the Kohn-Anomaly. This effect has been used to investigate the phonon dispersion close to **K** by Raman Spectroscopy [66, 67].

One generally refers to defects in graphitic materials as anything that breaks the symmetry of the infinite honeycomb lattice. Thus, different type of defects can exist such as grain boundaries, vacancies, implanted atoms and edges. Moreover, defects can be associated with a change in hybridization, for example from sp^2 to sp^3, or a change in the sp^2 configuration (clustering or conversion from rings into chains).

It is important to be able to probe defects in graphene because they strongly affect charge mobility. Furthermore, defect-engineering can allow one to tune the properties of graphene simply by introducing specific types and amounts of defects in the crystal lattice [68].

Defects were first analyzed in graphite in 1970 by Tuinstra and Koenig [27]: they investigated the evolution of the Raman spectrum of graphite with decreasing in-plane crystalline grains size (L_a), measured by X-ray diffraction. Here the defect is associated with the grain boundary, so the amount of disorder increases for decreasing grain size. As soon as disorder is introduced, they observed a new peak (the D peak) appearing in the Raman spectrum and its intensity incresing for increasing disorder. They concluded that the intensity ratio I(D)/I(G) could be used to quantify the in-plane grain size with the following relation: $I(D)/I(G) = C(\lambda)/L_a$ (TK relation), where $C(\lambda)$ is a constant which depends on the excitation energy. The higher the excitation energy, the smaller I(D)/I(G) [28].

This relation has been then extended to any type of defect, by replacing L_a with an average inter-defect distance. However, it is reasonable to expect I(D)/I(G) to depend also on the nature of the defect. This has been shown in disordered carbon [28], which is a material composed of a mixture of carbon atoms sp^2- and sp^3- bonded [28, 69]. For increasing sp^3 content, the rings start to get distorted and they cluster. Further increase of disorder forces the clusters to become smaller and smaller. At the same time, the rings, fewer and more distorted, start to open and convert into chains. In this case $I(D)/I(G) \sim La^2$, *i.e.* I(D)/I(G) is proportional to the number of hexagonal rings left in the structure [28]. When the amount of disorder is very high (sp^3 content $>80\%$), the D peak is no longer visible in the spectrum because all the rings have been converted into chains [28, 69]. This causes a strong change in the optical and electronic properties of the material [69, 70].

Based on the results obtained for disordered carbon, one should wonder about the possible use of the TK relation to quantify defects and disorder in graphene. Recently, a few works have shown a careful study of the evolution of the Raman spectrum of graphene with increasing number of defects, introduced by ion-bombardment [71, 72]. This treatment produces vacancies, typically 2 nm in size, with a controllable concentration. These works show that the TK relation cannot be applied to graphene and suggest a new relation between I(D)/I(G) and L_a. In any case, the evolution of the Raman spectrum is expected to depend not only on the number of defects, but also on their nature, as suggested by some preliminary results obtained in our group [73]. This makes Raman spectroscopy a powerful tool to fully characterise disorder in graphene.

3.6 Chemical derivatives of graphene

One of the main advantages of graphene, compared to graphite, is that its 2D-nature makes it easy to add, remove or move carbon atoms, *i.e.* to introduce only a specific type of defect. For example by gradually binding hydrogen or fluorine, it is possible to convert the sp^2-bonded carbon atom into sp^3 [74, 75]. Consequently, it is possible to fine tune the electronic properties of graphene from metallic to insulating simply by controlling the amount of bonded hydrogen or fluorine [74, 75]. This process can be performed on one-side only or on both-sides of graphene. In this second case, it is possible to achieve full hydrogenation or fluorination, leading to the creation of stoichiometric graphene-based derivatives, called graphane and fluoro-graphene, respectively [74, 75]. Note that, in contrast to graphite, graphene exhibits nm-scale corrugations. The associated strain and curvature can strongly influence the local surface reactivity, so the chemistry of graphene is expected to be different from the one of graphite [11]. This has been recently confirmed by electro-chemical measurements [76].

Raman Spectroscopy can easily detect hydrogen or fluorine bonding because the formation of the C–H sp^3 bonds breaks the translational symmetry of the crystal, *i.e.* the Raman spectrum will show defect-activated features. Figures 8a and 8b show the evolution of Raman spectra for supported and suspended graphene crystals from pristine (spectrum 1) to hydrogenated (spectrum 2) to annealed (spectrum 3). Hydrogenation results in the appearance of sharp D and D′ peaks, slight broadening and a decrease of the height of the 2D peak relative to the G peak, and the onset of a combination mode (D + D′) around 2950 cm^{-1}, which, unlike the 2D and 2D′ bands, requires a defect for its activation because it is a combination of two phonons with different momentum. Note that although the majority of carbon bonds in hydrogenated graphene are expected to acquire sp^3 hybridization, no Raman signature of C–C sp^3 bonds is expected because their cross section at visible light excitation is negligible as compared with that of the resonant C–C sp^2 bonds, as seen also in disordered carbons [28]. After annealing the D peak almost disappears due to desorption of the hydrogen. Figures 8c and 8d show schematic representations of pristine graphene (c) and graphane (d). The out-of plane atoms in graphane are the hydrogen atoms.

Another important derivative is graphene oxide (GO), obtained by chemical exfoliation of oxidized graphite [77]. There is a fundamental difference between hydrogenated/fluorinated graphene and GO. In the first case, the material contains only one type of defect, given by the change in hybridization. The number of defects directly depends on the amount of bonded hydrogen or fluorine. Graphene oxide, as synthesized, contains about 60% sp^3 content, but the disorder also affects the sp^2 network by introducing

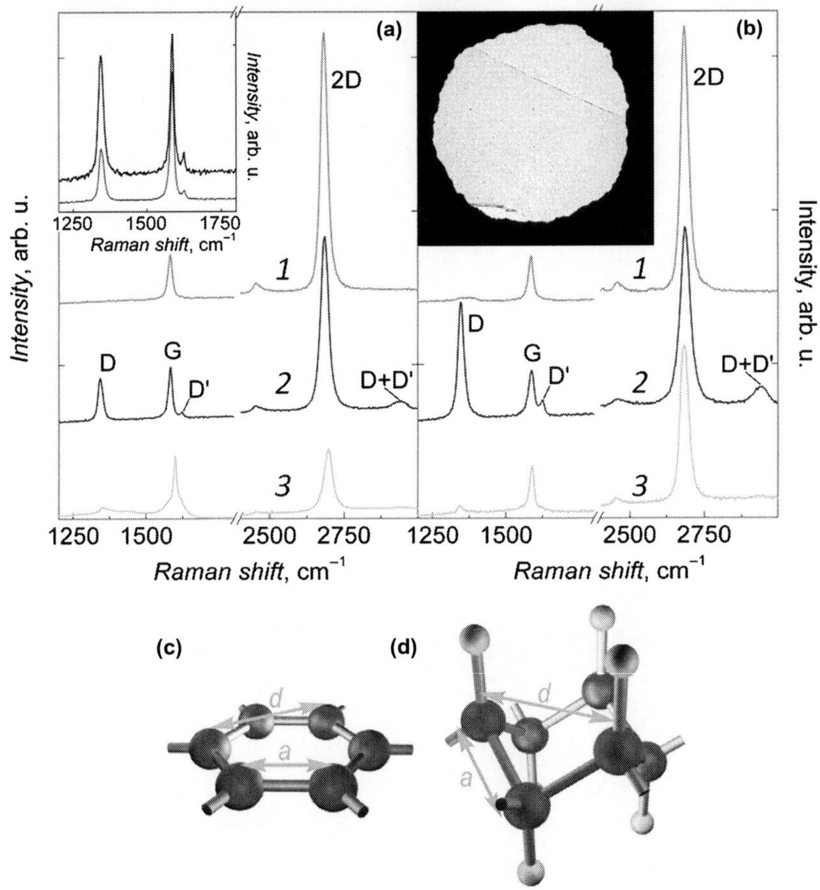

Fig. 8 (Colour online) Raman spectra of pristine graphene (spectrum 1) and graphane (spectrum 2) deposited on a substrate (a) and suspended (b). The D peak is activated by the binding event. Its intensity increases for increasing hydrogen content. The reaction with hydrogen is reversible, so that the D peak almost disappears after annealing of the sample (spectrum 3). c) Schematic representation of the crystal structure of graphene and d) theoretically predicted graphane. Adapted from Ref. 74.

lattice distortions, *i.e.* GO is the 2D-equivalent of disordered carbons. Chemical and thermal treatments have been used to reduce the number of defects in graphene oxide (called "reduced-graphene oxide", r-GO). However, it is not possible to achieve the perfect crystal lattice of graphene: r-GO still contains a wide range of defects, from topological defects to pentagon-heptagon pairs to holes, due to loss of carbon [78]. Because of the insulating behaviour and defective nature of GO and r-GO, these material are not suitable for use in field-effect transistors.

3.7 Graphene edges

It is common for mechanically exfoliated graphene not to have enough structural defects for the D peak to be clearly seen. This is indicative of the high crystallinity of graphene obtained by this technique. In this case, the D peak is only present at the edges, Fig. 9. An edge acts as defect on the Raman scattering process because it breaks the translation symmetry of the crystal.

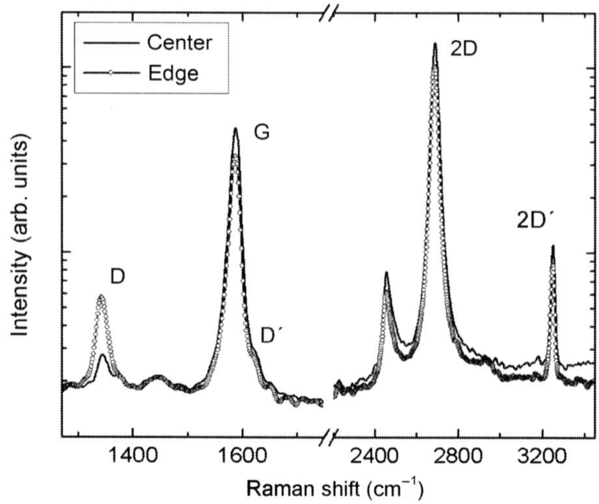

Fig. 9 Raman spectrum of graphene measured at the center of the sample and at the edge.

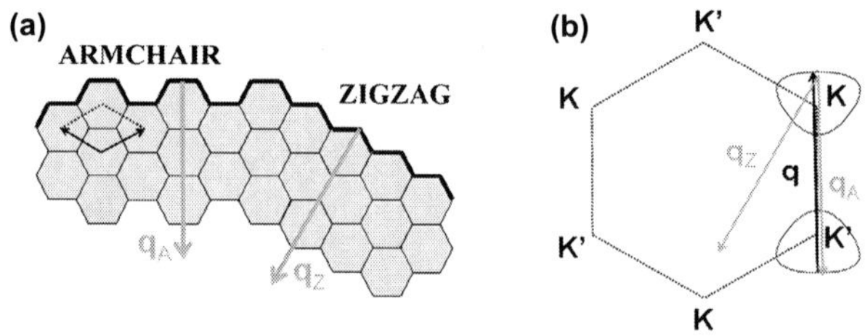

Fig. 10 (Colour online) a) Armchair and zigzag edges. The wavevector direction of electrons scattered by a zigzag, or armchair edge ($\mathbf{q}_z$, $\mathbf{q}_a$) is perpendicular to the edge direction. b) Schematic of the inter-valley process: only the exchanged momentum from armchair edges q_a can connect **K** and **K**′. Adapted from Ref. 64.

Graphene edges are very important in electronics: graphene shows a metallic behaviour making it difficult to create graphene-based field-effect transistors. A material with semiconducting properties is needed. One proposed way to change the electronic properties of graphene is to produce very narrow graphene nano-ribbons (GNRs) featuring perfectly controlled carbon atomic arrangements at the edges. A graphene edge can be disordered or it can perfectly follow a specific crystallographic direction. In the last case, we distinguish between zigzag and armchair edges. As shown in Fig. 2a, the unit cell of graphene contains two atoms A ad B. In an armchair edge there are two distinct carbon atoms for the two different sublattices. In a zigzag edge, there are only the carbon atoms belonging to the same sublattice, Fig. 10a. Because of this, the electronic properties of GNRs strongly depend on the type of edges [79]: a GNR with armchair edges can be either metallic or semiconducting, depending on its width, while a GNR with zigzag edges is metallic with peculiar edge states on both sides of the ribbon, no matter the width. A zigzag edge can have localized electronic states, in contrast to the armchair edges, where no edge states are possible [80, 81]. Superconductivity, ferromagnetism and Quantum Hall effect can appear from localized edge states [80, 81]. However, the perfect control of edges in GNRs is a very significant technological challenge. Thus, the investigation of edges has now become an important issue for future development of graphene-based electronics. In particular, this requires a simple and fast technique sensitive to the different atomic arrangements at the edges.

Raman Spectroscopy is a sensitive tool to probe the nature of graphene edges. The resonant theory suggests that momentum conservation upon reflection from an edge depends on the orientation and quality of the edge [64, 82, 83]. In particular, the wave-vector associated to the edge is strongly anisotropic because it has to be perpendicular to the edge direction [82], Fig. 10a. If we assume only zigzag or armchair edges, then a perfect zigzag edge cannot activate the D peak [82] because it does not satisfy inter-valley Raman scattering (*i.e.* it does not connect two inequivalent Dirac points), as shown in Fig. 10b.

This theory has been used to challenge the suggestion that perfect armchair or zigzag edges are routinely obtained in mechanically exfoliated graphene, even though the edges appear to follow defined directions on a large scale. However, Raman mapping on flakes with sharp edges forming angles of 90 and 180° with the use of circularly polarized light did now show any strong change in the D peak intensity [64]. Circular polarized light needs to be used in these measurements because the intensity of the D peak strongly depends on the light polarization: I(D) is maximum when the light polarization is parallel to the edge direction, while it is null for light polarization perpendicular to the edge direction [64, 82]. However, polarized Raman measurements at the edge have shown that I(D) is never null, even when using perpendicular polarized light. In addition, its residual intensity can change from edge to edge [64]. This is a clear fingerprint of the disordered nature of the edges of exfoliated graphene, in agreement with other experimental results [84].

The Raman theory has been extended to consider disordered edges [83]: by using polarized Raman spectroscopy it is possible to provide information on the nature of defects at the edge, for example if the edge is atomically rough or if it is given by many segments of perfect armchair and zigzag edges.

3.8 Raman spectroscopy of strained graphene- Grüneisen parameters

One speaks of strain if a material is compressed or stretched out of its equilibrium shape. This leads to a change in the atomic bond length along the direction of the applied strain. Raman spectroscopy is able to probe strain because this modifies the phonons' frequency of the crystal, usually resulting in mode softening (decrease of phonon frequency) for tensile strain and the opposite for compressive strain [85]. The magnitude of the change in the phonon frequency for a given change in the crystal volume, caused by the strain, is described by the Grüneisen parameter γ. This is defined as $\gamma = -\partial \ln\omega/\partial \ln V$, where ω is the frequency of the phonon and V the crystal volume.

Strain-engineering has been used in silicon and it can be applied to graphene too. For example it would be possible to make a complete circuit, where all the elements are made of graphene with different amounts and types of strain [86]. This can be easily achieved by making appropriate patterns, such as folds, trenches, and wells on the substrate [86]. Furthermore, the application of specific strain geometries can be used to induce pseudo-magnetic fields well above 100T [87]. This has been recently observed in graphene nano-bubbles with triangular shape [88]. Thus, it is very important to be able to apply strain and to probe different strain configurations in graphene.

Several works reported the use of Raman Spectroscopy on strained graphene [85, 89–95]. In most cases uniaxial strain was studied because the experimental setup is relatively easy to make: graphene needs to be transferred on a plastic substrate, which is then carefully bended, as shown in Fig. 11. Under uniaxial strain the following changes in the Raman spectrum have been observed:

i) Linear dependence of the peaks position with strain, where tensile strain produces a decrease in the peak positions, Fig. 11d.

ii) Splitting of the G peak for increasing tensile strain (Fig. 11d). This is understood in the following way: under uniaxial strain the hexagonal symmetry of the graphene lattice is broken. This causes the G peak, which is due to the doubly degenerate E_{2g}-mode, to split up into two components: one parallel to the strain and one perpendicular to the strain direction, which are termed G − and G + in analogy with carbon nanotubes. Because of it, the G + and G − intensities strongly depend on the direction of the light polarization [85].

iii) Broadening or splitting of the 2D peak for increasing strain (Fig. 11e): the 2D peak is activated by an inter-valley Raman scattering process. However, uniaxial strain causes a relative shift of the Dirac cones, depending on the direction of the strain, which in turn affects the Raman scattering process [89]. This makes very difficult to determine the Grüneisen parameter of the 2D peak.

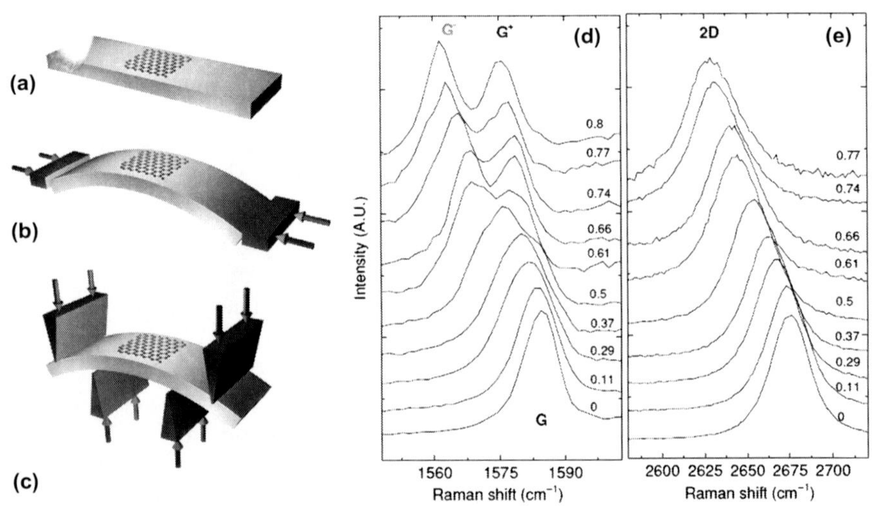

Fig. 11 a) Experimental setup used for uniaxial strain measurements; b) Raman spectra of graphene with increasing strain. Adapted from Ref. 85.

Despite the simplicity of the uniaxial experimental setup, different results have been obtained for the Grüneisen parameters.

Biaxial strain is more appropriate for the determination of the Grüneisen parameters because it is isotropic, *i.e.* it does not break the symmetry of the crystal. Furthermore, the Gruneisen parameters do not depend on the Poisson ratio, but only on the Raman peak shift and on the strain. However, biaxial strain is more difficult to apply experimentally, and only three works reported the study of biaxial strain by Raman Spectroscopy. Metzger *et al.* [93] deposited graphene over a shallow depression. The strain was directly measured with AFM. They observed the Raman spectrum to be rigidly down-shifted on the suspended part, compared to the supported graphene. Ding *et al.* [94] deposited graphene on a piezoelectric substrate, which was used to apply controlled strain. Similar to uniaxial strain, they observed a linear dependence of the peak position on the strain, were compressive strain results in a mode stiffening and tensile strain in a mode softening. They also did not observe strong changes in the 2D peak shape and no G peak splitting. However, in this work the strain was not directly measured, but derived from the G peak position, using the calculated G peak Grüneisen parameter $\gamma_G = 1.8$ from Ref. 85. This group also reported splitting of the D peak, which gives rise to a different Grüneisen parameter for D and 2D peaks, in contrast with the *ab-initio* calculations [85]. In a recent work, Zabel *et al.* [95] have used spherical graphene bubbles to determine the Grüneisen parameters: at the center of the bubble the strain is expected to be isotropic. Graphene bubbles are observed after the deposition of large graphene flakes on a oxidized silicon substrate [96]. The results obtained on graphene bubbles are in excellent agreement with *ab-initio* calculations of Ref. 85. These measurements also show that the coupling between graphene and the substrate can strongly affect the calculation of the Grüneisen parameters, introducing errors up to 30%. This could explain the discrepancy in the Grüneisen parameters previously reported. Further measurements on bilayer bubbles and balloons have shown that the Gruneisen parameters do not strongly change with the number of layers and that strain does not alter the AB-stacking configuration (up to $\sim$1% strain) [95].

4 Other spectroscopic techniques

Because of the atomic thickness of graphene, its relatively small size and weak optically visibility, most of the standard techniques used for thin films characterization can be hardly applied to graphene. Here, we describe the use of Rayleigh scattering and ellipsometry. Rayleigh scattering can be used to quickly map large area substrates in search of graphene flakes and can be used to explain the enhanced optical visibility of graphene when deposited on certain substrates [41]. Ellipsometry has been used to extract the optical constants of graphene and reconstruct its electronic dispersion near the K point [97]. Furthermore, a pronounced peak has been observed at 4.6 eV because of a Van Hove singularity in the density of states. The peak is however downshifted by 0.5 eV, probably because of excitonic effects [97].

4.1 Rayleigh scattering

Rayleigh spectroscopy provides a simple and quick way to really count the number of graphene layers on a substrate and to understand the optical visibility of graphene. In particular, by using interferometric modulation, it is possible to enhance the detection of graphene in the visible wavelength range.

Rayleigh scattering can be easily performed with a Raman spectrometer by removing the laser filter and replacing the CCD camera with an Avalanche Photodiode Detector (APD). In our case, we used a home-build system, composed by an inverted microscope, equipped with a piezo-electric stage, and two light sources: a He–Ne laser (633 nm) and a collimated white-light beam. Coherent white-light pulses are generated by pumping a photonic crystal fiber with the output of a Ti:Sa oscillator operating at 760 nm. The scattered light from the sample is collected in backscattering geometry, transmitted by a beam splitter and detected by the APD. Rayleigh images are obtained by raster scanning the sample with the piezoelectric scan stage. The acquisition time per pixel varies from a few milliseconds in the case of Rayleigh scattering to few minutes for Raman scattering. This empirically indicates that Rayleigh measurements are almost 5 orders of magnitude quicker than Raman measurements, being the Raman-scattered photons a minority compared to those elastically scattered. Thus, elastically scattered photons provide a very quick means to identify single and multilayer samples.

Fig. 5a shows an optical micrograph of a sample composed of 1, 2, 3, and 6 layers, as determined by combination of AFM and Raman Spectroscopy. These layers have a slightly different colour in the optical microscope. It appears that the darker colour corresponds to the thicker sample. Note, however, that the colour of much thicker layers (more than 10 layers) does not follow this trend and can change from blue to yellow to gray. Fig. 12a shows a Rayleigh map taken at 633 nm of this sample. The signal intensity in Fig.12(a) increases with the thickness. Thus, this technique can be used to directly count the number of layers. However, it has been shown that this

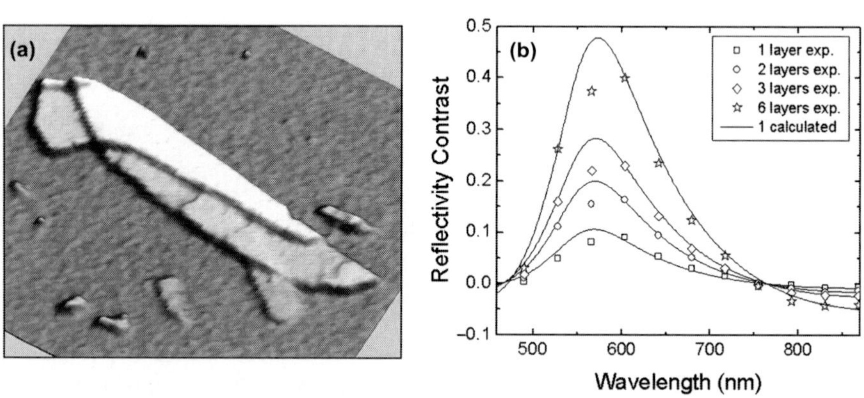

Fig. 12 a) Rayleigh map of the flake shown in Fig. 2a taken at 633 nm. b) The reflectivity contrast is proportional to the thickness and it strongly depends on the excitation wavelength. Adapted from Ref. 41.

effect strongly depends on the excitation wavelength, Fig. 12b: the reflectivity contrast is maximum at about 580 nm, while no enhancement is observed in the near IR or below 500 nm excitation wavelengths. Since the optical properties of graphene do not change in the visible range (Section 2), the modulation of the reflectivity contrast with the excitation wavelength has to be ascribed to the substrate. Graphene samples are usually deposited on a silicon substrate covered with an oxide layer, thus interference by multiple reflections is possible. By modelling the light modulation by multiple reflections with the recurrent matrix method for reflection and transmission of multilayered films, it has been possible to calculate the reflectivity contrast (the difference in reflectivity between graphene and the substrate) as a function of the excitation wavelength and the silicon oxide thickness [41]. The calculations are in very good agreement with our experimental results: for an oxide thickness of 300 nm, the contrast is maximum at 580 nm, Fig. 12b. The calculations also show that the maximum contrast is obtained only for a well chosen oxide thickness: ~300 nm and 90 nm oxide thicknesses need to be used to optically enhance the contrast of graphene on silicon [41].

Note that in order to calculate the reflectivity contrast, one needs to know the optical constants of graphene. The refractive index of graphene has been directly measured by ellipsometry: it has been shown that the refractive index of graphene does not significantly differ from the in-plane refractive index of graphite [97]. Thus, Rayleigh scattering is also a technique able to directly probe changes in the dielectric function of graphene.

Finally, note that interference by multiple reflections does not affect only the elastically scattered light, but also the Raman light. A strong dependence of the Raman signal with the oxide thickness and the excitation wavelength has been observed [98, 99]. In particular, when graphene is deposited on 300 nm oxide thickness, the Raman intensity is strongly enhanced by using an excitation wavelength of about 500 nm [98]. When using a 90 nm oxide thickness, the Raman signal is even higher and does not show a strong dispersion with the excitation wavelength [98]. The use of 90 nm oxide allows up to 8 times higher Raman intensity, depending on the excitation wavelengths, compared to the 300 nm oxide [98]. Because of this, one has to be careful in extrapolating information on the Raman scattering efficiency from the measured Raman intensities, when graphene is measured on an oxidized silicon substrate [98].

4.2 Ellipsometry

Graphene was first isolated by using micro-mechanical cleavage of graphite (Section 2). The single-layer flakes transferred on the substrate from the scotch tape were few and with typical size of $5 \times 10\,\mu m^2$ [35]. The technique has been strongly improved in the past few years and currently few flakes with lateral size of few hundred of µm can be easily obtained. Furthermore, studies on the optical properties of graphene and its optical visibility on a substrate [17, 41, 100] allowed the identification of graphene deposited on transparent substrates, such as quartz.

The use of large graphene flakes (size $> 200 \times 200\,\mu m^2$) deposited on quartz allowed measurements by a focused beam variable angle

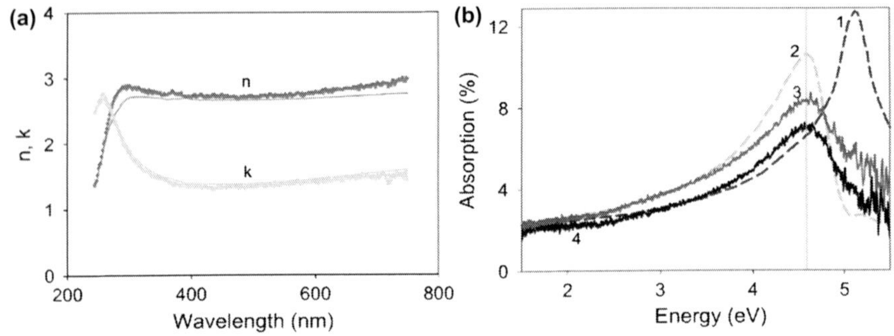

Fig. 13 a) Reconstructed optical constants of graphene. b) Absorption spectra of graphene. Solid curves 3 and 4 are experimental data, while dashed curves 1 and 2 are calculated. The symmetric peak at 5.2 eV is expected from non-interacting theory, whereas interaction effects should result in the asymmetric peak downshifted to 4.6 eV. Adapted from Ref. 97.

spectroscopic ellipsometer. Figure 13a shows the reconstructed optical constants of graphene using Fresnel theory, where graphene was modelled as a uniaxial anisotropic material with thickness of 0.335 nm [97]. Figure 13b shows the measured absorption spectra as a function of the incident light energy. This figure shows that the absorbance is constant in the visible range and equal to ~2.3%, in agreement with previous measurements [17]. However, the absorbance strongly increases for increasing incident energy, showing a prominent and asymmetric peak at 4.6 eV. This peak has been attributed to a Van Hove singularity in the density of states [97]. This behaviour has been observed in several samples, so the peak should not be related to any contamination effect [97]. Furthermore, a symmetric peak at 5.2 eV is expected from non-interacting theory; however, an excitonic effect should result in a down-shifted and asymmetric peak, as measured by ellipsometry [97].

Conclusions

Graphene is a material with outstanding properties. It can find applications in many areas, including transistors, field emission displays, capacitors, structural and conductive composites and photonic devices. In this chapter a short review of the Raman spectrum of graphene has been presented: it has been shown that the Raman scattering process in graphene involves resonant conditions, strong electron-phonon coupling effects and defect-activated processes. Thus, the Raman spectrum is sensitive to changes in the electronic structure and electron-phonon interactions: this means that Raman spectroscopy is able to identify graphene and probe doping. Defects and edges can be probed by defect-activated features such as the D and D′ peaks. Raman spectroscopy is also able to probe strain: uniaxial strain produces splitting of the G peak, while biaxial strain causes a rigid down-shift of the Raman spectrum for increasing tensile strain.

Other spectroscopic techniques can be used to investigate graphene: Rayleigh scattering is a quick technique for graphene's identification on large area substrates. This technique has been used to understand the optical visibility of graphene on a substrate. Ellipsometry has been used to reconstruct the optical constants of graphene and to study the absorbance in a wide range of energy, from visible to UV.

In conclusion, in this Chapter I have shown that Raman spectroscopy is a powerful, fast, non-destructive tool for graphene characterization.

Acknowledgments

The author acknowledges K. S. Novoselov, A. K. Geim, A. C. Ferrari, D. M. Basko, F. Mauri, M. Lazzari, A. Hartshuh, H. Quang, J. Robertson, S. Piscanec, S. Pisana, J. Zabel, A. Ott, P. Klar, A. Eckmann, T. Georgiou, R. R. Nair, D. C. Elias and L. Britnel. The use of Figure 3, 8, 11 and 13 have been authorized by A. C. Ferrari and K. S. Novoselov. Funding from the Humboldt Foundation in the framework of the Sofja Kovaleskaja Award, endowed by the Federal Ministry of Education and Research of Germany, is also acknowledged.

References

1 H. W. Kroto, J. R. Heath, S. C. O'Brien, R. F. Curl and R. E. Smalley, *Nature*, 1985, **318**, 162.
2 S. Iijima, *Nature*, 1991, **354**, 56.
3 K. S. Novoselov, A. K. Geim, S. V. Morozov, D. Jiang, Y. Zhang, S. V. Dubonos, I. V. Grigorieva and A. A. Firsov, *Science*, 2004, **306**, 666.
4 M. S. Dresselhaus, G. Dresselhaus, Ph. Avouris, and R. E. Smalley (eds.), *Carbon Nanotubes: Synthesis, Structure, Properties and Applications*, Springer-Verlag, New York, USA, 2008
5 A. Jorio, M. S. Dresselhaus, R. Saito, G. Dresselhaus, *Raman Spectroscopy in Graphene Related Systems*, Wiley-VCH; 2011.
6 S. Reich, C. Thomsen, J. Maultzsch, *Carbon Nanotubes: Basic Concepts and Physical Properties*, Wiley-VCH; 2004.
7 A. Jorio, G. Dresselhaus and M. S. Dresselhaus (eds.), *Carbon Nanotubes: Advanced Topics in the Synthesis, Structure, Properties and Applications.* Topics in Applied Physics 111, Springer-Verlag, New York, USA, 2008.
8 K. S. Novoselov, *Reviews of Modern Physics*, 2011, **83**, 837.
9 A. K. Geim, *Reviews of Modern Physics*, 2011, **83**, 851.
10 A. K. Geim and K. S. Novoselov, *Nat. Mater.*, 2007, **6**, 183.
11 A. K. Geim, *Science*, 2009, **324**, 1530.
12 A. S. Mayorov, R. V. Gorbachev, *et al.*, *Nano Lett.*, 2011, **11**, 2396.
13 S. V. Morozov, K. S. Novoselov, M. I. Katsnelson, F. Schedin, D. C. Elias, J. A. Jaszczak and A. K. Geim, *Phys. Rev. Lett.*, 2008, **100**, 016602.
14 X. Du, I. Skachko, A. Barker and E. Y. Andrei, *Nature Nanotech.*, 2008, **3**, 491.
15 C. R. Dean, A. F. Young, I. Meric, C. Lee, L. Wang, S. Sorgenfrei, K. Watanabe, T. Taniguchi, P. Kim, K. L. Shepard and J. Hone, *Nature Nanotechnology*, 2010, **5**, 722.
16 A. H. C. Neto, F. Guinea, N. M. R. Peres, K. S. Novoselov and A. K. Geim, *Rev. Mod. Phys.*, 2009, **81**, 109.
17 R. R. Nair, P. Blake, A. N. Grigorenko, K. S. Novoselov, T. J. Booth, T. Stauber, N. M. R. Peres and A. K. Geim, *Science*, 2008, **320**, 1308.
18 T. Ando, Y. Zheng and H. Suzuura, *J. Phys. Soc. Jpn.*, 2002, **71**, 1318.
19 V. P. Gusynin, S. G. Sharapov and J. P. Carbotte, *Phys. Rev. Lett.*, 2006, **96**, 256802.
20 A. B. Kuzmenko, E. van Heumen, F. Carbone and D. van der Marel, *Phys. Rev. Lett.*, 2008, **100**, 117401.

21 S. Piscanec, M. Lazzeri, F. Mauri, A. C. Ferrari and J. Robertson, *Phys. Rev. Lett.*, 2004, **93**, 185503.
22 J. M. Ziman, *Principles of the Theory of Solids*, Cambridge University Press, 1995, p. 155.
23 M. Lazzeri, C. Attaccalite and L. Wirtz, *Phys. Rev. B*, 2008, **78**, 081406.
24 P. Y. Yu, M. Cardona, *Fundamental of semiconductors*, Springer, 2005.
25 A. C. Ferrari, *Sol. State Comm.*, 2007, **142**, 47.
26 A. C. Ferrari, J. Robertson (eds.), *Raman Spectroscopy in Carbons: from Nanotubes to Diamond, Theme Issue*, Phil. Trans. Roy. Soc. A, 2004, **362**, 2267.
27 F. Tuinstra and J. L. Koenig, *J. Chem. Phys.*, 1970, **53**, 1126.
28 A. C. Ferrari and J. Robertson, *Phys. Rev. B*, 2000, **61**, 14095; ibid. 2001, **64**, 075414.
29 C. Thomsen and S. Reich, *Phys. Rev. Lett.*, 2000, **85**, 5214.
30 J. Maultzsch, S. Reich and C. Thomsen, *Phys. Rev. B*, 2004, **70**, 155403.
31 M. Kalbac, A. Reina-Cecco, H. Farhat, J. Kong, L. Kavan and M. S. Dresselhaus, *ACS Nano*, 2010, **10**, 6055.
32 C. Chen, C. Park, B. W. Boudouris, J. Horng, B. Geng, C. Girit, A. Zettl, M. F. Crommie, R. A. Segalman, S. G. Louie and F. Wang, *Nature*, 2011, **471**, 617.
33 D. M. Basko, *Phys. Rev. B*, 2008, **78**, 125418.
34 P. Venezuela, M. Lazzeri and F. Mauri, *Phys. Rev. B*, 2011, **83**, 235323.
35 K. S. Novoselov, D. Jiang, F. Schedin, T. J. Booth, V. V. Khotkevich, S. V. Morozov and A. K. Geim, *Proc. Natl Acad. Sci. USA*, 2005, **102**, 10451.
36 X. Li, W. Cai, J. An, S. Kim, J. Nah, D. Yang, R. Piner, A. Velamakanni, I. Jung, E. Tutuc, S. K. Banerjee, L. Colombo and R. S. Ruoff, *Science*, 2009, **324**, 1312.
37 S. Stankovich, D. A. Dikin, G. H. B. Dommett, K. M. Kohlhaas, E. J. Zimney, E. A. Stach, R. D. Piner, S-B. T. Nguyen and R. S. Ruoff, *Nature*, 2006, **442**, 282.
38 D. Li, M. B. Müller, S. Gilje, R. B. Kaner and G. G. Wallace, *Nature Nanotech*, 2008, **3**, 101.
39 Y. Hernandez, V. Nicolosi, M. Lotya, F. M. Blighe, Z. Sun, S. De, *et al.*, *Nature Nanotech.*, 2008, **3**, 563.
40 L. Dössel, L. Gherghel, X. Feng and K. Müllen, *Angewandte Chemie International Edition*, 2011, **50**, 2540.
41 C. Casiraghi, A. Hartschuh, E. Lidorikis, H. Qian, H. Harutyunyan, T. Gokus, K. S. Novoselov and A. C. Ferrari, *Nano. Lett.*, 2007, **7**, 2711.
42 A. C. Ferrari, J. Meyer, V. Scardaci, C. Casiraghi, M. Lazzeri, F. Mauri, S. Piscanec, D. Jiang, K. S. Novoselov, S. Roth and A. K. Geim, *Phys. Rev. Lett.*, 2006, **97**, 187401.
43 J. C. Meyer, A. K. Geim, M. I. Katsnelson, K. S. Novoselov, T. J. Booth and S. Roth, *Science*, 2007, **446**, 60.
44 A. Gupta, G. Chen, P. Joshi, S. Tadigadapa and P.C. Eklund, *Nano Lett.*, 2006, **6**, 2667.
45 S. Pisana, M. Lazzeri, C. Casiraghi, K. S. Novoselov, A. K. Geim, A. C. Ferrari and F. Mauri, *Nat. Mat.*, 2007, **6**, 198.
46 J. Yan, Y. Zhang, P. Kim and A. Pinczuk, *Phys. Rev. Lett.*, 2007, **98**, 166802.
47 M. Lazzeri and F. Mauri, *Phys. Rev. Lett.*, 2006, **97**, 266407.
48 A. Das, S. Pisana, B. Chakraborty, S. Piscanec, S. K. Saha, U. V. Waghmare, K. S. Novoselov, H. R. Krishnamurthy, A. K. Geim, A. C. Ferrari and A. K. Sood, *Nat. Nanotechnol.*, 2008, **3**, 210.
49 D. M. Basko, S. Piscanec and A. C. Ferrari, *Phys. Rev. B*, 2009, **80**, 165413.
50 C. Casiraghi, *Phys. Rev. B*, 2009, **80**, 233407.

51 C. Casiraghi, *Phys. Status Sol. (b)*, 2011, **248**, 2593.
52 W. Zhao, P. H. Tan, J. Liu and A. C. Ferrari, *J. Am. Chem. Soc.*, 2011, **133**, 5941.
53 C. Casiraghi, S. Pisana, K. S. Novoselov, A. K. Geim and A. C. Ferrari, *Appl. Phys. Lett.*, 2007, **91**, 233108.
54 C. Stampfer, F. Molitor, D. Graf, K. Ensslin, A. Jungen, C. Hierold and L. Wirtz, *Appl. Phys. Lett.*, 2007, **91**, 241907.
55 P. Blake, R. Yang, S. V. Morozova, F. Schedin, L. A. Ponomarenko, A. A. Zhukova, R. R. Nair, I. V. Grigorieva, K. S. Novoselov and A. K. Geim, *Sol. State Comm.*, 2009, **149**, 1068.
56 Y.-W. Tan, Y. Zhang, K. Bolotin, Y. Zhao, S. Adam, E. H. Hwang, S. Das Sarma, H. L. Stormer and P. Kim, *Phys. Rev. Lett.*, 2007, **99**, 246803.
57 Y. Y. Wang, Z. H. Ni, T. Yu, Z. X. Shen, H. M. Wang, Y. H. Wu, W. Chen and A. T. S. Wee, *J. Phys. Chem. B*, 2008, **112**, 10637.
58 I. Calizo, W. Bao, F. Miao, C. N. Lau and A. A. Balandin, *Appl. Phys. Lett.*, 2007, **91**, 201904.
59 S. Berciaud, S. Ryu, L. E. Brus and T. F. Heinz, *Nano Lett.*, 2009, **9**, 346.
60 Z. H. Hi, T. Yu, Z. Q. Luo, Y. Y. Whang, L. Liu, C. P. Wong, J. Miao, W. Huang and Z. X. Shen, *ACS Nano*, 2009, **3**, 569.
61 K. I. Bolotin, K. J. Sikes, Z. Jiang, M. Klim, G. Fudenberg, J. Hone, P. Kim and H. L. Stormer, *Solid State Comm.*, 2008, **146**, 351.
62 D. C. Elias, R. V. Gorbachev, A. S. Mayorov, S. V. Morozov, A. A. Zhukov, P. Blake, L. A. Ponomarenko, I. V. Grigorieva, K. S. Novoselov, F. Guinea and A. K. Geim, *Nature Physics*, 2011, **7**, 701.
63 A. S. Mayorov, D. C. Elias, M. Mucha-Kruczynski, R. V. Gorbachev, T. Tudorovskiy, A. Zhukov, S. V. Morozov, M. I. Katsnelson, V. I. Falko, A. K. Geim and K. S. Novoselov, *Science*, 2011, **333**, 860.
64 C. Casiraghi, A. Hartschuh, H. Qian, S. Piscanec, C. Georgi, A. Fasoli, K. S. Novoselov, A. K. Geim and A. C. Ferrari, *Nano Letters*, 2009, **9**, 1433.
65 M. A. Pimenta, G. Dresselhaus, M. S. Dresselhaus, L. G. Cancado, A. Jorio and R. Saito, *Phys. Chem. Chemical Phys.*, 2007, **9**, 1276.
66 P. H. Tan, L. An, L. Q. Liu, Z. X. Guo, R. Czerw, D. L. Carroll, P. M. Ajayan, N. Zhang and H. L. Guo, *Phys. Rev. B*, 2002, **66**, 245410.
67 L. M. Malard, J. Nilsson, D. C. Elias, J. C. Brant, F. Plentz, E. S. Alves, A. H. Castro Neto and M. A. Pimenta, *Phys. Rev. B*, 2007, **76**, 201401 R.
68 L. D. Carr and M. T. Lusk, *Nature nanotech.*, 2010, **5**, 316.
69 C. Casiraghi, A. C. Ferrari and J. Robertson, *Phys. Rev. B*, 2005, **72**, 085401.
70 J. Robertson, *Mater. Sci. Eng., R Rep.*, 2002, **37**, 129.
71 E. H. Martins Ferreira, M. V. O. Moutinho, F. Stavale, M. M. Lucchese, R. B. Capaz, C. A. Achete and A. Jorio, *Phys. Rev. B*, 2010, **82**, 125429.
72 L. G. Cançado, A. Jorio, E. H. Martins Ferreira, F. Stavale, C. A. Achete, R. B. Capaz, M. V. O. Moutinho, A. Lombardo, T. S. Kulmala and A. C. Ferrari, *Nano Letters*, 2011, **11**, 3190.
73 A. Eckman *et al.*, submitted.
74 D. C. Elias, R. R. Nair, T. M. G. Mohiuddin, S. V. Morozov, P. Blake, M. P. Halsall, A. C. Ferrari, D. W. Boukhvalov, M. I. Katsnelson, A. K. Geim and K. S. Novoselov, *Science*, 2009, **323**, 610.
75 R. R. Nair, W. Ren, R. Jali, I. Riaz, V. G. Kravets, L. Britnell, P. Blake, F. Schedin, A. S. Mayorov, S. Yuan, M. I. Katsnelson, H. Cheng, W. Strupinski, L. G. Bulusheva, A. V. Okotrub, I. V. Grigorieva, A. Grigorenko, K. S. Novoselov and A. K. Geim, *Small*, 2010, **6**, 2877.
76 A. T. Valota, I. A. Kinloch, K. S. Novoselov, C. Casiraghi, A. Eckmann, E. W. Hill, R. A. W. Dryfe, *ACS Nano*, DOI: 10.1021/nn202878f

77 D. A. Dikin, S. Stankovich, E. J. Zimney, R. D. Piner, G. H. B. Dommett, G. Evmenenko, S. B. T. Nguyen and R. S. Ruoff, *Nature*, 2007, **448**, 457.
78 K. P. Loh, Q. Bao, G. Eda and M. Chhowalla, *Nature Chemistry*, 2010, **2**, 1015.
79 K. A. Ritter and J. W. Lyding, *Nature Materials*, 2008, **8**, 235.
80 K. Nakada, M. Fujita, G. Dresselhaus and M. S. Dresselhaus, *Phys. Rev. B*, 1996, **54**, 17954.
81 K. Sasaki, J. Jiang, R. Saito, S. Onari and Y. Tanaka, *Phys. Soc. Jpn*, 2007, **76**, 033702.
82 L. G. Cançado, M. A. Pimenta, B. R. A. Neves, M. S. S. Dantas and A. Jorio, *Phys. Rev. Lett.*, 2004, **93**, 247401.
83 D. M. Basko, *Phys. Rev. B*, 2009, **79**, 205428.
84 A. K. Gupta, T. J. Russin, H. R. Gutierrez and P. C. Eklund, *ACS Nano*, 2009, **3**, 45.
85 T. M. G. Mohiuddin, A. Lombardo, R. R. Nair, A. Bonetti, G. Savini, R. Jalil, N. Bonini, D. M. Basko, C. Galiotis, N. Marzari, K. S. Novoselov, A. K. Geim and A. C. Ferrari, *Phys Rev. B*, 2009, **79**, 205433.
86 V. M. Pereira and A. H. Castro Neto, *Phys. Rev. Lett.*, 2009, **103**, 046801.
87 F. Guinea, M. I. Katsnelson and A. K. Geim, *Nature Phys.*, 2010, **6**, 30.
88 N. Levy, S. A. Burke, K. L. Meaker, M. Panlasigui, A. Zettl, F. Guinea, A. H. Castro Neto and M. F. Commie, *Science*, 2010, **329**, 544.
89 M. Huang, H. Yan, C. Chen, D. Song, T. F. Heinz and J. Hone, *PNAS*, 2009, **106**, 7304.
90 T. Yu, Z. Ni, C. Du, Y. You, Y. Wang and Z. Shen, *Phys. Chem. Lett. C*, 2008, **112**, 12602.
91 Z. H. Ni, T. Yu, Y. H. Lu, Y. Y. Wang, Y. P. Feng and Z. X. Shen, *ACS Nano*, 2008, **2**, 2301.
92 M. Huang, H. Yan, T. F. Heinz and J. Hone, *Nano Lett.*, 2010, **10**, 4074.
93 C. Metzger, S. Remi, M. Liu, S. V. Kusminskiy, A. H. Castro Neto, Anna K. Swan and B. B. Goldberg, *Nano Lett.*, 2010, **10**, 6.
94 F. Ding, H. Ji, Y. Chen, A. Herklotz, K. Dorr, Y. Mei, A. Rastelli and O. G. Schmidt, *Nano Lett.*, 2010, **10**, 3453.
95 J. Zabel, R. R. Nair, A. Ott, T. Georgiou, A. K. Geim, K. S. Novoselov and C. Casiraghi, *Nano Letters*, 2012, **12**, 617.
96 T. Georgiou, L. Britnell, P. Blake, R. V. Gorbachev, A. Gholinia, A. K. Geim, C. Casiraghi and K. S. Novoselov, *Appl. Phys. Lett.*, 2011, **99**, 093103.
97 V. G. Kravets, A. N. Grigorenko, R. R. Nair, P. Blake, S. Anissimova, K. S. Novoselov and A. K. Geim, *Phys. Rev. B*, 2010, **81**, 155413.
98 P. Klar, *et al.*, submitted.
99 Y. Y. Wang, Z. H. Ni, Z. X. Shen, H. M. Wang and Y. H. Wu, *Appl. Phys. Lett.*, 2008, **92**, 043121.
100 P. Blake, E. W. Hill, A. H. Castro Neto, K. S. Novoselov, D. Jiang, R. Yang, T. J. Booth and A. K. Booth, *Appl. Phys. Lett.*, 2007, **91**, 063124.

Solid-state NMR: a key tool for the understanding at a molecular level of well-defined heterogeneous catalysts and surface chemistry on top of oxide materials

David Gajan, Giuseppe Lapadula, Martin Schwarzwälder and Christophe Copéret*
DOI: 10.1039/9781849734899-00057

Catalysis is indeed one of the key disciplines for sustainable development, and will be a key technology in the future by enabling the efficient transformation of raw materials by helping us to convert from fossil fuels to sustainable sources. This domain has been traditionally divided in three sub-disciplines: homogeneous, heterogeneous and bio-catalysis. Each of those has advantages and disadvantages, and the choice is based according to the molecule or process desired. Heterogeneous catalysts show particular advantages for continuous processes, as they are more easily regenerated, recycled and separated from the reaction stream and therefore products. However, heterogeneous catalysts are typically complex and, in contrast to their homogeneous analogues based on well-defined systems: their structure(s) as well as their active sites are not very well-known making their development according to a powerful structure-reactivity approach often difficult, if not impossible.

Because of their advantages, considerable research effort has been oriented towards designing heterogeneous catalysts following a molecular approach,[1–7] which is known as Surface Organometallic Chemistry.[8,9] It is thus not surprising that solid state NMR has played an extremely important role in the characterization of those systems, along powerful complementary tools such as Vibrational Spectroscopy or X-Ray absorption spectroscopy (EXAFS). Furthermore associated with theoretical approaches, it allows a detailed understanding of the surface sites at the molecular scale.[10–17] Many groups are active in solid-state NMR directed at understanding heterogeneous catalysts and materials.[18–26] Here we mainly focus on well-defined heterogeneous catalysts based on controlled surface chemistry on oxide materials, in particular we discussed our own research development realized in collaboration with C-RMN.[21] Besides, we also discuss a new promising approach for the characterization of the surface sites, which should have a major impact, reaching far beyond the catalysis community. That technique has been called "Surface Enhanced Solid State NMR Spectroscopy" because it allows the NMR signatures of surface species to be preferentially enhanced.

ETH Zürich, Department of Chemistry, Wolfgang Pauli Strasse, 10, CH-8093 Zürich, Suisse; E-mail: ccoperet@inorg.chem.ethz.ch

Proton. Despite the 100% natural abundance and a good sensitivity ($\gamma_H = 267.5 \times 10^6$ rad $s^{-1}T^{-1}$), ^1H solid state NMR is often ignored in catalysis and material science. The reason lies in the high proton density of such systems, which induce tremendous signal broadening because of dipolar coupling. Note, however, that progresses in methodology (*e.g.* modern sequences of proton decoupling[27–29]) and technology (*e.g.* ultra fast[30,31] Magic Angle Spinning (MAS) with $\nu_{MAS} = 30–65$ kHz) have permitted considerable improvement.

In the particular case of the single-site surface complexes, obtained by grafting molecular complexes on strongly dehydroxylated surfaces, the proton network is typically not very dense, so the solid state NMR performed under standard conditions($\nu_{MAS} = 5–12$ kHz) without any decoupling (Scheme 1a), allows already good spectral resolution ($\Delta_{1/2} = 200–300$ Hz) so that different sites can be identified.[32,33] However, the resolution can be further improved by implementing optimized pulse sequences; *e.g.* by adding decoupling or/and spin echo (Scheme 1b – $\Delta_{1/2} < 100$ Hz).[34]

Resolution can be also significantly improved using ultra-fast MAS. For instance in the case of hybrid materials, which have a higher density of protons (at least one or more orders of magnitude), this technique is crucial to obtain high-resolution spectra. At a MAS spinning rate of 10 kHz it is still difficult to discriminate the different sites. Increasing the spinning rate to 60 kHz allows the resolution of all protons; the aromatic protons, the phenol and the ethoxy surface fragments are now all identified (Fig. 1). Thus this approach, combined with multidimensional techniques, allows the relatively easy assignment of the NMR proton signals of the surface species and the monitoring of surface reactions by proton NMR.

Carbon-13. Carbon-13 solid state NMR is particularly difficult on non-enriched samples because of the low density of the sites, in addition to the small natural abundance (1.1%) and weak sensitivity of carbon-13 ($\gamma_C/\gamma_H = 0.25$). The acquisition of NMR spectra is normally realized with sequences implying a polarization transfer from the protons to the carbons, cross polarization (CP, Scheme 2) under the Hartman-Hahn optimal conditions: $\omega_H = \gamma_H B_H = \gamma_C B_C = \omega_C$. This has the advantage to greatly enhance the intensity of the carbon signals. However, in the case of the characterization of surface species, it is still necessary to enrich samples with carbon-13 (10–100%). Yet, despite these problems, it has become a crucial technique for the identification of surface species.

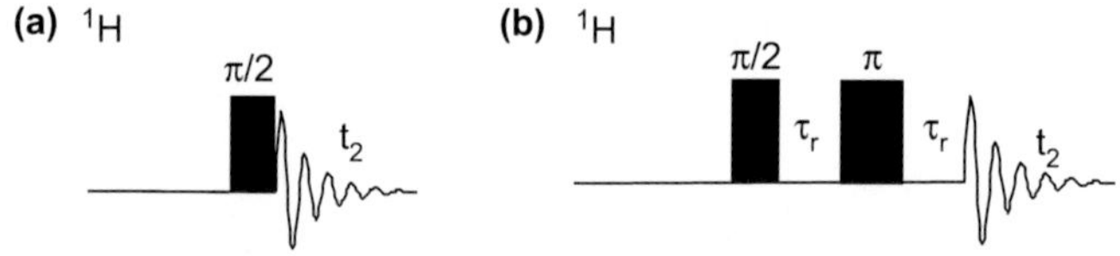

Scheme 1 Different pulse sequences for NMR experiments presented in the text are shown here. **a)** ^1H single pulse experiment. **b)** ^1H spin echo experiment. t_2 corresponds to the acquisition time, τ_r to a rotor period.

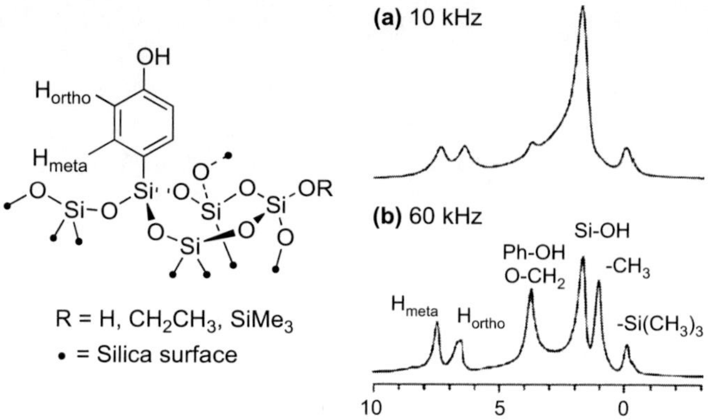

Fig. 1 Comparison of ^{1}H NMR spectra **a)** at 10 and **b)** at 60 kHz of a hybrid material containing phenol functionalities.[35] Adapted by kind permission of the authors.

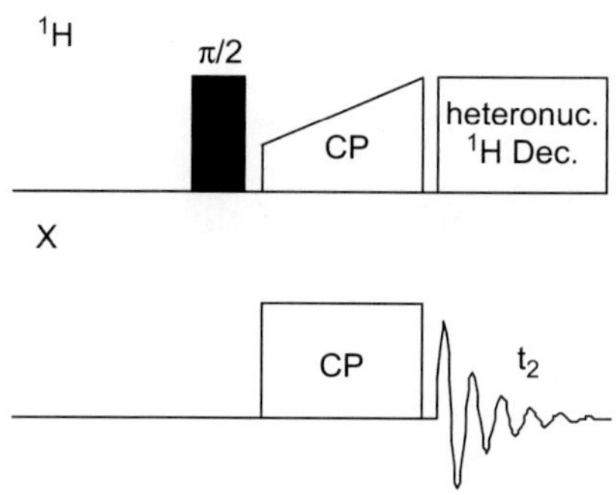

Scheme 2 ^{1}H-X cross polarization (CP) experiment. *heteronuc.^{1}H Dec.* is heteronuclear proton decoupling.

Chemical shifts associated with the carbon directly bound to the metal are very specific: it is relatively easy to distinguish the alkyls fragments (M–C, 40–100 ppm), the alkylidene (M=C, 150-250 ppm), or alkylidines (M=C, 250–300 ppm) (Table 1, Ti[36], Zr[37], Hf[38], Ta[14], Mo[14,32], W[14,39], Re[14]) as well as the metallacycles[40] and cyclopentadienyls derivatives, or aromatics.[41]

Table 1 Chemical shift of the ^{13}C in the perhydrocarbyne complexes (Ar = 2,6-*i*PrC$_6$H$_4$).

Complexes	M	Chemical shifts (δ_{Iso} ppm) M≡C	M=C	M–C
	Ti	–	–	113
(≡SiO)M(CH$_2$*t*Bu)$_3$	Zr	–	–	96
	Hf	–	–	106
(≡SiO)Ta(CH$_2$*t*Bu)$_2$(=CH*t*Bu)	Ta	–	245	96
(≡SiO)OS(CH$_2$*t*Bu)$_2$(≡C*t*Bu)	Os	263	–	22
(≡SiO)Re(CH$_2$*t*Bu)(=CH*t*Bu)(≡C*t*Bu)	Re	292	247	44
(≡SiO)M(CH$_2$*t*Bu)$_2$(≡C*t*Bu)	W	318	–	95
	Mo	329	–	90
(≡SiO)M(CH$_2$*t*Bu)(=CH*t*Bu)(≡NAr)	W	–	255	60
	M	–	279	56

Nitrogen-15. Nitrogen-15 NMR suffers from its low natural abundance (0.36%) as well as a weak sensitivity ($\gamma_N/\gamma_H = 0.10$). However, certain systems can be easily labeled, which allows identification of the surface compounds. For example the identification of the nature of the nitrogen ligands attached to tantalum resulting from the reaction of ammonia with tantalum hydride are easily accessed, thanks to the wide range of the chemical shift: coordinated amine ($\delta = -390$ ppm), amido ($\delta = -270$ ppm), imido ($\delta = -90$ ppm).[42] Finally, nitrogen-15 NMR is a convenient technique to evaluate Brönsted and Lewis acidity of the surface sites by adsorbing labeled nitrogen-15 pyridine as a probe molecule (Table 2).[43,44]

Table 2 ^{15}N chemical shift of chemisorbed pyridine on different supports function of the kind of interaction.[43,44]

Chemical shift (ppm)	Support examples	Interactions
− 65	silica	Physisorption
− 88	silica	Hydrogen bonds
− 110	zirconia	Lewis acidity
− 180	SO_3/Zirconia	Brönsted acidity

Phosphorus-31. Similarly to proton, 31-phosphorus has an ideal isotopic natural abundance (100%); it has, however, a lower sensitivity ($\gamma_P/\gamma_H = 0.40$). Phosphine oxides, in particular triethylphosphine oxide (TEPO) is a powerful probe to evaluate the acidity of surface sites (Drago scale, Table 3).[45] As an example, it has been used to reveal surface sites in supported lanthanides.[46]

Table 3 ^{31}P chemical shift of chemisorbed TEPO on different supports function of the acidity strength or basicity.[45]

Chemical shift (ppm)	Support examples	Interactions
50	silica	Physisorption
56	silica	Weak Brønsted acidity
75.6	H-M ordenite	Moderate Brønsted acidity
57	HZSM-5 site 2	Weak Brønsted acidity
89	HZSM-5 site 1	Strong Brønsted acidity
58.9	Sulfated silica gel	Strong Brøndted acidity
48.5	γ-Alumina	Brønsted basicity
48.6	NaOH	Brønsted basicity

This nucleus is also particularly interesting because organophosphorus compounds are classical ligands in organometallic chemistry. For instance, the addition of trimethylphosphine to Au(I) surface complexes allows two different sites to be detected thereby illustrating the heterogeneity of a silica surface: 30% of the sites can coordinate two phosphines, the remaining 70% only one.[47,48]

Beyond the characteristic chemical shift of the mono and di-phosphine adducts, the INADEQUATE (Incredible Natural Abundance DoublE QUAntum Transfer Experiment) 1D NMR of phosphorus-31 (Scheme 3a) has been essential for their attribution (Fig. 2).[48]

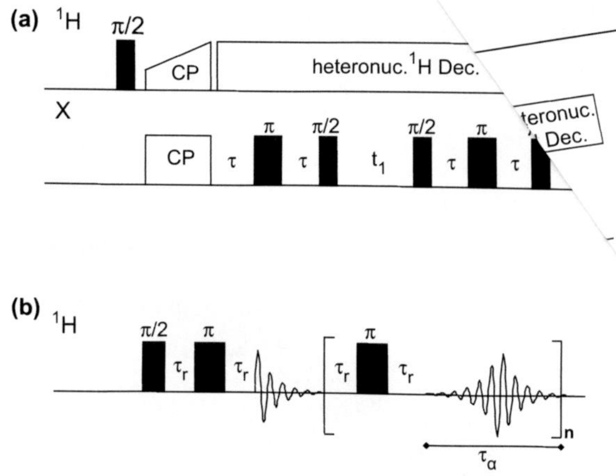

Scheme 3 **a)** ^{1}H-X cross polarized INADEQUATE sequence to detect X-X correlati
b) CPMG sequence (allowing a shorter experiment time for nuclei with a long relaxation time,
homonuc. ^{1}H Dec. corresponds to homonuclear proton decoupling.

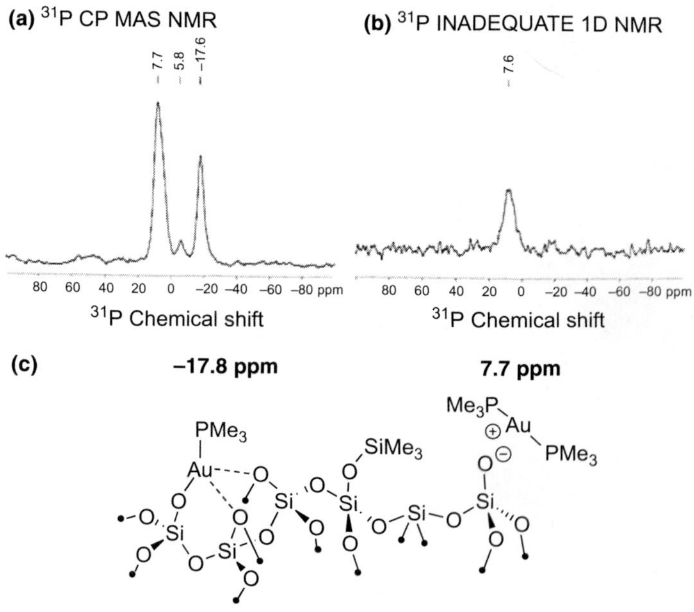

Fig. 2 **a)** ^{31}P CP/MAS and **b)** INADEQUATE 1D NMR of **c)** the surface Au(I) species after reaction with an excess of PMe_3. Adapted with permission of the authors.[48]

Silicon-29. Finally for silica based materials, silicon-29 NMR is particularly important, because it permits one to distinguish sites according to the number of oxygen atoms bound to the silicon, and if those silicon atoms are bound or not to other silicon atoms (Q_n and T_n sites, Table 4).[49] However, ^{29}Si suffers from its weak abundance (4.7%), sensitivity ($\gamma_{Si}/\gamma_H = 0.20$) as well as its very long relaxation times (frequently more than minutes under inert conditions); the latter further increasing experimental times.

Table 4 Diff… …nd T_n and their chemical shifts in ^{29}Si NMR.[49]

Q_4	…	Q_2	T_3	T_2	T_1
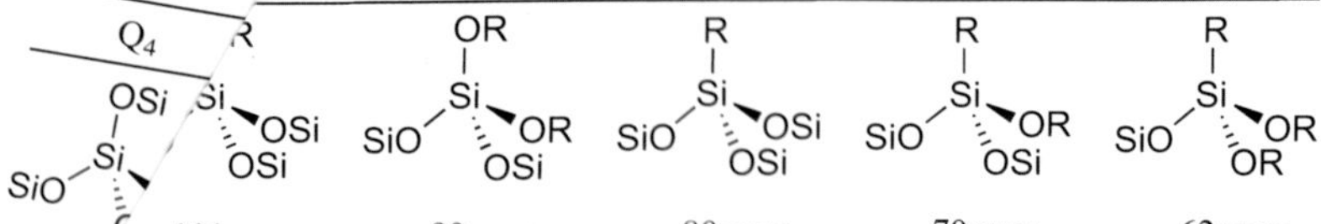					
– 11…	– 101 ppm	– 90 ppm	– 80 ppm	– 70 ppm	– 62 ppm

…d CPMG (Carr Purcell Meiboom Gill, Scheme 3b) measurements …he method of analysis to be accelerated, but this is accompanied a… loss of any quantitative information.[17,50–53] Silicon-29 NMR has …been used to study the level of fluorination of silica because the …emical shift is very sensitive to the number of fluorine atoms attached to …ilicon.[54]

Tin-119. Tin has three spin 1/2 isotopes: tin-115 (0.34% abundance, $\gamma_{Sn}/\gamma_H = 0.33$), tin-117 (7.68%, $\gamma_{Sn}/\gamma_H = 0.36$) and tin-119 (8.59%, $\gamma_{Sn}/\gamma_H = 0.37$). The latter isotope is mainly studied due to its higher natural abundance and sensitivity. It has proved possible to attribute specific chemical shifts of silica supported tin complexes as a function of its ligands (Table 5), *e.g.* types of alkyls (Me, Et, *i*Pr, Bu[55] or chiral ligands such as (–)-menthyl).[56] Moreover, tin-119 NMR allows probing on which kind of anchoring sites, $SnEt_4$, is grafted on cloverite, *i.e.* ≡GaO *vs.* ≡PO (Table 5).[57] Similarly, chemisorption of tin compounds was used to investigate the surface properties of H-Faujasite[58] zeolite and alumina.[59]

Table 5 ^{119}Sn chemical shifts (ppm) on different tin surface species.

Tin surface species	Chemical shift (δ ppm)
(silica)≡SiO–$SnMe_3$	136
(silica)≡SiO–$SnEt_3$	98
(silica)≡SiO–$SnBu_3$	106
(silica)≡SiO–Sn*i*Pr_3	34
(silica)≡SiO–$SnMe_2$[(-)-methyl]	103
(Alumina)(≡AIO)$SnBu_3$	80
(Alumina)H$SnBu_3$ (physisorbed)	– 77
(Alumina)(≡AIO)(≡AIOH)SnBu3	– 227
(H-Faujasite)(≡Si)(≡Al)O $SnMe_3$	260
(cloverite)≡PO–$SnEt_3$	175
(cloverite)≡GaO–$SnEt_3$	150

Fluorine-19. Despite interesting characteristics (100% abundance, $\gamma_F/\gamma_H = 0.94$), few fluorinated surface compounds have been studied. Fluorine-19 NMR has, however, been used as a powerful complementary technique to silicon-29 NMR to understand the surface sites of fluorinated silica particles (vide supra).[54]

Other spin 1/2 nuclei. Besides the typical spin 1/2 nuclei, cadmium and lead have been used in the context of metal-exchanged zeolites. They both have appreciable abundance and sensitivity (Cadmium-113: 12.2% abundance,

$\gamma_{Cd}/\gamma_H = 0.22$[60,61] and lead-207: 22.1% abundance, $\gamma_{Pb}/\gamma_H = 0.21$).[60,62] Both nuclei display chemical shifts, which highly depend on the coordination sphere and the hydratation of the metal site, thus providing information about the local structure of the zeolite.[60]

Spin 1/2 nuclei with low gamma. Of them, silver-109 (48.2% natural abundance, $\gamma_{Ag}/\gamma_H = 0.05$) and Yttrium-89 (100.0% abundance, $\gamma_Y/\gamma_H = 0.05$) have been studied. For instance, in hydrated Ag-exchange zeolites, a single sharp signal is observed, which corresponds to a highly mobile silver cation, and which upon dehydration is transformed partially into metallic silver.[63] For Yttrium-89 NMR, it has been mainly used to characterize the bulk of samples. Its chemical shift is characteristic of the coordinating ligands and the coordination number. However, the low ^{89}Y NMR sensitivity combined with long T_1 relaxation has so far precluded its use for supported species because of the low density of surface species.[64–69]

2 Chemical shift anisotropy for the molecular comprehension of the active sites (structure and dynamics)

Chemical shift is a tensorial value (second rank tensor), which implies that its intrinsic value depends on the orientation of the molecule in the magnetic field (B_0). If the molecule lies in its natural axe, the tensor is a diagonal matrix where the values on the diagonal are δ_{11}, δ_{22} and δ_{33} and the off-diagonal values are 0. The average of these values gives the isotropic chemical shift $\left(\delta_{iso} = \frac{1}{3}(\delta_{11} + \delta_{22} + \delta_{33})\right)$. There are two conventions to describe the asymmetry of the signal – the Herzfeld-Berger and the Haeberlen-Mehring-Spiess conventions. The Herzfeld-Berger convention uses the span breadth $\Omega = \delta_{11} - \delta_{33}$ to describe the broadness of the signal and the skew tensor $\kappa = \frac{[3(\delta_{22} - \delta_{iso})]}{\delta_{11} - \delta_{33}}$ to measure the amount of asymmetry. The Haeberlen-Mehring-Spiess convention uses $\delta_{aniso} = \delta_{33} - \delta_{iso}$ as a measure of signal breadth and the asymmetry parameter $\eta = \frac{(\delta_{22} - \delta_{11})}{(\delta_{33} - \delta_{iso})}$ indicating the difference in line shape from a signal with a axially symmetric tensor. In solution, the rapid movement of the molecules leads to averaging and to measuring directly the isotropic value. In contrast, for solids, under static conditions, the signal for an amorphous solid will be broad, and it will be "easy" to observe this anisotropy and δ_{11}, δ_{22} et δ_{33} (Fig. 3). Upon spinning the sample at the magic angle $\left(\arccos\left(\frac{1}{\sqrt{3}}\right) = 54.74°\right)$, the anisotropy will be averaged, and for a spinning rate greater than the anisotropy Ω, a unique signal will be found at the chemical shift value δ_{iso}. In the intermediate cases the NMR signal will be a series of sharp signals spaced by the rotational speed of the NMR rotor, which describe an envelope, which is the residual anisotropy (Fig. 3). Note that the actual dynamic of the molecules will also be able to average this signal so that the observed CSA can be a way to evidence dynamics.

Thus solid state NMR allows to access to the chemical shift anisotropy and the dynamic of a system. Since the chemical shift anisotropy is highly sensitive to hybridization, it can thus be used to explore slight electronic change at a specific site (Fig. 4).

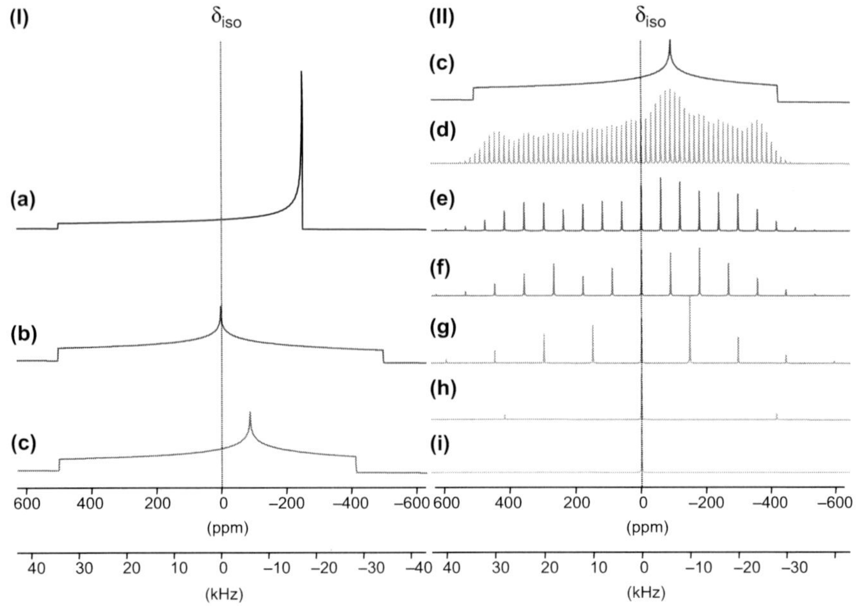

Fig. 3 Spectra simulations of spin 1/2; nucleus using dmfit software.[70]
I) Simulations under static conditions where $\delta_{iso} = 0$ ppm and $\delta_{aniso} = 500$ ppm are fixed while η varied. **a)** $\eta = 1$ ($\delta_{11} = -250$ ppm, $\delta_{22} = -250$ ppm, $\delta_{33} = 500$ ppm, $\Omega = 750$ ppm and $\kappa = -1$) under static condition. **b)** $\eta = 0$ ($\delta_{11} = -500$ ppm, $\delta_{22} = 0$ ppm, $\delta_{33} = 500$ ppm, $\Omega = 1000$ ppm and $\kappa = 0$) under static condition. **c)** $\eta = 0.65$ ($\delta_{11} = -412.5$ ppm, $\delta_{22} = -87.5$ ppm, $\delta_{33} = 500$ ppm, $\Omega = 912.50$ ppm and $\kappa = -0.29$).
II) Simulations where $\delta_{iso} = 0$ ppm, $\delta_{aniso} = 500$ ppm and $\eta = 0.65$ ($\delta_{11} = -412.5$ ppm, $\delta_{22} = -87.5$ ppm, $\delta_{33} = 500$ ppm, $\Omega = 912.50$ ppm and $\kappa = -0.29$) with MAS variations. **c)** under static conditions (MAS at 0 kHz) **d)** under MAS at 1 kHz. **e)** under MAS at 4 kHz. **f)** under MAS at 6 kHz. **g)** under MAS at 10 kHz (4 mm rotor). **h)** under MAS at 28 kHz (2.5 mm rotor). **i)** under MAS at 65 kHz (1.3 mm rotor).

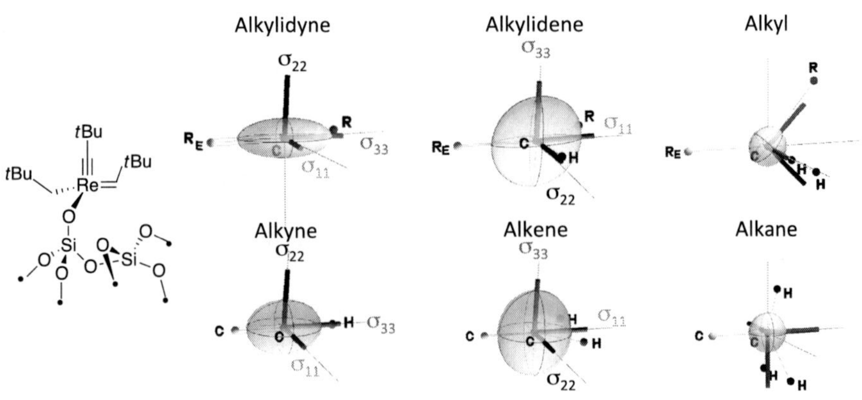

Fig. 4 Modelling of the chemical shift tensors orientation of organometallic fragments on a Re complex supported on silica: alkylidine (Re≡CR), alkylidene (Re=CHR) and alkyl (Re-CH_2R) compared to organic fragments: alkyne (C≡CH), alkene (C=CH_2) and alkane (C-CH_3).[14] σ_{11} in green, σ_{22} in black and σ_{33} in red.

Table 6 Determination of the C-H bond length and of the dipolar coupling constants C-H by NMR (experimental) and by DFT calculations for a rigid system.[14]

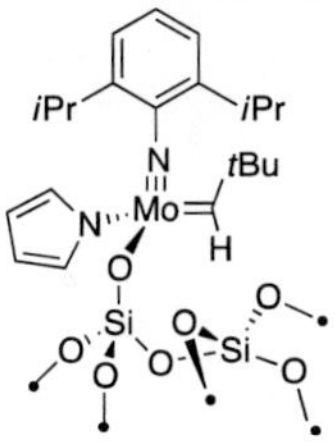

	Couplage dipolaire (kHz)		d_{CH} (Å)	
	Calculated	Experimental	Calculated	Experimental
M=*CR*–***H***	– 21.6	– 19.7	1.12	1.11
M–*CHR*–***H***	– 22.4	– 19.5	1.15	1.16

For instance, by measuring the chemical shift anisotropy of a series of surface alkylidene complexes and comparing with calculated values, monografted species on silica can be shown to be quite rigid, despite the attachment to the surface by a single $M–O_{surface}$ bond. This is particularly true for the complexes with bulky and aromatic ligands,[14] which is consistent with the fact that the loss of dynamics at room temperature was due to the Van der Waals interactions between the surface of the support and the organic fragments of the organometallic surface complex. For rigid systems, it has also been possible to use the measured dipolar coupling proton-carbon constant to evaluate the length of the bond, and the obtained value is very close to the calculated one (DFT) (Table 6).

Finally measuring chemical shift anisotropy of the carbons C_{sp2} of the fragment alkylidene could be exploited to probe the presence and the strength of agostic interactions between the C–H bond the alkylidene carbon and the metal center. Indeed, the principal components of the chemical shift tensor change orientation with the strength of the agostic interaction switching from δ_{33} (the most shielded) aligned with a perpendicular axis to the plane of the ligand alkylidene, in absence of the agostic interaction, to δ_{33} oriented along the M–C bond for a system which has a strong agostic interaction. This is in perfect agreement with the fact that carbon evolves from an sp^2 carbon to a carbon which has more and more sp character. This is associated with an increase of the strength of the agostic interaction, and this parallels the decrease of the coupling constant $J_{C\text{-}H}$ (Table 7), which is also a good reporter of hybridization.[71]

It is thus clear that a rigorous study of the chemical shift anisotropy, combining theory and experiment, is essential to understand at the molecular level the variation of coordination of a ligand on a metal center. This can in fact be used to understand the mode of absorption of carbon monoxide on metallic particles and metallic clusters.[72–79]

3 NMR of quadrupolar nuclei (spin > 1/2)

Quadrupolar nuclei ($I > 1/2$) display an anisotropic electric charge distribution (Scheme 4), in contrast to spin 1/2 nuclei, which have spherical distribution of their electric charge (isotropic).

Table 7 Comparison of the evolution of the chemical shift tensors and the coupling constant $J_{C\text{-}H}$ with the agostic interaction strength. σ_{11} in green, σ_{22} in black and σ_{33} in red.

Complexes	Tensors orientation	M=C–H Angle (°)	$^1J_{CH}$ (Hz)
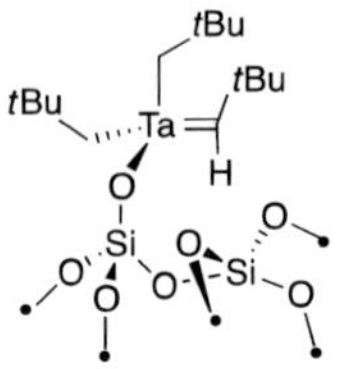		94.5	80
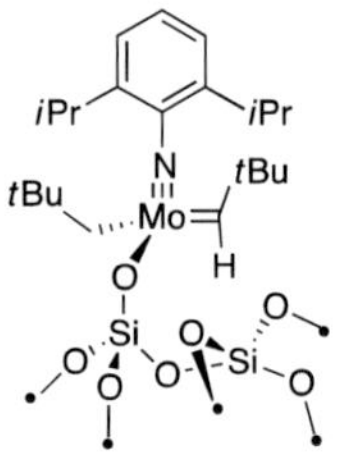	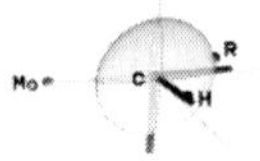	98.8	110
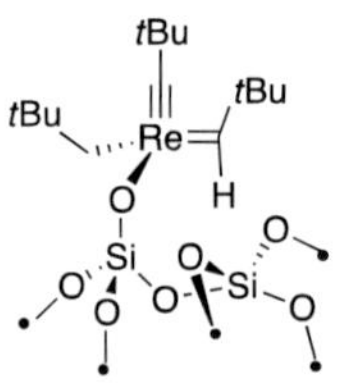	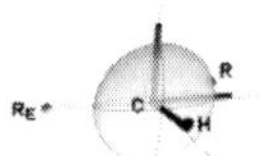	103.5	109
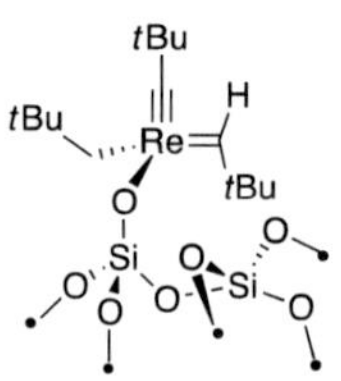	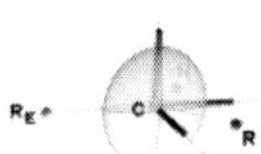	121.8	159

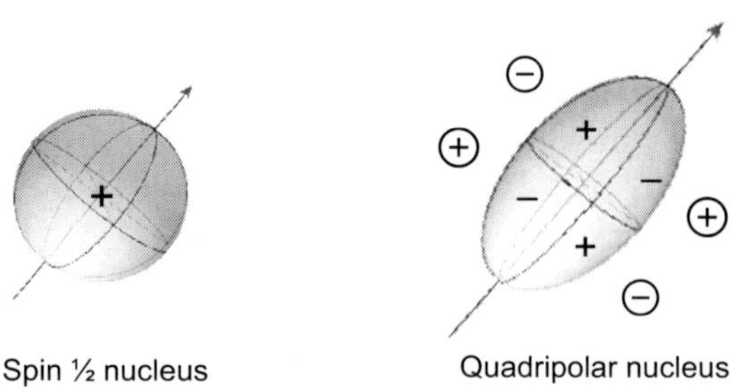

Scheme 4 Schematic representation of nuclei with spin 1/2; and quadrupolar +, − correspond to nuclear charge distribution and ⊕, ⊖ to the electric field gradients in the molecule.

The electric quadrupole moment is an intrinsic property of the nucleus: it does not depend on the environment. Quadrupolar nuclei interact with the electric field gradient (EFG) of the molecule, which depends on the distribution of charges in the molecule. Those gradients are described by a symmetric traceless tensor where $V_{33} > V_{22} > V_{11}$. It thus contributes in broadening the NMR spectrum (vide infra).

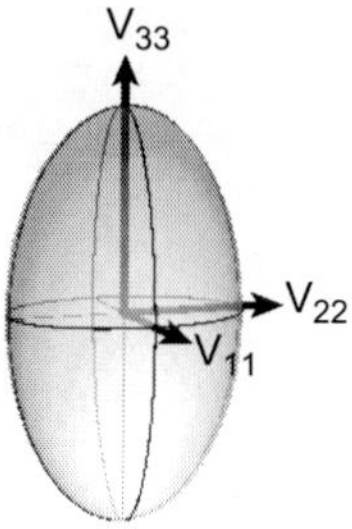

$$V = \begin{bmatrix} V_{xx} & V_{xy} & V_{xz} \\ V_{yx} & V_{yy} & V_{yz} \\ V_{zx} & V_{zy} & V_{zz} \end{bmatrix} \quad V^{PAS} = \begin{bmatrix} V_{11} & 0 & 0 \\ 0 & V_{22} & 0 \\ 0 & 0 & V_{33} \end{bmatrix}$$

For a quadrupolar nucleus in the center of a spherically symmetric molecule the resulting electric field gradient at the nucleus is close to zero, when the symmetry gets progressively broken the magnitude of this gradient grows, increasing the quadrupolar interaction.

The intensity of the interaction is described by the nuclear quadrupole coupling constant C_Q:

$$C_Q = \frac{eQV_{33}}{h}$$

where eQ is the nuclear electric quadrupole moment, a property of the nucleus, which does not change with the molecular environment.

The asymmetry is described by the asymmetry parameter η_Q:

$$\eta_Q = \frac{(V_{22} - V_{11})}{V_{33}}$$

$0 < \eta < 1$. If the asymmetry parameter is equal to zero the tensor is axially symmetric (Fig. 5).

It is noteworthy that the magnitude of quadrupolar coupling can largely exceed the range of chemical shift and can reach values of several MHz, making the detection (spreading of the signal), the distinction of sites and signal analysis particularly hard. This is even truer for surface species, where the density of sites is low making detection of isotropic chemical shift already a challenge in many cases. This probably explains why only a few studies, mainly devoted to deuterium, have been reported in surface chemistry. Yet, since the signal envelop of quadrupolar nuclei is highly sensitive to symmetry and thus coordination of the atom considered, it makes a powerful tool to identify sites and/or to evaluate their dynamics.

Deuterium. Despite its low natural abundance (0.015%) and its relatively weak sensitivity ($\gamma_D/\gamma_H = 0.15$), D-NMR ($I = 1$) has been particularly well studied. In particular, it is often easy to obtain 100% d-labeled samples; this is the case for metal-hydrides and metallic nanoparticles, which readily exchange their hydrogen for deuterium by H/D exchange with D_2 under mild conditions. For instance, the mobility of surface hydrides on Ru nanoparticles as a function of surface ligands has been studied in great details, clearly showing that the presence of donor ligands as stabilizing agent (phosphine in place of polyvinylpyrrolidone) freeze H-motion at the

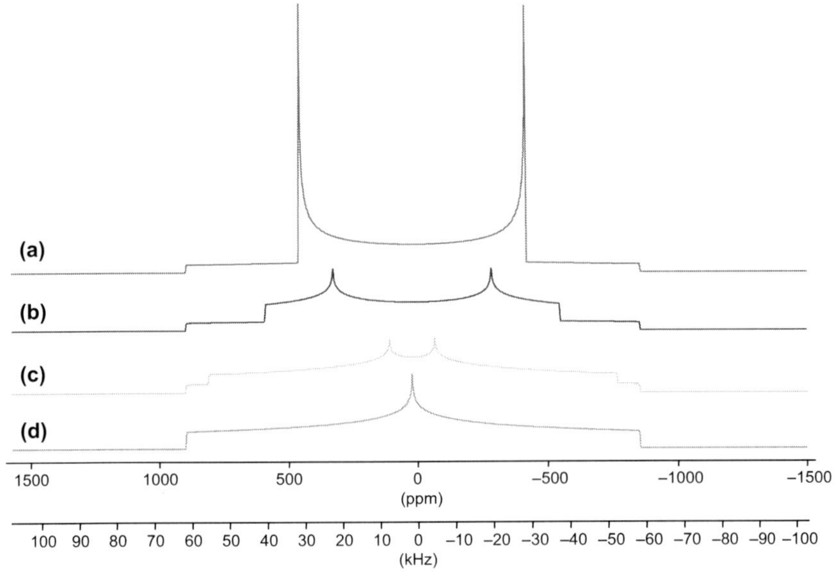

Fig. 5 Spectra simulations of a spin = 1 nuclei (such as Deuterium) using dmfit software. $^{70}\delta_{iso} = 0.00$ ppm, $\nu_Q = 120.00$ kHz (or $C_Q = 0.08$ MHz) are fixed, $V_{33} = 1.15 \times 10^{21}$ V · m^{-2} is calculated while η_Q varied: **a)** $\eta_Q = 0$, **b)** $\eta_Q = 0.3$, **c)** $\eta_Q = 0.8$ and **d)** $\eta_Q = 1$.

surface.[80] D-NMR can also be used to investigate the dynamics and the molecular motion of a model surface complex grafted on silica. For instance studying [–SiDMe$_2$] species grafted on various types of silica (non-porous *vs.* mesoporous SBA) using various pulse sequences {QE (quadrupole echo, as Scheme 1b for quadrupolar nuclei), QCPMG (quadrupole Carr-Purcell-Meiboom-Gill, as Scheme 3b for quadrupolar nuclei) and MAS spectra}, it was possible to shown that the dynamic of the surface species is more hindered on SBA materials showing that its surface is probably more rough than this of the corresponding amorphous silica nanoparticles.[81]

Oxygen-17 (I = 3/2). Despite the ubiquitous presence of oxygen in heterogeneous catalysts since the supports are typically metal oxides, studies related to surface species are rare. This is due to both low sensitivity ($\gamma_O/\gamma_H = 0.14$) and low natural abundance (0.038%) of oxygen-17. Additionally, oxygen-17 labeled compounds are relatively expensive, even for rather simple precursors such as di-oxygen or water as sources. Finally, oxygen-17 NMR is often associated with very broad signal unless it is in a fairly isotropic environment, which is often not the case for surface species. In silica obtained *via* a sol-gel process, the hydroxyl groups could be indirectly observed by ^{17}O solid-state MAS NMR. While the Si-O-Si and the Si-OH oxygens show similar chemical shifts, it is possible to distinguish and to quantify them by exploiting their very different T_1's (0.1 ms *vs.* 11 ms).[82] Oxygen-17 NMR was also used to distinguish the acid sites present in supercages *vs.* sodalite cages of zeolites.[83] While it could certainly provide information on how surface species interact with the oxygen functionality of the surface of the oxide supports, it has so far not been used (reported) in surface organometallic chemistry.

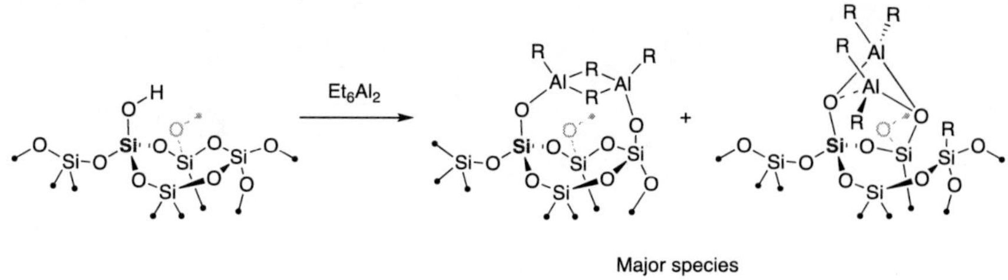

Scheme 5 Surface chemistry of triethylaluminium with silica.

Aluminium-27. Aluminum-27 NMR (I = 5/2, natural abundance: 100%, $\gamma_{Al}/\gamma_{H} = 0.26$) has also been an important tool to characterize heterogeneous catalysts since alumino-silicates and alumina are often used as supports or catalysts. In fact, the chemical shifts of aluminum sites is very sensitive to their coordination, the isotropic chemical shift being at *ca.* 70, 40 and 10 ppm for tetra-, penta- and hexa-coordinated aluminum sites, respectively. It has even been used to distinguish the position of aluminum sites in zeolites.[84] While this has been a powerful technique to distinguish the local geometry of bulk aluminum sites, the study of surface sites is more arduous, even if the use of indirect detection methods (CP, CPMG) can help to more or less selectively probes such sites.[85] A recent study of model surface aluminum species obtained by reaction of triethylaluminum on mesoporous silica has shown that the isotropic chemical shift is not sufficient to determine the local geometry at aluminum and that all NMR parameters must be considered, that is δ_{iso}, C_Q, and η_Q; the nature of the ligand bound to aluminum already greatly influences the chemical shift.[86] Using very high and different fields (17.6, 20.0 and 23.5 T), very high spinning rates (60–70 kHz) in combination with DFT calculations, it was, however, possible to clearly distinguish three different sites: two dimeric aluminum species, bound to the surface *via* two oxygen functionalities either both in terminal or bridging position along with even tetrahedral Al site (not shown), which arises from the integration of Al inside the silica matrix (Scheme 5). It is noteworthy that no monomeric tricoordinated Al species was detected.

Other quadrupolar nuclei. Recent studies in molecular organometallic chemistry have shown that zirconium-91 (I = −5/2) and chlorine-35 (I = 3/2) can be used to distinguish the first coordination sphere of metallocene complexes.[18,87] While not yet transposed to surface species, these studies make possible the investigation of the corresponding supported systems, even if obtaining good quality spectra (good S/N ratio) will be an issue.

4 Connectivity by multidimensional NMR

Multidimensional homonuclear Double-Quanta Proton-Proton NMR. 2D double-quanta NMR sequences (Scheme 6, sequencespost-C7[88] or BABA[89]) allow the selective observation of neighboring protons which interact through dipolar coupling.[90,91] Therefore, the 2D spectra provide a 2D map with the chemical shift scale in the F2 dimension (*e.g.* 0–13 ppm) and the double quanta scale in F1 (double scale, *e.g.* 0–26 ppm). Thus two protons

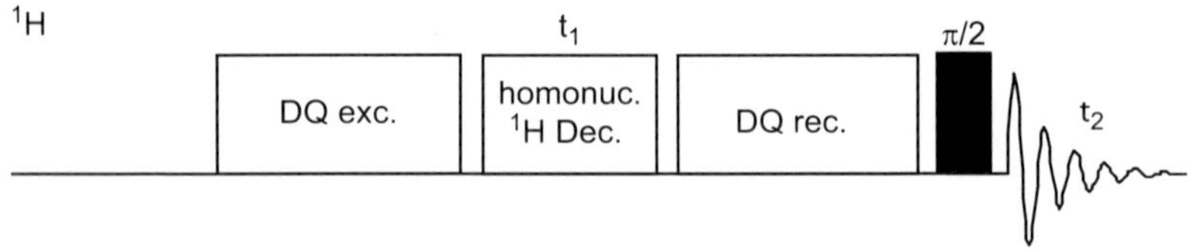

Scheme 6 ^1H-^1H double quanta experiment to investigate proton-proton connectivity. Where t_l denotes the evolution time, *DQ exc.* double quantum excitation and *DQ rec.* double quantum reconversion.

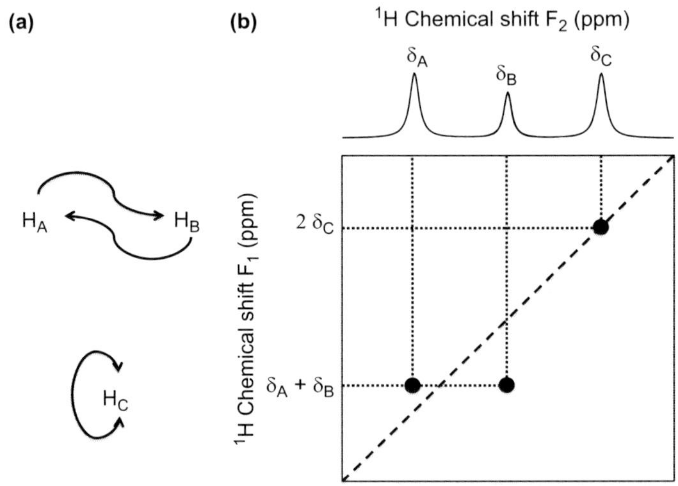

Fig. 6 Double quanta proton-proton multidimensional NMR. **a)** Dipolar interactions between the protons H_A, H_B and H_C. **b)** Expected correlation models of the peaks of the protons A and B, δ_A and δ_B, dipolarly coupled and the two protons C with the same chemical shift, δ_C.

H_A and H_B which have different chemical shift, δ_A and δ_B respectively, appear on F2 at their respective chemical shift, and in F1 at the sum of their chemical shifts ($\delta_A + \delta_B$). Protons which are not correlating with their neighbor will not appear in the second dimension (as they are excluded by the selection rule of double quanta), neighboring protons which are equivalent, and associated with same chemical shift, δ_C, will produce an auto-correlation peak at ($2 \times \delta_C$) in F1 (Fig. 6).

This sequence has been exploited to distinguish the surface sites of silica supported zirconium hydrides, which are known to catalyze hydrogenolysis[92] and the homologation of the alkanes[93] (Fig. 7):[94]

i) A zirconium monohydride, $(\equiv SiO)_3Zr\text{-}H$, close to a $[(\equiv SiO)_2SiH_2]$ fragment. Those species come from the opening of two siloxy bridges from a putative intermediate $(\equiv SiO)ZrH_3$ coming from the reaction of $(\equiv SiO)Zr(CH_2tBu)_3$ and H_2.

ii) A zirconium dihydride, $(\equiv SiO)_2Zr(H)_2$, close to a $[(\equiv SiO)_3SiH]$ fragment, which arise from the opening of only one siloxy bridge from the putative intermediate $(\equiv SiO)ZrH_3$.

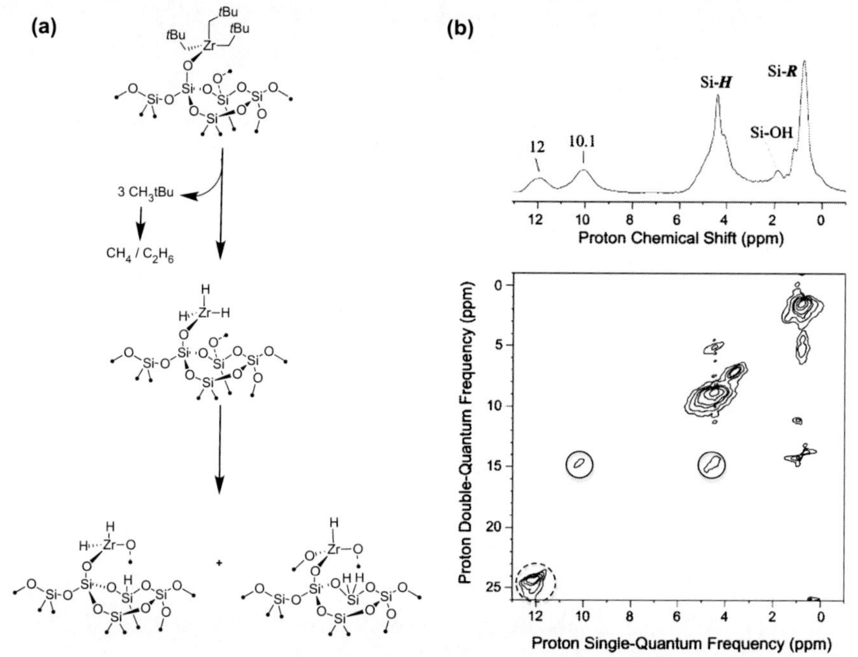

Fig. 7 **a)** Formation mechanism of different zirconium hydrides supported on silica **b)** Double quanta proton-proton multidimensional NMR allowing the identification of the two surface sites of the zirconium hydrides supported on silica, $(\equiv SiO)_2Zr(H)_2$ (auto correlation highlighted with dotted line) and $(\equiv SiO)_2Zr\text{-}H$ (correlation with surface SiH, full line). Adapted with permission from the authors.[94]

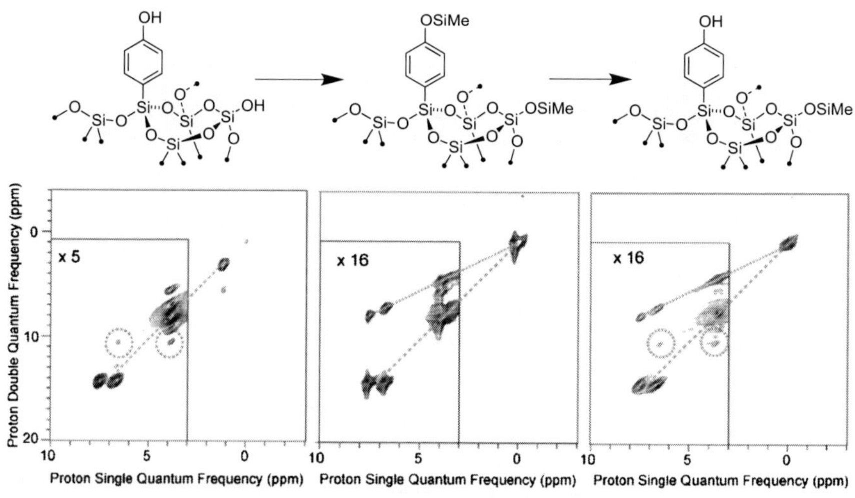

Fig. 8 Following the reaction by double quanta NMR of an exchangeable phenolic H on an hybrid material. Adapted with permission from the authors.[35]

Double-quanta NMR allows the recognition of sites and the monitoring of their reactivity, for example, the exchangeable proton of a phenol functionality of an hybrid material (Fig. 8) which disappears and re-appears upon selective post treatment.[35] The high proton density in this hybrid organic-inorganic material, leading to ^1H-^1H dipolar interactions, does not allow the resolution of all signals using standard MAS^1H Proton NMR (spinning at 14 kHz). Under ultra-fast MAS conditions (spinning at 67 kHz), most of the signals were resolved (vide supra) when combined with double quanta NMR. Several on- and off-diagonal correlations are obtained that allow the assignment of these otherwise unresolved signals.[95] We notice that other techniques (double CP from ^{29}Si or ^{31}P[96] and 2D heteronuclear correlations[97]) are rarely employed.

Finally it is also possible to refine the selection rules of this sequence to observe only correlations between three spins (triple quanta NMR (TQ), Scheme 7). By combining DQ and TQ NMR, it has been possible to determine the structure of the amides and imides deriving from the splitting of N_2 on silica supported Ta hydride.[42,98]

Heteronuclear multidimensional NMR. In the same way as in solution, heteronuclear NMR brings essential information to establish the connectivity of the supported organometallic fragments. In solid state NMR, the classic HETCOR (HETeronuclear CORrelation, Scheme 8) techniques use, however, through space dipolar interactions between neighbor atoms, which allows one to indirectly evaluate their connectivity (in place of scalar coupling in solution). Thus, by optimizing the contact time (short, 0.5–1 ms, or long, 5 ms), it is possible to « select » more or less distant couple of spins; short contact times favor interactions at short distances, *i.e.* between the closest atoms. Thank to these techniques, it has been possible to determine the connectivity of surface compounds and thus their structure (Re,[33] Mo,[34] W[16]). It has even been used to characterize the formation of reactive

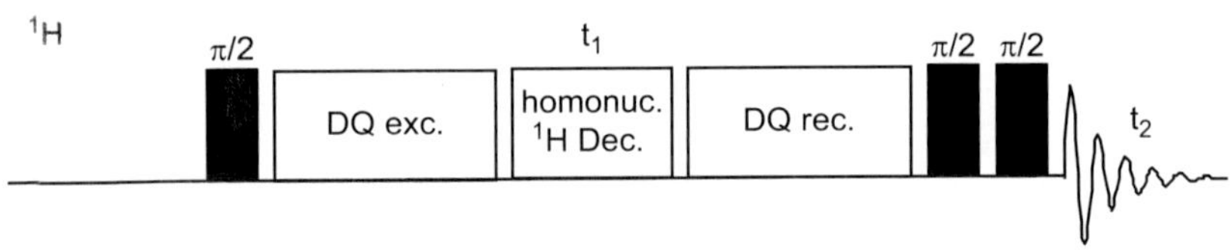

Scheme 7 ^1H-^1H triple quanta experiment to detect proton-proton connectivity.

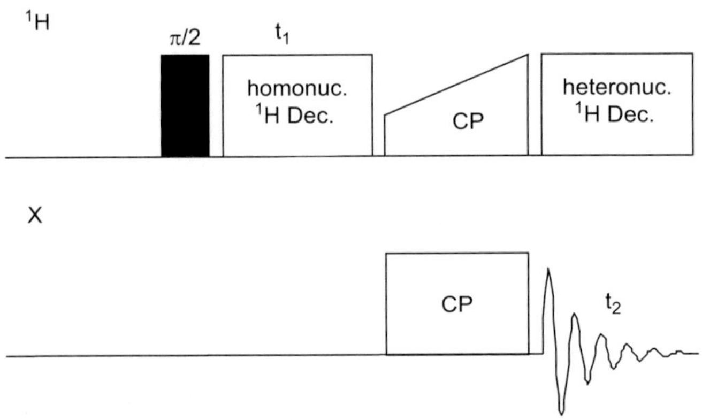

Scheme 8 ^1H-X HETCOR experiment to detect trough space nuclei connectivity.

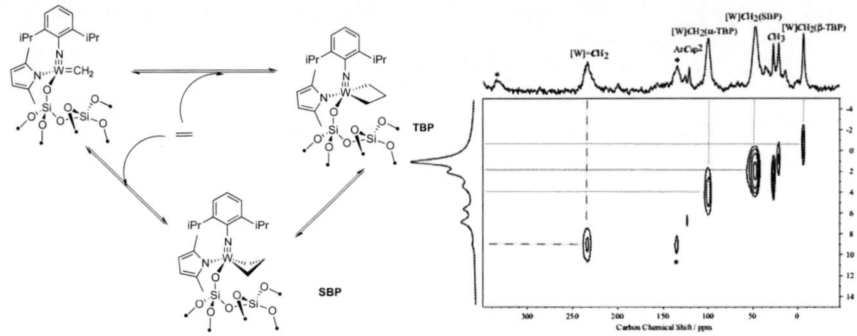

Fig. 9 Highlight of the intermediates of metallocyclobutane and alkylidene of the olefines metathesis with a W complex supported on silica. Adapted with permission from the authors.[99]

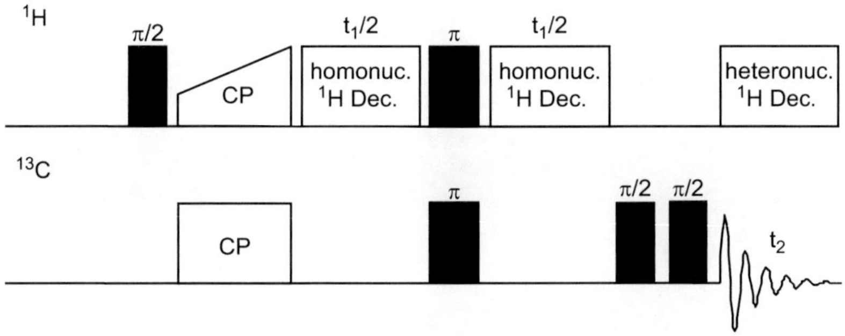

Scheme 9 Solid state J-resolved experiment to measure ^{1}H-X heteronuclear dipolar coupling J.

intermediates of well-defined alkene metathesis catalysts, *i.e.* metallocyclobutanes and alkylidene intermediates, and to demonstrate their role in the catalytic cycle of the reaction (Fig. 9).[99]

5 Measuring J coupling constants and structural considerations

Beyond the connectivity, the scalar coupling constant J is a characteristic fingerprint of the hybridization of a given nucleus, which allows you in certain cases to evaluate (M=C–H) bond angles. For example, the $J_{C\text{-}H}$ constant of the carbons of the alkylidene ligands, bound to a transition metal is very sensitive, and it correlates to the presence, or not, and to the degree of an agostic interaction (a three-center – two-electron bond). The coupling constant is related to the angle M=C–C(H) (more or less inferior to 120°)[100,101] by an empirical (quadratic) relation, determined on a series of molecular complexes. In solid state NMR, measuring this constant is not trivial because its value is much smaller than the corresponding dipolar coupling and chemical shift anisotropy. However, by optimizing the pulse sequences including decoupling[102] (Scheme 9), it is possible to extract it and thereby to evaluate the M–C–C bond angle of a surface species, thus providing detailed structural information. Such values have been measured for a series of surface complexes, and the experimental values are comparable with this obtained by DFT calculations (Table 8).[14]

Table 8 Comparison of the coupling constants $^{1}J_{CH}$ calculated from the M-C-H angles and experimentally determined by NMR.[14]

Complexes	Angle M=C–H (°)	$^{1}J_{CH}$ (Hz)	
		Calculated	Experimental
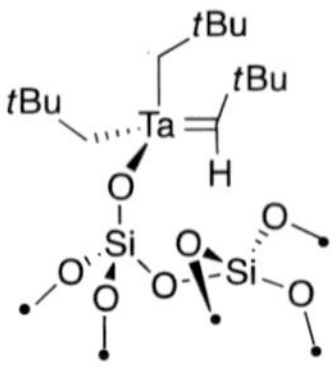	94.5	94	80
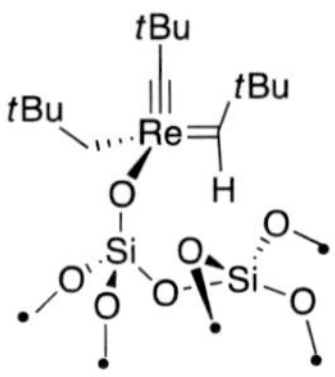	103.5	107	109
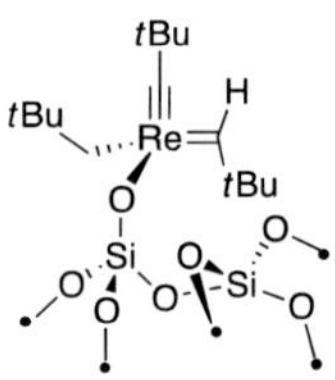	121.8	141	159
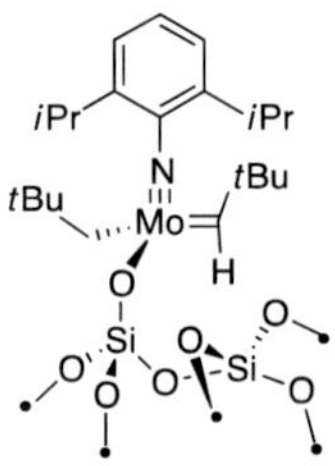	98.8	105	110
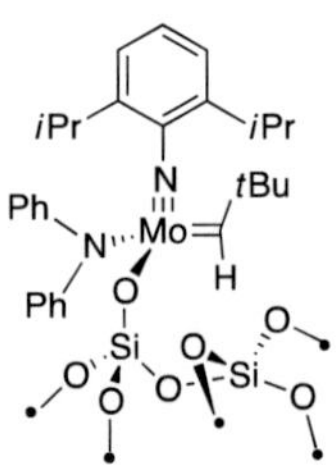	99.5	107	107
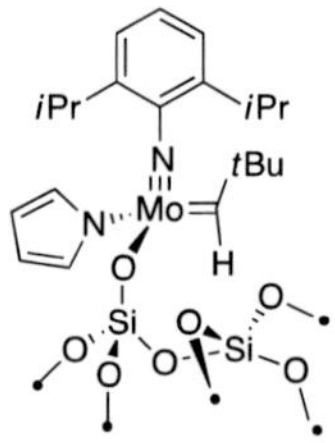	98.9	106	109

6 Surface enhanced solid-state NMR spectroscopy

So far, one of the major problems related to the NMR technique is its low sensitivity, which is even more serious in the case of surface species, because they correspond only to a (very) small fraction of the sample (1–10%): 0.1–2.0 mmol of a particular atom per gram of the sample which arrives to 1 to 20 μmol per gram of active nuclei for carbon-13. This is the reason why structural NMR studies of the surface species require often selective isotopic labeling (*e.g.* carbon-13, nitrogen-15, silicon-29, oxygen-17,...), which implies synthetic difficulties and high costs. Typically the enhancement of the signal for nuclei at low natural abundance is realized by cross-polarization, by exploiting the high gamma of protons and transferring its magnetization to their neighboring nuclei (vide supra). However, due to the weak abundance of the surface species, this technique allows only to obtain 1D spectra within reasonable acquisitions times (< 12–24 h for a carbon-13 spectrum), thus preventing the use of more advanced techniques (multi-dimensional NMR). Note also that sites featuring strong chemical shift anisotropy (CSA) are also hardly detected under these conditions. It is thus crucial to develop techniques, which further enhance the NMR signal.

One of the techniques, which has been developed, uses magnetic resonance of physisorbed ^{129}Xe. Xenon presents several advantages. First, it is an inert gas. Moreover the extremely polarizable electronic environment makes it an ideal probe, because the signal of xenon is sensitive to any interaction. The resulting electronic perturbation of ^{129}Xe isotope (spin 1/2), is directly transmitted to the surface nuclei with consequent effects on the chemical shift and enhancement; it has thus been exploited to a large extent to understand the surface sites of metal oxides and zeolites.[103] Another technique which can be used is producing a large nuclear spin polarization by employing the technique of para-hydrogen (the two spins of hydrogen are antiparallel, singlet state) induced polarization. The singlet state of the molecular hydrogen is transferred to a molecule *via* hydrogenation by contacting molecular hydrogen at a sufficient low temperature. By para-hydrogen assisted spectroscopy the sensitivity can be increased up to a factor of 31000. However this technique can be applied only to organometallic complexes, which can be hydrogenated, with both atoms of the para-hydrogen on the same metal atom. So far, this technique has not yet been exploited for surface species. Finally, another technique to enhance the NMR signal is the dynamic nuclear polarization (DNP). In CP NMR, the enhancement of the signal of a nucleus (i) is obtained by cross polarization usually with neighboring protons (^{1}H), and it depends on γ_H/γ_i. In DNP, the electrons, which are characterized by a very high gyromagnetic constant (γ_{e-}), 660 times higher than the protons, are thus better polarization « agents ».[104–106] This has been proposed since the first steps of NMR; DNP typically uses stable organic radicals as source of electron (Scheme 10). Technological advances in the domain of gyrotrons and solid state NMR allow today to enhance the NMR signal by two orders of magnitude.[107–116] It has been mainly used to characterize biological samples in frozen solutions. In 2010, this technique was transposed to selectively enhance the NMR signal of the surface functionalities of materials. One of the key steps of this technique has been the use of the Incipient Wetness Impregnation (IWI) method, corresponding to the wetting of material with a minimal quantity of solution, quantity that is approximately the same volume of the pore volume of the material. Then the material, appearing slightly wet – which reminds of the spring snow – is introduced in a zirconia rotor with thin walls or in a sapphire rotor, permeable to the microwaves, and the measurements are performed at *ca.*100 K under MAS (4–5 kHz).

The first positive results were obtained on hybrid organic-inorganic materials.[117,118] One of the challenges in this domain is to verify the integrity of the organic functionalities in the hybrid materials, and their interaction with the inorganic surface knowing that there is often less than one functionality per nm^2).

For instance, a hybrid material having *ca.* one phenol functionality per $nm^2$and prepared by direct synthesis,[35] requires *ca.* 20 hours of acquisition time for a low quality carbon-13 spectrum (low signal to noise ratio), and proper silicon-29 spectrum cannot be recorded: no signal specific for the organic fragment attached to the silica matrix can be detected even after 60 hours of acquisition, making it impossible to know how the silicon atom is bound to the organic fragment (sites T, Fig. 10). This is particularly true for the selected example below, where the silicon has no proton nearby.

After the impregnation of the material with an aqueous solution of TOTAPOL (Fig. 10), (optimal concentrations in the order of 10–15 mM), an enhancement of the signal in carbon-13 NMR[117] and silicon-29 NMR[118] of the order of 40–50 is observed, which allows to obtain spectra in record time, often in few minutes. It is noteworthy that the signal associated with the T sites of silicon are observed with a good signal-to-noise ratio in only few hours in contrast to obtaining a poor spectrum in 60 h with classical NMR techniques.[35] In particular it allows to evidence the effect of the preparative routes on the structure of the final materials. The material prepared *via* direct synthesis presenting principally the signal associated to the T_3 sites, along with T_2 sites in weaker intensity in contrast to the

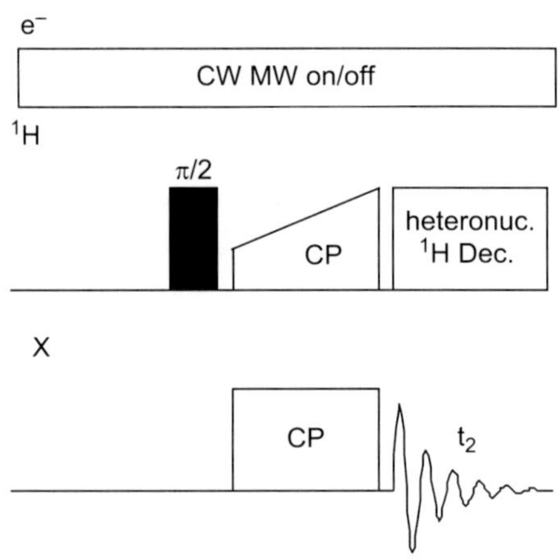

Scheme 10 ^{1}H-X cross polarization experiment with dynamic nuclear polarization (DNP). Continuous wave microwave (CW MW) can be turned on for excitation of the electrons or off.

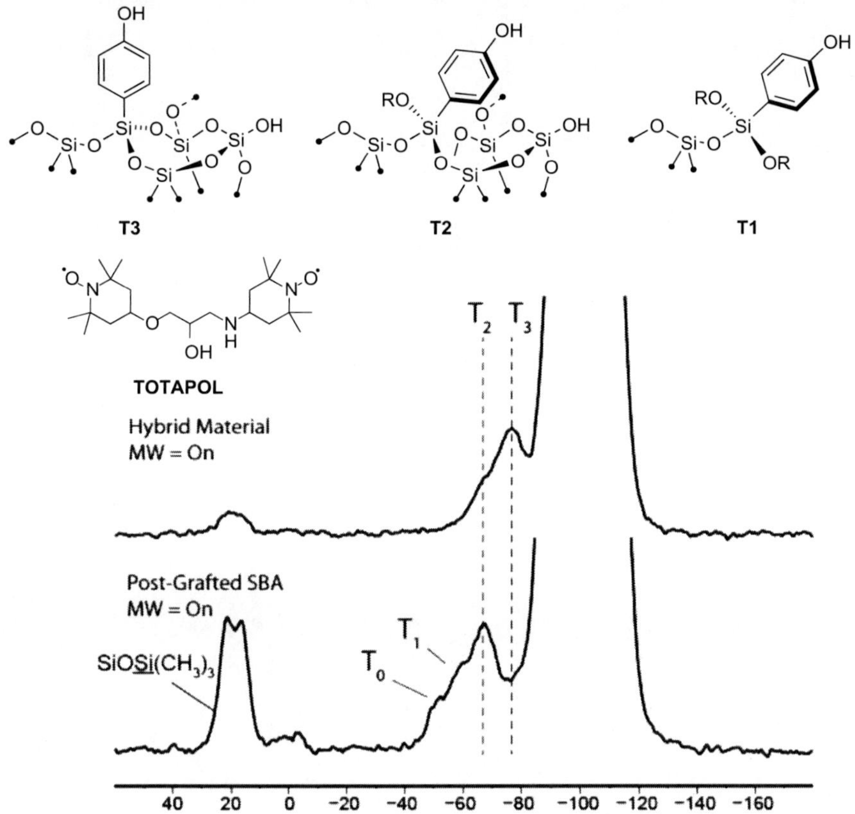

Fig. 10 T sites of a hybrid material and a post treatment SBA compared by NMR of ^{29}Si, enhanced by Dynamic Nuclear Polarization. Adapted with permission from the authors.[118]

material prepared post-synthesis, which shows no signal in the T_3 region, and mostly signals associated to T_2 and T_1 sites, showing the difficulty to integrate the fragment in the matrix with the "post-grafting approach".

Furthermore with the decrease of experimental time to record 1D spectra, SENS allows to perform multidimensional NMR experiences. Now proton-carbon NMR correlation (HETCOR experiment) allows the identifation without ambiguity of all the signals associated to the phenolic fragment of the unlabeled compounds (Fig. 11a). Similar proton-silicon correlation NMR spectroscopy on hybrid materials allows highlighting the interaction between the organic moiety with the surface for the systems characterized by flexible arms, and the absence of interaction for the rigid-arm systems. This allows further refinement in the understanding of the structure and the dynamics of the supported systems.

Finally SENS NMR is not limited to the use of aqueous solvents. Now a large range of organic non-protic solvents in particular chlorinated ones in combination with nitroxide radicals that are non-soluble in aqueous solutions can be used. For instance, use of a 10 mM solution of bTbK (bis-TEMPO-bisketal, Scheme 11a) in tetrachloroethane yields enhancements in the range of 36, making this technique compatible with the most sensitive systems (Fig. 11b)[119] and metal organic framework (MOF).[120] Using a

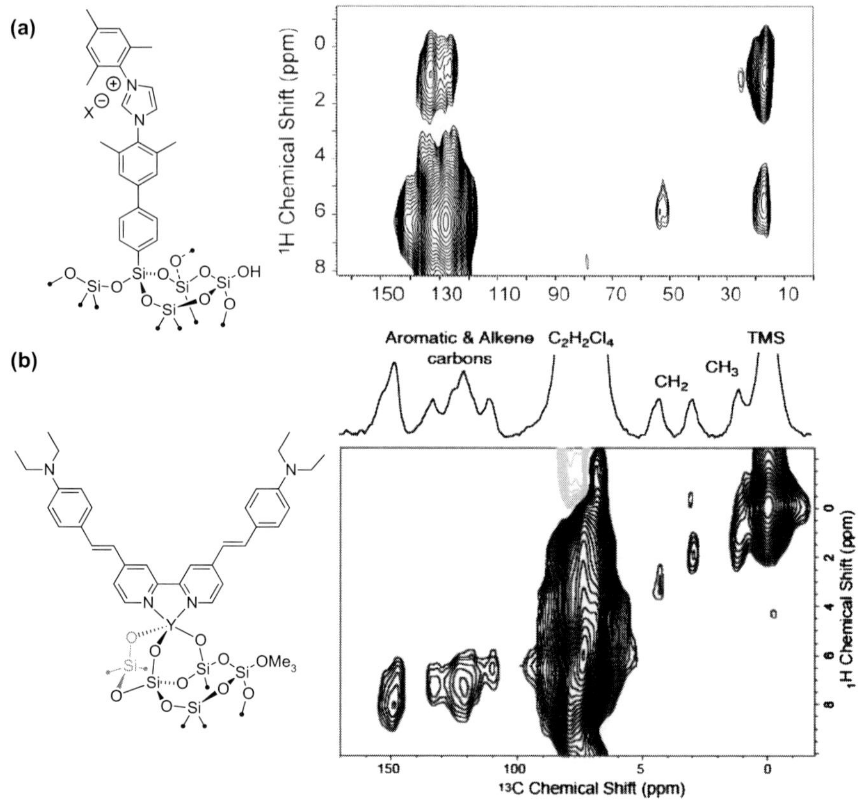

Fig. 11 Characterization by Dynamic Nuclear Polarization enhanced 2D NMR of the surface species **a)** of a hybrid material NMR HETCOR ^{1}H-^{29}Si[117] and **b)** of a surface organometallic complex, NMR HETCOR ^{1}H-^{13}C. Adapted with permission from the authors.[119]

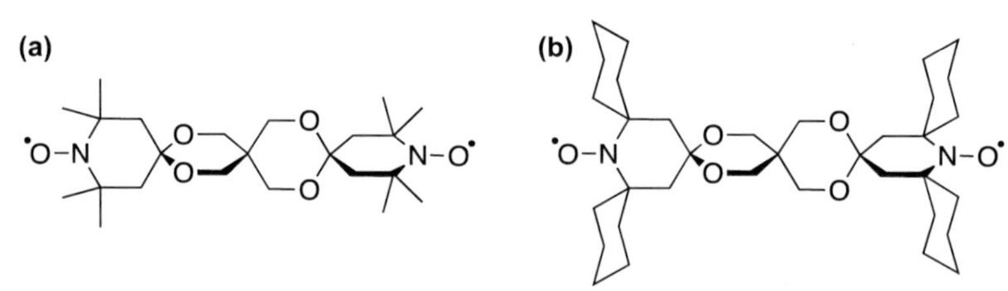

Scheme 11 Two biradicals soluble in organic solvents: **a)** bTbK and **b)** bCTbK.

radical with a longer T_{1e} (bCTbK, Scheme 11b), can further improve the signal enhancement, thus permitting to monitor by nitrogen-15 NMR the post-functionalisation of hybrid materials.[121] This technique is not limited to spin 1/2 nuclei, and recent results show that it can in principle be appliied to the study of quadrupolar nuclei such as Aluminium-27.[122]

These latter progresses introduce a new field of investigation thereby opening up the technique to a large range of solids, in particular water and oxygen sensitive materials such as heterogeneous catalysts, or surfaces functionalized by organometallic or coordination compounds derivatives.

7 Conclusions

This review is principally oriented towards the characterization of surface species of well-defined heterogeneous catalysts. We have illustrated that solid state NMR is essential for the comprehension at the molecular scale of structure and dynamics, therby making it possible to produce a molecular view of surface chemistry and heterogeneous catalysts. The progress in methods (pulse sequences more and more sophisticated) and technologies (high speed probes, cryo-probes, DNP NMR) have already revolutionized the world of the surface chemistry. Of those, SENS NMR, with its possibility to selectively increase the NMR signal of the surface sites, that is often difficult with other solid-state spectroscopy methods, should allow further progresses in the surface chemistry domain, including the production of structural information, which was not possible before. Time will tell!

References

1 D. G. H. Ballard, *Advan. Catal.*, 1973, **23**, 263.
2 Y. I. Ermakov and B. N. Kuznetsov, *Kinet. Katal.*, 1977, **18**, 1167.
3 T. J. Marks, *Acc. Chem. Res.*, 1992, **25**, 57.
4 J. Guzman and B. C. Gates, *Dalton Trans.*, 2003, 3303.
5 K. L. Fujdala, R. L. Brutchey and T. D. Tilley, *Top. Organomet. Chem.*, **16**, 69.
6 J. M. Thomas, R. Raja and D. W. Lewis, *Angew. Chem. Int. Ed.*, 2005, **44**, 6456.
7 M. Tada and Y. Iwasawa, *Coord. Chem. Rev.*, 2007, **251**, 2702.
8 J. M. Basset, J. P. Candy, C. Copéret, In *Comprehensive Organometallic Chemistry III*; R. H. Crabtree, D. Michael, P. Mingos (eds.), Elsevier: Oxford, 2007, p. 499.
9 C. Copéret, M. Chabanas, R. Petroff Saint-Arroman and J. M. Basset, *Angew. Chem. Int. Ed.*, 2003, **42**, 156.
10 J. Estephane, E. Groppo, J. G. Vitillo, A. Damin, D. Gianolio, C. Lamberti, S. Bordiga, E. A. Quadrelli, J. M. Basset, G. Kervern, L. Emsley, G. Pintacuda and A. Zecchina, *J. Phys. Chem. C*, 2010, **114**, 4451.
11 G. Kervern, G. Pintacuda, Y. Zhang, E. Oldfield, C. Roukoss, E. Kuntz, E. Herdtweck, J. M. Basset, S. Cadars, A. Lesage, C. Copéret and L. Emsley, *J. Am. Chem. Soc.*, 2006, **128**, 13545.
12 M. Hunger and J. Weitkamp, *Angew. Chem. Int. Ed.*, 2001, **40**, 2954.
13 R. Berthoud, N. Rendon, F. Blanc, X. Solans-Monfort, C. Copéret and O. Eisenstein, *Dalton Trans.*, 2009, 5879.
14 F. Blanc, J. M. Basset, C. Copéret, A. Sinha, Z. J. Tonzetich, R. R. Schrock, X. Solans-Monfort, E. Clot, O. Eisenstein, A. Lesage and L. Emsley, *J. Am. Chem. Soc.*, 2008, **130**, 5886.
15 J. Joubert, F. Delbecq, C. Copéret, J. M. Basset and P. Sautet, *Top. Catal.*, 2008, **48**, 114.
16 B. Rhers, A. Salameh, A. Baudouin, E. A. Quadrelli, M. Taoufik, C. Copéret, F. Lefebvre, J. M. Basset, X. Solans-Monfort, O. Eisenstein, W. W. Lukens, L. P. H. Lopez, A. Sinha and R. R. Schrock, *Organometallics*, 2006, **25**, 3554.
17 J. W. Wiench, Y. S. Avadhut, N. Maity, S. Bhaduri, G. K. Lahiri, M. Pruski and S. Ganapathy, *J. Phys. Chem. B*, 2007, **111**, 3877.
18 A. J. Rossini, R. W. Mills, G. A. Briscoe, E. L. Norton, S. J. Geier, I. Hung, S. Zheng, J. Autschbach and R. W. Schurko, *J. Am. Chem. Soc.*, 2009, **131**, 3317.

19 C. Bonhomme, C. Coelho, N. Baccile, C. Gervais, T. Azais and F. Babonneau, *Acc. Chem. Res.*, 2007, **40**, 738.
20 V. I. Bakhmutov, *Chem. Rev.*, 2011, **111**, 530.
21 F. Blanc, C. Copéret, A. Lesage and L. Emsley, *Chem. Soc. Rev.*, 2008, **37**, 518.
22 A. Lesage, *Phys. Chem. Chem. Phys.*, 2009, **11**, 6876.
23 D. Massiot, In *High Magnetic Fields: Applications in Condensed Matter Physics and Spectroscopy*; C. Berthier, L. P. Levy, G. Martinez (eds.), Springer-Verlag, Berlin, 2001; Vol. 595, p. 435.
24 J. W. Wiench and M. Pruski, *Solid State Nucl. Magn. Reson.*, 2004, **26**, 51.
25 A. Lupulescu, S. P. Brown and H. W. Spiess, *J. Magn. Reson.*, 2002, **154**, 101.
26 J. F. Haw, W. G. Song, D. M. Marcus and J. B. Nicholas, *Acc. Chem. Res.*, 2003, **36**, 317.
27 B. Elena, G. de Paepe and L. Emsley, *Chem. Phys. Lett.*, 2004, **398**, 532.
28 A. Lesage, D. Sakellariou, S. Hediger, B. Elena, P. Charmont, S. Steuernagel and L. Emsley, *J. Magn. Reson.*, 2003, **163**, 105.
29 D. Sakellariou, A. Lesage, P. Hodgkinson and L. Emsley, *Chem. Phys. Lett.*, 2000, **319**, 253.
30 A. Samoson, In *Encyclopedia of Nuclear Magnetic Resonance*; D. M. Grant, R. K. Harris (eds.), John Wiley & Sons Ltd.: Chichester, UK, 2002.
31 A. Samoson, T. Tuherm, J. Past, A. Reinhold, T. Anupold and N. Heinmaa, In *New Techniques in Solid-State Nmr*; J. Klinowski (ed.), Springer-Verlag, Berlin, 2005; Vol. 246, p. 15.
32 R. Petroff Saint-Arroman, M. Chabanas, A. Baudouin, C. Copéret, J. H. Basset, A. Lesage and L. Emsley, *J. Am. Chem. Soc.*, 2001, **123**, 3820.
33 M. Chabanas, A. Baudouin, C. Copéret, J. M. Basset, W. Lukens, A. Lesage, S. Hediger and L. Emsley, *J. Am. Chem. Soc.*, 2003, **125**, 492.
34 F. Blanc, C. Copéret, J. Thivolle-Cazat, J. M. Basset, A. Lesage, L. Emsley, A. Sinha and R. R. Schrock, *Inorg. Chem.*, 2006, **45**, 9587.
35 A. Roussey, D. Gajan, T. K. Maishal, A. Mukerjee, L. Veyre, A. Lesage, L. Emsley, C. Copéret and C. Thieuleux, *Phys. Chem. Chem. Phys.*, 2011, **13**, 4230.
36 F. Bini, C. Rosier, R. Petroff Saint-Arroman, E. Neumann, C. Dablemont, A. de Mallmann, F. Lefebvre, G. P. Niccolai, J. M. Basset, M. Crocker and J. K. Buijink, *Organometallics*, 2006, **25**, 3743.
37 R. Petroff Saint-Arroman, J. M. Basset, F. Lefebvre and B. Didillon, *Appl. Catal. A-Gen.*, 2005, **290**, 181.
38 G. Tosin, C. C. Santini, M. Taoufik, A. De Mallmann and J. M. Basset, *Organometallics*, 2006, **25**, 3324.
39 E. Le Roux, M. Taoufik, M. Chabanas, D. Alcor, A. Baudouin, C. Copéret, J. Thivolle-Cazat, J. M. Basset, A. Lesage, S. Hediger and L. Emsley, *Organometallics*, 2005, **24**, 4274.
40 F. Blanc, R. Berthoud, C. Copéret, A. Lesage, L. Emsley, R. Singh, T. Kreickmann and R. R. Schrock, *Proc. Natl. Acad. Sci. U. S. A.*, 2008, **105**, 12123.
41 F. Rataboul, M. Chabanas, A. De Mallmann, C. Copéret, J. Thivolle-Cazat and J.-M. Basset, *Chem. Eur. J.*, 2003, **9**, 1426.
42 P. Avenier, A. Lesage, M. Taoufik, A. Baudouin, A. De Mallmann, S. Fiddy, M. Vautier, L. Veyre, J.-M. Basset, L. Emsley and E. A. Quadrelli, *J. Am. Chem. Soc.*, 2007, **129**, 176.
43 J. F. Haw, I. S. Chuang, B. L. Hawkins and G. E. Maciel, *J. Am. Chem. Soc.*, 1983, **105**, 7206.
44 J. F. Haw, *Phys. Chem. Chem. Phys.*, 2002, **4**, 5431.
45 J. P. Osegovic and R. S. Drago, *J. Phys. Chem. B*, 2000, **104**, 147.

46 R. M. Gauvin, L. Delevoye, R. A. Hassan, J. Keldenich and A. Mortreux, *Inorg. Chem.*, 2007, **46**, 1062.
47 D. Gajan, K. Guillois, P. Delichere, J. M. Basset, J. P. Candy, V. Caps, C. Copéret, A. Lesage and L. Emsley, *J. Am. Chem. Soc.*, 2009, **131**, 14667.
48 D. Gajan, D. Levine, E. Zocher, C. Copéret, A. Lesage and L. Emsley, *Chem. Sci.*, 2011, **2**, 928.
49 S. L. Burkett, S. D. Sims and S. Mann, *Chem. Commun.*, 1996, 1367.
50 J. Trebosc, J. W. Wiench, S. Huh, V. S. Y. Lin and M. Pruski, *J. Am. Chem. Soc.*, 2005, **127**, 7587.
51 J. Trebosc, J. W. Wiench, S. Huh, V. S. Y. Lin and M. Pruski, *J. Am. Chem. Soc.*, 2005, **127**, 3057.
52 H. Y. Carr and E. M. Purcell, *Phys. Rev.*, 1954, **94**, 630.
53 S. Meiboom and D. Gill, *Rev. Sci. Instrum.*, 1958, **29**, 688.
54 G. Hartmeyer, C. Marichal, B. Lebeau, P. Caullet and J. Hernandez, *J. Phys. Chem. C*, 2007, **111**, 6634.
55 C. Nedez, A. Theolier, F. Lefebvre, A. Choplin, J. M. Basset and J. F. Joly, *J. Am. Chem. Soc.*, 1993, **115**, 722.
56 A. de Mallmann, O. Lot, N. Perrier, F. Lefebvre, C. C. Santini and J. M. Basset, *Organometallics*, 1998, **17**, 1031.
57 M. Adachi, J. Corker, H. Kessler, F. Lefebvre and J. M. Basset, *Microporous Mesoporous Mater*, 1998, **21**, 81.
58 X. X. Wang, A. de Mallmann, F. Bayard, F. Lefebvre and J. M. Basset, *Microporous Mesoporous Mater.*, 2003, **63**, 147.
59 C. Nedez, F. Lefebvre, A. Choplin, G. P. Niccolai, J. M. Basset and E. Benazzi, *J. Am. Chem. Soc.*, 1994, **116**, 8638.
60 A. Eldewik, J. M. Hook, N. K. Singh and R. F. Howe, *Magn. Reson. Chem.*, 1999, **37**, S63.
61 G. E. Pavlovskaya, C. D. Ren, M. Van Buskirk, C. Dybowski, D. R. Corbin, J. A. Reimer and A. T. Bell, *Catal. Lett.*, 2002, **80**, 19.
62 H. G. Niessen, M. Van Buskirk, C. Dybowski, D. R. Corbin, J. A. Reimer and A. T. Bell, *J. Phys. Chem. B*, 2001, **105**, 2945.
63 G. E. Pavlovskaya, C. F. Horton-Garcia, C. Dybowski, D. R. Corbin and T. Meersmann, *J. Phys. Chem. B*, 2004, **108**, 1584.
64 J. J. Wu, T. J. Boyle, J. L. Shreeve, J. W. Ziller and W. J. Evans, *Inorg. Chem.*, 1993, **32**, 1130.
65 H. Maekawa, K. Kawata, Y. P. Xiong, N. Sakai and H. Yokokawa, *Solid State Ionics*, 2009, **180**, 314.
66 N. J. Kim and J. F. Stebbins, *Chem. Mater.*, 2007, **19**, 5742.
67 H. Deters, J. F. de Lima, C. J. Magon, A. S. S. de Camargo and H. Eckert, *Phys. Chem. Chem. Phys.*, 2011, **13**, 16071.
68 H. Deters, A. S. S. de Camargo, C. N. Santos and H. Eckert, *J. Phys. Chem. C*, 2010, **114**, 14618.
69 S. E. Ashbrook, K. R. Whittle, G. R. Lumpkin and I. Farnan, *J. Phys. Chem. B*, 2006, **110**, 10358.
70 D. Massiot, F. Fayon, M. Capron, I. King, S. Le Calve, B. Alonso, J. O. Durand, B. Bujoli, Z. H. Gan and G. Hoatson, *Magn. Reson. Chem.*, 2002, **40**, 70.
71 K. W. Zilm, R. T. Conlin, D. M. Grant and J. Michl, *J. Am. Chem. Soc.*, 1980, **102**, 6672.
72 J. W. Gleeson and R. W. Vaughan, *J. Chem. Phys.*, 1983, **78**, 5384.
73 T. M. Duncan and T. W. Root, *J. Phys. Chem.*, 1988, **92**, 4426.
74 T. M. Duncan, A. M. Thayer and T. W. Root, *Phys. Rev. Lett.*, 1989, **63**, 62.

75 B. E. Hanson, *J. Am. Chem. Soc.*, 1989, **111**, 6442.
76 T. M. Duncan, *Colloids Surf.*, 1990, **45**, 11.
77 T. M. Duncan, A. M. Thayer and T. W. Root, *J. Chem. Phys.*, 1990, **92**, 2663.
78 L. Reven and E. Oldfield, *Inorg. Chem.*, 1992, **31**, 243.
79 D. Gajan, PhD, University of Lyon, 2010.
80 T. Pery, K. Pelzer, G. Buntkowsky, K. Philippot, H. H. Limbach and B. Chaudret, *ChemPhysChem*, 2005, **6**, 605.
81 J. Gath, G. L. Hoaston, R. L. Vold, R. Berthoud, C. Copéret, M. Grellier, S. Sabo-Etienne, A. Lesage and L. Emsley, *Phys. Chem. Chem. Phys.*, 2009, **11**, 6962.
82 E. R. H. van Eck, M. E. Smith and S. C. Kohn, *Solid State Nucl. Magn. Reson.*, 1999, **15**, 181.
83 L. Peng, H. Huo, Y. Liu and C. P. Grey, *J. Am. Chem. Soc.*, 2007, **129**, 335.
84 S. Sklenak, J. Dedecek, C. B. Li, B. Wichterlova, V. Gabova, M. Sierka and J. Sauer, *Angew. Chem. Int. Ed.*, 2007, **46**, 7286.
85 C. Magnenet, D. Massiot, I. Klur and J. P. Coutures, *J. Mater. Sci.*, 2000, **35**, 115.
86 R. N. Kerber, A. Kermagoret, E. Callens, P. Florian, D. Massiot, A. Lesage, C. Copéret, F. Delbecq, X. Rozanska and P. Sautet, *J. Am. Chem. Soc.*, 2012, **34**, 6767–6775.
87 E. Mazoyer, J. Trebosc, A. Baudouin, O. Boyron, J. Pelletier, J. M. Basset, M. J. Vitorino, C. P. Nicholas, R. M. Gauvin, M. Taoufik and L. Delevoye, *Angew. Chem. Int. Ed.*, 2010, **49**, 9854.
88 M. Hohwy, H. J. Jakobsen, M. Eden, M. H. Levitt and N. C. Nielsen, *J. Chem. Phys.*, 1998, **108**, 2686.
89 M. Feike, D. E. Demco, R. Graf, J. Gottwald, S. Hafner and H. W. Spiess, *J. Magn. Reson., Ser. A*, 1996, **122**, 214.
90 S. P. Brown and H. W. Spiess, *Chem. Rev.*, 2001, **101**, 4125.
91 S. P. Brown, I. Schnell, J. D. Brand, K. Mullen and H. W. Spiess, *J. Am. Chem. Soc.*, 1999, **121**, 6712.
92 C. Lecuyer, F. Quignard, A. Choplin, D. Olivier and J. M. Basset, *Angew. Chem.*, 1991, **103**, 1692.
93 C. Thieuleux, A. Maraval, L. Veyre, C. Copéret, D. Soulivong, J.-M. Basset and G. J. Sunley, *Angew. Chem. Int. Ed.*, 2007, **46**, 2288.
94 F. Rataboul, A. Baudouin, C. Thieuleux, L. Veyre, C. Copéret, J. Thivolle-Cazat, J. M. Basset, A. Lesage and L. Emsley, *J. Am. Chem. Soc.*, 2004, **126**, 12541.
95 G. Arrachart, C. Carcel, J. J. E. Moreau, G. Hartmeyer, B. Alonso, D. Massiot, G. Creff, J. L. Bantignies, P. Dieudonne, M. W. C. Man, G. Althoff, F. Babonneau and C. Bonhomme, *J. Mater. Chem.*, 2008, **18**, 392.
96 N. Folliet, C. Roiland, S. Begu, A. Aubert, T. Mineva, A. Goursot, K. Selvaraj, L. Duma, F. Tielens, F. Mauri, G. Laurent, C. Bonhomme, C. Gervais, F. Babonneau and T. Azais, *J. Am. Chem. Soc.*, 2011, **133**, 16815.
97 K. Mao, J. W. Wiench, V. S. Y. Lin and M. Pruski, *J. Magn. Reson.*, 2009, **196**, 92.
98 P. Avenier, M. Taoufik, A. Lesage, X. Solans-Monfort, A. Baudouin, A. de Mallmann, L. Veyre, J. M. Basset, O. Eisenstein, L. Emsley and E. A. Quadrelli, *Science*, 2007, **317**, 1056.
99 F. Blanc, J. Thivolle-Cazat, J. M. Basset and C. Copéret, *Chem.-Eur. J.*, 2008, **14**, 9030.
100 W. A. Nugent and J. M. Mayer, *Metal-Ligand Multiple Bonds: The Chemistry of Transition Metal Complexes Containing Oxo, Nitrido, Imido, Alkylidene, or Alkylidyne Ligands*; Wiley-Interscience: New York, 1988.
101 R. R. Schrock, *Acc. Chem. Res.*, 1979, **12**, 98.
102 A. Lesage, L. Emsley, M. Chabanas, C. Copéret and J. M. Basset, *Angew. Chem. Int. Ed.*, 2002, **41**, 4535.

103 J. L. Bonardet, J. Fraissard, A. Gedeon and M. A. Springuel-Huet, *Catal. Rev.-Sci. Eng.*, 1999, **41**, 115.
104 T. R. Carver and C. P. Slichter, *Phys. Rev.*, 1953, **92**, 212.
105 T. R. Carver and C. P. Slichter, *Phys. Rev.*, 1956, **102**, 975.
106 C. P. Slichter, *Phys. Chem. Chem. Phys.*, 2010, **12**, 5741.
107 A. B. Barnes, G. De Paepe, P. C. A. van der Wel, K. N. Hu, C. G. Joo, V. S. Bajaj, M. L. Mak-Jurkauskas, J. R. Sirigiri, J. Herzfeld, R. J. Temkin and R. G. Griffin, *Appl. Magn. Reson.*, 2008, **34**, 237.
108 L. R. Becerra, G. J. Gerfen, R. J. Temkin, D. J. Singel and R. G. Griffin, *Phys. Rev. Lett.*, 1993, **71**, 3561.
109 G. T. Debelouchina, M. J. Bayro, P. C. A. van der Wel, M. A. Caporini, A. B. Barnes, M. Rosay, W. E. Maas and R. G. Griffin, *Phys. Chem. Chem. Phys.*, 2010, **12**, 5911.
110 G. J. Gerfen, L. R. Becerra, D. A. Hall, R. G. Griffin, R. J. Temkin and D. J. Singel, *J. Chem. Phys.*, 1995, **102**, 9494.
111 D. A. Hall, D. C. Maus, G. J. Gerfen, S. J. Inati, L. R. Becerra, F. W. Dahlquist and R. G. Griffin, *Science*, 1997, **276**, 930.
112 T. Maly, L. B. Andreas, A. A. Smith and R. G. Griffin, *Phys. Chem. Chem. Phys.* 2010, **12**, 5872.
113 T. Maly, G. T. Debelouchina, V. S. Bajaj, K. N. Hu, C. G. Joo, M. L. Mak-Jurkauskas, J. R. Sirigiri, P. C. A. van der Wel, J. Herzfeld, R. J. Temkin and R. G. Griffin, *J. Chem. Phys.*, 2008, **128**, 19.
114 T. Maly, A. F. Miller and R. G. Griffin, *ChemPhysChem*, 2010, **11**, 999.
115 M. Rosay, L. Tometich, S. Pawsey, R. Bader, R. Schauwecker, M. Blank, P. M. Borchard, S. R. Cauffman, K. L. Felch, R. T. Weber, R. J. Temkin, R. G. Griffin and W. E. Maas, *Phys. Chem. Chem. Phys.*, 2010, **12**, 5850.
116 C. S. Song, K. N. Hu, C. G. Joo, T. M. Swager and R. G. Griffin, *J. Am. Chem. Soc.*, 2006, **128**, 11385.
117 A. Lesage, M. Lelli, D. Gajan, M. A. Caporini, V. Vitzthum, P. Mieville, J. Alauzun, A. Roussey, C. Thieuleux, A. Mehdi, G. Bodenhausen, C. Copéret and L. Emsley, *J. Am. Chem. Soc.*, 2010, **132**, 15459.
118 M. Lelli, D. Gajan, A. Lesage, M. A. Caporini, V. Vitzthum, P. Mieville, F. Heroguel, F. Rascon, A. Roussey, C. Thieuleux, M. Boualleg, L. Veyre, G. Bodenhausen, C. Copéret and L. Emsley, *J. Am. Chem. Soc.*, 2011, **133**, 2104.
119 A. Zagdoun, A. J. Rossini, D. Gajan, A. Bourdolle, O. Ouari, M. Rosay, W. E. Maas, P. Tordo, M. Lelli, L. Emsley, A. Lesage and C. Copéret, *Chem. Commun.*, 2012, **48**, 654.
120 A. J. Rossini, A. Zagdoun, M. Lelli, J. Canivet, S. Aguado, O. Ouari, P. Tordo, M. Rosay, W. E. Maas, C. Coperet, D. Farrusseng, L. Emsley and A. Lesage, *Angew. Chem. Int. Ed.*, 2012, **51**, 123.
121 A. Zagdoun, G. Casano, O. Ouari, G. Lapadula, A. J. Rossini, M. Lelli, M. Baffert, D. Gajan, L. Veyre, W. E. Maas, M. Rosay, R. T. Weber, C. Thieuleux, C. Coperet, A. Lesage, P. Tordo and L. Emsley, *J. Am. Chem. Soc.*, 2012, **134**, 2284.
122 V. Vitzthum, P. Mieville, D. Carnevale, M. A. Caporini, D. Gajan, C. Copéret, M. Lelli, A. Zagdoun, A. J. Rossini, A. Lesage, L. Emsley and G. Bodenhausen, *Chem. Commun.*, 2012, **48**, 1988.

Raman spectroscopy for solid oxide fuel cells

Robert C. Maher
DOI: 10.1039/9781849734899-00084

Introduction and overview

The solid oxide fuel cell (SOFC) is a high temperature fuel cell variant that has tremendous potential as a mechanism for the clean, efficient generation of energy [1]. A key advantage of the SOFC over other fuel cell technologies is its high operational temperature. This allows for efficient operation not only using hydrogen but also carbonaceous fuels such as methane without the use of expensive catalysts. As a result the SOFC represents a potential bridging technology between the current carbon dominated energy economy and a future clean energy economy fuelled by hydrogen. However, this potential can only be realised if SOFCs become economically competitive with incumbent technologies. While clean efficient SOFC systems have been available for a number of years, their cost is prohibitively high. Material and production costs must be lowered and lifetimes increased, which can only be accomplished by addressing specific problems. A detailed understanding of the physical processes occurring within operational SOFCs will allow for a much more targeted and effective development strategy, culminating in the rational design of material systems for application in SOFCs. New characterisation tools capable of providing real-time, *in-situ* characterisation of operational SOFCs are critical for this development process [2].

In recent years, Raman spectroscopy has emerged as an important tool for the characterisation of SOFCs. It has been shown to be capable of providing in depth information about component materials as well as the real-time physical and chemical processes occurring within operational SOFCs. This review focuses on studies that have applied Raman spectroscopy to the investigation of materials and processes relevant to SOFCs, primarily in the last 5 to 10 years. However, it is impossible to fully review each system completely in the space available due to the complexity of the SOFC in terms of the wide range of possible materials and system configurations. Instead, key results and review texts will be highlighted to allow the interested reader to investigate further. We will begin with a quick introduction to the SOFC leading onto a summary of results from investigations using *ex-situ* Raman spectroscopy split into the major component categories. A review of some significant results from *in-situ* studies will then be highlighted. Finally, the results and their relevance for SOFCs will be briefly summarised and the outlook for the future of Raman spectroscopy for SOFCs is discussed.

Imperial College, London; E-mail: robert.maher99@imperial.ac.uk

The solid oxide fuel cell

The SOFC is an electrochemical device, somewhat similar to a battery, which operates at high temperatures. The high operational temperatures allow systems to be designed to provide the combined generation of heat and power allowing for highly efficient power generation. SOFCs run on a continuous fuel stream and are not only efficient but also clean and allow for flexibility in fuel source. Potential fuels include the pure H_2 usually associated with fuel cells, as well as carbonaceous fuels such as methane, natural gas and biogas. The SOFC is of particular interest for distributed power generation, backup power, home heating and electrical power generation.

Fig. 1 shows a schematic representation of a single cell, which is typically a three layer structure consisting of porous anode and cathode electrodes separated by a dense ceramic electrolyte. Single cells are connected together in series into stacks to create useful systems for applications using interconnects. These components will now be quickly overviewed in order to highlight some of the most pressing issues related to each.

The most important SOFC component is the ceramic electrolyte which determines many of the operational parameters. The electrolyte must be highly ionically conductive whilst also electrically insulating to avoid short circuiting between the anode and cathode. It must also be stable in both oxidising and reducing environments at high temperatures for long periods of time. A three phase boundary exists within the anode where metal, electrolyte and gas all intersect to provide the fuel, ionic oxygen and a connection to the external electrical circuit necessary for electrochemical oxidation to occur. Oxygen is reduced on the cathode, adsorbed into the electrolyte and transported to the anode in order to provide a continuous oxide ion flux. Fuel such as H_2 is fed into the system on the anode side where it is then oxidised at the three phase boundary releasing electrons into the external circuit producing an electrical current.

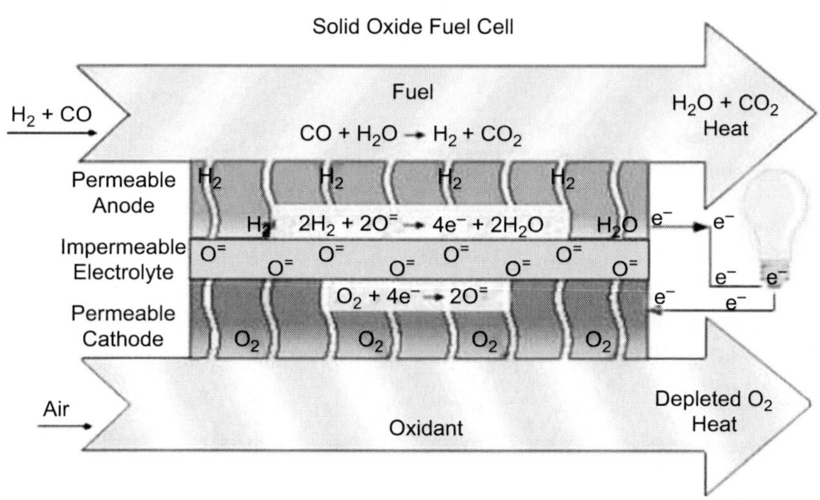

Fig. 1 Schematic of a SOFC consisting of a porous anode and cathode separated by a dense, impermeable ceramic oxide electrolyte. Figure reproduced from reference [4].

Ohmic loses within the electrolyte are a principal performance consideration and must be minimised for efficient operation. Electrolyte composition and design must be precisely controlled and operated at a sufficiently high temperature to achieve this. Doped Zirconia has traditionally been the electrolyte of choice, requiring operational temperatures between 800 and 1000 °C. Lower temperatures allow cheaper materials such as steel to be used for interconnects and supporting manifolds whilst increasing the long-term stability of the system. However, the development of intermediate temperature SOFCs (IT-SOFCs) presents serious challenges both in terms of design and optimisation of reaction kinetics. In recent years much of the research on SOFCs has focused on reducing operating temperatures to between 500 and 700 °C [3]. A great deal of effort has been expended on the development of new materials capable of high oxide ion conduction at lower temperatures. Although the search for optimised materials is on-going, doped ceria has emerged as the leading electrolyte material for these systems.

SOFC anodes are typically formed from a mixed composite of a ceramic electrolyte and metal (cermet). This most commonly is in the form of nickel and yttria stabilised zirconia (YSZ). Nickel is widely used as a catalyst for a variety of chemical reactions such as the hydrogenation of unsaturated compounds [5], production of carbon nanotubes [6–8], oligomerisation [9], polymerisation [10, 11] and is well suited for SOFC anodes [12]. In the case of electrolyte supported systems, anodes are prepared through the application of a cermet slurry consisting of electrolyte and NiO particles to the solid electrolyte surface through screen printing, tape casting or some other method. This results in the formation of a dense precursory layer of electrolyte-NiO composite. Significant dimensional changes occur within the anode during the initial reduction process creating the porous structure necessary for operation. The resulting strain generated within the structure must be properly managed to ensure mechanical stability.

The oxygen reduction reaction occurs at the cathode. Nearly all candidate cathode materials are based on perovskite oxides such as lanthanum strontium manganite (LSM). However, the oxygen reduction reaction is particularly affected by lower operating temperatures with kinetics and transport processes that are thermally activated. The development of materials with higher electrochemical activities for the oxygen reduction reaction is critical for the development of IT-SOFCs with higher efficiencies.

Design considerations can also have a large influence on operating conditions, degradation processes and production costs. For example, SOFCs must be designed with sufficient mechanical stability to enable economically viable operational lifetimes. Cells with thick electrolytes are typically robust, resistant to stresses induced by redox processes and allow for thin electrodes, increasing fuel access. However, ohmic losses in cells with thick electrolytes can be significant. Such losses can be dramatically reduced by using thinner electrolytes [13] as demonstrated by SOFCs with YSZ electrolytes operating below 600 °C using thin film technology [14, 15]. Numerous methods for creating electrolytes on the order of 10–30 μm thick have been developed where either the anode or cathode provides the necessary structural support. Drawbacks of the electrode supported cell configuration include limited gas transport through thick electrodes and increased redox stresses. The formation of a stable, dense, thermally compatible gas-tight electrolyte over a porous electrode also presents a significant challenge. Several studies have investigated the possibility of producing mechanically strong and thin electrolytes [16–18], and the electrode supported cell configuration is now the most established configuration and widespread for near commercial SOFC prototypes.

Individual cells are connected into useful large scale systems using interconnects. There is the potential for significant reductions in their cost as a result of lower operational temperatures. Interconnect materials must be stable at operational temperatures, in both oxidising and reducing environments, be electrically and thermally conductive whilst being ionically insulating. They must be resistant to attack from carbonaceous fuels, possess sufficient mechanical strength, and have a thermal expansion coefficient compatible with that of other component materials. Ceramics such as lanthanum chromite ($LaCaCrO_3$) have traditionally been used for interconnects in SOFCs operating at 1000 °C and higher [12, 19]. These are not ideal since they are typically poor electrical conductors, brittle and expensive due to their high rare-earth content and difficult fabrication. Lower operating temperatures (<800 °C) allows ceramic interconnects to be replaced with cheaper high-temperature steel alloys which are more easily fabricated and offer improved mechanical, electrical and thermal properties [20, 21]. Alloys with high chromium contents are typically preferred over other high-temperature alloys due to the relatively high conductivity of the Cr_2O_3 chromia scale. However, problems such as corrosion at high temperatures in hydrocarbon environments, oxide scale growth in oxygen containing atmospheres, and detrimental chromium migration to other components must be addressed [22].

Effective sealing of SOFC components is critical for efficient and effective operation and metallic, metal-ceramic composite and ceramic based seals are not ideal [23, 24]. Glass based sealants are typically recognised to be the best candidate due to their excellent electrical, mechanical and chemical stability in both oxidising and reducing environments properties at high temperature [25].

It is clear that SOFCs are demanding and complex systems involving a number of material systems. Issues such as carbon and sulphur deposition upon SOFC anodes, and a general improvement in material properties and stability under the harsh operational conditions for all components need to be achieved. Raman spectroscopy has become an increasingly important tool for the characterisation of SOFCs as will be demonstrated in the following sections.

Ex-situ characterisation studies

Raman spectroscopy has been used extensively in the *ex-situ* characterisation of potential materials for all the major SOFC components. It has also been used to investigate important degradation processes associated with long term SOFC operation such as phase separation, carbon deposition and sulphur poisoning. Key results for the important material systems are reviewed here.

The electrolyte

As already discussed, the electrolyte is the most important component of a SOFC, the properties of which determine critical parameters such as cell performance and temperature of operation. Because of this, much time has been devoted to developing and understanding the materials and their properties and Raman spectroscopy has been a key tool. This section summarises the key results of studies that have used Raman spectroscopy to investigate electrolytes for SOFCs and is split into three parts. The first focuses on Zirconia based materials; the conventional electrolyte of choice. The second will summarise the results of investigations on Ceria based electrolytes; the frontrunner electrolyte for intermediate temperature SOFCs. Other novel electrolytes which have some potential for reduced temperature operation will be summarised in the final section.

Zirconia. Zirconia (ZrO_2), primarily in the form of yttria-stabilised Zirconia (YSZ), has historically been the electrolyte of choice for SOFCs [12]. YSZ fulfils many of the criteria required for an effective electrolyte. It is chemically stable with excellent mechanical properties over a wide range of temperatures in a variety of atmospheres. However, while ZrO_2 doped with divalent or trivalent oxides has enhanced ionic conductivity, temperatures in excess of 800 °C are typically required for effective operation. This requirement places tough restrictions on other component materials. Physical properties such as the mechanical strength and oxygen ion conductivity are highly sensitive to the YSZ phase and such information can be extremely useful for optimisation of material preparation and of operational parameters. For example it is well known in the literature that the cubic phase of YSZ has the highest ionic conductivity.

Raman spectroscopy has been used extensively to distinguish between monoclinic, tetragonal and cubic phases of zirconia ceramics [26, 27]. Fig. 2 shows the Raman response of the three ZrO_2 phases as measured by Yashima *et al.* clearly demonstrating the sensitivity of the technique to the various phases [26]. The monoclinic phase of ZrO_2 is identified by 18 well defined and intense Raman peaks corresponding to the 9Ag and 9Bg modes expected from a consideration of the structural symmetry. The tetragonal phase has six broad Raman peaks resulting from the Raman active modes of A1g, 2B1g and 3Eg symmetry. The cubic phase usually has one broad band at around 640 cm^{-1} with a number of smaller broad background features. Although the ideal cubic phase should typically only have one Raman active mode with T_{2g} symmetry these additional features are associated with structural distortions. Ratios of the various peaks can be used to obtain information on the prevalence of the difference phases present in a particular sample.

Raman spectroscopy has many advantages over X-ray diffraction (XRD) for the characterisation of the structural phases of ZrO_2 [28–30]. It is generally much more sensitive to surface phase distributions and has been

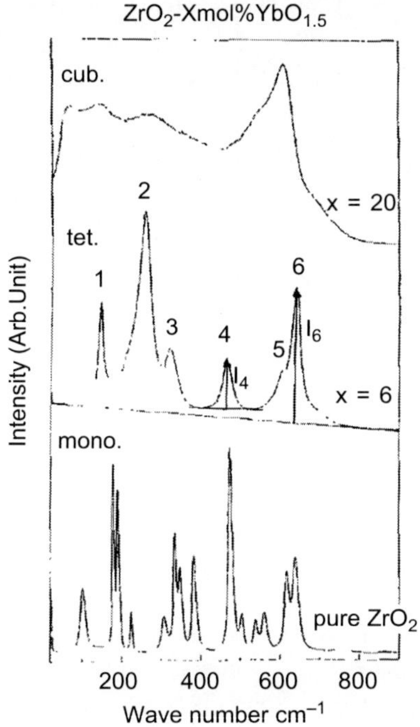

Fig. 2 Raman spectra of the cubic, tetragonal and monoclinic forms of ZrO_2 induced through doping. Figure reproduced from reference [26].

demonstrated to detect different phases for dense surfaces where XRD measurements suggest no variation. Raman spectroscopy can also characterise samples containing nanometre scale crystallinity for which XRD patterns are broad and ill defined. Confocal Raman microscopy systems allow phase distribution of electrolyte surfaces to be studied in detail. As a result, Raman spectroscopy has become a standard tool for the characterisation of ZrO_2 electrolytes produced by new methods [29].

Raman spectroscopy is also sensitive to the structural effects of doping in ZrO_2. Xie *et al.* used Raman spectroscopy to study the effect of yttria doping on the structure of $Sm_2Zr_2O_7$ [31]. Their samples were prepared as a function of Y doping through a solid state reaction process carried out at 1972 K for 10 hours in air. Fig. 3(a) shows the Raman spectra obtained from the ceramic as a function of Y dopant concentration. All peaks are clearly observed to broaden considerably as the dopant concentration is increased and this is emphasised in Fig. 3(b) where the FWHM of the Raman peaks arising from the O-Sm(Y)-O bending band at 307 cm^{-1} and the Zr-O stretch mode at 391 cm^{-1} is plotted as a function of Y doping. This broadening is suggestive of increasing oxygen disorder and structural disruption upon doping. Measurements of this type can be used in conjunction with other characterisation tools to optimise material composition for enhanced ion conductivity and stability.

Raman spectroscopy has also been used to study the structural properties of YSZ thin film electrolytes designed to operate at lower temperatures. Bernay *et al.* used Raman spectroscopy to characterise the structural

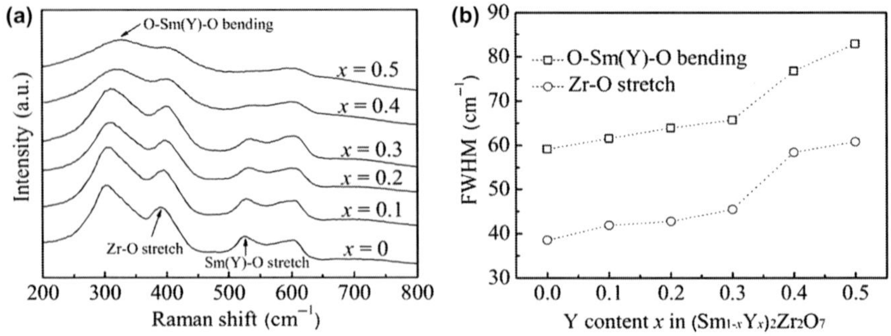

Fig. 3 (a) Raman spectra of $Sm_{1-x}Y_xZr_2O_7$ as a function of Y doping. (b) FWHM of Raman peaks arising from the O-Sm(Y)-O bending mode at 307 cm^{-1} and the Zr-O stretch mode at 391 cm^{-1} as a function of Y doping. Figure reproduced from [31].

properties of thin films of YSZ (<3 μm) prepared on typical anode and cathode materials using atomic layer deposition [32]. They were able to show that the thin electrolyte film was predominantly cubic in nature while the electrolyte component of the anode cermet was a mixture of tetragonal and cubic phases. Heiroth *et al.* characterised the phase dependence and strain effects of their thin film YSZ grown by pulsed laser deposition using Raman spectroscopy [30]. Their films, which ranged from 200 nm to 1.2 μm thick, were highly sensitive to annealing at relatively low temperatures and showed indications of strain induced effects [33]. Larrea *et al.* have developed a low cost method for reducing the effective thickness of YSZ electrolytes through laser ablation patterning [17]. Raman spectra collected from machined and pristine electrolyte surfaces confirmed the cubic structure of the YSZ electrolyte was unchanged by the laser. Laser ablation of selected areas of the electrolyte allowed operational temperatures to be reduced from 850 °C to 730 °C with scope for further development.

It is clear that Raman spectroscopy can provide detailed information on the structural effects of doping ZrO_2 relevant for the development of improved electrolytes. However, detailed analysis of the effects of actual SOFC operational conditions on the electrolyte materials is invaluable for monitoring not only the effects of doping, but also material stability. Such information allows for an improved understanding of the ionic properties of these materials and also provides a detailed understanding of the processes responsible for the degradation in performance over time. Such information can then be used to optimise not only material properties but also operational conditions. Many parameters affect the phase evolution of YSZ during operation such as sintering and annealing effects [28] as well as doping levels [29].

Hattori *et al.* investigated the aging effects of sintering and annealing on YSZ doped with between 8 mol % and 10 mol % by volume [34]. The electrical properties of the materials were monitored during the sintering and annealing processes. This was then correlated to changes in the crystal structure found from TEM, X-ray diffraction, and Raman microscopy. It is well known that the ionic conductivity of YSZ degrades over long periods of SOFC operation. To study this effect and the influence of doping and sintering conditions, samples were prepared as a function of doping, sintered for 10 hours at 1300, 1400 and 1500 °C and then annealed at the same temperatures for 1000 hours. The electrical measurements indicated that the conductivity of the samples doped below 9.5 mol% decreased significantly during the annealing process whereas those of 9.5 mol% and above were largely unchanged.

Fig. 4(a) shows the Raman spectra collected from the as sintered electrolyte surfaces for the different temperatures investigated. These clearly show the presence of ZrO_2 in the cubic phase as characterised by the dominant broad peak at 640 cm^{-1}. Spectra collected from the same surfaces after 1000 hours of thermal treatment are shown in Fig. 4(b). New bands are observed at approximately 260 and 460 cm^{-1} while the main 640 cm^{-1} is seen to split slightly. These new features are indicative of the formation of a significant amount of tetragonal phase Zirconia. The tetragonal phase of ZrO_2 is considerably less conductive than the cubic phase and its gradual formation over long periods of time explains the degradation in conductivity observed. These features were only observed for the samples doped below 9.5 mol %. Samples doped at 9.5 mol % and higher remained cubic in nature as evidenced by further Raman spectroscopy measurements. These measurements led Hattori *et al.* to suggest that the optimum composition of the YSZ system was 9.5 mol % Y_2O_3 based on its long-term stability and relatively high conductivity.

Similar studies have been performed in other Zirconia based electrolytes. Scandium stabilised zirconium (ScSZ) ceramics have been shown to have the highest oxide ion conductivity of Zirconia based electrolytes, potentially allowing for reduced operational temperatures [35]. Because of this, interest has grown in these electrolytes for reduced temperature SOFCs despite higher costs. The main difficulty in developing these materials as SOFC electrolytes arises from phase instability, which has important implications for the conductivity and mechanical properties of the material during long term SOFC operation. There is disagreement between authors on the precise origin of the phase instability stemming from the extremely complex

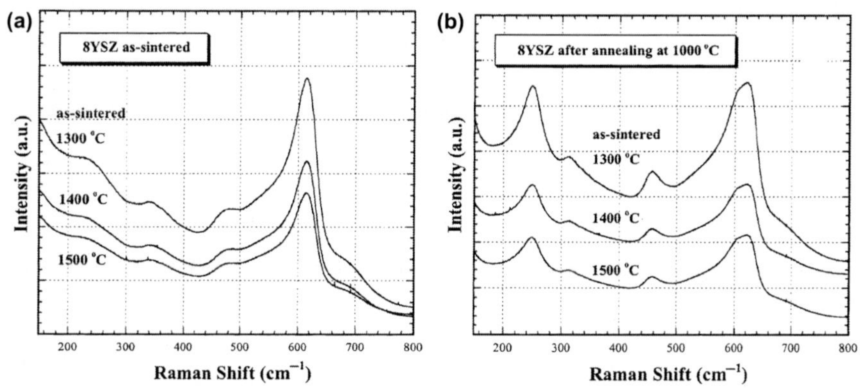

Fig. 4 (a) Raman spectra of 8YSZ as sintered at 1300, 1400 and 1500 °C. (b) Raman spectra of 8YSZ after thermal annealing at 1300, 1400 and 1500 °C for 1000 hours. Clear differences in the phase composition of the material are evident from the appearance of new peaks under long duration annealing. Figure reproduced from reference [34].

nature of the problem. Studies have shown that the process can be influenced by a wide range of parameters involved in both the preparation and operation of the electrolytes.

Bai *et al.* investigated the effect of doping concentration of Bi_2O_3 and sintering conditions on the structural properties of Scandia stabilised Zirconia using Raman spectroscopy [36]. Bi_2O_3 is typically used as an aid for reducing the temperatures required for effective sintering of the ScSZr electrolytes. Material phase was studied as a function of sintering temperature using Raman spectroscopy. Similarly, Zyryanov and co-workers used Raman spectroscopy to investigate the structure of the solid solution $Zr_{0.88}Sc_{0.1}Ce_{0.01}O_{1.955}$ prepared by different methods and sintered at 1620K in conjunction with X-ray diffraction [37]. The spectra suggested that the materials formed the same rhombohedral phase observed from solid solutions of ZrO_{2-x} $ScO_{1.5}$ [38].

Raman spectroscopy has been applied extensively to the investigation of aging in ZrO_2 electrolyte due to long term annealing and redox cycling [39–42]. Most recently, Shimazu *et al.* applied Raman spectroscopy to the investigation of ageing and phase separation effects in ZrO_2 co-doped with Sc_2O_3 and CeO_2 [42]. The ZrO_2 is co-doped with CeO_2 to stabilise the material in the cubic phase to optimise the conductivity. Unfortunately, the performance of Scandia stabilised Zirconia, like YSZ, has also been shown to decrease over time when exposed to high temperatures. This is typically attributed to the changing phase from the initial high ion conductivity cubic phase. Such processes have been shown to be accelerated in the presence of steam which is often present in fuels derived from biogas sources. Understanding the precise nature of these phase changes and the influence of H_2O on them is important for the development of electrolytes with improved properties. These effects were investigated in depth using Raman spectroscopy. Spectra were collected from electrolytes of cathode supported SOFCs before and after operations for a variety of times.

Raman spectra collected from the electrolytes before and after 100, 650 and 1000 hours of operation on wet hydrogen with a steam concentration of 3.0% by volume are shown in Fig. 5(a). Before operation three peaks were observed with the 600 cm^{-1} peak being assigned to the cubic of ZrO_2 phase and the weaker peaks at 350 and 700 cm^{-1} to the disordered structure of the oxygen vacancies. The spectral response from the electrolyte surface was found to be largely unchanged after 100 and 650 hours of operation. However, for longer times strong peaks appear at 150 and 260 cm^{-1} while the main 600 cm^{-1} peak broadens. These peaks are characteristic of the tetragonal phase of ZrO_2 and their appearance suggests a phase transformation of the ZrO_2 is occurring. Fig. 5(b) shows Raman spectra obtained from electrolyte surfaces after operation on wet hydrogen with a steam concentration of 28.5% by volume. Rapid phase transformation of the ZrO_2 was observed with indications of the formation of tetragonal ZrO_2 after only 100 hours. At longer timescales (>6000 hours), only peaks relating to the monoclinic phase were observed with increasingly intense signatures.

These results clearly show that the ZrO_2 electrolyte undergoes phase changes from cubic to tetragonal and then to monoclinic during operation.

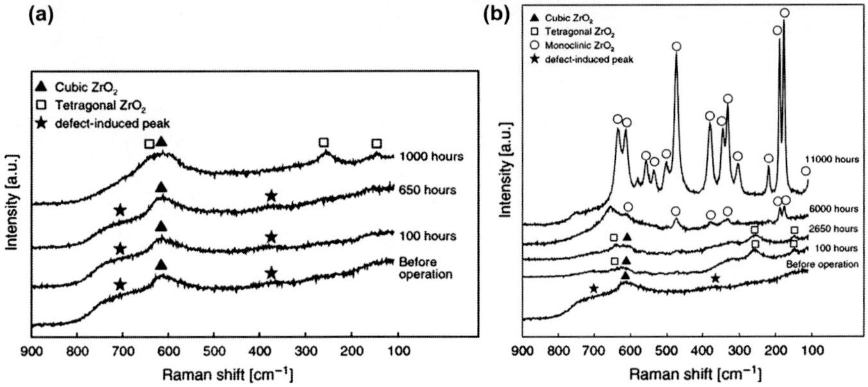

Fig. 5 (a) Raman spectra of SOFC electrolytes before and after 100, 650 and 1000 hours of operation in wet hydrogen with a steam concentration of 3.0% by volume. (b) Raman spectra of SOFC electrolytes before and after 100 2650, 6000 and 11000 hours operation on wet hydrogen with a steam concentration of 28.5% by volume. Figure reproduced from [42].

The speed with which these changes occur is strongly affected by the steam content in the fuel, accelerating as the steam concentration is increased. The authors suggest that the number of oxygen vacancies within the electrolyte structure maybe reduced in the presence of water leading to a destabilisation of the cubic phase. Such observations are extremely important given that fuels for use in real applications will often be complex mixtures including steam. While further work is required to fully understand the mechanisms involved in these processes, perhaps in conjunction with computational techniques such as density functional theory (DFT), the results clearly demonstrate the benefits of Raman spectroscopy for the study of these materials.

Ceria. Already an important material for a wide range of applications in catalysis [43] doped, ceria ceramics are emerging as the material of choice for IT-SOFCs electrolytes [44, 45]. This has primarily been driven by their high oxide ion conductivity in comparison to YSZ and other materials at reduced temperatures. Raman spectroscopy has been used extensively to characterise doped ceria produced using a wide range of preparation techniques [46–49]. This has usually been simply to confirm the formation of a uniform, single phase solid solution; something that Raman spectroscopy is particularly well suited to given its sensitivity to material phases [50–56]. However, several authors have utilised Raman spectroscopy to investigate these materials in much more detail. This has included the effects of a wide range of dopants, from the rare earths [57] to other metals such as iron [58, 59], cobalt [60] and copper [61]. The nature of oxygen defects within these materials and their dependence on preparation conditions such as sintering and annealing as well as doping concentrations [62] and exposure to different atmospheres [63, 64] have also been studied in detail.

Fig. 6 shows the Raman response of three different samples of gadolinium doped ceria (GDC) nanoparticles produced using reverse micelle processing with different production parameters [65]. The Raman spectrum is dominated by a single intense, well defined Raman peak at about 460 cm^{-1}, and

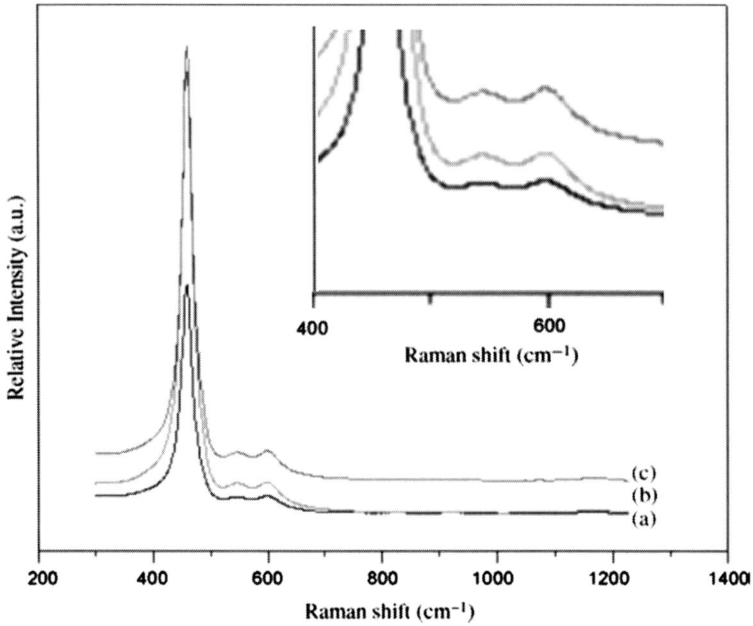

Fig. 6 Raman spectra of GDC doped with 10% Gd prepared using reverse micelle processing. The influence of the molar ratio of water to surfactant used in the production process on nanoparticle morphology and oxygen vacancy concentration was investigated. Clear changes in the oxygen vacancy peaks, seen at 550 and 600 cm^{-1}, of GDC were observed as the ratio was changed from (a) 4, (b) 6 and (c) 8 emphasising the effect of morphology on oxygen vacancies. Figure reproduced from reference [65].

is typical of doped cerium ceramics. This peak arises from the triply degenerate F_{2g} mode of the cubic phase of CeO_2 [57]. Its presence in conjunction with the lack of other peaks, such as the 360 cm^{-1} band of the cubic phase of Gd_2O_3, is generally taken as proof of the formation of a solid solution [66]. Further peaks are typically observed at 550 and 600 cm^{-1} as well second order features at around 1200 cm^{-1} [67].

In general, the spectral response of doped ceria ceramics is much simpler than the spectral response of doped zirconia. Despite this, detailed information can be obtained from the 550 and 600 cm^{-1} peaks shown in more detail in the inset of Fig. 6. These peaks are related to the oxygen vacancies present within the structure of the CeO_2 [68] which are introduced into the structure of CeO_2 through two different mechanisms. The first is an intrinsic effect related to the sample's surface to volume ratio and other structural defects. Oxygen ions are more mobile at grain surfaces and around defects and are thus able to leave the crystal lattice more easily than those within the bulk. Hence vacancy concentration is strongly related to the surface to volume ratio. The second mechanism is the extrinsic effect of doping, where substitution of Ce^{4+} ions from the lattice with lower valence ions such as Gd^{3+} leads to a loss of O^{2-} ions to preserve the charge neutrality. Nakajima *et al.* assigned the 550 cm^{-1} to extrinsic oxygen vacancies and the 600 cm^{-1} to intrinsic oxygen vacancies. It is not unusual for these peaks to be extremely broad and thus unresolvable as individual peaks. However, even in this case a great deal of information can still be extracted from the relative intensity of this single broad peak in relation to the F_{2g} peak.

Doped ceria have been shown to form solid solutions that maintain the fluorite structure of CeO_2 for low doping concentrations relevant for application [57]. As the doping concentration is increased, the F_{2g} peak typically shifts to lower wavenumbers, broadens and becomes increasingly asymmetric with a low frequency tail. The oxygen vacancy peak also appears and increases as the doping levels are increased.

The Raman response is highly dependent on particle and grain size due to the significant influence of grain boundaries on the concentration of oxygen vacancies and other effects such as phonon confinement [50, 67, 69, 70]. This has been used to characterise precursor GDC nano-particles used to create electrolytes, as well as the dense electrolytes, and the effect of processes such as sintering and annealing. Fig. 7 shows the Raman spectra of a CeO_2 nano-powder in comparison to a mono-crystalline reference [69]. The F_{2g} is observed to broaden as the particle size reduces to the point where it obscures the vacancy peaks.

Ceria based electrolytes are typically produced through the compression of nano-powders into dense pellets, which are then sintered in order to fuse the particles together. Askrabic *et al.* studied the phase stability and oxygen vacancy concentrations using Raman spectroscopy as a function of thermal treatment [68]. The F_{2g} and vacancy peaks were observed to significantly sharpen after thermal treatment indicating particle growth and fusing. A slight phase separation of the CeO_2 and Gd_2O_3 was also observed suggesting a degree of instability of the system.

The strong dependence of the Raman features of doped ceria systems on their oxygenation allows oxidation states to be investigated in detail. Mineshige *et al.* studied the influence of annealing atmosphere composition on the oxygenation levels of Samaria doped ceria (SDC) [63]. SDC pellets

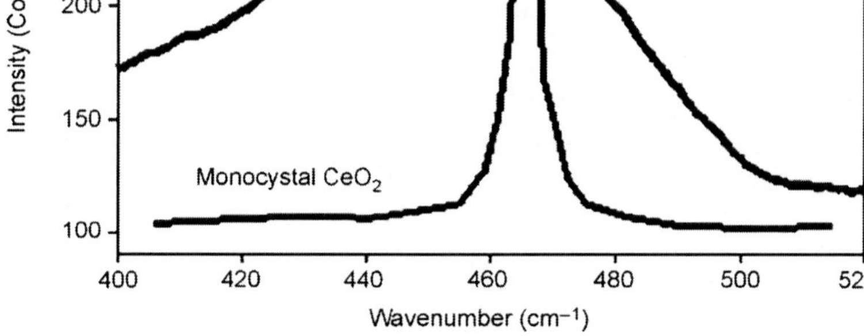

Fig. 7 Raman spectra of mono-crystalline and nanopowdered ceria with a diameter of 7.5 nm. Figure reproduced from reference [69].

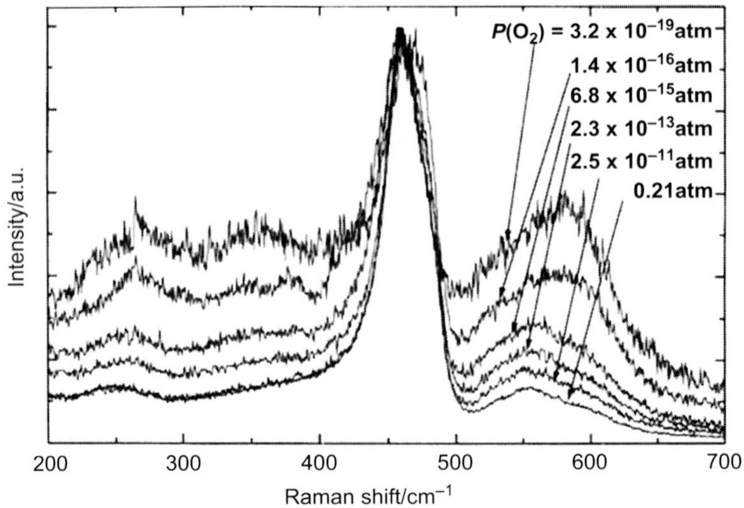

Fig. 8 Raman spectra obtained from $Ce_{0.8}Sm_{0.2}O_{2-\delta}$ samples annealed at 1273 K in atmospheres of varying partial oxygen pressure. δ indicates the level of oxygen depletion resulting from exposure to the oxygen depleted atmosphere and will increase as the $P(O_2)$ decreases. Figure reproduced from reference [63].

were annealed at 1273 K for several hours in a wide range of atmospheres with different partial oxygen pressures ($P(O_2)$). The oxidation state of the pellet will change significantly as a result of the annealing process, resulting in drastic changes to its optical properties. Fig. 8 shows the Raman response of SDC pellets annealed in atmospheres of different $P(O_2)$. Spectra were normalised by the intensity of the F_{2g} peak to allow for them to be directly compared. As the surface is reduced it will become more reflective reducing scattering volume probed by the laser and consequently the observed intensity will also decrease. This is manifested by the degradation in signal to noise ratio in the spectra as the oxygen concentration of the annealing environment is reduced.

It is clear that there is a substantial increase in the intensity of the oxygen defect peaks in the range of 550 to 600 cm^{-1} relative to the F_{2g} peak as the annealing environment becomes increasingly reducing. This clear and robust relationship allowed Mineshige and co-workers to determine the effective $P(O_2)$ profile through the cross-section of an SDC pellet exposed to a $P(O_2)$ gradient.

In order to achieve this, a 4 mm thick SDC pellet was exposed to a simulated fuel cell environment where the fuel side had a $P(O_2)$ of 4×10^{-18} and the oxygen side 0.21 atm. The pellet was then annealed for 5 hours at 1273 K, quenched in liquid N_2 and fractured. Raman spectra were collected every 50–400 μm through the fractured cross-section as a function of distance from the fuel side. Fig. 9(a) shows a selection of the Raman spectra obtained from the fractured SDC pellet with the position of each indicated. The intensity of the vacancy peaks are clearly seen to decrease significantly relative to the F_{2g} peak close to the side of the pellet exposed to air. Fig. 9(b) shows the effective $P(O_2)$ concentration profile through the electrolyte pellet as measured by Raman spectroscopy in comparison to an empirical

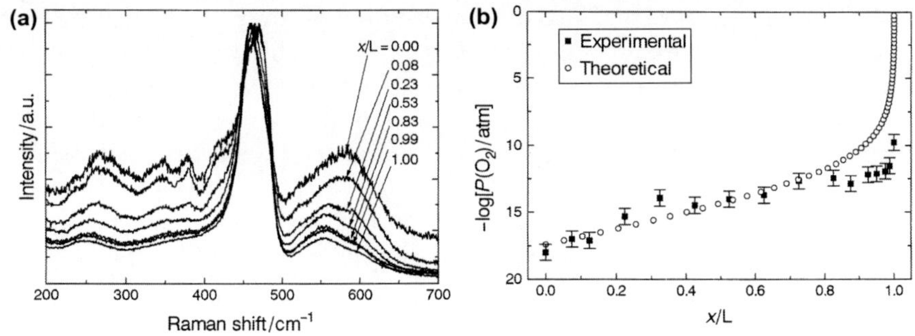

Fig. 9 (a) Raman spectra obtained from the fractured cross-section of a 4 mm thick Samarium doped Ceria electrolyte pellet exposed to a partial oxygen pressure differential of 4×10^{-18} and 0.21 atm of the fuel and air sides respectively at 1273 K. Spectra were collected as a function of distance (x) from the fuel side surface. Spectra have been labelled according to x normalised by the total thickness, L, of the pellet. (b) Experimental and theoretical profiles of the $P(O_2)$ concentration through the electrolyte pellet calculated from the experimental Raman data and a empirical model of the ion diffusion through the pellet. Figure reproduced from reference [63].

model of ion diffusion through the pellet.The experimental and theoretical profiles agree well indicating that most of the SDC pellet was a mixed ionic/electronic conductor under these conditions, with only a thin layer of the electrolyte on the air side remaining purely ionically conducting.

As with zirconia, ohmic losses can be significantly reduced by decreasing the thickness of the electrolyte. Burninskas and co-workers used Raman spectroscopy to characterise GDC electrolytes formed from solid state annealing of sputtered Gd_2O_3/CeO_2 multilayers [71]. Solid solutions of Gd doped ceria were observed to form when multilayer structures were annealed at a temperature of 900 °C. Films had to be annealed for more than one hour and prepared with a bilayer period of less than 70 nm in order to produce a well-ordered and uniformly doped phase.

Other electrolytes. Although zirconia and ceria based electrolytes are the focus for commercial applications there remain strong reasons to continue the search for improved oxide ion electrolytes for use in IT- SOFCs [72]. A number of investigations have focussed on the combination of zirconia and ceria in the hope of creating ceramics with improved properties for a variety of applications including SOFCs. Raman spectroscopy is ideal for the characterisation of these materials given its sensitivity to the various phases of ZrO_2 and CeO_2 as demonstrated by the results in the previous sections. The structure and phase of Ce-Zr-Y based ceramics has been characterised as a function of composition and sintering to confirm the formation and stability of solid solutions for a variety of different production methods [73–76].

A full understanding of the evolution of the phase and microstructure is essential if these composites are to be developed for application in SOFCs, as phase separation will clearly be an issue for long term operation. Recently, Boaro *et al.* have studied the effects of redox cycling on the structural, chemical, and electrical properties for a range of Zr/Ce electrolyte compositions [77]. Raman spectroscopy was used to study the effects of up to three reduction oxidation cycles at temperatures between 1073 K and

1440 K using temperature programmed reduction in 5%H_2-Ar and a two hour oxidation procedure in pure O_2 at 873 K. This was carried out for three different compositions: primarily zirconia ($Ce_{0.2}Zr_{0.8}O_2$), equal proportions of Ce and Zr ($Ce_{0.5}Zr_{0.5}O_2$) and primarily ceria ($Ce_{0.8}Zr_{0.2}O_2$). Raman measurements showed that the redox treatment initiated the formation of cation ordered phases in the 50:50 Ce:Zr composition leading to higher electronic conductivity. While some phase changes induced by the redox treatments were observed, these were not correlated to increases in their electrical conductivity. Although this investigation concentrated on relatively short exposures to simulated environments it does suggest that compositions of equal proportions of Ce and Zr will be unsuitable for long term SOFC operation.

Various oxide-ion conducting ABO_3 perovskite oxides have generated significant interest as alternative electrolyte materials for potentially reducing SOFC operating temperatures [72]. The properties of these systems can be easily modified through doping, which allows for a wide range of applications. Raman spectroscopy has been used to study the structural and phase properties of these materials for a range of compositions, and as stated earlier, with a greater sensitivity than XRD. He *et al.* used Raman spectroscopy to investigate the phase composition of $LaInO_3$ doped with Sr as a function of concentration and thermal treatment [78]. The structure of doped lanthanum orthogallate ($LaGaO_3$), synthesised using a metal nitrate glycine gel route, has been characterised at various preparation stages to demonstrate the formation of a single phase material [79]. Similarly, Farhadi and Sepahvand used Raman spectroscopy to confirm the formation of single phase $LaCoO_3$ produced from solid state microwave decomposition of a $La[Co(CN)_6]_5H_2O$ precursor [80]. However, despite the demonstrated sensitivity of Raman spectroscopy to material phase in the perovskite system, there has been no investigation of the aging of perovskite electrolytes during SOFC operation similar to those performed in the more traditional Zirconia and Ceria systems, to the best of my knowledge.

In recent years interest has grown in the potential of apatite-type lanthanum silicate ($La_{10}Si_6O_{27}$) for SOFC electrolytes [81–83]. Raman spectroscopy is well suited to the structural characterisation of silicates and has found some recent application to the study of potential SOFC electrolytes [84, 85]. Kim *et al.* studied the structural effects of high temperature sintering on LaSiO [84]. The ionic conductivity of LaSiO increases as the sintering temperature is increased as a result of induced structural changes. Such changes were observed in the Raman response of these materials through the appearance of a new Raman peak which was assigned to local structural distortions which provided additional oxide ion conduction pathways. The Raman response of LaSiO when doped with Ca, Sr and Ba has been found to broaden and shift suggesting the increased conductivity induced by doping is a result of increased interstitial oxygen [85].

Although oxide ion conducting oxides have traditionally been used as the separating electrolyte, interest in proton conducting oxides for SOFC electrolytes has been growing [86, 87]. Perovskites such as barium cerate have been shown to be ideal for candidate materials and have been the subject of intense research [88]. Although a full discussion of these materials is beyond the scope of this review, it is worth mentioning that Raman spectroscopy has also found application in the study of structural and conduction properties of candidate materials as a function of doping and thermal processing [89–93]. Interested readers are referred to these references as well as the review article by Kreuer [87], for further information.

The electrodes

The anode and cathode electrodes are key SOFC components where the critical electrochemical processes of fuel oxidation and oxygen reduction occur respectively [94]. Each component presents challenges which must be met for cost effective efficient, long term SOFC operation. Raman spectroscopy has been applied extensively to the study of SOFC anodes where Raman cross-sections of important material systems and processes are particularly strong. This allows material phase, oxidation states and processes such as carbon deposition relevant to anodes to be studied in detail relatively easily. On the other hand, the characterisation of cathode materials and processes has been hampered due to the low cross-sections of the oxygen reduction reaction. Progress in the application of Raman spectroscopy to the study of cathode materials has been made through the application of Surface Enhanced Raman Spectroscopy (SERS) but it is still at an early stage. Recent results on the application of Raman spectroscopy to the study of SOFC anodes and cathodes are reviewed here, and important references highlighted for the interested reader.

The anode. Raman spectroscopy has been extensively applied to the study of SOFC anodes, particularly for the characterisation of carbon formation on anodes operated on carbonaceous fuels. The ability of SOFCs to operate efficiently on a variety of hydrocarbon fuels without the use of expensive catalysts is a key advantage. However, progressive carbon build up within the anode leads to SOFC performance degradation through diminished fuel access and eventual catastrophic failure *via* anode cracking and delamination. Currently hydrocarbon fuels are typically converted into a mixture of hydrogen, CO, steam and CO_2 before being introduced into the system through steam reforming and partial oxidation. Although, this minimises carbon formation, reforming adds to the balance of plant component of the full SOFC system, greatly increasing complexity and cost. The direct electrochemical oxidation of hydrocarbon fuels on the anode would remove this additional complication and ultimately reduce costs. However, for this to be an effective cost reduction strategy anode material systems and SOFC operational parameters must be optimised to minimise carbon formation. An understanding of the dynamics of carbon formation within operational SOFCs is important for developing operational methodologies to minimise its effects. For example, amorphous carbon deposits are more easily removed through electrochemical oxidation. If operational parameters can be tuned to favour the growth of amorphous carbon rather than graphitic then SOFC lifetimes may be extended through the use of a periodic regeneration procedure designed for optimised carbon removal. A full understanding of the dynamics of the processes leading to carbon formation will greatly help with this process and Raman spectroscopy has been shown to be a powerful tool for the study of carbonaceous species [95].

Many operational parameters have a strong influence on carbon deposition such as the precise fuel composition, the electrolyte/metal composition used as the anode, and the properties of the underlying electrolyte. Researchers based at AIST in Japan have investigated the dynamics of carbon deposition and its dependence on anode morphology in a series of papers [96–98]. Most recently, Yoshinaga investigated the dynamics of carbon deposition on GDC-Ni, ScSZ-Ni and YSZ-Ni anode composites using Raman spectroscopy [98]. The anode composites were prepared by sputtering metal films onto oxide substrates, which were then annealed at 800 °C. This ensured that a well-defined surface was formed for each of the three electrolytes with comparable properties. The composites were then exposed to a simulated SOFC fuel environment of 4.9% CH_4/Ar at 600 °C for 20 minutes, cooled, and characterised using Raman spectroscopy.

Fig. 10 (top) shows the Raman spectra collected from areas of Ni and electrolyte for GDC-Ni and YSZ-Ni composites. The D and G peaks characteristic of carbonaceous material are observed at approximately 1350 and 1580 cm^{-1} respectively [95]. The relative intensity, position and widths of these peaks can provide a great deal of information regarding the precise nature of the carbon deposits formed in terms of how amorphous or graphitic they are. For example, broad Raman features in combination with an intense D peak relative to the G peak are suggestive of amorphous carbon. Conversely, sharp intense features with a weak D peak intensity relative to the G peak can be related to graphitic carbon. The specific position of the D and G peaks is also intimately and strongly related to the type and size of carbon and its situation as seen by the difference in position of the carbon features seen for the GDC-Ni and YSZ-Ni surfaces shown in Fig. 10. They also act, as in this case, as a direct indication of the amount of carbon present within a particular measurement. Yoshinaga and co-workers utilised Raman mapping to assess the distribution of carbon deposition across anodes for the different compositions. White light images of the areas mapped are shown in Fig. 10 (middle) with the mapped areas and the different materials indicated. Raman maps generated from the carbon G peak are shown in Fig. 10 (bottom). These show that the distribution of carbon deposits on the different anodes is significantly different. It is clear from these and similar measurements on the ScSZ-Ni anode composition that significant amount of graphitic carbon forms on the exposed nickel surfaces of the YSZ-Ni and ScSZ-Ni systems. Little carbon formation was observed on the GDC system. These differences likely occur from the differences in oxide conductivity combined with possible differences in water mediated interactions.

Su *et al.* also studied carbon deposition and its effect on cell performance and degradation for a range of anode compositions [99]. As part of this study, Raman spectroscopy was used to investigate the relative propensity for the various anode composites to react with CO. Powders of the candidate anode materials SDC-Ni, ScSZ-Ni and $La_{0.8}Sr_{0.2}Sc_{0.2}Mn_{0.8}O_3$ (LSSM) were exposed to pure CO for 5 hours at 600, 850 and 900 °C. Raman spectra were then collected from the various powders using a 514 nm laser in order to compare the amount and type of carbon formed under the different conditions. A significant amount of carbon was observed to form on the

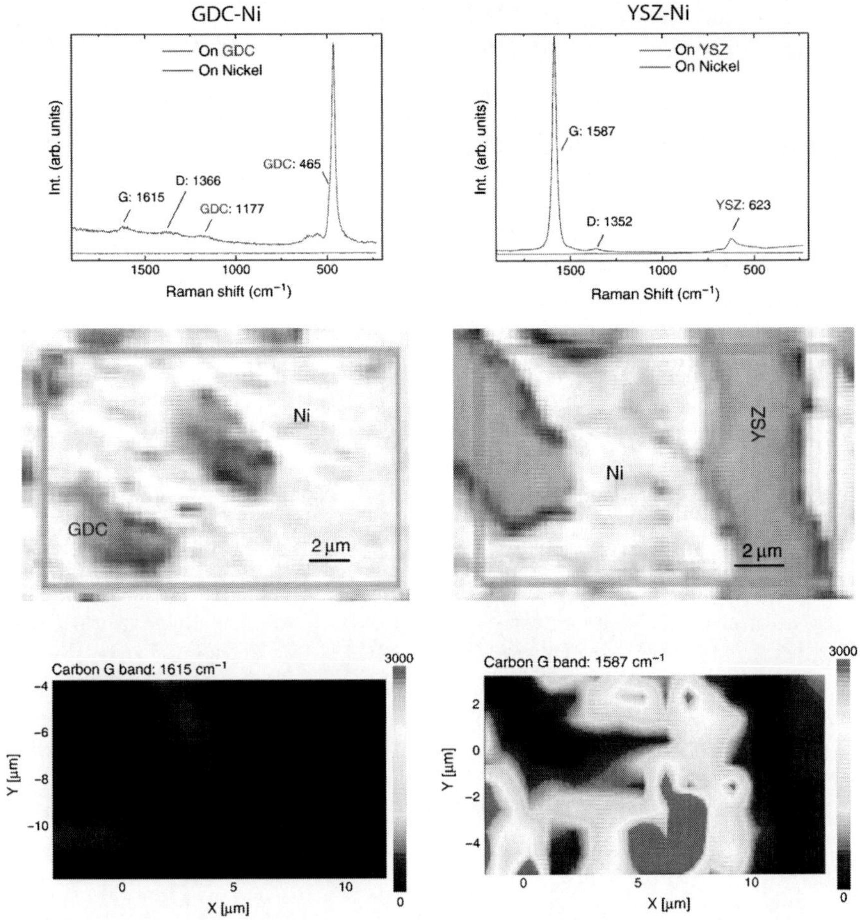

Fig. 10 Raman characterisation of carbon distributions of (left) GDC-Ni and (right) YSZ-Ni cermet anodes. Clear differences in the Raman spectra (top) are observed showing a pronounced difference in the deposition characteristics between the two anode compositions. The optical images (middle) correspond to the areas mapped using Raman spectroscopy. Intensity maps generated from the G peak of carbon (bottom) show there is there much less carbon deposited on the GDC-Ni composite compared to the YSZ.-Ni Figure reproduced from reference [98].

SDC-Ni and ScSZ-Ni powders. Higher temperatures resulted in increased deposition with deposits becoming increasingly graphitic. On the other hand, LSSM showed no sign of carbon formation suggesting that LSSM would be an ideal material for SOFC anodes operating with carbonaceous fuels. However, LSSM was found to be unstable under CO operation at open circuit potential (OCP) in operational systems, highlighting the importance of *in-situ* measurements to fully understand interactions. *In-situ* test measurements performed on SOFCs with anodes formed of the three compositions suggested that ScSZ-Ni is the most suitable for carbon resistant anodes.

Carbon formed on the anode can also strongly influence electrochemical interactions [100]. Li *et al.* investigated how the formation of carbon on the YSZ-Ni anode of a functional SOFC affected these electrochemical reactions using SEM, EDS, XPS and Raman spectroscopy [101]. They used a commercial SOFC with an YSZ electrolyte and ScSZ-Ni anode running on CH_4 for 30 minutes at OCP, 0.7V and under direct carbon fuel cell discharging. In order for carbon to participate in the electrochemical reactions occurring on the anode, the reactive carbon must have access to the electrolyte for O^{2-} and be electrically connected to the external circuit to allow the electrons to be released. From their measurements they concluded that carbon deposited on the YSZ, Ni, and at the three phase boundary can all participate in electrochemical reactions albeit with differing activity. Carbon formed at the three phase boundary was found to be the most reactive, followed by that adsorbed to the YSZ and finally on the Ni. This goes some way to explaining the observation of Yoshinaga and co-workers that GDC is much more resistant to carbon deposition given that GDC is partially reduced under operational conditions. Under these conditions GDC is a mixed conductor allowing surface carbon to be electrochemically active through providing access to both O^{2-} and the external circuit [102].

The group of Zongping Shao based at Nanjing University in China has also been active in using Raman spectroscopy to understand carbon deposition for heavier carbon fuels [103, 104] and for the evaluation of catalyst layers for *in-situ* fuel reforming [105, 106]. They have investigated, using *ex-situ* Raman spectroscopy, the temperature dependent carbon formation on YSZ-Ni catalysts suitable for SOFC anodes which were exposed to pure dimethyl ether (DME) and a DME-CO_2 mix for 30 minutes between 650 and 850 °C. DME was found to be readily decomposed to a mix of CH_4, CO and H_2 over the catalyst. Analysis of the characteristic D and G carbon peaks indicated that carbon deposits become increasingly graphitic at higher temperatures while the presence of CO_2 in the fuel was shown to significantly decrease the graphitic nature of the deposits formed.

Raman spectroscopy has also been applied to characterising carbon formed on mixed LiLaNi-Al_2O_3 and copper catalyst layers for *in-situ* reforming, to study the effects of doping concentrations and operational temperatures [105]. Copper is of interest for SOFC anodes due to its enhanced resistance to carbon formation. Carbon formation was found to be reduced for pure LiLaNi-Al_2O_3 and 50:50 LiLaNi-Al_2O_3/Cu exposed to pure methane for 5 minutes at 850 °C. The Cu composite also encouraged formation of a more amorphous carbon, as indicated by a larger D/G ratio. Hornes *et al.* have also used Raman spectroscopy as part of a larger study on carbon formation on Cu/Ni and Cu/Fe doped Ceria composites [61, 107].

Sumi *et al.* investigated the effect of the crystal structure of YSZ and ScSZ in nickel based anodes on the dynamics of carbon deposition and oxidation [108]. Anodes formed from YSZ-NiO and ScSZ-NiO were first characterised with Raman spectroscopy as a function of Ni content after calcination at 1673 K in air and at 1073 K in 10% H_2/N_2 mixed atmosphere. The Raman spectra suggested a mix of primarily cubic phase zirconia with a small contribution from other phases. The prevalence of cubic phase of ScSZ was found to be stronger than other phases compared to the YSZ based materials. The low solubility of Ni in ScSZ was observed to generate fine Ni particles on NiO reduction. The susceptibility of the different anode compositions to carbon deposition was then investigated by exposure to a 10% CH_4/N_2 mixture at 1073 and 1273 K for 10 and 60 minutes. Graphitic carbon growth was found to initiate from the interface between Ni and oxide particles for both the YSZ and ScSZ electrolytes. The finer Ni particles formed during reduction of the ScSZ composite resulted in rod shaped carbon formations, demonstrating the strong dependence of carbon formation morphology and crystallinity on the Ni morphology. The dynamics of carbon deposition and oxidation on operational anodes formed of the two composites were then investigated. Anodes were exposed to a fuel mixture of 10% CH_4-N_2 at 1273 K for 1 hour without current loading, followed by 1 hour at 0.3 A/cm^2. The YSZ-Ni anode was found to be much more susceptible to amorphous carbon formation than the ScSZ-Ni at OCP. As previously mentioned, the more amorphous the deposit the more easily it is removed through electrochemical oxidation. Thus, much of the carbon formed on the YSZ anode was removed under load with graphitic carbon remaining. This confirms that the more graphitic the carbon, the more difficult it is to remove through any regenerative procedure.

Nickel oxidation is a key issue for SOFC operation, where the reduction and oxidation (redox) tolerance of the anodes is an important practical consideration [109]. Many studies have characterised the reduction of pure NiO to Ni using H_2, although there is a great deal of inconsistency regarding the reported activation energy E_a. Richardson *et al.* reviewed the E_a values reported in the literature and found that values ranged between 17 and 133 kJ mol^{-1}. They concluded that the measured E_a was strongly dependent on the type of sample and its preparation [110]. One might expect the process to be significantly modified in complex anode composites and worthy of further study. This is particularly true given the long term effects of repeated redox cycling on SOFC anodes. While such structural changes are required during the initial reduction of SOFC anodes, the dimensional changes associated with subsequent redox cycles and the associated sintering of Ni leads to a reduction in surface area and the formation of cracks within the anode and/or electrolyte [111]. As such, high redox stability would greatly enhance the lifetime, stability and operation of SOFCs and ultimately reduce cost. Although Raman spectroscopy has not been extensively applied to the investigation of the redox dynamics of nickel within anode composites, it would be ideally suited. Raman spectroscopy is reasonably sensitive to the presence of NiO which has a Raman fingerprint at high temperatures dominated by a single broad intense peak at approximately 1100 cm^{-1} arising from the 2LO 2 phonon scattering mode [112]. This has already been demonstrated by Pomfret *et al.* in their 2006 application of *in-situ* Raman spectroscopy to the study of various processes relevant to SOFC operation including the NiO redox process [113]. However, there is significant scope for further investigation.

Despite the progress made with anode characterisation, there is much more scope for further investigation particularly *in-situ*. While most studies have focussed on simplified fuels such as methane, which is the main component of natural gas as well as coal or bio derived gases, further work is required to understand the chemical processes occurring within more complicated fuel compositions.

The cathode. Potential cathode materials must satisfy a range of requirements. They must be highly catalytically active for the oxygen reduction reaction, and both highly electronically and ionic conductive. As with other material components they must also be mechanically and chemically compatible with other component materials. Perovskite materials fulfil many of these requirements and dominate the list of candidate cathode materials. Many of their properties can be precisely optimised for particular application through doping with a wide range of metal oxides. Lanthanum strontium manganite (LSM) has traditionally been the material of choice for high temperature SOFC cathodes. However; at lower temperature operation this material no longer has sufficient activity or electronic conductivity [114]. The oxygen reduction reaction is particularly affected by electrochemical loss of activity since the electrode kinetics and transport processes are thermally activated. In recent years numerous material compositions have been investigated for cathodes for IT-SOFCs [115].

In general there have been relatively few studies that have utilised Raman spectroscopy to study cathode materials beyond basic characterisation [80, 116, 117]. This is primarily because of the low Raman intensity associated with the processes of interest for cathodes such as the oxygen reduction reaction. In recent years there has been some notable progress on the application of Surface Enhanced Raman Spectroscopy (SERS) to the investigation of SOFC cathodes driven by Mielin Liu's group based at Georgia Tech [118, 119]. Although promising, further work will be required to further refine the use of SERS for investigation of SOFC materials.

Interconnects

The replacement of traditional ceramic interconnects with metallic components to create SOFC stacks offers significant improvements in SOFC performance along with reduced cost. However, understanding the degradation processes in terms of corrosion and chromium migration is critical for developing cost effective SOFCs. Raman spectroscopy is particularly sensitive to the relevant oxides and corrosion processes.

Zeng *et al.* studied the corrosive effect of carbonaceous gases on fourteen different high temperature alloys using Raman spectroscopy [120]. These alloys were exposed to a simulated SOFC fuel environment to investigate the process of metal dusting which attacks metals exposed to high activity carbon species at high temperature [121, 122]. The exact details of the process depend on the specific composition of the alloy in question. For example in iron based alloys the bulk is thought to be broken into small particles of iron and carbon through the decomposition of metastable Fe_3C which is formed on the surface in the carbon atmosphere [123]. On the other hand carbon particles are thought to grow into the nickel metal causing its disintegration into small particles.

Fig. 11(a) shows example spectra obtained from two different compositions of stainless steel alloy. The two peaks relate to the surface oxides formed – either Cr_2O_3 scale or spinel (such as Fe_3O_4) type oxides – at 550 and 690 cm^{-1} respectively. Each alloy was exposed to various compositions of H_2, CO, H_2O and CO_2 at 593 °C for 1000 hours at low and high pressure (15 atm). Deep pits were observed to form on the exposed surfaces with

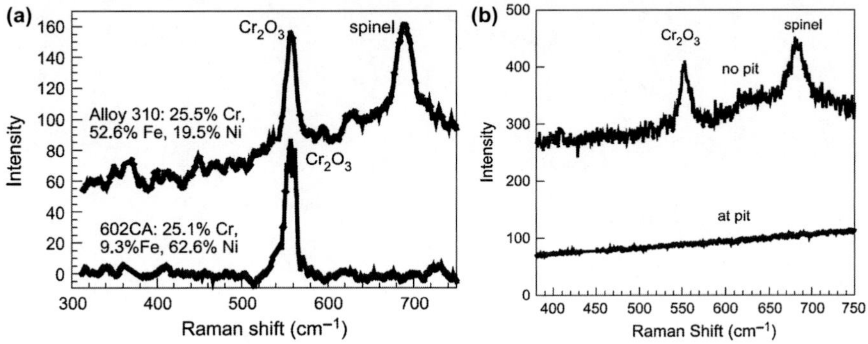

Fig. 11 (a) Raman spectra of two different stainless steel alloys. (b) Raman spectra of stainless steel exposed to a mixed hydrogen and carbonaceous fuel atmosphere from normal and damaged areas. Figure reproduced from reference [120].

material disintegrating into a mixture of carbon, metal and carbide particles. Fig. 11(b) shows Raman spectra collected from the damaged and undamaged areas of the surface. Oxide scales were clearly identifiable in undamaged regions and absent from damaged areas indicating that the oxides protect the alloys from the metal dusting process. Alloys with an oxide scale predominately consisting of the Cr_2O_3 oxide phase were protected to a greater degree and suffered less damage compared to those with a high proportion of spinel at the surface. From this it was concluded that Ni based alloys with scales containing less spinel perform better than Fe based alloys. High water vapour contents were also found to reduce the damage by increasing the oxygen partial pressure resulting in a reduced carbon activity and consequently an increased stability of the spinel oxide phase. Higher pressures were found to significantly accelerate the degradation process.

On the cathode side, long term exposure to oxidising atmospheres can result in the deterioration of SOFC performance due to the growth of oxide scale surface layers. The resistance of the interconnect gradually increases as the oxide scale layer thickens and mechanical stability is reduced. Several methods to mitigate this process have been under investigation including surface doping either to slow or prevent oxide growth or to improve the conductivity of the scale and minimise the effect on SOFC performance.

Numerous studies have investigated coating the interconnect with a dense protective oxide layer to reduce surface oxidation and chromium contamination of other key components [124, 125]. Perovskite materials such as $LaCrO_3$ typically used for higher temperature ceramic interconnects are ideal as their conductivity and thermal expansion coefficient can be tailored through doping to ensure compatibility and optimise performance.

Several deposition techniques have been investigated such as laser ablation, pulsed laser deposition, electrostatic spray and sputtering [127]. Orlovskaya and co-workers utilised Raman spectroscopy in conjunction with SEM, XRD, and energy dispersive X-ray analysis to study the structural properties of a 200 nm thick $LaCrO_3$ protective thin film deposited on high temperature steels by radiofrequency magnetron sputtering [126]. Fig. 12(a) shows the Raman spectrum of the polished stainless steel surface

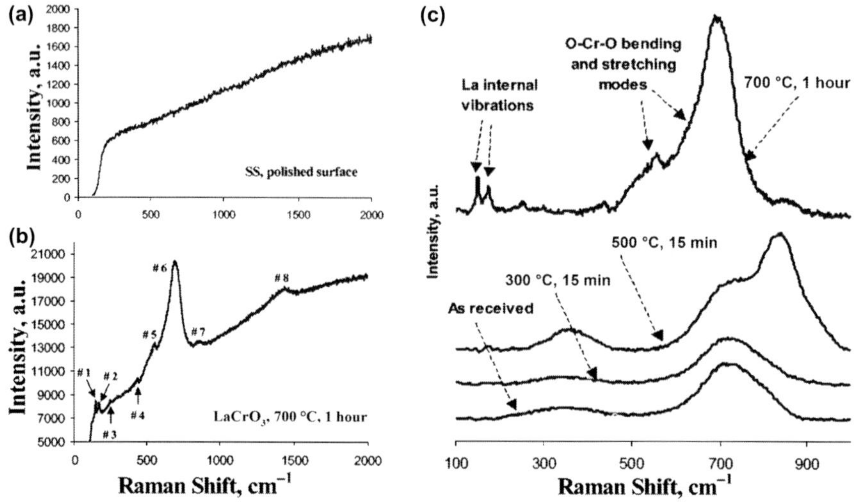

Fig. 12 Raman spectra of stainless steel (a) before treatment and (b) after treatment and annealing at 700 °C with $LaCrO_3$ perovskite thin film. (c) Raman spectra of $LaCrO_3$ coatings annealled at various temperatures. Figure reproduced from reference [126].

prior to thin film deposition. No surface oxides are apparent and no features from Cr_2O_3 or spinel were observed. Fig. 12(b) shows the Raman response of the surface after coating with the $LaCrO_3$ perovskite thin film and subsequent annealing at 700 °C for an hour confirming the formation of a dense perovskite surface layer.

Fig. 12(c) shows the Raman response of the coating before and after thermal treatments in air at different temperatures. Raman features observed in the coating at 320, 730 and 1430 (not shown) cm^{-1} are associated with amorphous La-Cr-O and no changes were observed on annealing at 300 °C. Several weak features between 150 and 180 cm^{-1} appear along with new features at 350 and 855 cm^{-1} on annealing at 500 °C. The 350 and 855 cm^{-1} peaks are associated with Cr in the higher oxidation state of the $LaCrO_4$ phase [128, 129]. Their appearance at 500 °C is most likely due to the transformation of amorphous to the monazite $LaCrO_4$ which occurs between 495 and 530 °C [130]. Orlovskaya *et al.* suggested that the weak features maybe lanthanum vibrations.

The Raman spectrum obtained from the film after annealing at 700 °C for an hour is dominated by the characteristic signature of perovskite $LaCrO_3$ [131–133]. The sharp peaks at 150 and 170 cm^{-1} arise from lanthanum vibrations while the stronger peaks at 580 and 695 cm^{-1} are assigned to O-Cr-O bending and stretching vibrations [134]. A slight 855 cm^{-1} peak is indicative of small amount of residual monazite $LaCrO_4$ phase. The ratio of these peaks can give an indication of the proportions of the different phases.

Unprotected steel was exposed to similar conditions to evaluate the effectiveness of the protective oxide layer. Several oxide phases were observed to form with the majority being hematite. On the other hand, no oxide species were observed for protected steel indicating a good level of protection. More recently, Lair *et al.* have studied the crystal structure of thin film samaria doped ceria protective layers grown on interconnect steel as a function of doping using Raman spectroscopy in combination with SEM and XRD [135]. Thin ceria films were electrochemically deposited on a steel substrates with 0 to 50% samaria doping at room temperature and annealed at 600 °C. Samaria was found stabilise the ceria in the cubic phase in films with up to 5% doping as inferred by increased F_{2g} peak intensity. Further doping resulted in decrease in the F_{2g} peak intensity and appearance of peaks characteristic of Sm_2O_3, suggesting a reduced crystalinity and phase separation.

Sealing materials

Effective sealing of SOFC components to ensure gas separation are critical for efficient long term operation. Exposure to both oxidising and reducing environments places stringent requirements on potential materials. Glass based sealants have excellent properties at high temperatures which make them ideally suited. Raman spectroscopy is a useful characterisation tool for studying the structural and network properties of silicate glasses [136–140] and has been applied to investigate potential SOFC sealing glasses [141, 142].

Mahapatra and Bodnar used Raman spectroscopy to study the change in the network structure of the novel $SrO\text{-}La_2O_3\text{-}Al_2O_3\text{-}B_2O_3\text{-}SiO_2$ (SABS) based glass as a function of composition [142]. Fig. 13(a) shows Raman spectra as a function of increasing $B_2O_3{:}SiO_2$ ratio. A great deal of information on the network structure can be obtained from the Raman response of these glasses which is of significant interest given their influence on the thermal and mechanical properties. It is clear from these spectra that large-scale changes are occurring within the glass structure as their composition is changed. Several extremely broad peaks are observed which result from the convolution of numerous peaks related to different glass substructures.

A great deal of structural information can be extracted from such Raman spectra given the sensitive dependence of these peaks on the various structural glass components [137]. Fig. 13(b) shows the deconvolution of the peaks for the glass with no B_2O_3 using symmetric Gaussian functions based on the procedures developed by Mysen *et al.* [137]. The peaks located between 400 to 700 cm^{-1} are related to delocalised vibrations of the Si-O-Si system [140]. The broad peak centred at 1000 cm^{-1} is a convolution of a series of 4 major regions related to different silicate structural

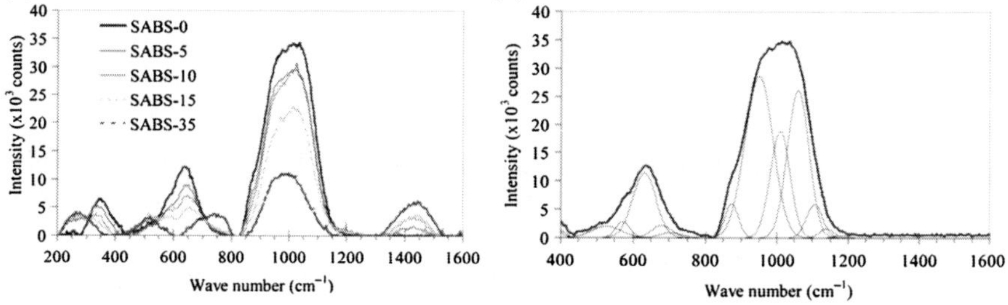

Fig. 13 a) Raman spectra of the different SABS glasses investigated by Mahapatra and Bodnar and b) the deconvoluted Raman spectrum for the SABS-0 glass. Deconvolution of all the spectra using Gaussian peaks allows for the effect of the $B2O_3{:}SiO_2$ ratio on the structural properties of the glass. See text for further details. Figure reproduced from reference [142].

components [138]. The 850 cm^{-1} peak arises from the orthosilicate composition with zero bridging oxygen atoms. The 900 cm^{-1} peak is related to the pyrosilicate composition with one bridging oxygen atom. The stretching motion of metasilicate composition with two bridging oxygen atoms is described by the 950–1000 cm^{-1} bands, while the 1050–1100 cm^{-1} region describes the stretching motion of the metasilicate composition with two bridging oxygen atoms. The peak around 1400 cm^{-1} is directly attributed to B_2O_3 arising from the B-O stretching mode. Its intensity clearly increases with increasing B_2O_3 proportion in Fig. 13(a) and it is absent from the standard silicate glass without B_2O_3 shown in Fig. 13(b). Although not shown here, it can also be deconvoluted into specific components relevant to specific structural elements.

By analysing the Raman spectra in detail, Mahapatra and Bodnar were able to show that the micro-hetrogeneity and the degree of non-bridging oxygen atoms increased with the B_2O_3 content. This was shown to reduce glass stability for long exposures to high temperatures. In an earlier study Mahapatra and Lu studied the influence of nickel on the glass network structure of the same SrO-La_2O_3-Al_2O_3-B_2O_3-SiO_2 based glass system and its effect on the thermal properties [141]. As already discussed, nickel is a key component for SOFC anodes and as a result is a possible modifier of potential glass sealing materials. Through a similarly detailed study of the Raman response of a range of glass compositions they were able to show that presence of nickel leads to decreased glass network connectivity, increased non-bridging oxygen atoms (thus decreasing thermal stability), and glass devitrification.

In-situ characterisation studies

The preceding sections demonstrate the progress made in the understanding of SOFC materials and processes that has been made through *ex-situ* Raman spectroscopy characterisation. However, the implicit assumption that the physical processes leave unique surface "signatures" which then remain unchanged during cooling can result in important details being misconstrued or missed altogether. Questions such as whether different fuels form carbon deposits on a surface at different rates or whether compounds change phase or migrate onto critical surfaces during the cooling process will always be a concern, not only for Raman spectroscopy but any *ex-situ* characterisation. For these reasons there has been a drive in recent years to develop novel *in-situ* techniques including Raman spectroscopy as well as IR thermal imaging [143, 144], interferometry [145], and XRD [146].

In-situ optical characterisation of SOFC components represents a significant technical challenge given the extreme temperatures and atmospheres involved [148]. IR spectroscopy can often be rendered completely ineffective by blackbody radiation from SOFC surfaces at high temperatures. Raman characterisation of materials exposed to a single atmosphere as a function of temperature can be achieved relatively simply through the use of commercial optical cells such as those provided by Linkam [149]. Raman characterisation of a fully operational SOFC represents much more of a challenge given the need for a two atmosphere setup with electrical

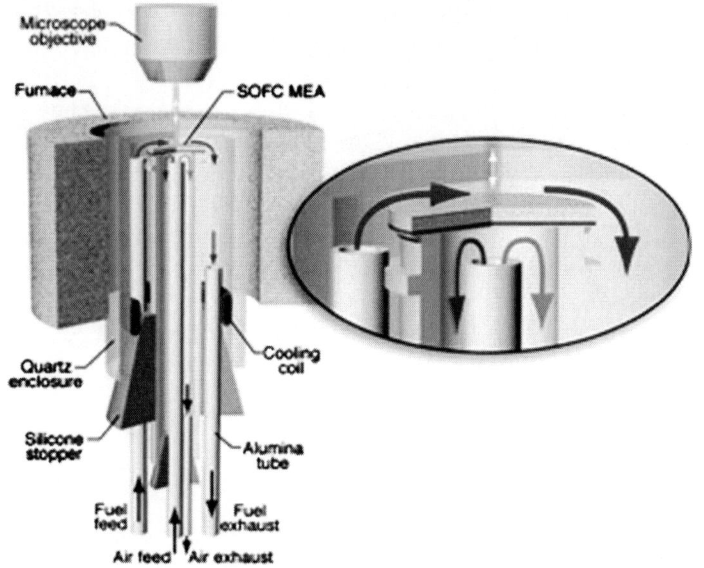

Fig. 14 Schematic representation of the SOFC rig developed by Rob Walker's group. This setup allows for combined Raman spectroscopy and electrochemical characterisation of operational SOFCs. Figure reproduced from [147].

connections. Currently no commercial SOFCs with suitable optical access are available and groups have developed systems in-house. A typical *in-situ* Raman spectroscopy setup, as developed by Rob Walker's group, is shown in Fig. 14 [147]. In this case, a full scale bench furnace has been adapted to accept a SOFC test pellet within a quartz enclosure. In both single atmosphere and full SOFC systems the laser is focused onto the SOFC surface through an optical window using a long working distance objective through which scattered light is also collected, as in a standard Raman microscope system. Such a setup allows for a much more detailed investigation of component materials and their interactions. Significant results that demonstrate the importance of *in-situ* Raman characterisation of SOFCs are reviewed here.

Sulphur poisoning of the anode

Sulphur is a well-known poison of catalysts that presents a serious problem for SOFC anodes [150]. Many potential carbonaceous SOFC fuels such as natural gas and biogas contain significant concentrations of sulphurous compounds which can lead to a severe degradation in SOFC performance. While it is possible to remove the majority of the sulphur from the fuel prior to its introduction to the cell, this increases the complexity and cost of the entire system. In order to remove the need for fuel processing, many investigations have focused on developing sulphur tolerate anodes with the aim of reducing the complexity and cost of integrated systems.

To this end, interest has grown in the application of Raman spectroscopy for the identification and detection of sulphurous surface species. However, problems can arise for nickel based catalysts whose complicated range of surface sulphide species can lead to ambiguous Raman responses [151, 152].

As a result there is some disagreement in the Raman fingerprints of sulphides reported in the literature. The group of Meilin Liu have focussed on the application of Raman spectroscopy for the study of sulphur poisoning of SOFC anodes and other nickel based alloys [153–157]. They have concentrated on resolving these issues through careful comparison of room temperature and *in-situ* measurements at high temperature.

Ni_3S_2 is one such species important for SOFCs that exists as the mineral heazlewoodite at room temperature. Several groups have reported conflicting Raman spectra from the rhombohedral heazlewoodite form of Ni_3S_2 [154, 158, 159]. To address this in relation to SOFCs, Cheng *et al.* performed a detailed study of Ni_3S_2 formation on YSZ-Ni cermets at room and high temperature [154]. Raman measurements were correlated to factor group analysis and XRD, to characterise the room temperature rhombohedral form as well as the cubic phase prevalent at high temperatures.

YSZ-Ni (60% YSZ, 40% Ni) composites were prepared with large grained Ni particulates to allow for detailed analysis of known surfaces. These were then heated to temperature in H_2 and exposed to flowing H_2/H_2S gas with a concentration of H_2S of 100 ppm for 12 hours. Cheng *et al.* were able to clearly identify the Ni_3S_2 through normal white light imaging of the surface during the measurement. This allowed *in-situ* Raman spectroscopy to be collected both before and after Ni_3S_2 formation and *ex-situ* analysis of the deposits to be performed after cool down.

Fig. 15(a) shows the *in-situ* Raman spectra collected from a typical Ni grain before and after the formation of Ni_3S_2. A clear signature is observed with 6 peaks consistent with those expected from group factor analysis for the rhombohedral form of Ni_3S_2. These spectra were found to be

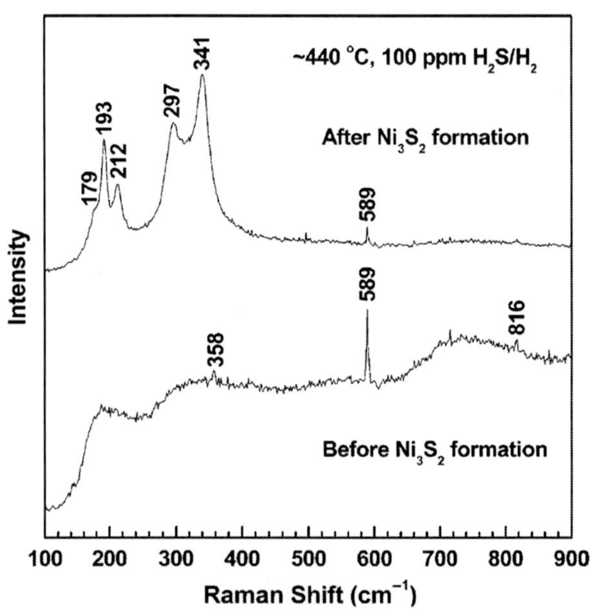

Fig. 15 *In-situ* Raman spectra obtained from a Ni region of a YSZ-Ni composite before and after exposure to 100 ppm H_2S/H_2 at 440 °C. Figure reproduced from reference [154].

significantly different from previous reports in the literature [158, 159] including previous work by the same authors [153]. Liu and coworkers speculated that such discrepancies could be due to the presence of other unexpected stoichiometries such as Ni_7S_6.

Further confusion may occur at higher temperatures where sulphur species may undergo significant phase changes. Raman spectra at temperatures close to the transition temperature ($565 \pm 5\,°C$ [151, 152]) between the low temperature rhombohedral and high temperature cubic phases of Ni_3S_2 are shown in Fig. 16. Below the transition temperature the Raman spectra closely resembles those measured at room temperature. Features related to the rhombohedral phase disappear above the transition temperature and reappear upon cooling. The broad feature at $320\,cm^{-1}$, found to remain above the transition temperature, was assigned to the F_{2g} mode of the cubic phase of Ni_3S_2. The authors note that this broad feature would be difficult to use as a spectral fingerprint for Ni_3S_2 during *in-situ* observation and it may be necessary to cool samples to confirm the existence of Ni_3S_2. This is not ideal as it almost defeats the point of performing *in-situ* measurements. Indeed Cheng and Liu showed that artefacts can be introduced

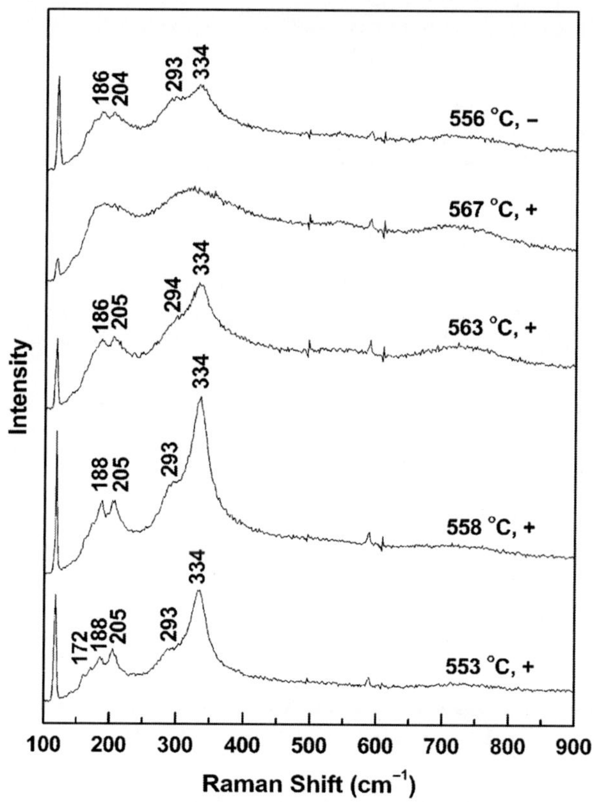

Fig. 16 Raman spectra of Ni_3S_2 around the temperature of the transition between the low temperature rhombohedral and the high temperature cubic phases. Spectra marked with a + indicate spectra taken on heating and those with a – were taken during cooling. The sharp peaks at 120 and $590\,cm^{-1}$ are due to stray light and adsorbed hydrogen respectively. Figure reproduced from reference [154].

simply as a result of differences in the cooling rate of the system [155]. This work in particular demonstrates the importance of *in-situ* characterisation of SOFC processes and emphasises the importance of careful consideration of all experimental parameters.

Carbon deposition

As already discussed, carbon deposition is a significant problem for SOFC anodes operated on carbonaceous fuels. Deposits fill the open microstructure of the anode leading to a reduction in performance over time from restricted fuel access as well as metal dusting corrosion and catastrophic failure *via* cracking and delamination. The group of Rob Walker, previously based Maryland and now in Montana, has greatly advanced the used of *in-situ* Raman spectroscopy, particularly for the investigation of carbon deposition [113, 147, 160–162].

In their most recent study Eigenbrodt *et al.* reported a detailed investigation of carbon deposition on YSZ-Ni anodes operating in YSZ electrolyte based SOFC run on methanol and methane using *in-situ* near infrared imaging and Raman spectroscopy [162]. Time dependent Raman spectra collected from SOFC anodes which were held at OCP at 715 °C and exposed to methanol and methane are shown in Fig. 17(a) and (b) respectively. The characteristic D and G peaks of carbon are observed to appear for the anode exposed to methanol while only the G peak is observed for the anode exposed to methane. It is clear from the evolution of these peaks that carbon forms on the anode exposed to methanol much more quickly than that exposed to methane. This is emphasised in Fig. 17(c) where spectra were

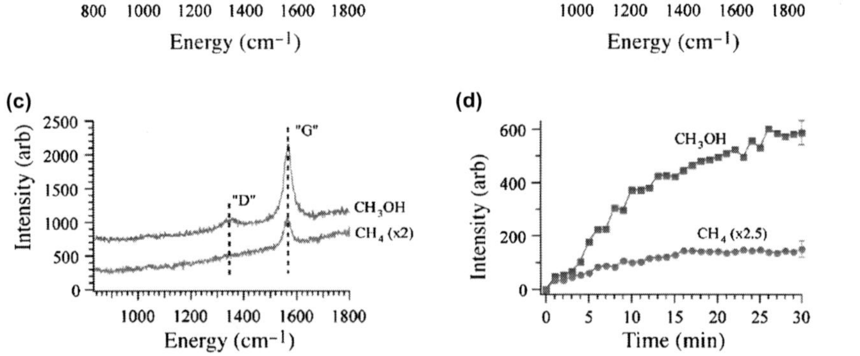

Fig. 17 *In-situ* Raman monitoring of carbon formation on YSZ-Ni anodes operated carbonaceous fuels at OCP and 715 °C. (a) Carbon formation as a function of time for operation with CH_3OH fuel. (b) Carbon formation as a function of time for operation with CH_4 fuel. (c) Raman spectra of anode surfaces after 30 minutes of operation on CH_3OH (blue) and CH_4 (red). (d) G peak intensity for both CH_3OH (blue) and CH_4 (red) as a function of time. Figure reproduced from reference [162].

collected from the anodes after 30 minutes of exposure to the different environments.

The sharp G features observed from both surfaces are indicative of highly graphitic carbon. The presence of the small D peak in the spectrum from the anode exposed to the methanol fuel suggests that the carbon formed differs slightly from that formed from methane. A more detailed study of the time dependent dynamics of the carbon formation process for the two anodes reveals differences in the deposition process associated with the two fuels. Fig. 17(d) shows the integrated peak areas of the G peak for the two fuels as a function of time. Growth initially proceeds at the same rate for both fuels until about 5 minutes after exposure. After this point the growth rate for carbon formation in the methane environment slows significantly and eventually stops indicating that graphite formation from methane fuel is self-limiting. Conversely, substantial carbon deposition in the methanol environment continues for the duration of the measurement. This is a direct result of the presence of the OH group within the methanol which is in general much more reactive than methane and leads to carbon deposition through multiple reaction pathways. Such insights are important for optimising operational conditions to minimise carbon deposition and perhaps even inform the development of strategies for its eventual removal and regeneration of degraded anodes.

Oxidation states

The *ex-situ* studies of electrolyte materials highlighted in previous sections clearly demonstrate the detailed information Raman spectroscopy is capable of providing regarding oxidation states and material phase that is relevant for SOFCs. Perhaps the most pressing issue facing SOFCs today is the development of materials with improved redox stability characteristics. Detailed information regarding material phases and oxygen concentration under operational conditions over long periods of time will be key to optimising materials, their fabrication, and conditions of operation for long term operation.

We have investigated the use of *in-situ* Raman spectroscopy for the characterisation of the thermal properties and structural properties of IT-SOFC electrolyte GDC exposed to simulated SOFC conditions [163]. The strong, well defined Raman response of GDC is ideal for the study of important processes. Subtle changes occurring within the Raman features can be resolved in on short timescales which can then be related to changes occurring within the material structure and composition in real time.

Fig. 18 shows the integrated area (top) and position (bottom) of the F_{2g} peak of GDC as a function of time when exposed to a strongly reducing 97% H_2 3%H_2O atmosphere. After a short period of time during which H_2 is filling the system, the intensity of the F_{2g} peak is observed to significantly decrease with a step like profile. The peak position also shifts in the same profile. This is directly related to the formation of a reduced surface layer where the oxygen ion content is greatly decreased. Such detail can be used in conjunction with computational calculations to predict and better understand material stoichiometries formed under different experimental conditions. In this case we investigated the thermodynamic phase stability of ceria

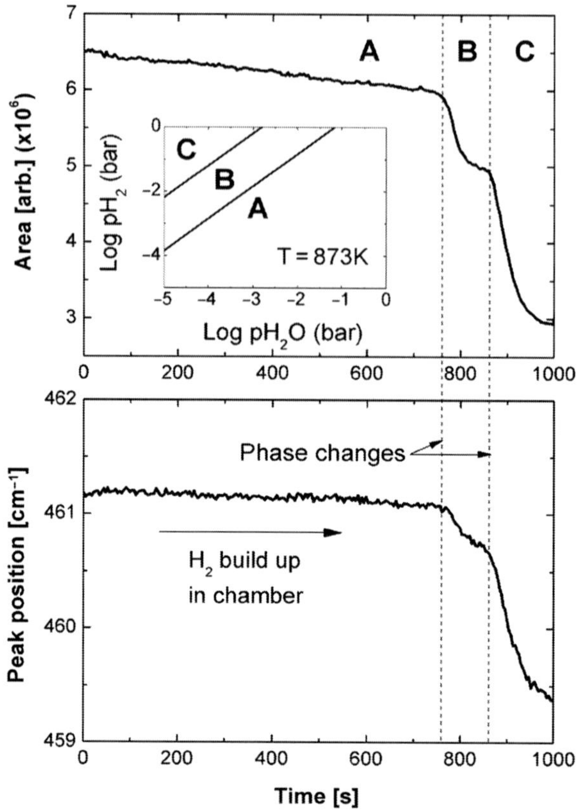

Fig. 18 Peak area (top) and position (bottom) of the F_{2g} 460 cm^{-1} peak of GDC when exposed to 97%H_2 3%H_2O at 873K as a function of time. Top inset shows the predicted oxidation state of ceria as a function of pH_2 and pH_2O at 873 K. Figure reproduced from reference [163].

as a function of pH_2 and pH_2O using HSC Chemistry version 5.1. The results of this analysis are shown in the inset of Fig. 18. These calculations suggest that three possible phases of Ceria might be expected under these conditions: (A) CeO_2, (B) $CeO_{1.85}$ and (C) $CeO_{1.72}$. Although these calculations ignore the possible influence of gadolinium doping, the three phases predicted would appear to support the Raman observations.

More recently, Raman spectroscopy has been used to investigate similar effects in YSZ in greater detail. Fig. 19(a) shows a cross-sectional schematic of the SOFC setup used by Eigenbrodt and Walker for their surface impedance and Raman oxide mapping study of YSZ [164]. A top-down view showing the electrode orientation and the rough locations investigated for the position dependent Raman scattering is shown in Fig. 19(b).

Normalised Raman spectra of the YSZ surface exposed to oxidising (air) and reducing (H_2 in Ar) atmospheres at 715 °C are shown in Fig. 19(c). Both spectra were collected with the same integration time and the characteristic F_{2g} peak of cubic YSZ is clearly observed in both. The spectral features are almost identical, indicating that the lattice structure of the YSZ is the same in both atmospheres. However, there is a clear decrease in the intensity of

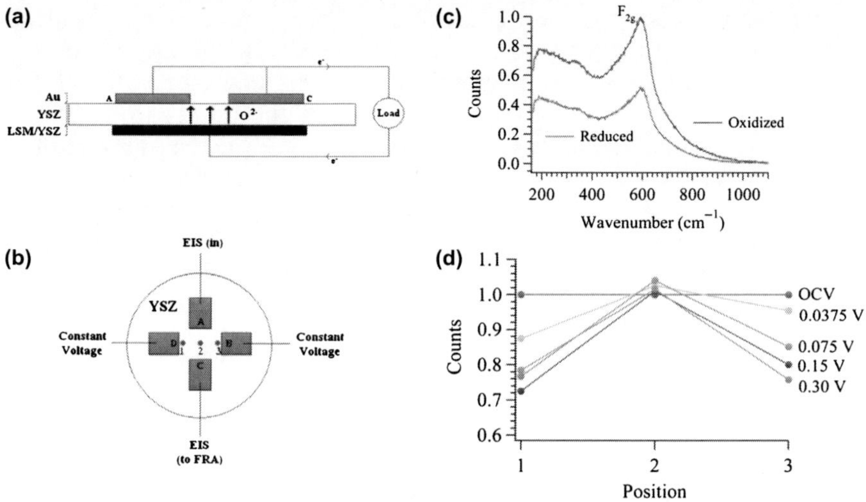

Fig. 19 Schematic (a) cross-section and (b) overview of electrode orientation on YSZ electrolyte pellet. (c) Normalised spectra collected from the YSZ electrolyte under oxidising and reducing conditions at 715 °C. (d) Normalised intensity of the F_{2g} YSZ electrolyte peak for various electrode potentials for the surface positions indicated in (b). Figure reproduced from reference [164].

the F_{2g} peak upon exposure to a reducing environment. This reduced intensity was attributed to the formation of a surface reduced YSZ phase extending 5 nm into the bulk. The formation and removal of this surface reduced phase of YSZ was found to be almost completely reversible on a timescale of about 15 minutes for multiple reduction oxidation cycles. Such a reduced surface phase of YSZ is of interest for its potential influence over the catalytic properties of SOFC anodes, as well as electrolyte oxide ion concentrations.

The sensitivity of Raman spectroscopy to the local chemical condition of the YSZ surface allowed Eigenbrodt and Walker to investigate relative oxide concentrations across the electrolyte surface in response to cell polarisation. Fig. 19(d) shows the normalised F_{2g} peak intensity as a function of cell potential for the three positions. The positions 1 and 3 were approximately 10 μm from each electrode, whilst position 2 was in the centre of the pellet approximately 2 mm from each electrode. Under operational conditions at OCP the pellet is exposed to a reducing atmosphere resulting in the formation of the partially reduced surface phase of YSZ. *In-situ* Raman spectroscopy clearly shows that the electrolyte surface becomes increasingly reduced as the potential is increased. This is somewhat surprising given that higher over potentials will result in greater oxygen ion flux to the anode. From this we might expect the YSZ in proximity to the anode to show evidence of electrochemical oxidation with an increased F_{2g} peak intensity. Further measurements showed that surface oxide depletion extended up to 10 μm away from the anode-electrolyte triple phase boundary. They suggested that the reduction in oxide ion concentration within 10 μm of the anode is a result of the increasingly strong drive of oxygen to the electrochemically active three phase boundary.

Specific data relating to the chemical conditions prevalent at the surface can be correlated to electrical measurements in order to inform our fundamental understanding of interactions occurring under operational conditions. For example, the lateral impedance of the cell was observed to decrease as the cell polarisation was increased. The increasingly reduced state of the YSZ surface in response to increased cell polarisations will result in higher electronic conductivity contributing to this effect. Such insights cannot be obtained from any other characterisation tools currently available emphasising the utility of Raman spectroscopy for the study of SOFCs under operational conditions.

Conclusion and outlook

Over the past decade Raman spectroscopy has increasingly become more important for the characterisation of the materials and processes vital for SOFCs. A full understanding of these is critical for the development of SOFCs if they are ever to achieve their full potential as a competitive, clean and efficient energy generation system. *Ex-situ* Raman spectroscopy characterisation has been extensively applied to the investigation of all major SOFC components and processes such as carbon and sulphur poisoning of anodes, as well as oxygen ion concentrations within electrolytes. The recent development of *in-situ* Raman spectroscopy has opened a wide range of physical processes to detailed study under realistic operational conditions.

In the near future we can expect the development of Raman spectroscopy for the characterisation of SOFCs to continue apace. Progress in the application of *in-situ* Raman spectroscopy in conjunction with more conventional electrochemical characterisation can be expected to be particularly important given the wealth of information it provides. The application of *in-situ* Raman spectroscopy to the investigation of sulphur and carbon poisoning of anodes as well as *in-situ* determination of oxygen concentrations within electrolyte materials has been highlighted here. However, these represent only a small selection of SOFC relevant processes for which *in-situ* Raman spectroscopy is ideally suited. Given this potential it is clear that Raman spectroscopy has a bright future for *in-situ* SOFC characterisation.

References

1 S. C. Singhal and K. Kendall (eds.), High-temperature Solid Oxide Fuel Cells: Fundamentals, Design and Applications, 1st edn, Elsevier, Oxford, 2003.
2 D. J. L. Brett, A. R. Kucernak, P. Aguiar, S. C. Atkins, N. P. Brandon, R. Clague, L. F. Cohen, G. Hinds, C. Kalyvas, G. J. Offer, B. Ladewig, R. Maher, A. Marquis, P. Shearing, N. Vasileiadis and V. Vesovic, *Chemphyschem*, 2010, **11**(13), 2714–2731.
3 D. J. L. Brett, A. Atkinson, N. P. Brandon and S. J. Skinner, *Chemical Society Reviews*, 2008, **37**(8), 1568–1578.
4 http://www.seca.doe.gov.
5 M. P. Sohier, G. Wrobel, J. P. Bonnelle and J. P. Marcq, *Applied Catalysis a-General*, 1992, **84**(2), 169–186.
6 K. P. De Jong and J. W. Geus, *Catalysis Reviews-Science and Engineering*, 2000, **42**(4), 481–510.

7 P. Serp, M. Corrias and P. Kalck, *Applied Catalysis a-General*, 2003, **253**(2), 337–358.
8 S. Helveg, C. Lopez-Cartes, J. Sehested, P. L. Hansen, B. S. Clausen, J. R. Rostrup-Nielsen, F. Abild-Pedersen and J. K. Norskov, *Nature*, 2004, **427**(6973), 426–429.
9 R. Malgas, S. F. Mapolie, S. O. Ojwach, G. S. Smith and J. Darkwa, *Catalysis Communications*, 2008, **9**(7), 1612–1617.
10 L. K. Johnson, C. M. Killian and M. Brookhart, *Journal of the American Chemical Society*, 1995, **117**(23), 6414–6415.
11 S. D. Ittel, L. K. Johnson and M. Brookhart, *Chemical Reviews*, 2000, **100**(4), 1169–1203.
12 N. Q. Minh, *Journal of the American Ceramic Society*, 1993, **76**(3), 563–588.
13 D. Beckel, A. Bieberle-Hutter, A. Harvey, A. Infortuna, U. P. Muecke, M. Prestat, J. L. M. Rupp and L. J. Gauckler, *Journal of Power Sources*, 2007, **173**(1), 325–345.
14 P. C. Su, C. C. Chao, J. H. Shim, R. Fasching and F. B. Prinz, *Nano Letters*, 2008, **8**(8), 2289–2292.
15 A. C. Johnson, A. Baclig, D. V. Harburg, B. K. Lai and S. Ramanathan, *Journal of Power Sources*, 2010, **195**(4), 1149–1155.
16 P. Bruschi, A. Diligenti, A. Nannini and M. Piotto, *Thin Solid Films*, 1999, **346**(1–2), 251–254.
17 A. Larrea, D. Sola, M. A. Laguna-Bercero, J. I. Pena, R. I. Merino and V. M. Orera, *Journal of the Electrochemical Society*, 2011, **158**(10), B1193–B1197.
18 C. D. Baertsch, K. F. Jensen, J. L. Hertz, H. L. Tuller, S. T. Vengallatore, S. M. Spearing and M. A. Schmidt, *Journal of Materials Research*, 2004, **19**(9), 2604–2615.
19 J. W. Fergus, *Solid State Ionics*, 2004, **171**(1–2), 1–15.
20 W. Z. Zhu and S. C. Deevi, *Materials Science and Engineering a-Structural Materials Properties Microstructure and Processing*, 2003, **348**(1–2), 227–243.
21 J. W. Fergus, *Materials Science and Engineering a-Structural Materials Properties Microstructure and Processing*, 2005, **397**(1–2), 271–283.
22 K. Hilpert, D. Das, M. Miller, D. H. Peck and R. Weiss, *Journal of the Electrochemical Society*, 1996, **143**(11), 3642–3647.
23 J. W. Fergus, *Journal of Power Sources*, 2005, **147**(1–2), 46–57.
24 P. A. Lessing, *Journal of Materials Science*, 2007, **42**(10), 3465–3476.
25 M. K. Mahapatra and K. Lu, *Journal of Power Sources*, 2010, **195**(21), 7129–7139.
26 M. Yashima, K. Ohtake, M. Kakihana, H. Arashi and M. Yoshimura, *Journal of Physics and Chemistry of Solids*, 1996, **57**(1), 17–24.
27 N. R. Harlan, R. I. Merino, J. I. Peña, A. Larrea, V. M. Orera, C. González, P. Poza and J. Llorca, *Journal of the American Ceramic Society*, 2002, **85**(8), 2025–2032.
28 X. S. Xin, Z. Lu, Z. H. Ding, X. Q. Huang, Z. G. Liu, X. Q. Sha, Y. H. Zhang and W. H. Su, *Journal of Alloys and Compounds*, 2006, **425**(1–2), 69–75.
29 Q. Yuan, L. L. Li, S. L. Lu, H. H. Duan, Z. X. Li, Y. X. Zhu and C. H. Yan, *Journal of Physical Chemistry C*, 2009, **113**(10), 4117–4124.
30 S. Heiroth, R. Frison, J. L. M. Rupp, T. Lippert, E. J. B. Meier, E. M. Gubler, M. Dobeli, K. Conder, A. Wokaun and L. J. Gauckler, *Solid State Ionics*, 2011, **191**(1), 12–23.
31 X. L. Xia, J. H. Ouyang, Z. G. Liu, S. A. Gao and S. Li, *Journal of the Electrochemical Society*, 2010, **157**(4), B470–B476.
32 C. Bernay, A. Ringuede, P. Colomban, D. Lincot and M. Cassir, *Journal of Physics and Chemistry of Solids*, 2003, **64**(9–10), 1761–1770.

33 M. Tanaka, M. Hasegawa, A. F. Dericioglu and Y. Kagawa, *Materials Science and Engineering: A*, 2006, **419**(1–2), 262–268.
34 M. Hattori, Y. Takeda, J. H. Lee, S. Ohara, K. Mukai, T. Fukui, S. Takahashi, Y. Sakaki and A. Nakanishi, *Journal of Power Sources*, 2004, **131**(1–2), 247–250.
35 Y. J. Leng, S. H. Chan, K. A. Khor, S. P. Jiang and P. Cheang, *Journal of Power Sources*, 2003, **117**(1–2), 26–34.
36 B. Bai, N. M. Sammes, A. L. Smirnova and G. Tompsett, *Journal of Fuel Cell Science and Technology*, 2010, **7**(2).
37 V. Zyryanov, N. Uvarov, A. Ulikhin, V. Kostrovskii, B. Bokhonov, V. Ivanov, V. Sadykov, A. Titov and K. Paichadze, *Inorganic Materials*, 2009, **45**(1), 90–98.
38 H. Fujimori, M. Yashima, M. Kakihana and M. Yoshimura, *Journal of the American Ceramic Society*, 1998, **81**(11), 2885–2893.
39 H. Kishimoto, N. Sakai, T. Horita, K. Yamaji, Y. P. Xiong, M. E. Brito and H. Yokokawa, *Solid State Ionics*, 2008, **179**(35–36), 2037–2041.
40 W. Araki, T. Koshikawa, A. Yamaji and T. Adachi, *Solid State Ionics*, 2009, **180**(28–31), 1484–1489.
41 S. Boulfrad, E. Djurado and J. Fouletier, *Solid State Ionics*, 2009, **180**(14–16), 978–983.
42 M. Shimazu, T. Isobe, S. Ando, K. Hiwatashi, A. Ueno, K. Yamaji, H. Kishimoto, H. Yokokawa, A. Nakajima and K. Okada, *Solid State Ionics*, 2011, **182**(1), 120–126.
43 A. Trovarelli, *Catalysis Reviews-Science and Engineering*, 1996, **38**(4), 439–520.
44 H. Inaba and H. Tagawa, *Solid State Ionics*, 1996, **83**(1–2), 1–16.
45 B. C. H. Steele, *Solid State Ionics*, 2000, **129**(1–4), 95–110.
46 G. Q. Xie, M. F. Luo, M. He, P. Fang, J. M. Ma, Y. F. Ying and Z. L. Yan, *Journal of Nanoparticle Research*, 2007, **9**(3), 471–478.
47 E. Rossinyol, E. Pellicer, A. Prim, S. Estrade, J. Arbiol, F. Peiro, A. Cornet and J. R. Morante, *Journal of Nanoparticle Research*, 2008, **10**(2), 369–375.
48 P. Bocchetta, M. Santamaria and F. Di Quarto, *Journal of Applied Electrochemistry*, 2009, **39**(11), 2073–2081.
49 D. H. Prasad, H. R. Kim, J. S. Park, J. W. Son, B. K. Kim, H. W. Lee and J. H. Lee, *Journal of Alloys and Compounds*, 2010, **495**(1), 238–241.
50 J. E. Spanier, R. D. Robinson, F. Zhang, S.-W. Chan and I. P. Herman, *Physical Review B*, 2001, **64**(24), 245407.
51 Y. Ji, J. Liu, T. M. He, J. X. Wang and W. H. Su, *Journal of Alloys and Compounds*, 2005, **389**(1–2), 317–322.
52 S. K. Tadokoro and E. N. S. Muccillo, *Journal of the European Ceramic Society*, 2007, **27**(13–15), 4261–4264.
53 T. Yuzhakova, V. Rakic, C. Guimon and A. Auroux, *Chemistry of Materials*, 2007, **19**(12), 2970–2981.
54 S. P. Li, J. Q. Lu, P. Fang and M. F. Luo, *Materials Letters*, 2009, **63**(20), 1689–1692.
55 C. H. Liang, Z. Q. Ma, H. Y. Lin, L. Ding, J. S. Qiu, W. Frandsen and D. S. Su, *Journal of Materials Chemistry*, 2009, **19**(10), 1417–1424.
56 L. D. Jadhav, M. G. Chourashiya, A. P. Jamale, A. U. Chavan and S. P. Patil, *Journal of Alloys and Compounds*, 2010, **506**(2), 739–744.
57 J. R. Mcbride, K. C. Hass, B. D. Poindexter and W. H. Weber, *Journal of Applied Physics*, 1994, **76**(4), 2435–2441.
58 B. Matovic, Z. Dohcevic-Mitrovic, M. Radovic, Z. Brankovic, G. Brankovic, S. Boskovic and Z. V. Popovic, *Journal of Power Sources*, 2009, **193**(1), 146–149.

59 W. D. Cai, F. Chen, X. X. Shen, L. J. Chen and J. L. Zhang, *Applied Catalysis B-Environmental*, 2010, **101**(1–2), 160–168.
60 M. P. Woods, P. Gawade, B. Tan and U. S. Ozkan, *Applied Catalysis B-Environmental*, 2010, **97**(1–2), 28–35.
61 A. Hornes, G. Munuera, A. Fuerte, M. J. Escudero, L. Daza and A. Martinez-Arias, *Journal of Power Sources*, 2011, **196**(9), 4218–4225.
62 X. Q. Sha, Z. Lu, X. Q. Huang, J. P. Miao, Z. H. Ding, X. S. Xin and W. H. Su, *Journal of Alloys and Compounds*, 2007, **428**(1–2), 59–64.
63 A. Mineshige, T. Taji, Y. Muroi, M. Kobune, S. Fujii, N. Nishi, M. Inaba and Z. Ogumi, *Solid State Ionics*, 2000, **135**(1–4), 481–485.
64 A. Mineshige, T. Yasui, N. Ohmura, M. Kobune, S. Fujii, M. Inaba and Z. Ogumi, *Solid State Ionics*, 2002, **152**, 493–498.
65 J. Chandradass, B. Nam and K. H. Kim, *Colloids and Surfaces a-Physicochemical and Engineering Aspects*, 2009, **348**(1–3), 130–136.
66 M. J. Godinho, R. F. Goncalves, L. P. S. Santos, J. A. Varela, E. Longo and E. R. Leite, *Materials Letters*, 2007, **61**(8–9), 1904–1907.
67 W. H. Weber, K. C. Hass and J. R. McBride, *Physical Review B*, 1993, **48**(1), 178–185.
68 S. Askrabic, Z. D. Dohcevic-Mitrovic, M. Radovic, M. Scepanovic and Z. V. Popovic, *Journal of Raman Spectroscopy*, 2009, **40**(6), 650–655.
69 M. Stojmenovic, S. Zec, S. Boskovic, B. Babic, B. Matovic, D. Bucevac, Z. Dohcevic-Mitrovic and F. Aldinger, *Journal of Alloys and Compounds*, 2010, **507**(1), 279–285.
70 I. Kosacki, T. Suzuki, H. U. Anderson and P. Colomban, *Solid State Ionics*, 2002, **149**(1–2), 99–105.
71 S. Burinskas, V. Adomonis, J. Dudonis, G. Laukaitis, V. Minialga and D. Milcius, *Solid State Ionics*, 2011, **184**(1), 14–18.
72 J. B. Goodenough, *Annual Review of Materials Research*, 2003, **33**, 91–128.
73 C. H. Liang, J. S. Qiu, Z. L. Li and C. Li, *Nanotechnology*, 2004, **15**(7), 843–847.
74 Y. C. Hu, *Journal of the American Ceramic Society*, 2006, **89**(9), 2949–2951.
75 C. L. Li, X. Gu, Y. Q. Wang, Y. J. Wang, Y. G. Wang, X. H. Liu and G. Z. Lu, *Journal of Rare Earths*, 2009, **27**(2), 211–215.
76 S. Somacescu, V. Parvulescu, P. Osiceanu, J. M. Calderon-Moreno and B. L. Su, *Journal of Colloid and Interface Science*, 2011, **363**(1), 165–174.
77 M. Boaro, S. Desinan, C. Abate, M. Ferluga, C. de Leitenburg and A. Trovarelli, *Journal of the Electrochemical Society*, 2011, **158**(2), P22–P29.
78 H. P. He, X. J. Huang and L. Q. Chen, *Journal of Physics and Chemistry of Solids*, 2001, **62**(4), 701–709.
79 A. M. Azad, M. Ramachandran and N. Schweitzer, *Solid State Ionics*, 2007, **178**(25–26), 1476–1486.
80 S. Farhadi and S. Sepahvand, *Journal of Alloys and Compounds*, 2010, **489**(2), 586–591.
81 S. Nakayama, H. Aono and Y. Sadaoka, *Chemistry Letters*, 1995, **6**, 431–432.
82 S. Nakayama and M. Sakamoto, *Journal of the European Ceramic Society*, 1998, **18**(10), 1413–1418.
83 H. Yoshioka and S. Tanase, *Solid State Ionics*, 2005, **176**(31–34), 2395–2398.
84 D. K. Kim, S. H. Jo and P. Muralidharan, *Electrochimica Acta*, 2009, **54**(28), 7495–7501.
85 S. P. Jiang, L. Zhang, H. Q. He, H. W. Wu and C. Z. Li, *International Journal of Hydrogen Energy*, 2011, **36**(11), 6862–6874.
86 T. Norby, *Nature*, 2001, **410**(6831), 877–878.
87 K. D. Kreuer, *Annual Review of Materials Research*, 2003, **33**, 333–359.

88 H. Iwahara, H. Uchida, K. Ono and K. Ogaki, *Journal of the Electrochemical Society*, 1988, **135**(2), 529–533.
89 I. Ahmed, C. S. Knee, M. Karlsson, S. G. Eriksson, P. F. Henry, A. Matic, D. Engberg and L. Borjesson, *Journal of Alloys and Compounds*, 2008, **450**(1–2), 103–110.
90 A. Slodczyk, P. Colomban, S. Willemin, O. Lacroix and B. Sala, *Journal of Raman Spectroscopy*, 2009, **40**(5), 513–521.
91 C. S. Tu, R. R. Chien, V. H. Schmidt, S. C. Lee, C. C. Huang and C. L. Tsai, *Journal of Applied Physics*, 2009, **105**(10).
92 S. Ricote, N. Bonanos, M. C. M. de Lucas and G. Caboche, *Journal of Power Sources*, 2009, **193**(1), 189–193.
93 A. Subramaniyan, J. H. Tong, R. P. O'Hayre and N. M. Sammes, *Journal of the American Ceramic Society*, 2011, **94**(6), 1800–1804.
94 E. V. Tsipis and V. V. Kharton, *Journal of Solid State Electrochemistry*, 2011, **15**(5), 1007–1040.
95 A. C. Ferrari and J. Robertson, *Physical Review B*, 2000, **61**(20), 14095–14107.
96 K. Yamaji, H. Kishimoto, Y. M. Xiong, T. Horita, N. Sakai, M. E. Brito and H. Yokokawa, *Solid State Ionics*, 2008, **179**(27–32), 1526–1530.
97 M. Yoshinaga, H. Kishimoto, M. E. Brito, K. Yamaji, T. Horita and H. Yokokawa, *Journal of the Ceramic Society of Japan*, 2011, **119**(1388), 307–309.
98 M. Yoshinaga, H. Kishimoto, K. Yamaji, Y. P. Xiong, M. E. Brito, T. Horita and H. Yokokawa, *Solid State Ionics*, 2011, **192**(1), 571–575.
99 C. Su, Y. Z. Wu, W. Wang, Y. Zheng, R. Ran and Z. P. Shao, *Journal of Power Sources*, 2010, **195**(5), 1333–1343.
100 S. McIntosh, J. M. Vohs and R. J. Gorte, *Journal of the Electrochemical Society*, 2003, **150**(4), A470–A476.
101 C. Li, Y. X. Shi and N. S. Cai, *Journal of Power Sources*, 2011, **196**(2), 754–763.
102 M. Mogensen, N. M. Sammes and G. A. Tompsett, *Solid State Ionics*, 2000, **129**(1–4), 63–94.
103 C. B. Wang, C. C. Lee, J. L. Bi, J. Y. Siang, J. Y. Liu and C. T. Yeh, *Catalysis Today*, 2009, **146**(1–2), 76–81.
104 C. Su, R. Ran, W. Wang and Z. P. Shao, *Journal of Power Sources*, 2011, **196**(4), 1967–1974.
105 W. Wang, C. Su, R. Ran, H. J. Park, C. Kwak and Z. P. Shao, *International Journal of Hydrogen Energy*, 2011, **36**(9), 5632–5643.
106 W. Wang, C. Su, R. Ran and Z. P. Shao, *Journal of Power Sources*, 2011, **196**(8), 3855–3862.
107 A. Hornes, D. Gamarra, G. Munuera, A. Fuerte, R. X. Valenzuela, M. J. Escudero, L. Daza, J. C. Conesa, P. Bera and A. Martinez-Arias, *Journal of Power Sources*, 2009, **192**(1), 70–77.
108 H. Sumi, P. Puengjinda, H. Muroyama, T. Matsui and K. Eguchi, *Journal of Power Sources*, 2011, **196**(15), 6048–6054.
109 D. Sarantaridis and A. Atkinson, *Fuel Cells*, 2007, **7**(3), 246–258.
110 J. T. Richardson, R. Scates and M. V. Twigg, *Applied Catalysis a-General*, 2003, **246**(1), 137–150.
111 M. Ettler, H. Timmermann, J. Malzbender, A. Weber and N. H. Menzler, *Journal of Power Sources*, 2010, **195**(17), 5452–5467.
112 M. J. Massey, N. H. Chen, J. W. Allen and R. Merlin, *Physical Review B*, 1990, **42**(13), 8776–8779.
113 M. B. Pomfret, J. C. Owrutsky and R. A. Walker, *Journal of Physical Chemistry B*, 2006, **110**(35), 17305–17308.
114 M. J. Jorgensen and M. Mogensen, *Journal of the Electrochemical Society*, 2001, **148**(5), A433–A442.

115 H. Zhao, Q. Li and L. P. Sun, *Science China-Chemistry*, 2011, **54**(6), 898–910.
116 W. Zhou, J. Sunarso, Z. G. Chen, L. Ge, J. Motuzas, J. Zou, G. X. Wang, A. Julbe and Z. H. Zhu, *Energy & Environmental Science*, 2011, **4**(3), 872–875.
117 Z. H. Zhu, W. Zhou, J. Sunarso, J. Motuzas, F. L. Liang, Z. G. Chen, L. Ge, S. M. Liu and A. Julbe, *Chemistry of Materials*, 2011, **23**(6), 1618–1624.
118 H. W. Abernathy, E. Koep, C. Compson, Z. Cheng and M. L. Liu, *Journal of Physical Chemistry C*, 2008, **112**(34), 13299–13303.
119 K. S. Blinn, H. W. Abernathy, and M. L. Liu, Surface enhanced Raman spectroscopy for investigation of sofc cathodes, in *Advances in Solid Oxide Fuel Cells V*, N. P. Bansal and P. Singh (eds.), Amer Ceramic Soc: Westerville, 2010, pp. 65–73.
120 Z. Zeng and K. Natesan, *Solid State Ionics*, 2004, **167**(1–2), 9–16.
121 G. H. Meier, W. C. Coons and R. A. Perkins, *Oxidation of Metals*, 1982, **17**(3), 235–262.
122 H. J. Grabke, *Materials and Corrosion-Werkstoffe Und Korrosion*, 1998, **49**(5), 303–308.
123 R. F. Hochman, Catastrophic deterioration of high-temperature alloys in carbonaceous atmospheres, in *Properties of High-Temperature Alloys*, A. Foroulis and F. S. Pettit (eds.), The Electrochemical Society, Princeton, 1977.
124 X. Chen, P. Y. Hou, C. P. Jacobson, S. J. Visco and L. C. De Jonghe, *Solid State Ionics*, 2005, **176**(5–6), 425–433.
125 Z. G. Yang, G. G. Xia, G. D. Maupin and J. W. Stevenson, *Surface & Coatings Technology*, 2006, **201**(7), 4476–4483.
126 N. Orlovskaya, A. Coratolo, C. Johnson and R. Gemmen, *Journal of the American Ceramic Society*, 2004, **87**(10), 1981–1987.
127 D. J. Jan, C. T. Lin and C. F. Ai, *Thin Solid Films*, 2008, **516**(18), 6300–6304.
128 D. H. Cho, S. D. Yim, G. H. Cha, J. S. Lee, Y. G. Kim, J. S. Chung and I. S. Nam, *Journal of Physical Chemistry A*, 1998, **102**(41), 7913–7918.
129 K. Azegami, M. Yoshinaka, K. Hirota and O. Yamaguchi, *Materials Research Bulletin*, 1998, **33**(2), 341–348.
130 C. Johnson, R. Gemmen and N. Orlovskaya, *Composites Part B-Engineering*, 2004, **35**(2), 167–172.
131 V. B. Podobedov, A. Weber, D. B. Romero, J. P. Rice and H. D. Drew, *Solid State Communications*, 1998, **105**(9), 589–593.
132 Y. Aoki, H. Konno, H. Tachikawa and M. Inagaki, *Bulletin of the Chemical Society of Japan*, 2000, **73**(5), 1197–1203.
133 M. Popa, L. Van Hong and M. Kakihana, *Physica B-Condensed Matter*, 2003, **327**(2–4), 233–236.
134 K. H. Kim, J. Y. Gu, H. S. Choi, G. W. Park and T. W. Noh, *Physical Review Letters*, 1996, **77**(9), 1877–1880.
135 V. Lair, L. S. Zivkovic, O. Lupan and A. Ringuede, *Electrochimica Acta*, 2011, **56**(12), 4638–4644.
136 W. L. Konijnendijk and J. M. Stevels, *Journal of Non-Crystalline Solids*, 1975, **18**(3), 307–331.
137 B. O. Mysen, L. W. Finger, D. Virgo and F. A. Seifert, *American Mineralogist*, 1982, **67**(7–8), 686–695.
138 P. McMillan, *American Mineralogist*, 1984, **69**(7–8), 622–644.
139 A. K. Hassan, L. M. Torell, L. Borjesson and H. Doweidar, *Physical Review B*, 1992, **45**(22), 12797–12805.
140 E. I. Kamitsos, J. A. Kapoutsis, H. Jain and C. H. Hsieh, *Journal of Non-Crystalline Solids*, 1994, **171**(1), 31–45.
141 M. K. Mahapatra and K. Lu, *Journal of Power Sources*, 2008, **185**(2), 993–1000.

142 M. K. Mahapatra, K. Lu and R. J. Bodnar, *Applied Physics a-Materials Science & Processing*, 2009, **95**(2), 493–500.
143 D. J. L. Brett, P. Aguiar, R. Clague, A. J. Marquis, S. Schottl, R. Simpson and N. P. Brandon, *Journal of Power Sources*, 2007, **166**(1), 112–119.
144 M. B. Pomfret, D. A. Steinhurst, D. A. Kidwell and J. C. Owrutsky, *Journal of Power Sources*, 2010, **195**(1), 257–262.
145 H. Guo, G. Iqbal and B. S. Kang, *International Journal of Applied Ceramic Technology*, 2010, **7**(1), 55–62.
146 J. S. Hardy, J. W. Templeton, D. J. Edwards, Z. G. Lu and J. W. Stevenson, *Journal of Power Sources*, 2012, **198**, 76–82.
147 M. B. Pomfret, J. C. Owrutsky and R. A. Walker, *Analytical Chemistry*, 2007, **79**(6), 2367–2372.
148 M. B. Pomfret, J. C. Owrutsky and R. A. Walker, *Annual Review of Analytical Chemistry*, 2010, **3**, 151–174.
149 http://www.linkam.co.uk/.
150 Y. Matsuzaki and I. Yasuda, *Solid State Ionics*, 2000, **132**(3–4), 261–269.
151 H. Fjellvag and A. Andersen, *Acta Chemica Scandinavica*, 1994, **48**(4), 290–293.
152 S. Stolen, F. Gronvold, E. F. Westrum and G. R. Kolonin, *Journal of Chemical Thermodynamics*, 1991, **23**(1), 77–93.
153 J. Dong, Z. Cheng, S. W. Zha and M. L. Liu, *Journal of Power Sources*, 2006, **156**(2), 461–465.
154 Z. Cheng, H. Abernathy and M. L. Liu, *Journal of Physical Chemistry C*, 2007, **111**(49), 17997–18000.
155 Z. Cheng and M. L. Liu, *Solid State Ionics*, 2007, **178**(13–14), 925–935.
156 J. H. Wang, Z. Cheng, J. L. Bredas and M. L. Liu, *Journal of Chemical Physics*, 2007, **127**(21).
157 S. H. Choi, J. H. Wang, Z. Cheng and M. Liu, *Journal of the Electrochemical Society*, 2008, **155**(5), B449–B454.
158 U. S. Ozkan, L. P. Zhang, S. A. Ni and E. Moctezuma, *Energy & Fuels*, 1994, **8**(4), 830–838.
159 D. W. Bishop, P. S. Thomas and A. S. Ray, *Materials Research Bulletin*, 2000, **35**(7), 1123–1128.
160 M. B. Pomfret, C. Stoltz, B. Varughese and R. A. Walker, *Analytical Chemistry*, 2005, **77**(6), 1791–1795.
161 M. B. Pomfret, J. Marda, G. S. Jackson, B. W. Eichhorn, A. M. Dean and R. A. Walker, *Journal of Physical Chemistry C*, 2008, **112**(13), 5232–5240.
162 B. C. Eigenbrodt, M. B. Pomfret, D. A. Steinhurst, J. C. Owrutsky and R. A. Walker, *Journal of Physical Chemistry C*, 2011, **115**(6), 2895–2903.
163 R. C. Maher and L. F. Cohen, *Journal of Physical Chemistry A*, 2008, **112**(7), 1497–1501.
164 B. C. Eigenbrodt and R. A. Walker, *Analytical Methods*, 2011, **3**(7), 1478–1484.

Integrated analytical techniques for analysing individual environmental particles

S. Potgieter-Vermaak,*[a, b] R. Van Grieken[b] and J.H. Potgieter[c]
DOI: 10.1039/9781849734899-00123

Introduction

Environmental particles of various origins are frequently analysed with more than one analytical technique to obtain their chemical composition, structure and morphology. The knowledge thus obtained often facilitates the understanding of their reactivity, sources, transport and changes of chemical species. In this way predictions regarding their impact on the environment and health can be made, aiding preventive conservation of cultural heritage, setting of pollution guideline values, legislation for pollution control measures, source apportionment, and climate change elucidation, to name but a few. The analysis techniques for environmental particles can broadly be divided into two groups, namely bulk (*e.g.* water-soluble ionic content by means of ion chromatography for particulate matter, elemental concentrations by means of elemental techniques such as X-ray fluorescence spectrometry for all environmental particles, chemical structural information by means of X-ray diffraction for larger environmental particles, like sediments and sands, *etc.*) and micro-analytical techniques, with which the character of any single particle can be probed. The latter, single particle analysis (SPA), permits one to obtain more unambiguous and detailed information than the former and addresses the chemically and morphologically heterogeneous nature of environmental particles; albeit local detection limits are often in the 1000 ppm range. Our concern in this chapter will be on the latter, and will focus on environmental particles in the fine to medium-fine range (aerodynamic diameter 0.2–16 µm). These particles are typically in the respirable size range, with a heterogeneity that varies from pure inorganic – pure organic (primary) to mixed particles (binary and ternary). Because of their diverse nature the use of a single technique to characterise them, is inadequate.

Among the variety of micro-analytical techniques used, electron probe X-ray micro-analysis (EPXMA) and computer controlled scanning electron microscopy with energy-dispersive X-ray detection (CC-SEM/EDX) are most commonly used. Both can be used in fully automated mode, and in combination with cluster analysis and/or multivariate techniques. They are ideally suited for the analysis of representative numbers (300–1000

[a] *School of Science and the Environment, Division of Chemistry and Environmental Sciences, Manchester Metropolitan University, Oxford Road, Manchester, M1 5GD, UK; E-mail: S.potgieter@mmu.ac.uk*
[b] *Department of Chemistry, University of Antwerp, Universiteitsplein 1, B-2610 Antwerp, Belgium*
[c] *School of Research, Enterprise and Innovation, Manchester Metropolitan University, Oxford Road, Manchester, M1 5GD, UK*

particles per sample) of single particles in the size range mentioned earlier. The methodology is based on automated particle recognition and characterisation following three sequential steps: localising, sizing and chemical characterising. The use of hierarchical, non-hierarchical or principal component analysis, allows one to discern particle classes. With appropriate collection methods and substrate optimization, quantification of the elemental concentrations, area, perimeter, and diameter of an individual particle with a size of 0.2 μm and larger, is achieved. During the early stages of development this methodology enabled only analysis of high-Z ($Z =$ atomic number) elements. The upgrading to "thin window" detection allows low-Z element detection (TW-EPXMA). Because of the size, shape and matrix effects in environmental particles, conventional quantification procedures cannot straightforwardly be applied. Therefore a home-made quantification method was used which is based on a modified version of the single scattering Casino Monte Carlo program. It allows the simulation of electron trajectories in spherical, hemi-spherical and cubic low-Z particles on a smooth surface, and calculates the characteristic and continuous background X-ray flux produced by the particle and the substrate. The measured X-ray intensities and the geometrical parameters are set as input for the quantification procedure. In an iterative way, the weight fraction or concentration of a certain element in particles is calculated until the simulated and measured X-ray intensities, including the background, are in good agreement; normally this is within 5–15%, even for low-Z elements. More information on this quantification procedure is given *e.g.* by Ro *et al.*[1] and Szaloki *et al.*[2] Analyses at liquid nitrogen cooling conditions retain the integrity of more volatile compounds, such as ammonium sulphate and nitrate, ubiquitous in airborne particulate matter frequently analysed by the group of Rene Van Grieken, based in the University of Antwerp, Belgium[1–7] using this methodology. Furthermore, analyses of surface and in-depth layers of single particles can be achieved by using variable energy electron excitation followed again by quantification calculations. This approach has been used in the analysis of various types of environmental particles *e.g.* road dust[8] (and refs. therein).

Although the clustering results of the elemental concentrations classify the particles into representative groups, they do not provide structural information. The sensitivity of micro-Raman spectroscopy (MRS) (which enables one to investigate the composition, phase, crystallinity, crystal orientation, and, in some cases, doping of materials on a micrometer-scale), makes it an ideal tool to characterise individual heterogeneous particles in the fine and course particle size ranges. As a stand alone technique it is indeed used extensively for the analysis of environmental particles from various sources (soils,[9] sediments,[10,11] atmospheric aerosols,[5,12–15] minerals[11,16,17] and bio-compounds).[18,19] Some of the very first publications on Raman spectroscopy and aerosol particles were those of Fung and Tang,[20–22] who reported on an extended study of artificial inorganic salt particles. A sophisticated set-up was built, in which levitated single particles of known composition in solid and liquid states could be analysed by Raman. A valuable contribution has been made by this study in that the result of mixed salts, typically found in aerosols, were characterized for their respective Raman spectra. Since the early 1990's, various studies on levitated particles exposed to different conditions and their consequent heterogeneous reactions were published, some with the use of Raman spectroscopy.[23–25] Wu *et al.*[26] used MRS to observe the effect of potassium nitrate on the phase transitions of ammonium nitrate and came to the conclusion that ammonium nitrate may exist in phase form III at lower temperatures in the presence of potassium nitrate and therefore this phase transition should be considered possible during modelling. Apart from these carefully controlled laboratory investigations (a review of which has been written by Schweiger[12]) various studies on carbonaceous content of airborne particulate matter were published.[27] Finally, a preliminary investigation on the use of MRS on particulate matter collected on various substrates was reported by Potgieter-Vermaak and Van Grieken.[28] The authors were able to identify various compounds, apart from soot, and discussed the practical challenges of analysing particulate matter collected by impaction. In addition, mention was made of the ambiguous characterization of conglomerates and the necessity of complementing the molecular characterization with an elemental characterization of the same particle. It is evident that the addition of this micro-analytical technique to the wealth of information obtained by EPXMA makes it a powerful two tier approach and its benefit to elucidate the origin and fate of particulate matter, and environmental particles in general, seems obvious. Consequently, this approach has been utilised by various researchers in the last decade[11,16,29–33] and its application to environmental particles is ever growing. Of late, the group of Chul-Un Ro at Inham University, Incheon, Korea[34–38] has used attenuated Total Reflection Fourier Transform-Infrared Imaging Techniques (ATR-FT-IR-IT) instead of MRS to obtain structural information. In addition, EPXMA could be interfaced with MRS or micro-X-ray fluorescence spectrometry (μXRFS) and Auger spectrometry, avoiding the difficulties discussed in Potgieter-Vermaak, *et al.*[29] such as particle relocation.

This chapter will start off by giving a historical overview of the use of EPXMA/CC-SEM/EDX and TW-EPXMA applied to environmental particles, concentrating on airborne particulate matter. The more recent advances where TW-EPXMA is combined with MRS and ATR-FT-IR-IT to elucidate the chemical and structural nature of airborne particulate matter will then be discussed. The few examples where airborne particulate matter has been analysed by SEM/EDX interfaced with MRS (SEMSCA, illustrated in Fig. 1) will be summarised and the chapter will conclude with a view of potential future developments and requirements for further extensions to current practices.

First era: CCSEM/EDX or EPXMA for SPA

In an earlier review by Jambers *et al.*,[39] a number of applications of EPXMA/EDX analyses to various environmental SPA were reported, among them aerosol particulate matter from Antartica, Siberia, the Amazon basin, Hawaii, Chile, Sudan, Antwerp in Belgium, the English Channel and the North Sea.

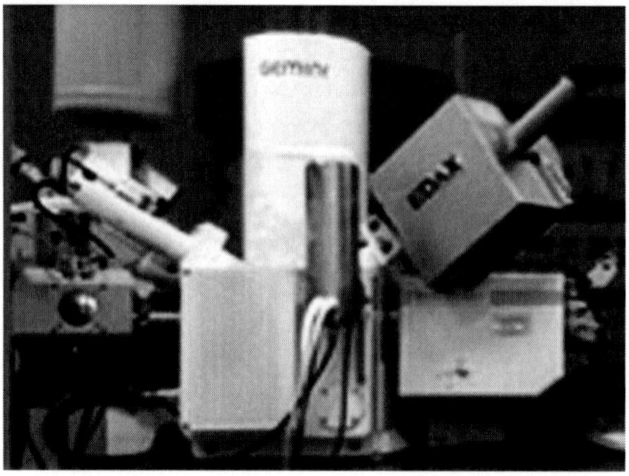

Fig. 1 SEM/EDX interfaced with MRS (SEMSCA).

In follow-up work, marine aerosols, sampled along the coast of central California, USA, by employing an aircraft (C131), were analysed in the framework of the Monterey area Ship Tracks Experiment (MAST).[40] The main aim was to investigate the processes involved in the anthropogenic modification of cloud albedo (cloud reflectivity). Such individual cloud droplet residual, ship track and ship plume particles were analysed by CC-SEM/EDX. Using the methodology described elsewhere in this review, it was discerned that the aerosols from the ship plumes played an active role in the formation of ship tracks (due to the identification of Si-rich particles found only in the below-cloud plume and ship track residual particles) and thus on cloud modification. The clear drawback of the instrumentation used, is that only elements with atomic number higher than 10 could be analysed and the "organic fraction" had to be inferred as belonging to those particles with a very low sum of X-ray intensities. Figs. 2a and b illustrate the % abundance of the 13 particle clusters identified for the below-cloud aerosol samples, collected during 8 research flights and investigated by single particle SEM-EDX analysis. The results for marine boundary layer considered to be clean (C) and intermediately polluted (IP), are shown. Some 77% of the analysed particles in this sample appeared to be pure sea salt. This publication also described a particle cluster called "Transformed NaCl" where its composition is ascribed to SO_2 condensation on the sea salt particle (22% abundance in Fig. 2a). The other sample classified as 'clean' reveals besides sea salt and transformed sea salt particles also a number of particles (9%) identified as marine crystallisation products (C.P. in Fig. 2) which are produced by the crystallisation of successive salts ($CaCO_3$, dolomite, $CaSO_4$, NaCl …) upon evaporation of sea water drops. The remaining particles, characterised in this sample (11%), seemed to be composed of organic or other low-Z material, exclusively associated with a characteristic Cl-peak in the X-ray spectrum. The authors state, however, that they have no further evidence that these particles are in fact truly organic or whether they are elemental carbon or other low Z-elements.

Further work on samples collected in the North Sea[41] using a combination of cluster and factor analysis approaches, concluded that human expert

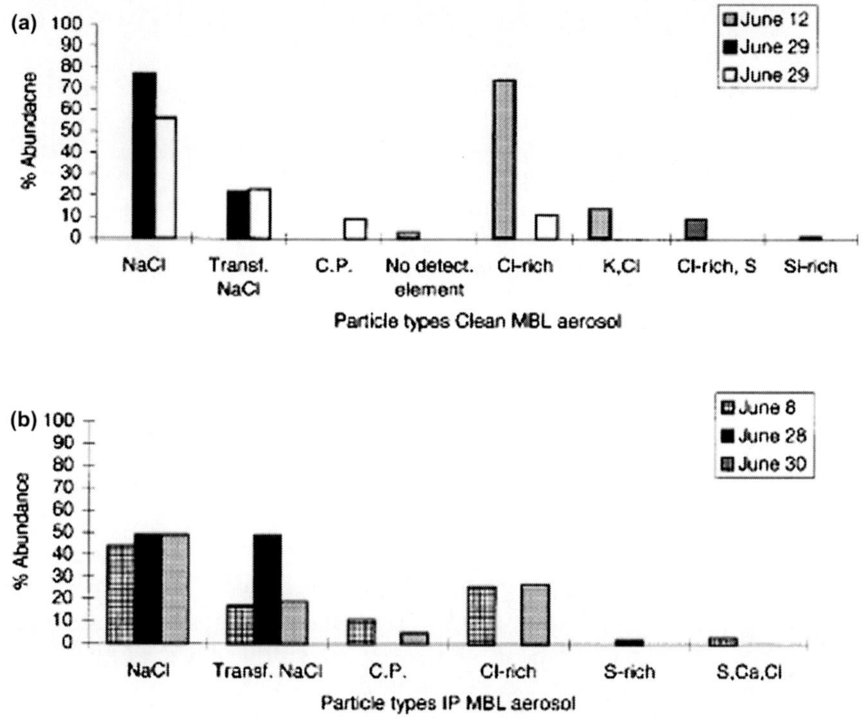

Fig. 2 a & b. Particle clusters identified in marine boundary layer aerosols analysed by SPA with CC-EPXMA.

knowledge remains critically important in selecting the best conditions and combinations to obtain the correct structural analysis in large data sets of environmental particulate matter. Wegrzynek *et al.*[42] described an investigation into the analysis of sediment cores from the Rhone river suspended matter in the Gulf of Lion in order to study the input of Sahara aerosols to these sediments and the amount of matter transported by the river itself. Individual particle analyses were conducted on approximately 15,000 relevant individual particles and the authors identified several typical particle groups in the sediment samples, *e.g.* Si-rich and aluminosilicate groups constituting approximately 70% of the particulate matter, Ca-Si and Ca-Si-Mg particles making up about 20% of the core samples, with the remaining 10% composing of Fe-S, Fe-Si-Al, Ti-Si and Ti-Si-Fe-Al. The authors concluded that aerosol input from the Sahara was hardly noticeable in the deposited sediments analysed. In a paper published by de Miranda *et al.*[43] aerosol samples collected in the Sao Paulo Metropolitan area in Brazil were analysed by a number of techniques to elucidate their elemental composition. The EPXMA analyses were performed in automated mode by employing a commercial software program that located every particle on the filter, determined its size, acquired secondary electron (SE) and backscattered electron (BE) images thereof and then analysed it by EDX. By using a combination of techniques ((EPXMA, Particle Induced X-ray Emission (PIXE) and gravimetry) in conjunction with hierarchical cluster analysis of the EPXMA results, the authors were able to characterise the aerosol particles into well-defined classes and types.

Due to the limitations posed by the inability to analyse low-Z elements, especially when dealing with marine aerosol particles, further developments in this field shifted to modifications like thin-window and windowless X-ray instruments, as well as incorporation of a low temperature stage to help preserve easily volatilised compounds, *e.g.* ammonium salts. This trend, which is particularly obvious in reports published towards 2000 and later on years, marks the onset of the second era in single particle analysis.

Second era: low-Z and beam sensitive particles

The seminal work to develop semi-quantitative measurement of low Z-elements by using a combination of a liquid-nitrogen cooled stage and appropriate automated particle recognition, analyses and classification software, was described early in the 21st century by the Van Grieken group in Antwerp.[3,44] An EPXMA instrument with EDX detection was used in conjunction with Si(Li) detectors and without a beryllium window (windowless) or with an ultra-thin polymer foil (thin window) to observe light and heavy elements simultaneously. The group evaluated different substrates such as beryllium, aluminium, silicon and silver for sample/particles collection, while the size, shape and position of particles were obtained from SE or BE images. Instrumental parameters were optimised to yield low electron energy, minimum X-ray background, minimum beam damage, and automated particle recognition. The huge amount of spectral and morphological information that were obtained was processed with a modified reverse Monte Carlo simulation procedure. This allowed the semi-quantitative composition calculation of particles containing light elements (*e.g.* O, N and C). Chemometric algorithms were employed to refine the obtained data sets into main particle types in each sample. Cluster Analysis (CA), Factor Analysis (FA) and Principal Component Analysis (PCA) were chosen as the main classification algorithms to identify the various particle types and classes. The team performed an exercise on comparing synthetically produced typical marine particles and naturally collected ones and found an excellent correlation between their compositions when employing this newly developed approach.

In one of the first papers reporting the application of a thin window EPXMA to measure light elements ($Z<11$), Osán *et al.*[45] used the reverse Monte Carlo method developed at the University of Antwerp to calculate the concentrations of light elements in individual atmospheric aerosol particles collected over Lake Balaton in Hungary. Approximately 25,000 individual particles were analysed by CC-EPXMA and classified into various groups and types using hierarchical and non-hierarchical cluster analyses. PCA was applied to the combined set of single particle EPXMA and bulk XRF results to determine potential sources of the aerosol particles. The composition of the aerosol particles did not display typical seasonal variations, but could be correlated to air mass movements of marine origin, continental origin and Eastern European origin.

The anthropogenic contribution to the particulate matter composition above the North Sea was probed by relating the meteorological conditions

Fig. 3 Position of the Berner impactor sampler on the upper deck of the vessel.

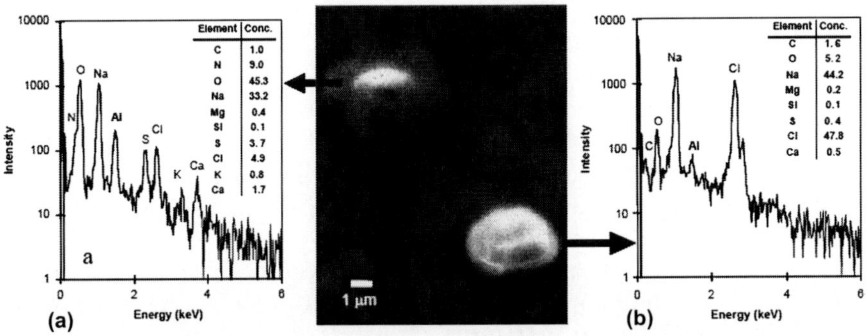

Fig. 4 Image, spectrum and calculated elemental concentrations (%w) for two particles: (a) aged sea salt; (b) sea salt.

with the chemical composition of individual atmospheric particles collected on a research vessel (R/V Belgica, illustrated in Fig. 3).[46]

Particles were analysed with TW-EPXMA, as described above and classified into 13 groups using a non-hierarchical clustering algorithm. The impact of marine and continental influences on the particulate matter could be studied by determining the air mass backward trajectories (obtained from the Hybrid Single-Particle Lagrangian Integrated Trajectory – HYSPLIT model). Samples were thus divided into 2 groups; those influenced by marine air and those influenced by the continent. The results showed that the particulate matter influenced by the continent had significant quantities of "aged sea salt", in contrast to those influenced mainly by marine air. Aged sea salt is evidence of chemical reactions of sodium and magnesium chloride with nitric and sulphuric acid, for instance, forming the nitrate and sulphate salts. With TW-EPXMA it is possible to detect and quantify nitrogen and oxygen, and sodium nitrate could be discerned from sodium-nitrogen-oxygen-rich clusters. This is illustrated in Fig. 4, where it is evident that two particles appearing the same under the electron beam proved to be quite different in composition.

In addition, a distinction between organic sulphur particles and ammonium sulphate, previously all characterised as sulphur-rich, could be made. It was then also possible to identify organic and biogenic (higher relative phosphorus or potassium content) particles, of which the highest relative abundances were found in the continentally influenced samples (indicative of anthropogenic sources such as traffic, industrial processes, biomass burning, *etc.*).

The usefulness of these latest developments in analyses and data handling/processing, is further illustrated in a paper by Osán *et al.*[47] in which an investigation was launched into a mine pollution event occurring in the Tisza river in Hungary as a result of a mine dumping incident in neighbouring Romania. Sediment samples taken at six different river sites in Hungary indicated serious heavy metal contamination and pollution of the Tisza river. During this work, the abundance of particle classes obtained from the bulk elemental concentrations measured by XRF and those from the single particle EPXMA analyses were combined and PCA applied to it. The polluted samples containing Cu, Zn and Pb from the pyrite and zinc sulphide particles were clearly identifiable as a distinct group. The additional X-ray Diffraction (XRD) analyses that were performed on the same samples, clearly pointed to the need for structural analysis in addition to chemical compositional profiles and can therefore be considered a prediction of the next stage of development and application in the field of SPA.

In another study underlining the necessity of recording both chemical composition of the surrounding atmosphere as well as the observed decay patterns and occurrences as a result thereof, Torfs and Van Grieken[48] described an investigation into the chemical relationships between atmospheric aerosols and stone decay layers on historic buildings at the Mediterranean coast at sites in Greece, Malta, Italy and Spain. The study indicated a clear relationship between the chemical composition of the weathered stone layers and the chemical composition of the surrounding atmosphere, which is further exacerbated by continuous wet-dry cycles. Severe decay was observed in parts of churches and inside cathedrals at Cadiz (Spain) and Bari (Italy), as well as in a church in Malta. At this stage it would have become pretty obvious to the experienced observer and worker in this field that the next phase in the saga on SPA would involve some form and/or technique capable of also providing structural/molecular composition information. Thus providing a better understanding of the composition and potential effects of single particles/atmospheric aerosols on the environment, cultural heritage and human health.

Third era: the addition of complementary techniques

In the early 21st century the information obtained from CCSEM/EDX or EPXMA of individual particles were often not sufficient to elucidate their impact on the environment and human health. As a consequence various other micro-analytical techniques are nowadays combined (in a parallel, sequential or interfaced manner) with them to obtain structural, chemical, phase and morphological information on an individual particle basis. Some developmental work on substrate selection, particle recognition, methodology and spectral analysis for the characterisation of single particles of environmental origin by parallel or sequential/interfaced SEM/EDX (or EPXMA) and MRS has been published, details of which can for example be found in Potgieter-Vermaak *et al.*,[29] Worobiec *et al.*,[49,50] Batonneau *et al.*[32] The additional advantage of using MRS is that lower concentrations could be probed, where EPXMA despite advances over the decades have a detection limit in the order of 0.1% due to high Bremsstrahlung.

This approach found application in various areas, one of which is cultural heritage and conservation. Works of art are under constant threat of degradation due to the deposition of atmospheric particles. Particulate matter can induce and intensify surface damage, particularly because of its potential to serve as a centre for moisture condensation and adsorption of gaseous pollutants. SPA provides essential information to investigate the potential harm that these particles can cause. By combining EPXMA with MRS, additional information on the chemical structure can be obtained, and this could therefore aid in preventive conservation. The challenge is of course to be able to analyse the same particle by both techniques. During an investigation in the Rubens House museum in Antwerp, a coordinate system was used (Transmission Electron Microscopy (TEM) grid), upon which individual particles were transferred by means of nano-manipulation, elegantly described in Godoi *et al.*[5] To illustrate the power of such a combined approach, the analysis of a particle of interest (one collected inside a showcase containing the famous artist's chair, illustrated in Fig. 5) will be discussed. The elemental composition of this particle showed weight percentage concentrations of 34% carbon, 28% oxygen, 15% aluminium, 17% calcium and 7% chlorine. Classification of this particle, based only on the elemental concentrations, would be ambiguous. The Raman spectra indicated the presence of a mono-substituted benzene ring and an aluminium hydroxyl vibration at the lower wavenumbers, suggesting that it was a particle with mixed composition. Penetration depth of the laser beam depends of course on the nature of the particle, but it could be assumed that the whole particle was analysed as its particle size was in the 0.5–1 μm range and penetration depth probably of the same order. Using variable keV (5–15) the particle could be analysed at variable penetrations depths and the results confirmed the heterogeneous character of the particle. It was made-up of an aluminium–oxygen–rich core (500 nm), surrounded by a

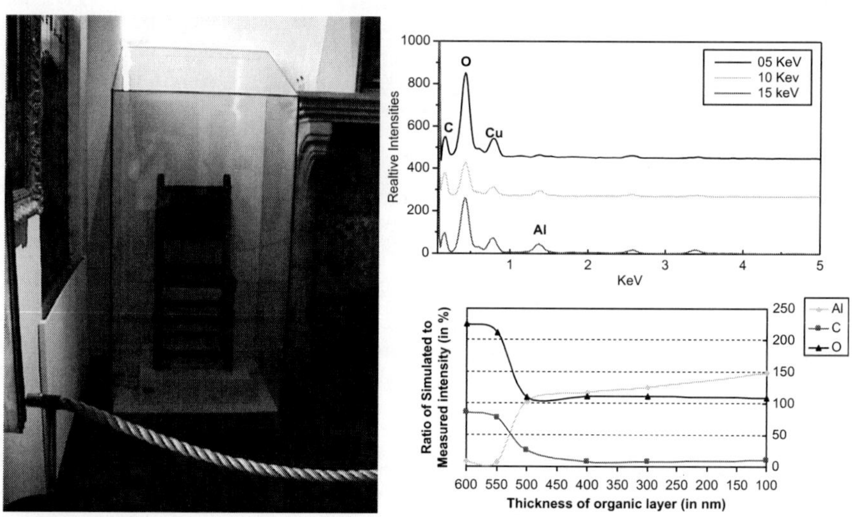

Fig. 5 Showcase housing Rubens' chair, with EDX and simulated data to determine the layered composition of atmospheric particles.

carbon-oxygen-rich outer layer, as illustrated in Fig. 5. The showcase was also sampled and analysed for its gaseous air pollutants and excessively high toluene was reported. It is therefore quite probable that an inorganic core could have outer layers of adsorbed organic components, probably emitted from the wood of the chair or that of the showcase, indicating deterioration.

In a more recent study,[51] size segregated airborne particulate matter (collected *via* a May impactor, carefully regulating time to ensure that particles collected in the various aerodynamic ranges could be analysed by CC-EPXMA, *i.e.* were well-separated) at the Alhambra's Ambassador Hall was studied with EPXMA and MRS performed on the same sample. In this case, MRS data was obtained on 50–100 particles per sample stage to ensure representative results. MRS data elucidated and confirmed the structural composition of many of the particle classes discerned from EPXMA data, using manual classification. Examples are the dominance of $NaNO_3$ in the 4–8 μm fraction, the presence of dolomitic particles and Ca-Mg-carbonates in the presence of sulphates, and mixed ammonium salts in the fine fraction ($d_a = 0.5$ to 1 μm). The MRS data also indicated that the particle types suggested by the EPXMA results were not pure and existed rather as mixed particles, for example mixed iron oxide-sea salt particles. These intertwined mineral and sea salt particles strongly suggest the presence of iberulites, *i.e.* microspherulites which are exclusively found in the southern Iberian Peninsula. These vortex-like particles (60–90 μm) consist of minerals and sea salts, formed during a Saharan dust outbreak. When collected on an impactor plate, iberulites splash into numerous smaller particles, which were recognised by MRS as mixed particles. A good example of this complementary approach is illustrated in Fig. 6. Fig. 6(a) gives typical abundance plots for the discerned particle classes, obtained from EPXMA data, where one can observe a pure carbonaceous, a sea salt, a secondary

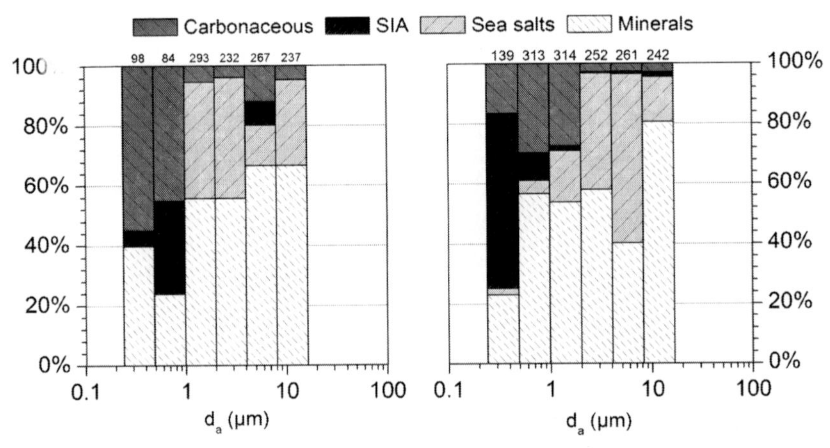

Fig. 6a Abundance plots constructed for the 4 classes of particles identified by EPXMA analysis for summer (left) and winter (right) airborne particulate matter collected at the Alhambra museum. Aerodynamic diameter (d_a) size fractions between 0.25 and 16 μm correspond to the stages of the May impactor. For each size fraction in the abundance plot, the total number of analysed particles (N) is indicated on top of the corresponding column. SIA indicates secondary inorganic aerosol particles.

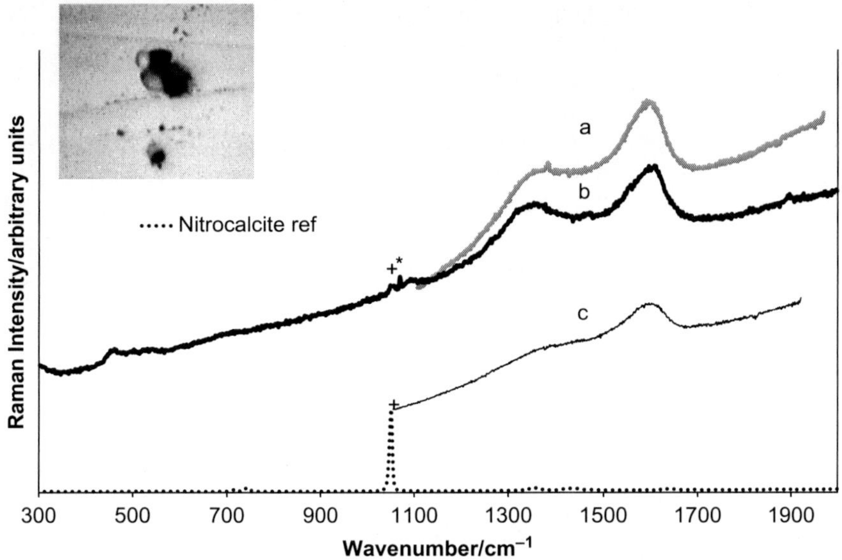

Fig. 6b Raman spectra of amorphous carbon [soot + ? (a); soot + sodium nitrate + nitrocalcite (b); soot (c)] identified on the stage 5 sample during summer. (* sodium nitrate band (1067 cm^{-1}) + nitrocalcite band (1050 cm^{-1})).

inorganic aerosol (typically ammonium salts formed due to reactions between gases and aerosols) and a mineral class. The carbonaceous fraction (pure according to EPXMA) was shown to have various inorganic salts adsorbed onto it (illustrated in Fig. 6(b)), which is of particular concern in terms of degradation potential.

Another area where this approach proved to be extremely suitable is for investigations involving climatology changes. A recent paper [52](and refs. therein), utilised a suite of micro-analytical techniques to investigate the mixing state of African dust, which is important in radiative budget calculations. Individual particles were identified based on their elemental compositions obtained by CCSEM/EDX with low-Z analysis due to an ultra thin window and collection of airborne particulate matter on boron substrates. Use of transmission electron microscopy (TEM) and electron energy loss spectroscopy (EELS) enabled the collection of information on carbon structural variations. The G (graphitic) and D (disordered) bands of amorphous carbons were determined by Raman spectroscopy. Deconvolution of these bands revealed a close resemblance of the carbonaceous airborne particulate matter with those of reference soot samples. The combined results enabled the investigators to conclude that volatile and semi-volatile organic compounds probably adsorb onto dust particles and this will influence the hygroscopic behaviour and thus the radiating properties of the airborne particulate matter.

An area of increasing concern that greatly benefit from SPA following a multiple technique protocol is the toxicological fate of environmental particles (especially those of thoracic size). A group at the University of Sciences and Technology of Lille in France[15,32,33,53–56] used Raman mapping, mostly in conjunction with SEM/EDX and environmental SEM

(ESEM/EDX) to characterise size-segregated atmospheric particles. Various sophisticated multivariate analysis techniques were used to extract intensities of pure species from a matrix of up to 2400 spectra. Only in the papers by Choël *et al.*[55] and Uzu *et al.*[56] was use made of individual particle analysis and/or CCSEM/EDX. This group has sampled dust emissions at a battery recycling plant during three periods, as well as particulate matter at some distance from the smelter during two periods in the late 1990's and early 2000's. In all the articles the collected industrial dust was characterised by using this approach, sometimes in conjunction with other techniques, but ultimately with the goal to determine the elemental associations of the heavy metals, which is important to assess their health impact on the industrial work force. Uzu *et al.*[57] of the same group, recently published results on SPA and bulk analysis of airborne particulate matter from the same lead smelter. Special attention was paid to the various lead phases (characterised by XRD, Extended X-ray Absorption Fine Structure (EXAFS), MRS, ESEM), of which only the ESEM and MRS results were on single particles. After some analogue phytoavailability tests to determine metal exchangeability and calculated doses deposited in the lung (using particulate matter mass concentrations and the inhalable convention) the potential risk to human health could be estimated. A similar approach has been used to characterise the toxic and carcinogenic components in the respirable inhalable fraction of road dust collected from a heavy-trafficked road in central Manchester, reported by Potgieter-Vermaak *et al.*[58] (accepted subject to revision in Environmental Geochemistry and Health). SPA was performed on size-segregated particles using CC-EPXMA and MRS (point analysis) and bulk analysis was performed with EDXRFS and Inductively Coupled Plasma Mass Spectrometry (ICP-MS). The results indicated that the finest fraction was enriched in Cr and Pb and that the Cr was often present in Cr(VI) oxidation state, mostly as lead chromate. As the structural information showed that the inhalable fraction contains toxic and carcinogenic components that could potentially be mobile in body fluids when ingested or inhaled, these fractions were leached with artificial body fluid to determine bioaccessibility. The results indicated a time-dependent increase in concentration of both Cr and Pb.

This two-tier approach was also used extensively by the group of Chul-Un Ro[34–38] with the difference that MRS was replaced by Attenuated Total Reflection Fourier Transform-Infrared Imaging Technique (ATR-FT-IR-IT), and is therefore treated separately in this chapter. Although imaging instruments improved to a spatial resolution of 3.1 μm at 1726 cm^{-1}, it still is not adequate for environmental SPA. However, in conjunction with EPXMA this proved to be feasible for the analysis of atmospheric and other environmental particles in the micron size range. Ryu and Ro[34] thus investigated artificially generated atmospheric particles to illustrate the power of this approach and applied it to some exemplar indoor atmospheric particles to demonstrate its practical feasibility. Under the leadership of Chul-Un Ro various publications have since seen the light, ranging in the analysis of the externally heterogeneous mineral particles, mineral dust aerosols, soil and dust originating from soil, indoor aerosols and Arctic aerosols. Mineralogical, chemical and structural characterisations on an individual particle basis, as well as consequent source apportionment, were the main achievements in these papers.

Finally, a discussion of the contribution of interfaced techniques would be beneficial to the reader. One of the first published results of individual particle analysis, using SEM/EDX interfaced with MRS (using the Scanning Electron Microscopy Structural Chemical Analyser – SEM SCA, patented by Renishaw, illustrated in Fig. 1), was in the article by Potgieter-Vermaak[16] on heavy mineral sands. Samples of the non-magnetic fraction of a heavy mineral concentration plant were analysed. The samples originated specifically from the non-conducting fraction (Zr-rich) and were taken at various stages in its clean-up process. A sample that was supposed to be Ti-rich and practically free of Zr, was analysed by MRS and SEMSCA. MRS analyses showed that the Ti-rich part comprised of anatase and rutile. These particles were covered in places with a red substance that appeared amorphous, potentially indicating secondary alteration, and proved to be anatase in some instances, but it also fluoresced to such an extent in another case that it could not be identified. Performing SEM/EDX and MRS on the same particle, indicated that, apart from the anatase present, this red layer was Al rich (illustrated in Fig. 7). This maybe the reason for inefficient magnetic separation of these concentrates in the mineral industry. Work on airborne particulate matter up to this stage has been limited to artificially generated particles, such as iron-rich welding dust and ammonium and sodium salts.[49,50]

An important aspect that was not yet reported on is the Raman behaviour of electron beam sensitive particles during the analyses with interfaced SEM/EDX/MRS. In unpublished preliminary investigations a number of phenomena were observed during the analyses of semi-volatile environmental particles such as ammonium sulphate, ammonium nitrate and sodium nitrate. These particle types were chosen because of their importance in the study of environmental pollution, their sources, processes and transportation. An example of an ammonium sulphate particle analysed by

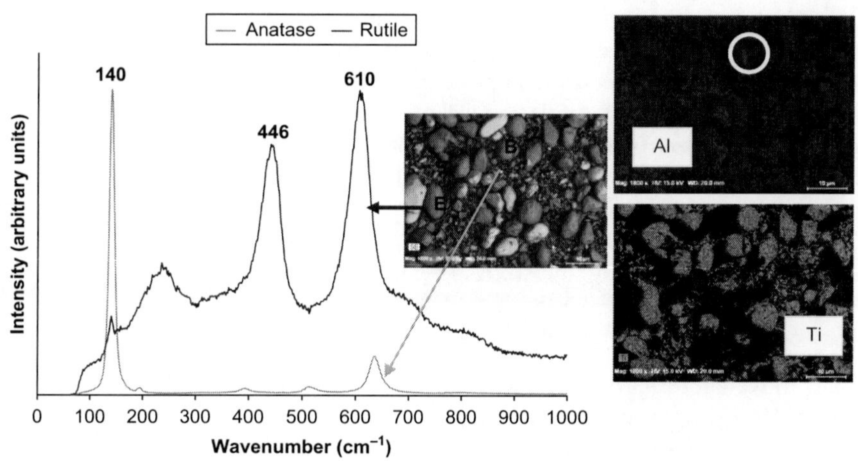

Fig. 7 Corresponding Raman spectra of the titania polymorphs found in the sample, and the aluminium association.

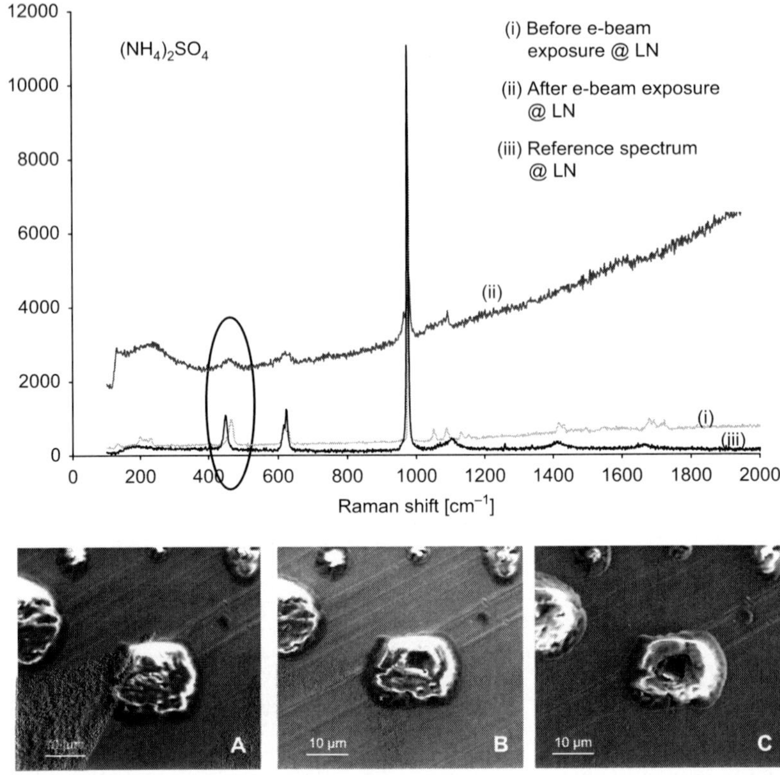

Fig. 8 The SE image of an ammonium sulphate particle A indicates that the beam sensitive particle stays stable at liquid-Nitrogen (LN) temperatures, with some degradation in B after electron beam exposure and further degradation in C after the Raman spectrum was taken. A good spectrum was obtained (i) for the particle prior to electron beam exposure, but the second spectrum (ii) showed band broadening, high background and poor spectral quality.

Field Emission Gun with Energy Dispersive X-ray detection (FEG/EDX) interfaced with MRS is illustrated in Fig. 8, where it is observed that the particle has undergone changes after exposure to the electron beam. In addition, Raman spectra obtained after exposure to the electron beam showed band broadening, higher background and lower spectral quality. To this end, the importance of a robust protocol to be used for the analysis of environmental particles (especially of the finer fraction) with an interfaced two tier approach has been illustrated. It also seems from these preliminary observations that the protocol may have to be particle type specific. In addition, particle behaviour when exposed to the interfaced instrument may not mirror its behaviour when analysed under stand-alone conditions.

Expectations and challenges for the future

The above sections do not intend to give a complete literature overview about SPA for environmental particles or not even about electron induced X-ray techniques in this field. Rather we have emphasized mostly our own contributions (because we know these best) and those of the research groups we had most contact with. There are of course other teams that are active in this field and produce excellent and innovative results, like *e.g.* the ones around Alexander Laskin in Richland, WA, USA, Peter Buseck in Tempe, AZ, USA and Stephan Weinbruch at the Technical University of Darmstadt in Germany.

A caveat:- after more than two decades of SPA by EPXMA and SEM, this should still not be considered as a routine technique for analysing environmental particles. To get results for a representative number of particles, automating the searching and sizing and analysing of the particles on the holder is desirable, but many artefacts might occur in computer controlled techniques, *e.g.* with geometrically overlapping particles. The available commercial software for the handling of the huge amounts of data is often inadequate, and the in-house versions that several labs developed have usually not been commercialised and are not very user-friendly. The interpretation of data implies some prior knowledge to correctly predict particle classes, *e.g.* for aerosol data which particles are likely to be present or relevant, and which properties are of environmental importance.

In addition to the numerous elemental characterisations of ambient particles, we have recently seen more applications in fundamental physico-chemistry. This includes laboratory studies on gas-particle reactions, *e.g.* sea salt[59] or soil[60] dust with nitric acid. Environmental SEM (ESEM) or environmental TEM (ETEM) does offer excellent future prospects in this field. In ESEM and ETEM, it is possible to obtain high quality images and to do some chemical analysis while the gaseous environment around the sample is controlled (*i.e.* vacuum in the neighbourhood of the sample is not necessary, the relative humidity can be varied and a temperature around $-30\,^{\circ}C$ can be maintained). Ice nucleation properties of individual atmospheric particles have also been studied recently. Both ESEM and ETEM are relatively new techniques, of which the potential has not fully been explored.

In the past, a combination of several analysis techniques has often been applied to environmental aerosols because one technique can normally not give all requested answers. *E.g.* bulk elemental analysis by XRF or particle-induced X-ray emission (PIXE) or inductively-coupled mass or emission spectrometry (ICP-MS/ICP-OES) is often combined with ion chromatography (IC) to assess the soluble ions, gas chromatography mass spectrometry (GC-MS) for some organics, aethalometry for ambient soot determinations and gravimetry for aerosol mass per m^3.

There is also a trend now to combine several micro-analysis techniques as well, either as stand alone or interfaced instruments. In the previous section, much was said about the combination of SEM/EDX with MRS, while a few other combinations were mentioned. A first problem in this and other combined or interfaced micro-techniques is that it is not always easy to ascertain that the same single particles are being assessed by both measurements. A second problem is in general that the particle volume that is probed can vary drastically. An extreme example is in the novel JEOL JAMP-9500F microprobe, which combines in one interfaced instrument: advanced field-emission EPXMA, micro-XRF (by irradiation of a secondary target) and Auger electron spectrometry (AES). The high resolution SEM enables one to obtain high quality images of both submicrometer and nano-scale size with their fine detail revealed, but the micro-XRF technique cannot be reasonably applied to single aerosol particle analysis because its spatial resolution does not match the size of relevant atmospheric particles, and the Auger electron escape depth does not exceed 2 nm, while the sharply focused electron beam has a lateral diameter of 2–3 nm, thanks to the superbly performing electron-optical column. The local AES technique can be successfully applied for the composition test of particles of any size and for obtaining the element distributions along a line (line profile), in an area (Auger mapping) and in depth with the help of ion etching (depth profile). In preliminary analyses of urban samples,[61] AES indicated *e.g.* a strong enrichment of N in the surface (nanometers!) layers, undoubtedly the result of interactions with gaseous NOx in the air. This could not be seen by the other techniques which do not have such extreme surface sensitivities, but quantification of the results is not trivial.

Recently, several articles have appeared on the combined use of different other microanalysis techniques for atmospheric aerosols. Let us mention one. Hopkins *et al.*[62] successfully used CC SEM/EDX, time-of-flight secondary ion mass spectrometry (TOF-SIMS) and scanning transmission X-ray microscopy/near-edge X-ray absorption fine structure analysis (STXM/NEXAFS) to study quantitatively methanesulphonate ($CH_3SO_3^-$) and non-sea salt sulphate in individual marine aerosol particles. This indicated *e.g.* that $CH_3SO_3^-$ salts were the dominant source of non-seasalt sulphur in large particles, while sulphate was more common in smaller particles. Quantitative assessment of these two forms is important for kinetic modelling concerning the pathways of natural dimethylsulphide oxidation and the impact on the number and size of cloud condensation nuclei in the marine environment.

In the future, more emphasis will necessarily be on advanced characterisation of more complex molecules in air particulate matter. The further development of these methods and their applications will be challenges for future research.

References

1 C.-U. Ro, J. Osán and R. Van Grieken, *Anal. Chem.*, 1999, **71**, 1521–1528.
2 I. Szalóki, J. Osán, C.-U. Ro and R. Van Grieken, *Spectrochim. Acta B*, 2000, **55**, 1017–1030.
3 I. Szalóki, J. Osán, A. Worobiec, J. de Hoog and R. Van Grieken, *X-ray Spectrom.*, 2001, **30**, 143–155.
4 A. Worobiec, J. de Hoog, J. Osán, I. Szalóki and R. Van Grieken, *Spectrochim. Acta B*, 2003, **58**, 479–495.
5 R. H. M. Godoi, S. Potgieter-Vermaak, J. De Hoog, R. Kaegi and R. Van Grieken, *Spectrochim. Acta B*, 2006, **61**, 375–388.
6 F. Delalieux, R. Van Grieken and J. H. Potgieter, *Mar. Pollut. Bul.*, 2006, **52**, 606–611.
7 D. Avigo Jr., A. F. L. Godoi, P. R. Janissek, Y. Makarovska, A. Krata, S. Potgieter-Vermaak, B. Alföldy, R. Van Grieken and R. H. M. Godoi, *Anal. Bioanal. Chem.*, 2008, **391**, 1459–1468.
8 M. Murakami, F. Nakajima, H. Furumai, B. Tomiyasu and M. Owri, *Chemosphere*, 2007, **67**, 2000–2010.

9 J. Jehlička, H. G. M. Edwards, S. E. J. Villar and J. Pokorny, *Spectrochim. Acta A*, 2005, **61**, 2390–2398.
10 C. D. Moody, S. E. Jorge Villar, H. G. M. Edwards, D. A. Hodgsonc, P. T. Doran and J. L. Bishop, *Spectrochim. Acta A*, 2005, **61**, 2413–2417.
11 E. A. Stefaniak, A Worobiec, S. Potgieter-Vermaak, A. Alsecz, S. Török and R. Van Grieken, *Spectrochim. Acta B*, 2006, **61**, 824–830.
12 G. Schweiger, *J. Aerosol Sci.*, 1999, **21**, 483–509.
13 A. K. Y. Lee and C. K. Chan, *Atmos. Environ.*, 2007, **41**, 4611–4621.
14 A. Sengupta, M. L. Laucks, N. E. Dildine, Drapal and E. J. Davis, *Aerosol Sci.*, 2005, **36**, 651–664.
15 S. Sobanska, G. Falgayrac, J. Laureyns and C. Brémard, *Spectrochim. Acta A*, 2006, **64**, 1102–1109.
16 S. Potgieter-Vermaak, In: The Proceedings of: *The 6th International Heavy Minerals Conference 'Back to Basics'*, The Southern African Institute of Mining and Metallurgy, 2007.
17 G. A. Hope, R. Woods and C. G. Munce, *Min. Eng.*, 2001, **14**, 1565–1577.
18 A. Guedes, N. Ribeiro, H. Ribeiro, M. Oliveira, F. Noronha and I. Abreu, *Aerosol Sci.*, 2009, **40**, 81–86.
19 J. R. Baena and B. Lendl, *Curr. Opin. Chem. Biol.*, 2004, **8**, 534–539.
20 K. H. Fung and I. N. Tang, *J. Colloid Inter. Sci.*, 1989, **130**, 219–224.
21 K. H. Fung and I. N. Tang, *J. Aerosol Sci.*, 1992, **23**, 309–307.
22 I. N. Tang and K. H. Fung, *J. Aerosol Sci.*, 1989, **20**, 609–617.
23 C. L. Aardahl, W. R. Foss and E.J. Davis, *J. Aerosol Sci.*, 1996, **27**, 1015–1033.
24 M. Trunk, J. Popp and W. Kiefer, *Chem Phys. Lett.*, 1998, **284**, 377–381.
25 J. Musick and J. Popp, *Phys. Chem. Chem. Phys.*, 1999, **1**, 5497–5502.
26 H. B. Wu and C. K. Chan, *Atmos. Environ.*, 2008, **42**, 313–322.
27 R. Escribano, J. J. Sloan, N. Sidique, S.-K. Sze and T. Dudev, *Vib. Spectros.*, 2001, **26**, 179–186.
28 S. S. Potgieter-Vermaak and R. Van Grieken, *Appl. Spectroc.*, 2006, **60**, 39–47.
29 S. Potgieter-Vermaak, A. Worobiec, L. Darchuk and R. Van Grieken, *Fundamentals and Applications in Aerosol Spectroscopy*, Taylor and Francis Ed Ruth Signorell, Jonathan P Reid, 2010, ISBN: 9781420085617.
30 A. Worobiec, E. A. Stefaniak, V. Kontozova, L. Samek, P. Karaszkiewicz, K. Van Meel and R. Van Grieken, *e-Preservation Science*, 2006, 63–68.
31 A. Worobiec, E. A. Stefaniak, S. Kiro, M. Oprya, A. Bekshaev, Z. Spolnik, S. S. Potgieter-Vermaak, A. Ennan and R. Van Grieken, *X-Ray Spectrom.*, 2007, **36**, 328–335.
32 Y. Batonneau, J. Laureyns, J.-C. Merlin and C. Brémard, *Anal. Chim. Acta*, 2001, **446**, 23–37.
33 Y. Batonneau, S. Sobanska, J. Lauryens and C. Brémard, *Environ. Sci. Technol.*, 2006, **40**, 1300–1306.
34 J. Ryu and C.-U. Ro, *Anal. Chem.*, 2009, **81**, 6695–6707.
35 Y.-C. Song, J. Ryu, M. A. Malek, H.-J. Jung and C.-U. Ro, *Anal. Chem.*, 2010, **82**, 7987–7998.
36 M. A. Malek, B. W. Kim, H.-J. Jung, Y.-C. Song and C.-U. Ro, *Anal. Chem.*, 2011, **83**, 7970–7977.
37 S. Maskey, T. H. Kang, H.-J. Jung and C.-U. Ro, *Indoor Air*, 2011, **21**, 12–24.
38 S. Maskey, H. Geng, Y.-C. Song, H. Hwang, Y.-J. Yoon, K.-H. Ahn and C.-U. Ro, *Environ. Sci. Technol.*, 2011, **45**, 6275–6282.
39 W. Jambers, L. De Bock and R. Van Grieken, *Analyst*, 1995, **120**, 681–692.
40 L. A. De Bock, P. E. Joos, K. J. Noone, R. A. Pockalny and R. E. Van Grieken, *J. Atmos. Environ.*, 2000, **37**, 299–329.

41 L. De Bock, B. Treiger, L. Van der Auwera and R. Van Grieken, *Mikrochim. Acta*, 1998, **128**, 191–200.

42 D. Węgrzynek, W. Jambers and R. Van Grieken, *Marine Chem.*, 1997, **57**, 41–53.

43 R.M. de Miranda, M. Andrade, A. Worobiec and R. Van Grieken, *Atmos. Environ.*, 2002, **36**, 345–352.

44 J. Osán, I. Szalóki, C.-U. Ro and R. Van Grieken, *Mikrochim. Acta*, 2000, **132**, 349–355.

45 J. Osán, B. Alföldy, S. Kurunczi, S. Török, L. Bozó, A. Worobiec, J. Injuk and R. Van Grieken, Időjárás *(Quartely Journal of the Hungarian Meteorological Service)*, 2001, **105**, 146–156.

46 J. De Hoog, J. Osán, I. Szalóki, K. Eyckmans, A. Worobiec, C.-U. Ro and R. Van Grieken, *Atmos. Environ.*, 2005, **39**, 3231–3242.

47 J. Osán, S. Kurunczi, S. Török and R. Van Grieken, *Spectrochim. Acta B*, 2002, **57**, 413–422.

48 K. Torfs and R. Van Grieken, *Atmos. Environ.*, 1997, **31**, 2179–2192.

49 A. Worobiec, S. Potgieter-Vermaak, A. Brooker, L. Darchuk, E. Stefaniak and R. Van Grieken, *Microchem. J.*, 2009, doi:10.1016/j.microc.2009.09.003.

50 A. Worobiec, L. Darchuk, A. Brooker, J. H. Potgieter and R. Van Grieken, *J. Raman Spectrosc.*, 2011, **42**, 808–814.

51 S. Potgieter-Vermaak, B. Horemans, W. Anaf, V. Kontozova, C. Cardell and R. Van Grieken, In: Book of Abstracts, RAA2011 (6th International Conference on the application of Raman spectroscopy on Art and Archaeology, Parma, Italy), September 2011.

52 K. Deboudt, P. Flament, M. Choël, A. Gloter, S. Sobanska and C. Colliex, *J. Geophys. Res.*, 2010, **115**, D24207.

53 N. Y. Dupuy and Y. Batonneau, *Anal. Chim. Acta*, 2003, **495**, 205–215.

54 G. Falgayrac, S. Sobanska, J. Laureyns and C. Brémard, *Spectrochim. Acta A*, 2006, **64**, 1095–1101.

55 M. Choël, K. Deboudt, P. Flament, G. Lecornet, E. Perdrix and S. Sobanska, *Atmos. Environ.*, 2006, **40**, 4439–4449.

56 G. Uzu, S. Sobanska, Y. Aliouane, P. Pradere and C. Dumat, *Environ. Pollut.*, 2009, **157**, 1178–1185.

57 G. Uzu, S. Sobanska, G. Sarret, J. J. Sauvain, P. Pradere and C. Dumat, *J. Haz. Mat.*, 2011, **186**, 1018–1027.

58 S. Potgieter-Vermaak, G. Rotondo, V. Novakovic and K. G. Taylor, In: Book of Abstracts SEGH2011 (International Conference on Environment & Health incorporating the SEGH 28th European Conference, Ormskirk, UK), April 2011.

59 Y. Liu, J. P. Cain, H. Wang and A. Laskin, *J. Phys. Chem.*, 2007, **A111**, 10026–10043.

60 A. Laskin, T. W. Wietsma, B. J. Krueger and V. H. Grassian, *J. Geophys. Res.-Atmospheres*, 2003, **111**(D10), D10208, 430–434.

61 R. Van Grieken, S. Potgieter-Vermaak, L. Darchuk and A. Worobiec, *XRF Newsletter*, December 2009, N° **18**, 9–13.

62 R. J. Hopkins, Y. Desyaterik, A. V. Tivanski, C. M. Berkowitz, T. Tyliszczak, M. K. Gilles and A. Laskin, *J. Geophys. Res.-Atmospheres*, 2008, **113**(D4), D04209, 15.

Materials sciences using synchrotron infrared light sources

Michael C. Martin*[a] and Paul Dumas[b]
DOI: 10.1039/9781849734899-00141

Synchrotron Infrared spectroscopy has witnessed several important applications in Materials Science over the recent years. This chapter is aimed at highlighting the most recent studies that could inspire new studies from readers. Soft matter (in particular polymer science), catalysis and microscopic ellipsometry have achieved important steps forward in their applications recently, while well-established studies in semiconductors and high pressure studies have generated important results and findings. The field is evolving quickly towards new directions, mainly in the production of intense THz beams that are opening new research directions, in time resolved studies, in fast imaging and in near field infrared microscopy. The recent advances are reported in this chapter.

1 Introduction

The goal of this chapter is to provide the reader an overview of recent developments across various scientific disciplines, which were achieved by using synchrotron light sources around the world with their high-brightness infrared beams. Synchrotron-based micro-spectroscopy and spectroscopy in the mid- and far-IR have had a number of important applications over recent years several of which we will highlight. While users working on biological and biomedical applications often have the largest fraction of beamtime at most synchrotron infrared beamlines [1], materials science research continues to be a strong player pushing technical developments and generating exciting and high-profile results.

A synchrotron light source is most typically associated with, and is primarily built for delivering, hard and soft x-rays for a wide array of science. Their broad bending magnet spectrum extends from the x-ray all the way through the UV, visible and infrared, and their unique properties also make synchrotrons an exciting source in the infrared. The most commonly exploited property of these lights beams is their high brightness – how much IR light can be focussed to a small spot. For mid- and far-IR wavelengths the size of the beam can be focussed down to the diffraction-limit, yielding a spatial resolution that is only limited by the wavelength of the light, and providing approximately 1000 times more light per unit area than a conventional thermal FTIR source. Therefore many scientific applications take advantage of this brightness to make high-resolution spectral microscopy much faster or even possible at all. Other properties that are exploited by some researchers are the fast-pulsed nature of the synchrotron source, and greater overall flux at longer wavelengths, with a

[a] *Advanced Light Source Division, Lawrence Berkeley National Laboratory, 1 Cyclotron Road, B6R2100, Berkeley, CA 94720; E-mail: MCMartin@lbl.gov*
[b] *Synchrotron SOLEIL, L'Orme des Merisiers, BP48 - Saint Aubin, 91192 Gif-sur-Yvette Cedex, France; E-mail: dumas@synchrotron-soleil.fr*

special operating mode, allowing extremely high flux at very long wavelengths.

After introducing the synchrotron as an infrared light source in greater detail, this review will discuss some interesting ways in which these properties of infrared synchrotron radiation have been used for studying samples important to solid state physics, materials science, materials under extreme conditions, catalysis, and soft matter such as polymer science. We will also provide citations for readers interested in reviews that provide greater information and details about uses of synchrotron radiation for biological, environmental, and other sciences. Finally we will mention a few future directions that are being explored, and how they may provide even more opportunities for the broad field of materials science. For more details and to discuss how synchrotron infrared techniques may play a role in your research, we encourage you to contact an infrared beamline scientist at a synchrotron light source near you [2].

2 Synchrotron infrared sources

Infrared (IR) synchrotron radiation has become much more popular over the last decade. Almost all synchrotron facilities worldwide have an operational, under construction or planned IR beamline (Table 1 in [2] and [3]). Synchrotron infrared radiation is generated by electrons travelling at relativistic velocities, that are accelerated by a constant magnetic field causing the electrons to travel along a curved path {*i.e.* bending magnet (BM) radiation [4]} or when the electrons' trajectories encounter a change in magnetic field, *e.g.*, at the edges of bending magnets {called edge or transition radiation (ER) [5, 6]}. In short, as magnetic fields are used to bend the electrons' trajectory into a closed orbit, synchrotron radiation (SR) is produced at each of these bending magnets. The emitted radiation spans an extremely broad spectral domain, extending from the X-ray regime to the very Far Infrared (FIR) region. The SR from the constant magnetic field provides circularly polarized light, left-handed and right-handed for the radiation extracted above and below the storage ring plane, respectively. The degree of polarization depends on the details of the radiation-extraction scheme. By contrast, the edge radiation is linearly polarized radially around the direction of propagation. The flux and brightness for these two types of IR synchrotron emission are almost equivalent, but the opening angle of the ER is narrower than that of a BM [7, 8].

The crucial parameter for why synchrotrons make appealing infrared sources is the effective size of the electron source for the light. For newer storage rings, it is of the order of 100 nm or less, well below the size of the wavelengths of infrared (and even visible) light, so the light is emitted into a narrow range of angles and the resulting brightness is increased by 2–3 orders of magnitude, above the brightness of a thermal infrared source, which is typically mm's in size [8]. In the X-ray, the SR source offers many orders of magnitude more flux than a coventional laboratory source. In the MIR energy range, however, this advantage has diminished to only 1–2 orders of magnitude compared to a thermal source, but the brightness (flux per unit area) advantage remains. At longer far-IR wavelengths the synchrotron can again beat laboratory sources by several orders of magnitude in both flux and brightness.

Because of the very small effective source size, the spatial resolution of SR-FTIR for micro-spectroscopy is usually only limited by the wavelength, λ, of the IR light [9]. Schwarzschild microscope objectives are used to focus the broad spectrum of IR light onto the sample, and these objectives typically have relatively large numerical apertures in the range 0.5–0.6. While diffraction will naturally limit the spot size on the sample [10], one can also use a single aperture before the sample to exactly define the illuminated sample region. In such a situation, the diffraction-limited spatial resolution is approximately $2\lambda/3$ [11], which corresponds to 1.7 μm (at 4000 cm^{-1}) and 13 μm (at 500 cm^{-1}) the two extremes of typical mid-IR measurements.

Other microscopes operate in the so-called confocal arrangement, where a second aperture is also placed after the sample so as to limit light collection to the region of interest of the sample. In this geometry, the two apertures are placed in conjugates planes of the two IR objectives. This method, discussed by G.L. Carr [8, 11] improves spatial resolution to $\lambda/2$. In addition, the confocal arrangement also reduces the Schwarzschild's first-order and higher-order diffraction rings, resulting in improved image contrast [8]. In reflection geometry only half of the objective is typically used for illumination and the other half for collection. This worsens the spatial resolution in one direction, but can be overcome and full resolution recovered by using the full objective for illumination and collection by replacing the half covered mirror with a beamsplitter [12]. More details about typical instrument and beamline characteristics are available [7, 8]. Lastly,another exciting development in sources for longer wavelengths is exploiting coherent synchrotron radiation (CSR) for extremely high power THz sources. CSR is produced when the electron bunch length is physically shorter than the wavelengths of light being emitted. When this happens, the electric fields of the emitted photons coherently add which means that instead of the total intensity of the synchrotron radiation being proportional to the total number of electrons in the bunch, the intensity of the CSR is proportional to the total number of electrons squared. As the number of electrons in a bunch is on the order of 10^{10}, having an additional factor of 10^{10} in intensity is truly exciting. The high power terahertz radiation of CSR was demonstrated in 2002 [13], and the first scientific experiments using CSR from a synchrotron source to measure the Josephson plasma resonance in the high temperature superconductor $Bi_2Sr_2CaCu_2O_8$ was published in 2004 [14]. We will briefly touch on how this novel mode of operation is still being used for previously difficult problems in the THz frequency domain.

3 Synchrotron IR opportunities in polymer science

Vibrational spectroscopy is a well-accepted and fundamental tool for the characterization of polymeric materials. Indeed, infrared spectroscopy is probably the most widely used analytical characterization technique in this research field today. It can provide a rich source of information on diverse aspects of polymeric materials, from their composition, to molecular configuration and conformation, to orientational properties of polymer chains. Hyphenated FTIR approaches are often used to approach the study of polymers, and the combination of IR spectroscopy and microscopy has been well established for many years. Within the polymer community a wide number of applications of the micro-spectroscopic analysis of polymeric systems have been reviewed in the literature [15]. Despite the fact that the community is well versed in the use of vibrational micro-spectroscopy, the full potential of synchrotron infrared spectroscopy and microscopy in polymer science is not yet fully realized.

However, over the last decade, important advances have been made in IR micro-spectroscopy with a synchrotron source [16], which can provide new opportunities and motivation for the study of polymeric materials. The most utilized key synchrotron parameter for this community is the ability, thanks to the source brightness, to differentiate the chemical nature of the constituents in multicomponent polymeric systems. When the sizes of the different domains are in the range of the IR wavelength (micron scale), the spatial differentiation and study of each of these domains is possible using SR-FTIR micro-spectroscopy. Such analytical tools allow the study of various aspects related to the chemical composition, structure and morphology of the polymeric materials. Some of the areas that have benefited from synchrotron infrared over the last four years are reviewed in this section.

3.1 Polymer blends and composites

The blending of immiscible polymers *via* conventional routes is rife with technical issues such as chemical compatibilizer kinetic barriers, thermal degradation, inadequate mixing, undesirable side reactions with a solvent, and dispersed-phase domain coarsening during thermal shaping. The solid-state approach provides an opportunity to blend immiscible polymers that are otherwise incompatible *via* conventional means and to better understand the factors that control blending. Retention of dispersed-phase domain sizes during thermal shaping is the typical proof of compatibility, *i.e.* coarsening of the dispersed phase leads to macroscopic phase separation and degradation of material properties. In fact, fine (micron or sub-micron scale) dispersions evolve from reduced interfacial tensions and large interfacial thicknesses. The reduction of particle size and crystallinity by Neat Net-Shaping Manufacturing (NNSM) associated to a mechanical attrition followed by consolidation, using hot isostatic pressing in the solid state favours the intimate mixing of polymers [17 and reference therein]. As the mixing domains lie in the micron to sub-micron size domains, a technique like synchrotron-based infrared microscopy is a relevant approach.

A better understanding of NNSM-induced morphological changes has been initiated by employing synchrotron-based infrared micro-spectroscopy. In this recent study [17], authors have chosen three common semi-crystalline and amorphous polymers for blending: Hydrophobic polymers polystyrene (PS), polyethylene terephthalate (PET), and poly(methyl-methacrylate) (PMMA) labelled hereafter PS/PET and PET/PMMA. The results of such studies clearly show the potential of the synchrotron technique to detect the transition region between the two polymers, and provide evidence for the fluctuations in blend concentration for PS/PMMA.

There is clearly a room for improvement, since well optimized infrared beamlines are able to record very good quality spectra at lateral resolution down to 3 microns, as demonstrated in previous studies [16 and references therein].

3.2 Crystalline morphology and phase transformation

Laser microperforated biodegradable microbial polyhydroxyalkanoate substrates have been the subject of highly active research during recent years. Such substrates have important applications in tissue repair strategies [18]. The substrates are based on synthetic polymers and they serve as a temporary support for cells while they synthesize their own extracellular matrix during tissue regeneration and repair. The use of biocompatible polymers that can undergo degradation is highly desirable since these are potential candidates for therapeutic approaches. Among degradation pathways, the hydrolysis of labile bonds (*e.g.*, ester and amide) leads to nontoxic and non-immunogenic byproducts [19]. In this context, a recent study by G. Ellis *et al.* concentrated on the copolymer poly(3-hydroxy-butyrate-co- 3-hydroxyvalerate) (PHBHV) with different molar ratios [18]. Selective feeding of bacteria such as Ralstonia Eutropha [20] are used in the process to obtain such substrates. Substrate surfaces play a key role in the determination of the interactions between cells and biomaterials. G. Ellis *et al.* [18] used synchrotron infrared (IR) microspectroscopy and imaging for obtaining local information about the molecular structure of samples of interest with the highest lateral resolution possible, fully exploiting the high brightness of a synchrotron source for FTIR micro-spectroscopy.

HaCaT cells were seeded on the substrate after laser perforation. These cells were genetically modified with retroviral vectors introducing the green fluorescence gene into the cell genome (EGFP), which confers a green colour to the epithelial cells facilitating their detection on the PHBHV surfaces.

Fig. 1 shows the FTIR spectra recorded at a synchrotron infrared microscopy beamline, using $6 \times 6\text{-}\mu m^2$ confocal apertures and mapping at the positions indicated in the image.

The two spectra shown are representative of the bulk biopolymer film and the edge of the micropore. False colour maps shown superimposed on the image were constructed based on the variations in (a) the peak intensity of the carbonyl band at $1723\,cm^{-1}$ and (b) the $1228/1453\,cm^{-1}$ intensity ratio representing the crystallinity. A colour scale bar is given for each of these false colour maps in the figure. Both the carbonyl intensity and the crystallinity are clearly seen to diminish as the pore edge is approached. The dramatic reduction in the integrated intensity of the bands associated with the carbonyl stretch is accompanied by a reduction in intensity of the bands in the C–O–C stretching region between 1330 and $1230\,cm^{-1}$ co-located with bands associated with crystalline conformations at around 1289, 1279, and $1263\,cm^{-1}$. However, beyond these changes, no clear evidence for structural modification of the chain induced by laser ablation was observed in the spectra at the pore edge.

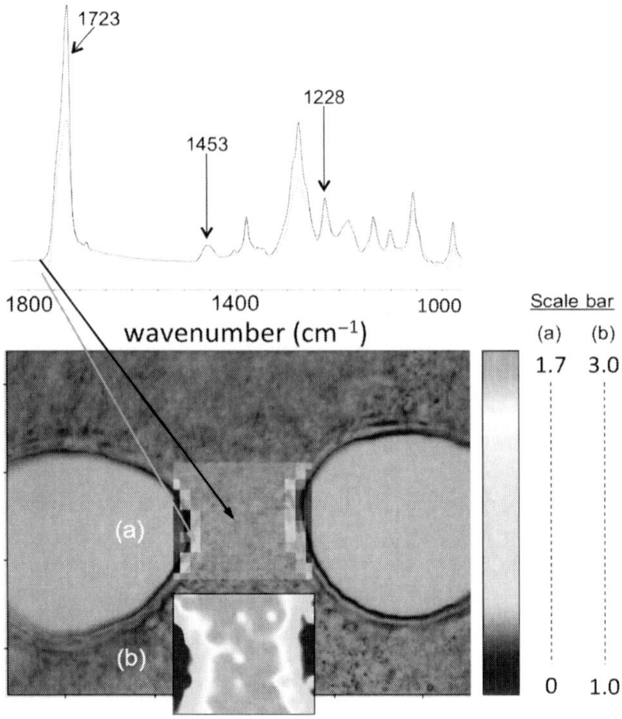

Fig. 1 Spectra and false colour maps corresponding to (a): the carbonyl band intensity and to (b): the 1228/1453 cm^{-1} band ratio (intensity scale bars shown) from the region between two micropores. Reprinted with permission from [18]. Copyright 2011 Springer.

While no clear evidence for thermal degradation products were observed at the micropore edge, the lower crystallinity observed is particularly important at the surface. Increased surface amorphicity at the pore edges may facilitate cell adhesion and be employed as a strategy to promote templated growth and migration of autologous or genetically engineered cells for regenerative and therapeutic medicine.

High spatial resolution and high spectral quality were essential for this study, demonstrating the need for using a synchrotron, and we hope this type of approach will motivate new studies in this research area.

3.3 Orientational properties

Exploring orientation properties in polymers films often requires a high spatial resolution and a high degree of polarisation of the probing beam. Synchrotron radiation has well known polarisation properties; the emission follows strict polarisation properties, as detailed in [21].

One recent work illustrates a good example of an orientational polymer study [21]. The molecular mechanisms involved in the residual stress in relation to shape memory effects in glassy amorphous starch has been investigated using a combination of synchrotron-based Wide Angle X-ray Scattering (WAXS) and polarized SR infrared micro-spectroscopy. The aim of this study was to analyse and better understand the structural anisotropy revealed by polarized light microscopy. A local molecular orientation was unearthed by WAXS and IR dichroism. The residual stress observed in the temporary shape was found to be due to local molecular orientation in the amorphous starch resulting from the flow of matter induced by the deformation process. This molecular orientation disappears when the sample relaxes and recovers its initial shape.

The orientation properties polymer are relevant to many lines of inquiry about these materials, and we expect an increase in this field of research and its use of synchrotron techniques in the coming years.

3.4 Microstructured polymeric materials

Infrared synchrotron micro-spectroscopy is also an appropriate method for identifying and visualizing the existence of localized water at buried interfaces, particularly between multilayers of polymers. It was recently shown that water inclusions can be imaged at the buried interface of solid-contact-ion-selective electrodes (SC-ISEs) [22]. In this study a poly(methyl methacrylate)-poly(decyl methacyrlate) [PMMA-PDMA] copolymer was used. Since the PMMA-PDMA copolymer is known to be water repellent and unsuitable for water sorption at measurable levels in the bulk membrane, the detection (or non-detection) of water by reflectance SR-FTIR is symbolic of the presence (or absence) of localized zones of water at the buried interface of a solid-contact ISE employing PMMA-PDMA as the sensing membrane. In fact, SR-FTIR revealed the presence of micrometer-sized inclusions of water at the gold-to-membrane interface, whereas coupling a hydrophobic solid contact of poly(3-octylthiophene 2,5-diyl) (POT) prevented the accumulation of water at the buried interface (Fig. 2) [22].

The aforementioned outcomes from SR-FTIR present valuable insights into the physical behaviour of technologically important materials such as

Fig. 2 *Left:* Optical image of a region of interest in a SE system, comprising PMMA-PDMA calcium selective membrane, coated on a conductive polymer layer on a gold substrate after being exposed to solution. *Right:* Three-dimensional intensity map of OH bending mode of water in the same region. Reprinted with permission from [22]. Copyright (2010) American Chemical Society.

ISE membranes exposed to fluid media for prolonged periods. Parallels between the physicochemical information obtained *via* SR-FTIR micro-spectroscopy and the chemical behaviour of the functional material can be drawn to achieve an enhanced understanding of the material systems under consideration.

3.5 Polymers in forensic science

Glitter is ubiquitous in everyday life as a form of decoration: it is in many cosmetics, is used for decoration on clothing, added to inks and paints, incorporated into fiberglass, and is widely used by both young and old in arts and crafts. Because glitter is small and light, it is readily transferred and retained on contact, often without our knowledge. For these reasons, glitter may be important associative evidence in criminal investigations [23]. The variety of glitter is astounding and may be found in an assortment of colours, shapes, and sizes. Thus, the more ways a glitter particle may be characterized, the smaller the subclass of evidence it will fall into and the greater its potential value as associative evidence.

Although there may be one or more metallized (aluminium) layers, most of the layers of glitter consist of polymers. In addition to the general shape, size, colour, thickness, and specific gravity of a glitter particle, the number of layers and their thicknesses and composition are important characteristics. Because some of the layers may be quite thin ($<10\,\mu m$), obtaining an infrared spectrum of an individual layer may not be possible with an ordinary FT-IR microscope system. In criminal trials, trace (or transfer) evidence may assist in showing an association between the victim/suspect and crime scenes. Most glitter particles are composed of three and sometimes many more distinct layers. Measurements of cross sections that determine the number of layers, the thickness of each layer, and the chemistry (infrared spectrum) of each layer can help to place a questioned glitter particle into a smaller subclass of trace evidence (it would be statistically less likely that glitter particles having all these same properties would by chance be found at the crime scene, on the victim, and on the suspect, for example).

Fig. 3 illustrates the analysis of such a multi-layered glitter particle using SR-FTIR micro-spectroscopy [24].

An estimate of rarity is still necessary to properly assess the value of glitter as associative evidence. It will be necessary to both create and maintain a searchable database showing both the variety and the frequency of occurrence of specific glitter types before this is used in criminal court cases.

This synchrotron infrared micro-spectroscopy study serves as a proof of concept, and should help motivate further work in this field.

4 Synchrotron IR studies in catalysis

The full potential of using a bright infrared beam in catalysis, a traditional tool for catalytic research by vibrational spectroscopy, is only beginning to be seen. We describe a very good example of such an application in the following.

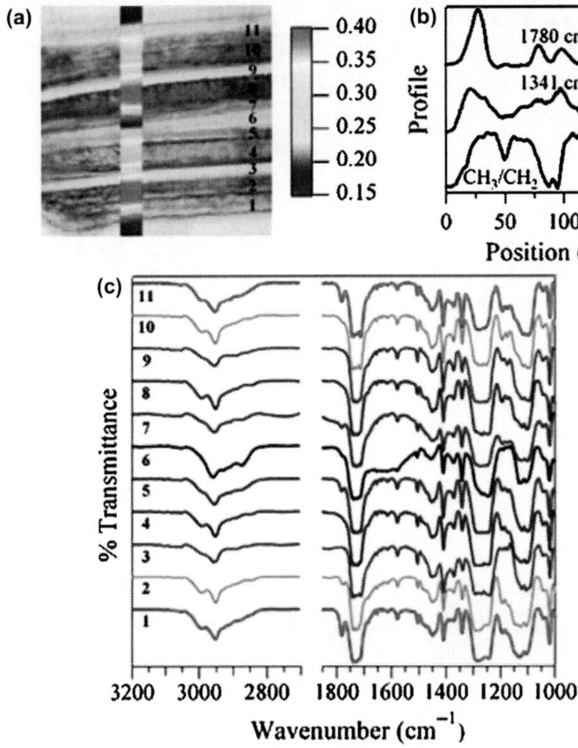

Fig. 3 (a) Image of the distribution of the ratio of CH_3 (at 2990 cm^{-1}) to CH_2 (at 2960 cm^{-1}), superimposed on an optical image of a glitter sample. The color code represents the value of the ratio (b): Profiles of the 1780 cm^{-1} and the 1341 cm^{-1} bands, as well as the CH_3/CH_2 ratio across the glitter cross-section. (c): synchrotron infrared micro-spectra across the 11 layers, colour coded so that similar spectra have the same colour. Reprinted with permission from [24]. Copyright (2011) Elsevier.

Conventional spectroscopic tools deal insufficiently with many of the scientific challenges pertaining to the understanding of heterogeneous catalysis at a very fundamental level. There are several aspects that limit their potential, and in particular the size of the probed area. Fundamental studies in surface science have convincingly demonstrated that distinct areas of the catalytic particles, for instance, crystalline faces of different orientations can exhibit varying catalytic activities and/or selectivities. [25, 26]. While model catalysts, such as well-defined metal or metal oxide surfaces, have been successfully deployed, the real-world systems often possess structural features having nanometer to micrometer sizes. Therefore, a strong need exists for *in-situ* spectroscopy at these length scales in order to record dynamic information on the different regions of a catalyst to understand their structure–performance, which, ultimately will provide a more complete picture of the overall catalytic process [27, 28]

Vibrational methods are one of the most powerful techniques in the arsenal of the catalysis researcher thanks to the direct chemical information that can be extracted from IR (and/or Raman) spectra [29]. These reveal the nature of chemical bonding, from which the determination of the chemical structure of the reaction intermediates and products can be identified.

In parallel, the state of the catalyst, in terms of chemical composition, hydration degree and redox state can be probed under *in situ* conditions. In addition, deactivation of the catalysts, often originating from carbonaceous deposits, can be easily identified since they exhibit strong features in the vibrational spectra.

L.R. Aramburo *et al.* investigated the porosity, acidity, and reactivity of de-aluminated Zeolite ZSM-5 at the single particle level. This study used a combination of approaches including synchrotron IR micro-spectroscopy, Atomic Force Microscopy (AFM), High Resolution Scanning Electron Microscopy (HRSEM), Focused Ion Beam - Scanning Electron Microscopy (FIB-SEM) milling and imaging, X-ray Photoelectron Spectroscopy (XPS), Confocal Fluorescence Microscopy (CFM), and UV-Vis spectroscopy [30] . ZSM-5 Zeolites were chosen because their physical properties and catalytic behaviours are mainly determined by the number and nature of their acid sites, while their location and distribution affects the effective diffusion paths of reactant molecules. In addition, diffusion limitations play a crucial role in their catalytic performance as well as in their eventual deactivation. In order to improve the molecular diffusion of zeolite materials, different approaches have been studied, *e.g.* exfoliating layered zeolites [31, 32] or synthesis of the zeolite nanocrystals [33]. To shed detailed insight into the variation of the catalytic properties it has been essential to understand the synergisms between the acidic properties, the internal and external diffusion barriers and micro- and meso-pore accessibility at the level of an individual zeolite crystal.

Three different ZSM-5 crystal types have been studied by synchrotron infrared micro-spectroscopy, namely parent zeolite (ZSM-5-P), mildly treated zeolite (ZSM-5-MT) and severely treated zeolite (ZSM-5-ST). The goal is to compare the influence of different dealumination conditions on the molecular diffusion and reactivity of individual ZSM-5 crystals. Synchrotron-based IR micro-spectroscopy [34] was used in combination with pyridine adsorption. Pyridine, having a molecular dimension of 0.57 nm, is able to diffuse throughout the micropore system of ZSM-5 and allows the detection of all acidic sites present within the zeolite material [35]. The sorption of pyridine was introduced at room temperature and the samples were subsequently heated up to 573 K and 673 K. The Brønsted acid sites, giving rise to IR spectra of adsorbed pyridine at around 1545 cm^{-1}, allow the visualization of differences in the nature and strength of the acid sites in the zeolite crystals under investigation. Fig. 4 shows the IR spectra of the adsorbed pyridine collected at room temperature, 573 K and 673 K for the ZSM-5-P, ZSM-5-MT and ZSM-5-ST crystals.

Fig. 4 shows clearly that the steaming treatment conditions have a considerable impact on the variation of the acidic properties of ZSM-5 zeolites. Notably, ZSM-5-ST only possesses residual Brønsted acidity. ZSM-5-MT exhibits a lower number of pyridine molecules chemisorbed on Brønsted acid sites at room temperature, compared to ZSM-5-P. As steaming post-treatment extracts part of the framework aluminium species (acting as Brønsted acid sites) and converts these into extra-framework aluminium, (acting as Lewis acid sites), a decrease in the number of Brønsted acid sites is expected. A close look to the band at 1545 cm^{-1} of ZSM-5-MT at 573 K and 673 K reveals a higher number of pyridine molecules attached to

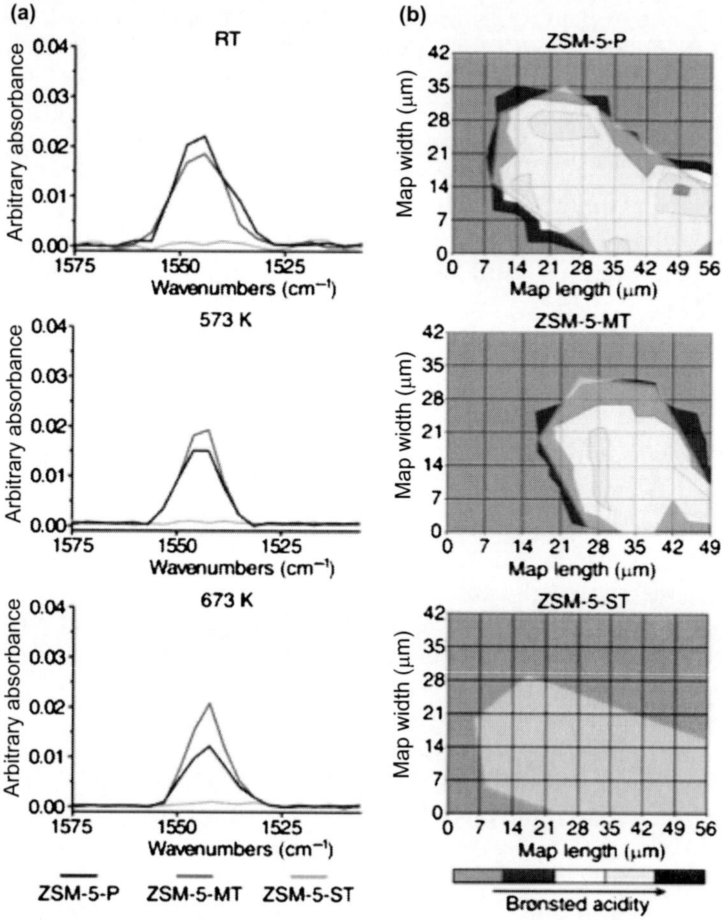

Fig. 4 (a) Synchrotron-based IR spectra of zeolites with adsorbed pyridine at room temperature, 573 K and 673 K. The IR bands at $1545\,cm^{-1}$ correspond to pyridine interacting with Brønsted acid sites. In black, spectra for the parent sample (ZSM-5-P), in red for mildly treated sample (ZSM-5-MT) and green for the severely treated (ZSM-5-ST) sample. (b) Spatially resolved Brønsted acid site maps revealing the amount of chemisorbed pyridine in ZSM-5-P, ZSM-5-MT and ZSM-5-ST at 523 K. Maps are overlaid with the footprints of the scanned crystals depicted in grey. No signal was detected during recording the ZSM-5-ST map. Reprinted with permission [30] f. Copyright © 2011 WILEY-VCH.

Brønsted acid sites compared to ZSM-5-P. Importantly, the relatively constant value of Brønsted acid sites detected in ZSM-5-MT at different temperatures compared to the progressive decrease of the amount of Brønsted acid sites observed for ZSM-5-P indicates an enhanced acid site strength distribution resulting from the steaming post-treatment. A likely explanation is that the observed enhancement involves the coupling of two effects taking place simultaneously, the extraction of aluminium from the zeolite framework (which increases the acidity of the remaining framework aluminium) and the synergism between framework and extra-framework aluminium species (the so-called "super acidity" effect) [36].

IR spatially resolved 2D maps help visualize the spatial distribution of the Brønsted acid sites within an individual zeolite crystal. As shown in Fig. 4b, a nearly homogeneous Brønsted acid site distribution can be found in ZSM-5-P and ZSM-5-MT after being heated up to 523 K. A small variation between the crystal edges and the main body does seem to occur possibly due to the twinning of the ZSM-5 crystals, however aluminium content is uniformly distributed in 2-D within the ZSM-5-P crystal.

Furthermore, it is important to stress that the presence of a nearly homogeneous distribution will most probably not cause differences in the susceptibility towards steaming within an individual zeolite crystal. It is clear that the steaming conditions have a considerable impact on the variation of the acidic properties of ZSM-5 zeolites.

Such a study highlights the potential synchrotron infrared microspectroscopy has for catalytic-related research.

5 Synchrotron IR studies in extreme conditions

Exploring the properties of matter beyond the usual room temperature, pressure, and/or magnetic field allows us to survey complex phase diagrams and to study substances in constrained environments. Ultra-high pressures, often together with high temperature, experiments are a new challenge being met by synchrotron radiation techniques. Samples mounted in a diamond anvil cell (DAC) can be investigated using both IR and X-ray synchrotron beamlines, since the diameter of the sample within the gasket of the DAC nicely matches the spot size of the SR source.

Several disciplines are benefiting over recent years, from high-pressure synchrotron infrared studies (*e.g.*, hydrogen-rich solids [37] and related planetary materials, minerals of the Earth's crust, mantle, glasses and melts [38], whole-rock samples [39], and pioneering studies of pressure-induced phase transitions in oxides and strongly correlated systems [40]).

Among recent studies, some researchers have focused on fluid and matter interactions in the extreme conditions found in the interiors of the Earth and other planets. *In-situ* observations by using diamond anvil cells allow obtaining data yielding surprising new insights into their molecular structure. For example, the water content in minerals localized in the (pressurized) mantle transition zone of the Earth affects several physical properties (*e.g.*, molar volume, thermal expansion, elastic moduli, and heat transfer). Water is likely to affect the mantle dynamics, so the interpretation of seismic data directly benefits from enhanced knowledge on the impact of water. The mantle's X-discontinuity, usually attributed to seismic velocity reflectors in the 260–330 km depth range, is not explained in terms of simple mineralogical phase transformations. Using a combination of synchrotron IR, synchrotron X-ray diffraction and Raman microscopy, S.D. Jacobsen *et al.* [41], have shown that concentrations of only 1300 ppm of water content in $MgSiO_3$ crystals can change its low-pressure structure (clinoenstatite) to its high-pressure structure. This suggests that a minor change of water content can drastically change the transformation boundary, *e.g.* the depth of the mantle X-discontinuity.

Water carriers in the Earth's mantle, in particular in the subducting slabs has remained a significant research area. Using synchrotron infrared microscopy, two phases of hydrous magnesium silicates (named Phase D and E) have been found to be stable at pressure up to about 40 GPa [42]. The synchrotron infrared technique was exploited in this study since, thanks to the small dimension of the probing beam, only part of the gasket hole between the two diamond windows was filled with sample, leaving the space beside the sample for accurate background measurement and subtraction. Organic molecules are found in carbonaceous and ordinary chondrites. Massive amounts fell on the early Earth, but the processes by which these materials became the earth's hydrosphere, atmosphere and biosphere are unknown. Montgomery *et al.* [43] observed the formation either of new organo-silicates or new complex organic molecules *via* templating of the crystalline structure of the minerals, at pressures associated with the upper mantle and planetary assembly. Contributing to the earth's primordial carbon budget is coronene ($C_{24}H_{12}$), a polycyclic aromatic hydrocarbon which is present in the interstellar medium and meteorites. Synchrotron IR microscopy has shown that Coronene, which is known to have survived atmospheric entry during the earth accretion, is able to survive through geological processes such as subduction and silicate melting of the rock cycle [44]. A reversible phase transition has been found between 2 and 3.3 GPa at ambient temperature (Fig. 5).

Phase transitions in solids are a subject of active infrared studies. It was recently observed using a synchrotron infrared microscopic study in both the far- and mid-IR regions, that there is a new behaviour of the optical modes in the rocksalt ZnO. High pressure induces a decrease of the LO – TO splitting resulting in a small reduction of ionicity, as deduced from the pressure coefficient of both the transversal and longitudinal effective charges. In this polymorph, the rate at which the ionicity decreases is larger than in wurtzite ZnO [45]. A similar investigation has been carried out on non-stoichiometric wurtzite at high pressure [39]. A localized mode was observed and assigned to the presence of vacancies in the crystal structure. The behaviour of this mode, seen by infrared spectroscopy, at high pressure has suggested pressure-induced charge localization near the vacancy sites in

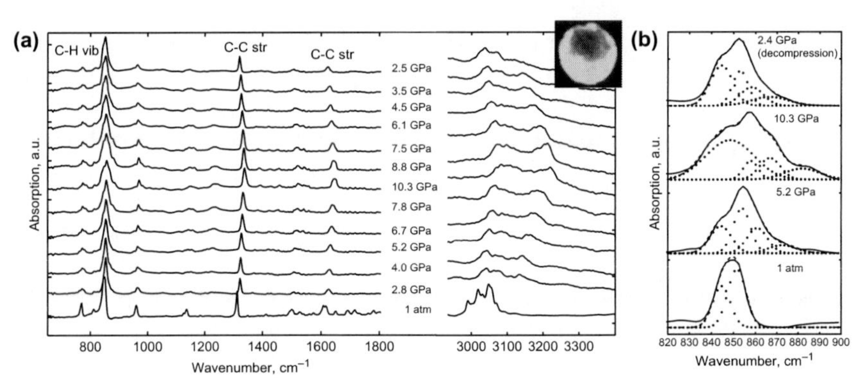

Fig. 5 (a) Normalized infrared spectra of coronene at 300 °C, and pressures up to 10.3 GPa (compression and decompression are shown) over the spectral range of interest. Micrograph of the *in situ* sample is shown at upper right. Peak formation is visible in the 840–880 cm^{-1} region, while peaks at 3000 cm^{-1} show Fermi resonances affected by pressure. (b) Gaussian peak fits to ambient temperature compression data in the range 820–900 cm^{-1}. Reprinted with permission from [44]. Copyright (2010) American Chemical Society.

$Fe_{0.91}O$. Accordingly, the pressure dependencies of the dielectric properties of wurtzite have been quantified.

X-ray diffraction, most often using synchrotron radiation sources, is the main technique for studying structural phase transitions in solids. Often such studies are complemented by synchrotron infrared micro-spectroscopy to further explore the pressure-dependent IR optical density of the samples, as exemplified by the recent work on structural change of multiferroic $BiFeO_3$ [46] and multiferroic hexagonal RMnO3 (R = Y,Ho,Lu) [47].

Experiments at very high pressure (up to 100 GPa, to above 300 GPa) are exceedingly difficult but are being reached with DAC cells, but sample volumes become very small (typical sample volumes are approximately less than 20 microns in diameter and less than 10 microns thick). Synchrotron X-ray diffraction and spectroscopy, synchrotron infrared micro-spectroscopy and optical techniques (*e.g.* Raman and Brillouin scattering) are the main tools for such investigation. At these extreme pressures, the pursuit of superconductivity and novel quantum behaviour in light-element systems is of great experimental and theoretical interest, fuelled largely by predictions of metallization and high-temperature superconductivity in hydrogen, in both paired and monatomic states [48]. Metallization of hydrogen is a very challenging research area, where synchrotron infrared microscopy could play a crucial role in demonstrating the metallicity *via* optical conductivity measurements. The search for the means to convert molecular hydrogen to a metal under static conditions at high pressure has been recently reviewed with emphasis on selected recent developments in both experimental studies and theoretical approaches [49]. There are on-going studies using large working distance synchrotron IR microscopy, to collect decent signals through 8 microns or less gasket holes at pressures over 300 Gpa, both at room and liquid nitrogen temperatures [50].

It was also suggest that hydrogen-dominant metallic alloys could become high temperature superconductors [51]. By this argument, covalent hydrides could exhibit metallization at significantly reduced pressures compared to pure hydrogen due to "chemical precompression". That is, covalent hydrides should possess a larger average valence electron density compared to bulk hydrogen at equivalent pressures and could also exhibit features important for phonon-mediated high-temperature superconductivity: wide overlapping bands, high density of states, and strong electron-phonon interaction. Silane (SiH_4) was extensively studied up to 150 GPa by a combination of Raman spectroscopy, optical microscopy, synchrotron infrared reflectivity, optical absorption, and synchrotron x-ray diffraction measurements. Chen *et al.* [52] have measured the IR reflectivity of SiH_4 between 12 and 70 GPa, and found that, on one hand at pressure below 40 GPa the reflectivity spectra were characterized by phonon reflection of the bending mode at around 850 cm^{-1}, consistent with an insulating molecular solid, and on the other hand with increasing pressure (>55 GPa) a marked increase in the reflectivity over the wide frequency domain is observed. These observations led to the suggestion that SiH_4 undergoes an insulator-to-metal transition at 60 GPa. From the published theoretical [53] and experimental studies on silane [54], it appeared that the pressure behaviour of this compound remains not fully understood. More recently, it

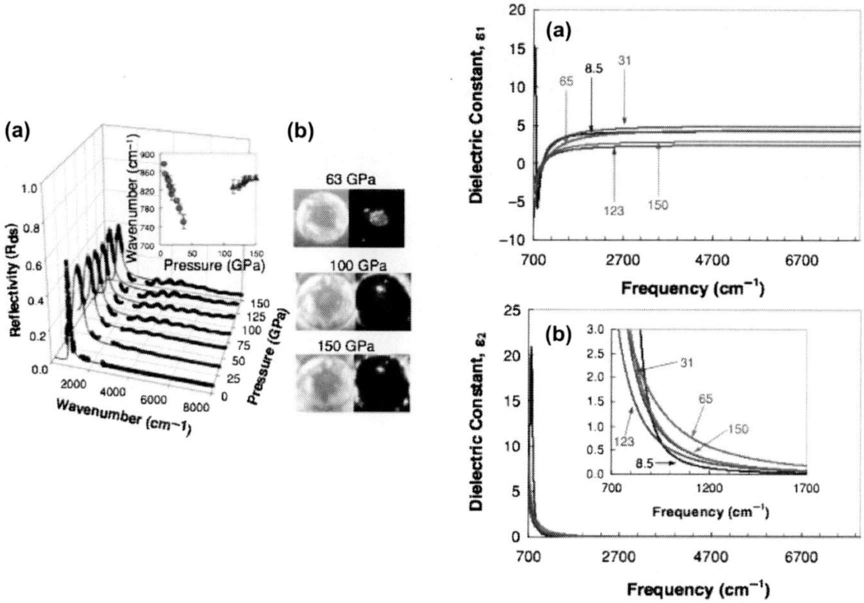

Fig. 6 *Left:* (a) Synchrotron IR reflectivity at the diamond-silane interface as a function of pressure. Lines show best-fit regressions to the experimental data (points). The nitrogen defect impurity and diamond multiphonon absorption regions are removed. Oscillations in the spectra are due to interference fringes between the two parallel diamond surfaces. Reproducibility of reflectivity from repeat measurements was within 5%. The inset shows the frequency of maximum reflectivity in the v4 region as a function of pressure. (b) Photomicrographs obtained in reflection (left) and transmission (right). Integration time on the CCD camera was several seconds. Measurable transmission of visible light was found through the sample at 150 GPa. *Right:* (a) Real and (b) imaginary parts of the dielectric function for SiH_4 at various pressures (indicated on figure in GPa). Reprinted with permission from [55]. Copyright 2011 by the American Physical Society.

was suggested that previous reports of silane metallization are likely a consequence of decomposition and that superconductivity may originate from silicon or hydrogen-doped silicon that is not in fact silane [55]. In the absence of decomposition, silane remains partially transparent and nonmetallic to at least 150 GPa with a band gap constrained to the range of 0.6 to 1.8 eV.

IR reflectivity results are reported in Fig. 6.

6 Synchrotron IR ellipsometry

Ellipsometry is an efficient method to determine the optical properties of matter. It has been largely employed with grating spectrometers in the visible, UV and near-infrared ranges for the characterization of thin films, surfaces and interfaces. In the mid- and far-infrared, attempts using conventional infrared (IR) light sources, such as globars or mercury arc lamps, were limited to relatively large samples. For example, ellipsometric studies of the cuprate high T_c superconductors (HTSC), using a laboratory source, could be performed only on thin films or on polycrystalline samples whose quality and homogeneity remained a major concern. With the use of a synchrotron source, these studies have been made possible in the far- to mid- IR region [56] on small single crystals of high quality.

Other important applications of infrared spectroscopic ellipsometry (IRSE) are the determination of the structure, orientation and homogeneity of thin layers on metallic and semiconductors surfaces. With the use of synchrotron sources, the probed area can be reduced to below 1 mm^2, whereas with a laboratory source the probed area is around 50 mm^2. This advance has opened the way to map samples and generate images of the distribution of optical functions on the surface. This has been recently illustrated in the study of biosensing surfaces, specifically for the absorption of cysteine modified CHL peptide on gold surfaces, and cytosine and guanine films on Si(111) [57]. In both cases, mapping revealed that the peptide film is inhomogeneous in thickness and structure, while the cytosine film on silicon is homogeneous for a thin (54 nm) film, but presents inhomogeneous lateral molecular structures for thicker (125 nm) films. This study demonstrates that synchrotron mapping ellipsometry made it possible to detect the structural inhomogeneity of biomolecular films that could not be observed directly in lab measurements that can only analyse a larger area.

Another illustration of the new capabilities provided by synchrotron infrared ellipsometry imaging is the study of a polymer brush [58]. A binary polymer brush consisting of weak polyelectrolytes was investigated *in-situ* under the influence of different aqueous solutions. The thickness of the brush layer in its dry state was ~15 nm. The brush, consisting of poly(ethylene glycol) and poly(acrylic acid)-b-poly(styrene) in a 50/50 composition, was switched between two different states by changing the pH of the solution. The results show strong chemical changes in the brush layer due to COOH - COO- conversion of the PAA's carboxylic groups.

The recent ellipsometry study of air exposed graphene flakes [59] is another very interesting example. The two-dimensional (2D) material graphene exhibits unusual infrared (IR) characteristics that make it the focus of intense technical and scientific interest [60 and references therein]. High-quality exfoliated graphene flakes tend to be restricted to dimensions of a few hundred micrometers at the largest, while chemically deposited graphene, which can be fabricated in extended sheets, suffers from large numbers of defects and purity issues. Measuring the ellipsometric angles of small graphene flakes in the IR region is beyond the capabilities of standard table top equipment since focusing causes a large spread of the incidence angle, resulting in an unacceptable level of depolarization. With the synchrotron source, ellipsometric measurements with defined optical conditions (incidence angle) of ultrathin films are readily feasible with submillimeter lateral resolution. In [59] the authors used a unique micro-focus-mapping ellipsometer [61] to measure the polarized reflection properties of single layer graphene and ultrathin graphite (Fig. 7). The ability to locate and characterize individual flakes with lateral dimensions of *ca.* $150 \times 380\,\mu m^2$ was demonstrated. In addition, the ellipsometric parameters were recorded from which the complete complex dielectric functions were calculated without resorting to Kramers-Kronig analysis. This work has revealed drastic differences between the dielectric functions of a graphene

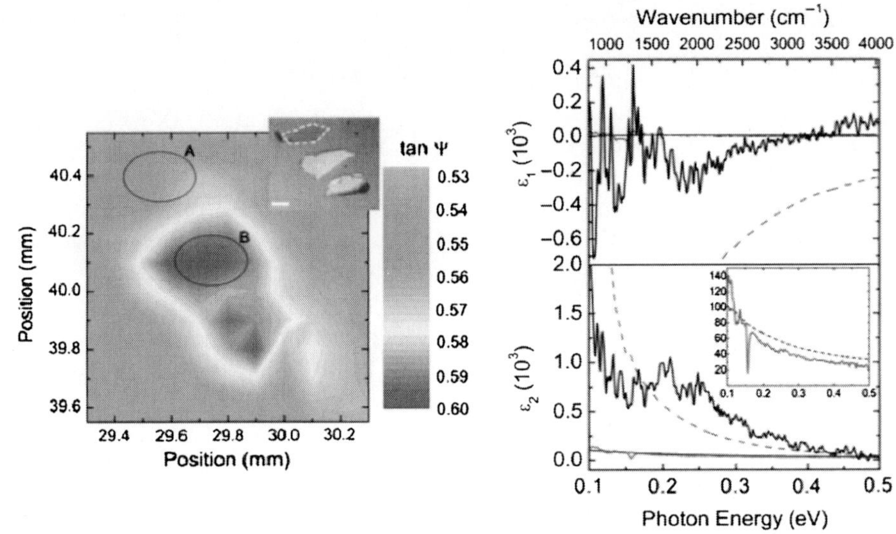

Fig. 7 *Left:* Map of tan Ø at 2992 cm^{-1} showing the two graphite flakes and the spot size and locations of scans A and B. The inset shows a digital photograph of the graphite flakes (middle and bottom) and the graphene flake (white outline) which is visible by eye. *Right:* Real (top) and imaginary (bottom) parts of the dielectric functions of the graphene (thick solid black lines) and graphite (thin solid red lines) flakes measured at positions A and B, respectively. Also shown are the measured curves for gold (green line). The inset shows a zoomed region comparing the thin graphite and bulk Highly Ordered Pyrolytic Graphite (HOPG) (dashed blue line) curves. The boxed area denotes a region of values that are influenced by a substrate phonon feature. Reprinted with permission from [59]. Copyright 2011, American Institute of Physics.

monolayer versus thin graphite (ε_2 of the graphene flake being about an order of magnitude higher than for the thin graphite).

7 Novel condensed matter compounds

Synchrotron infrared beamlines have been used for studying a number of exciting condensed matter systems from high temperature superconductors, to VO_2, to semiconductors, to graphene. Here we overview some recent work about the fascinating optical properties of graphene and the use of coherent synchrotron radiation as a novel and powerful far-infrared source.

7.1 Bilayer graphene

Synchrotron-based infrared spectroscopy has revealed unusual tuneable electronic structure and optical behaviour in electrically gated bilayer graphene. In a dual-gate bilayer graphene device, researchers were able to control the carrier doping semiconductor bandgap independently by using different combinations of the top and bottom gate voltages [62]. This unique tuneable bandgap can lead to many new physical phenomena. One example is an unusual phonon-exciton Fano resonance when the electronic bandgap is tuned to match the phonon vibration energy [63]. Excitation pathway control can also be achieved in graphene with electrostatic doping revealing quantum interference between different Raman pathways [64]. Finally, plasmon behaviour in graphene is important for the understanding of many body interactions of 2D Dirac fermions and it underlies potential graphene applications in ultrahigh speed THz electronics [65].

The tuneable plasmon excitations observed in the Terahertz range are remarkably strong, which demonstrate the exotic behaviour of 2D massless Dirac electrons and shed new light on graphene's application in THz metamaterials [65]. Plasmons describe collective oscillations of electrons and play a fundamental role in dynamic responses of electron systems and form the basis of research into optical metamaterials. Researchers explore plasmon excitations in engineered graphene micro-ribbon arrays by detecting the transmission of polarized light with THz frequencies. They observed plasmon resonances with remarkably large oscillator strengths, as manifest in prominent room-temperature optical absorption peaks. Also, they demonstrated that graphene plasmon resonances can be tuned over a broad terahertz frequency range by changing the micro-ribbon width and *in situ* electrostatic doping (Fig. 8).

This combination of electrical tuning and structural engineering in one material makes graphene a truly unique platform for plasmonic

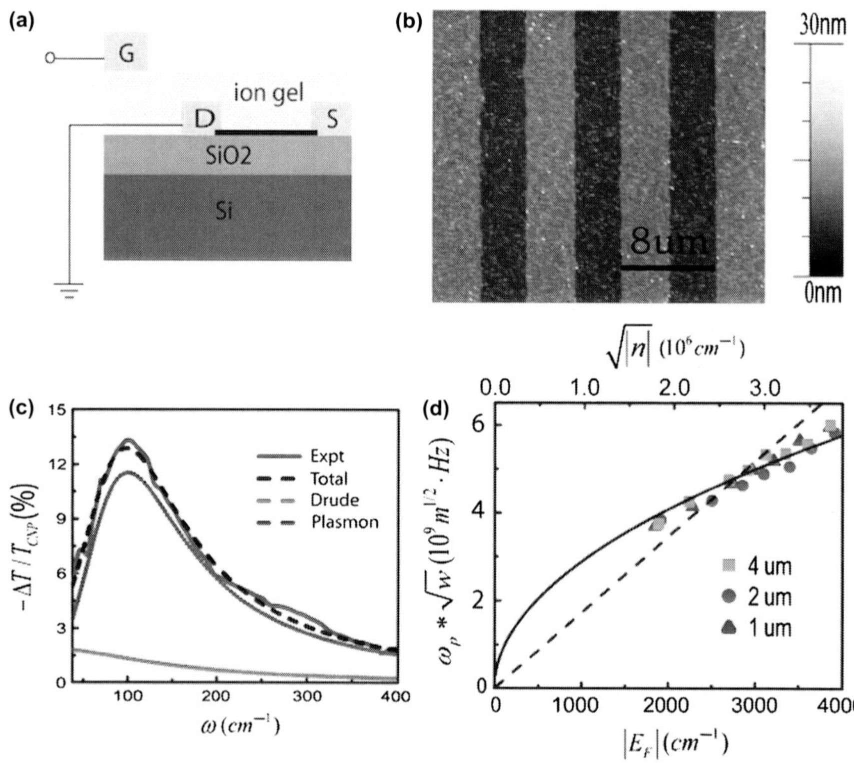

Fig. 8 (a) Side view of the graphene-based device in the experiment. (b) AFM image of graphene micro-ribbon array. (c) Plasmon absorption spectrum of graphene micro-ribbon array in the THz range (1 THz = 33.3 cm^{-1}). Over 13% of incident light is absorbed by the atomically thick graphene ribbon array. (d) Scaling laws of plasmon frequency as a function of charge density and ribbon width. Solid lines represent the scaling law obeyed by massless electrons in graphene. Dash lines represent that of ordinary massive electrons. Reprinted by permission from Macmillan Publishers from [65] with permission.

metamaterials. The observed unusual scaling behavior of plasmon frequency with ribbon width and carrier concentration also bears the signature to 2D Dirac electrons in graphene. The results represent a first look at light-plasmon coupling in graphene and point to potential graphene-based terahertz metamaterials.

7.2 Superconductivity with CSR

Coherent Synchrotron Radiation (CSR) has opened up the ability to perform measurements within the so-called THz-gap using this novel high-power THz and sub-THz source [13, 14]. Several facilities now offer a few days a year where the accelerator is run in a mode to specifically produce the short electron bunches required for CSR. One scientific area where this has been specifically exploited is the low-frequency study of novel superconductors. A high-flux synchrotron source allows a high accuracy in the detection of small effects in the reflectivity spectra and overcomes the problem of the very low intensity transmitted by these systems.

After an initial demonstration of a measurement along the physically small c-axis of a cuprate high temperature superconductor [14], a number of novel materials have been studied. Recently V_3Si was studied using CSR and the optical data showed a two-band framework within this material [66]. Two optical gaps were identified with $2\Delta/kT_c$ values of 1.8 and 3.8. This two-gap superconductivity behaviour has been of increasing interest since the discovery of two distinct gaps in MgB_2 [67].

Boron-doped diamond has also been recently discovered to be a superconductor [68], and its low-energy optical properties were elucidated using CSR [69]. The reflectivity of such samples were measured in the traditionally difficult sub-THz region and fit to a BCS model showing the opening of an optical gap below Tc, and a charge-lattice interaction involving optical phonons of pure diamond together with lower frequency modes induced by doping. Together these results point to a consistent dirty limit BCS picture of superconducting diamond, and are another demonstration of the use of CSR to greatly improve the signal-to-noise ratio of a conventional Michelson interferometer in the sub-THz spectral region.

8 Some future directions

The infrared spectral region remains under active development at accelerator based sources. Some of the areas which are being pursued include using the high power CSR pulses to provide high field time-domain sample pumps for pump and probe spectroscopies, careful tuning of the coherent THz pulse properties, increasing the speed and power of spectral micro imaging, and working to go beyond the diffraction limit and into the nano scale.

8.1 Coherent THz pumps & probes

Terahertz technologies are receiving increasing attention throughout the world [70], and time-domain THz spectroscopy is becoming well known for studying the dynamic behaviour of photo-excited carriers with time resolutions as fast at 10 fs using visible or near-visible laser pulses as pumps and synchronized THz pulses as probes. Broadening the spectral range for both the pump and probe pulses so that either can be anywhere from the THz all the way up to the hard x-ray is a major goal of ultrafast science with specific targeted property control being a grand challenge. Ultra-fast and extremely powerful x-ray sources are being developed using Free Electron Laser (FEL) technologies, and coupling these advances with the ability to make high-power tailored THz pulses *via* CSR processes is being pursued.

The first high-power THz CSR demonstrations were performed at an infrared FEL [13], showing that short electron bunches can naturally produce significant THz powers. More recently, a time-resolved THz pump and VUV probe beamline has been built at the FLASH facility that is the world's first soft x-ray FEL [71]. This beamline uses a THz undulator in line with the VUV lasing undulator source and overlaps the two pulses at the sample with an adjustable time-delay. The success of this beamline has led to plans and explorations for THz beamlines at the FERMI@ELETTRA FEL in Trieste, Italy, as part of the SwissFEL project plans in Switzerland, as well as others.

The UVSOR facility has recently demonstrated [72] the ability to produce tuneable quasi-monochromatic CSR THz pulses using a laser slicing technique to produce modulations on the electron beam and these modulations then radiate coherently [73–76], see Fig. 9. Based on this ability, a new THz-pump and VUV photoemission probe spectroscopy beamline has been designed and is under commissioning [77]. It will use THz CSR and VUV coherent harmonic generation from the same electron bunches in the storage ring to help elucidate the electronic structure relating to physical properties of solids by using selective excitations of low energy electronic and vibrational structures. Mode selective excitation of vibrational modes has been shown to be a powerful way to control a phase of matter in manganites [78] and combining this control ability with a wide array of probe techniques and wavelengths will further our understanding of complex materials.

8.2 Faster spectral imaging

One advance over the past decade in conventional FTIR microspectroscopy is the advent and availability of focal plane array detectors. These larger

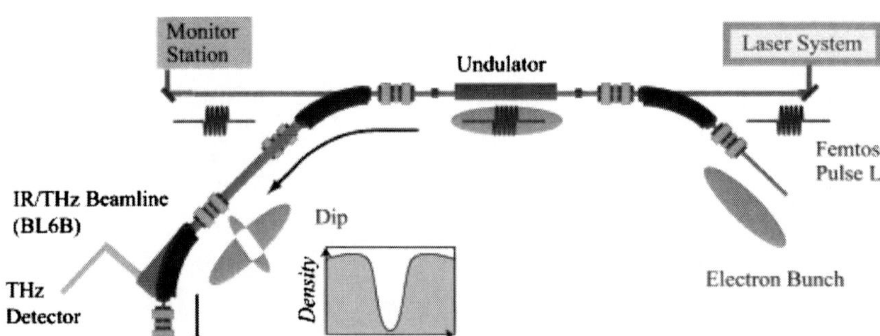

Fig. 9 Schematic view of the experimental setup for modulating an electron beam *via* laser slicing and monitoring the coherent THz emission from the density dip(s) created in the bunch. THz work at UVSOR. Reprinted with permission from [76]. Copyright (2011) Elsevier.

area pixelated detectors allow parallel spectral acquisition across a larger area of a sample. These couple to low-brightness thermal IR sources as these cannot be focussed to small diffraction-limited spots. A synchrotron source yields stable, broadband and high-brightness IR radiation making them excellent for FTIR microspectroscopy, as exemplified above. However coupling such a high-brightness source to a wide-field array detector essentially throws away the brightness advantages of a synchrotron.

At the SRC in Wisconsin, a new beamline has been built to find a compromise between high brightness and wide field by using multiple (12) synchrotron beams to cover a wide-field array detector and increase the speed of such spectral imaging [79]. The improvement in acquisition time opens the way for faster high resolution imaging at the diffraction-limit and is already being used for several biological and environmental sciences applications.

Continued development of pixelated detectors optimized for the high-brightness of synchrotron sources and further improvements beamline sources should yield ever more capable wide-field rapid imaging facilities and instrumentation.

8.3 Beating the diffraction limit

Science on the nanoscale is characterized by new physical and chemical properties of matter that emerge at dimensions intermediate between those of individual atoms or molecules and the homogeneous bulk. Protein and DNA folding, micelle formation, lipid bilayer membranes, supramolecular assemblies, and phase-separated polymers are examples of intra- or inter-molecular self-assembly associated with an increase in the internal degree of organization of the system. The chemical specificity of infrared vibrational spectroscopy has early on generated the desire for simultaneous spatial resolution to determine the microscopic composition of a sample. The coupling of FTIR with an IR microscope thus allows for micro-spectroscopy and mapping the chemical composition of a sample. Synchrotron infrared beamlines have provided the ability to obtain high quality spectral images at the diffraction-limit, but with the mid-infrared wavelengths between 3 and 20 microns the diffraction limited resolution achieved is at best 1.5–10 microns [10]. Optical near-field techniques are thus needed to reach beyond the far-field diffraction limit.

Photothermal techniques have been used with a synchrotron [80] resulting in some spatial resolution improvement. Greatly improved spatial resolution down to the nanometer scale has been realized using monochromatic laser sources and taking advantage of the local-field enhancement and scattering of nanoscopic metal probe tips scanned in close proximity to the sample by means of atomic-force or scanning-tunneling microscopy [81]. However broad-band spectroscopy is needed to realize the desire for full chemical analysis of FTIR coupled with nanometer spatial resolution (Fig. 10).

A first step towards such a nano-FTIR system has recently been taken using a thermal IR sources and a scanning infrared antenna [82]. These initial measurements were limited in spectral range to about 750–1250 cm^{-1}, but they show better than 100 nm resolution and more importantly that the idea of broadband nano-FTIR could indeed work. Several

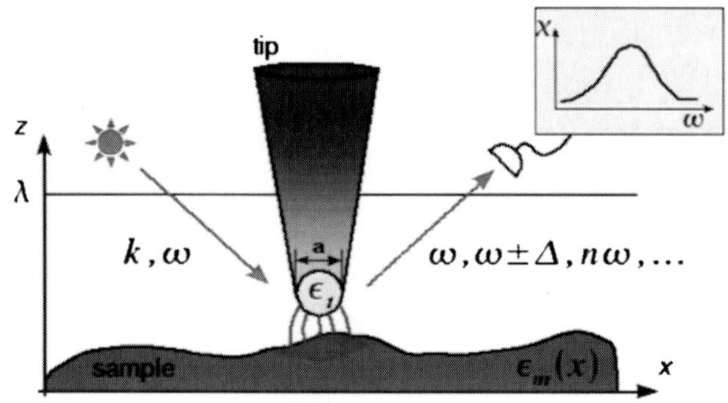

Fig. 10 General concept of the tip-enhanced region for scattering-type near-field optical microscopy (s-SNOM) for spectroscopic imaging. (From [80] with permission).

groups are currently working to couple a scanning nano imaging system to the high-brightness and broadband synchrotron IR source with the goal of achieving true nanospectrosopy. A first step towards this coupling was recently reported from SPring-8 [83], and plans are underway at ANKA [84], LNLS, the ALS, and elsewhere. Hopefully in the coming few years nanoscale spectroscopy will be realized at a synchrotron user facility and be available for users across the world.

Conclusions

The world has a growing number of synchrotron infrared beamlines and their use has also rapidly grown across a wide number of scientific applications. This review is far from comprehensive, but we hope it has highlighted several of the exciting recent developments in the field and the application of these sources to a wide variety of materials science. And in the near future we anticipate several advances, which will further increase the uses and capabilities of synchrotron, based infrared spectroscopies and microscopies. To find out more about how synchrotron infrared techniques may play a role in your research, we encourage you to contact one of the many friendly infrared beamline scientists at a synchrotron light source near you [2].

References

1 L. M. Miller and P. Dumas, *Current Opinion in Structural Biology*, 2010, **20**, 649–656.
2 See list of synchrotron infrared beamlines worldwide at http://infrared.als.lbl.gov/content/web-links/45-srir
3 M. C. Martin, U. Schade, P. Lerch and P. Dumas, *Trends in Analytical Chemistry*, 2010, **29**, 453–463.
4 W. Duncan and G. P. Williams, *Appl. Opt.*, 1983, **22**, 2914.
5 R. A. Bosch and O. Chubar, *Am. Inst. Phys. Conf. Proc.*, 1997, **417**, 35.
6 Y. L. Matthis, P. Roy, B. Tremblay, A. Nucara, S. Lupi, P. Calvani and A. Gerschel, *Phys. Rev. Lett.*, 1998, **80**, 1220.
7 P. Dumas, L. M. Miller and M. J. Tobin, *Acta Phys. Pol.*, 2009, **A115**, 446.

8 G. L. Carr, O. Chubar and P. Dumas, In *Spectrochemical Analysis Using Infrared Detectors*. Edited by Bahrgava IWLR, Blackwell Publishing, 2006.
9 L. M. Miller and P. Dumas, *Biochim. Biophys. Acta*, 2006, **1758**, 846.
10 E. Levenson, P. Lerch and M. C. Martin, *J. Synchrotron Radiation*, 2008, **15**, 323–328.
11 G. L. Carr, *Rev. Sci. Instrum.*, 2001, **72**, 1613–1619.
12 H. A. Bechtel, M. C. Martin, T. E. May and P. Lerch, *Rev. Sci. Instrum.*, 2009, **80**, 126106.
13 G. L. Carr, M. C. Martin, W. R. McKinney, K. Jordan, G. R. Neil and G. P. Williams, *Nature*, 2002, **420**, 153–156.
14 E. J. Singley, M. Abo-Bakr, D. N. Basov, J. Feikes, P. Guptasarma, K. Holldack, H. W. Hübers, P. Kuske, M. C. Martin, W. B. Peatman, U. Schade and G. Wüstefeld, *Phys. Rev. B*, 2004, **69**, 092512.
15 N. G. Everall and J. Chalmers (eds.), *Vibrational Spectroscopy of Polymers: Principles & Practice*, 2nd edn, John Wiley & Sons, Chichester, 2007.
16 G. Ellis, G. Santoro, M. A. Gómez and C. Marco, Synchrotron IR micro-spectroscopy: Opportunities in polymer science, *IOP Conf. Series: Materials Science and Engineering*, 2010, **14**, 012019.
17 R. J. Schexnaydre and B. S. Mitchell, *Materials Letters*, 2007, **61**, 2151–2155.
18 G. Ellis, P. Cano, M. Jadraque, M. Martin, L. Lopez, T. Nunez, E. DeLa Pena, C. Marco and L. Garrido, *Anal. Bioanal. Chem.*, 2011, **399**, 2379–2388.
19 F. Causa, P. A. Netti and L. Ambrosio, *Biomaterials*, 2007, **28**, 5093–5099.
20 T. Freier, *Adv Polym. Sci.*, 2006, **203**, 1–61.
21 C. Vechambre, A. Buleon, L. Chaunier, F. Jamme and D. Lourdin, *Macromolecules 2010*, 2010, **43**, 9854–9858.
22 J. P. Veder, K. Patel, G. Clarke, E. Grygolowicz-Pawlak, D. S. Silvester, R. De Marco, E. Pretsch and E. Bakker, *Anal. Chem.*, 2010, **82**, 6203–6207.
23 K. Aardahl, S. Kirkowski and R. D. Blackledge, *Science and Justice*, 2005, **45**, 7–12.
24 L. Vernoud, H. A. Bechtel, M. C. Martin, J. A. Reffner and R. D. Blackledge, *Forensic Science International*, 2011, **210**, 47–51.
25 G. A. Somorjai, Y. Li, *Introduction to Surface Chemistry and Catalysis*, New York, Wiley, 2010.
26 G. Ertl, *Reactions at Solid Surfaces*, Edited by: Wiley, 2009.
27 B. M. Weckhuysen, *Chem. Commun.*, 2002, 97.
28 B. M. Weckhuysen, *Phys. Chem. Chem. Phys.*, 2003, 4351.
29 B. M. Weckhuysen: *In-situ Spectroscopy of Catalysts*: American Scientific Publishers, 2004.
30 L. R. Aramburo, L. Karwacki, P. Cubillas, S. Asahina, D. A. de Winter, I. L. C. Buurmans, E. Stavitski, D. Mores, M. Daturi, *et al.*, *Chemistry - A European Journal*, 2011, on-line first.
31 A. Corma, V. Fornes, S. B. Pergher, T. L. Maesen and J. G. Buglass, *Nature*, 1998, **396**, 353–356.
32 A. Corma, V. Fornes, J. Martinez-Triguero and S. B. Pergher, *J. Catal.*, 1999, **186**, 57–63.
33 L. Tosheva and V. P. Valtchev, *Chem. Mat.*, 2005, **17**, 2494–2513.
34 E. Stavitski and B. M. Weckhuysen, *Chem. Soc. Rev.*, 2010, **39**, 4615–4625.
35 F. Thibault-Starzyk, I. Stan, S. Abello, A. Bonilla, K. Thomas, C. Fernandez, J. P. Gilson and J. Perez-Ramirez, *J. Catal.*, 2009, **264**, 11–14.
36 V. Gruver and J. J. Fripiat, *J. Phys. Chem.*, 1994, **98**, 8549–8554.
37 T. A. Strobel, M. Somayazulu and R. J. Hemley, *Phys.Rev.Lett.*, 2009, **103**, 065701.

38 V. V. Struzhkin, A. F. Goncharov, R. Caracas, H.-K. Mao and R. J. Hemley, *Phys. Rev. B*, 2009, **77**, 165133.
39 C. T. Seagle, W. Zhang, D. L. Heinz and Z. Liu, *Phys. Rev. B*, 2009, **79**, 014104.
40 A. Perucchi, C. Marini, M. Valentini, P. Postorino, R. Sopracase, P. Dore, P. Hansmann, G. Jepsen, O. Sangiovanni, A. Toschi, *et al.*, *Phys. Rev. B*, 2009, **80**, 073101.
41 S. D. Jacobsen, Z. Liu, T. B. Ballaran, E. F. Littlefield, L. Ehm and R. J. Hemley, *Physics of the Earth and Planetary Interiors*, 2010, **183**, 234–244.
42 S. R. Shieh, T. S. Duffy, Z. Liu and E. Ohtani, *Physics of the Earth and Planetary Interiors*, 2009, **175**, 106–114.
43 W. Montgomery, J. Tuff, S. C. Kohn and R. L. Jones, *Chemical Geology*, 2011, **283**, 171–176.
44 E. Jennings, W. Montgomery and P. Lerch, *J. Phys. Chem.*, 2010, **114**, 15753–15758.
45 J. Pellicer-Porres, A. Segura, V. Panchal, A. Polian, F. Decremps and P. Dumas, *Phys. Rev. B*, 2011, **84**, 5202–5207.
46 R. Haumont, P. Bouvier, A. Pashkin, K. Rabia, S. Franck, B. Dkhil, W. A. Crichton, C. A. Kuntscher and J. Kreisel, *Phys. Rev. B*, 2009, **79**, 184110.
47 P. Gao, Z. Chen, T. A. Tyson, T. Wu, K. H. Ahn, Z. Miu, R. Tappero, S. B. Kim and S. W. Cheong, *Phys.Rev.B.*, 2011, **83**, 224113.
48 E. Babaev, A. Sudbø and N. W. Aschroft, *Nature*, 2004, **431**, 666.
49 D. D. Klug and Y. Yao, *Phys. Chem. Chem. Phys.*, 2011, **13**, 16999–17006.
50 F. Occelli, S. Lefrancois, P. Loubeyre and P. Dumas, *In preparation.*
51 N. W. Ashcroft, *Phys. Rev. Lett.*, 2004, **92**, 187002.
52 X. J. Chen, V. V. Struzhkin, Y. Song, A. Goncharov, M. Ahart, Z. Liu, H. K. Mao and R. J. Hemley, *Proc. Nat. Acad. Phys.*, 2008, **105**, 20–23.
53 Y. Yao and D. D. Klug, *Proc. Nat. Acad. Phys.*, 2010, **107**, 20893–20898.
54 T. A. Strobel, M. Somayazulu and R. J. Hemley, *Phys. Rev. Lett.*, 2009, **103**, 065701.
55 T. A. Strobel, A. F. Goncharov, C. T. Seagle, Z. Liu, M. Somayazulu, V. V. Struzhkin and R. J. Hemley, *Phys.Rev.B.*, 2011, **83**, 144102.
56 C. Bernhard, J. Humlıcek and B. Keimer, *Thin Solid Films*, 2004, **455–456**, 143–149.
57 G. Sun, D. M. Rosu, X. Zhang, M. Hovesta, S. Pop, U. Schade, D. Aulich, M. nsch, B. Ay, H. G. Holzhutter, *et al.*, *Phys. Status Solidi B*, 2010, **247**, 1925–1931.
58 D. Aulich, O. Hoy, I. Luzinov, K. J. Eichhorn, M. Stamm, M. Gensch, U. Schade, N. Esser and K. Hinrichs, *Phys. Status Solidi C*, 2010, **7**, 197–199.
59 J. W. Weber, K. Hinrichs, M. Gensch, M. C. M. van de Sanden and T. W. H. Oates, *Applied Physics Letters*, 2011, **99**, 061909.
60 F. Bonaccorso, Z. Sun, T. Hasan and A. C. Ferrari, *Nat. Photonics*, 2010, **4**, 611.
61 K. Roodenko, Y. Mikhaylova, L. Ionov, M. Gensch, M. Stamm, S. Minko, U. Schade, K. J. Eichhorn, N. Esser and K. Hinrichs, *Appl. Phys. Letters*, 2008, **92**, 103102.
62 Y. Zhang, T. T. Tang, C. Girit, Z. Hao, M. C. Martin, A. Zettl, M. F. Crommie, Y. R. Shen and F. Wang, *Nature*, 2009, **459**, 820–823.
63 T. T. Tang, Y. Zhang, C. H. Park, B. Geng, C. Girit, Z. Hao, M. C. Martin, A. Zettl, M. F. Crommie, S. G. Louie, Y. R. Shen and F. Wang, *Nature Nanotechnology*, 2009, **5**, 32–36.
64 C.-F. Chen, C.-H. Park, B. W. Boudouris, J. Horng, B. Geng, C. Girit, Al. Zettl, M. F. Crommie, R. A. Segalman, S. G. Louie and F. Wang, *Nature*, 2011, **471**, 617–620.

65 L. Ju, B. Geng, J. Horng, C. Girit, M. C. Martin, Z. Hao, H. A. Bechtel, X. Liang, A. Zettl, Y. R. Shen and F. Wang, *Nature Nanotechnology*, 2011, **10**, 630–634.
66 A. Perucchi, D. Nicoletti, M. Ortolani, C. Marini, R. Sopracase, S. Lupi, U. Schade, M. Putti, I. Pallecchi, C. Tarantini, M. Ferretti, C. Ferdeghini, M. Monni, F. Bernardini, S. Massidda and P. Dore, *Phys. Rev. B*, 2010, **81**, 092509.
67 M. Kortus, I. I. Mazin, K. D. Belashchenko, V. P. Antropov and L. L. Boyer, *Phys. Rev. Lett.*, 2001, **86**, 4656.
68 E. A. Ekimov, V. A. Sidorov, E. D. Bauer, N. N. Mel'nik, N. J. Curro, J. D. Thompson and S. M. Stishov, *Nature*, 2004, **428**, 542–545.
69 M. Ortolani, S. Lupi, L. Baldassarre, U. Schade, P. Calvani, Y. Takano, M. Nagao, T. Takenouchi and H. Kawarada, *Phys. Rev. Lett.*, 2006, **97**, 097002.
70 M. Tonouchi, *Nature Photonics*, 2007, **1**, 97–105.
71 M. Gensch, *AIP Conference Proceedings*, 2010, **1214**, 19.
72 S. Bielawski, C. Evain, T. Hara, M. Hosaka, M. Katoh, S. Kimura, A. Mochihashi, M. Shimada, C. Szwaj, T. Takahashi and Y. Takashima, *Nature Physics*, 2008, **4**, 390.
73 J. M. Byrd, Z. Hao, M. C. Martin, D. S. Robin, F. Sannibale, R. W. Schoenlein, A. A. Zholents and M. S. Zolotorev, *Phys. Rev. Lett.*, 2006, **96**, 164801.
74 Y. Shen, X. Yang, G. L. Carr, Y. Hidaka, J. B. Murphy and X. Wang, *Phys. Rev. Lett.*, 2011, **107**, 204801.
75 J. M. Byrd, M. C. Martin, W. R. McKinney, D. V. Munson, H. Nishimura, D. S. Robin, F. Sannibale, R. D. Schlueter, W. G. Thur, J. Y. Jung and W. Wan, *Infrared Physics & Technology*, 2004, **45**, 325–330.
76 N. Yamamoto, M. Shimada, M. Adachi, H. Zen, T. Tanikawa, Y. Taira, S. Kimura, M. Hosaka, Y. Takashima, T. Takahashi and M. Katoh, *Nucl. Instrm. Meth. A*, 2011, **637**, S112–S115.
77 S. Kimura, E. Nakamura, M. Hosaka, T. Takahashi and M. Katoh, *AIP Conf. Proc.*, 2010, **1234**, 63–65.
78 M. Rini, R. Tobey, N. Dean, J. Itatani, Y. Tomioka, Y. Tokura, R. W. Schoenlein and A. Cavalleri, *Nature*, 2007, **449**, 72–74.
79 M. J. Nasse, M. J. Walsh, E. C. Mattson, R. Reininger, A. Kajdacsy-Balla, V. Macias, R. Bhargava and C. J. Hirschmugl, *Nature Methods*, 2011, **8**, 413–416.
80 L. Bozec, A. Hammiche, M. J. Tobin, J. M. Chalmers, N. J. Everall and H. M. Pollock, *Meas. Sci. Technol.*, 2002, **13**, 1217.
81 M. B. Raschke, L. Molina, T. Elsaesser, D. H. Kim, W. Knoll and K. Hinrichs, *Chemphyschem*, 2005, **6**, 2197–2203.
82 F. Huth, M. Schnell, J. Wittborn, N. Ocelic and R. Hillenbrand, *Nature Materials*, 2011, **10**, 352–356.
83 Y. Ikemoto, T. Moriwaki, T. Kinoshita, M. Ishikawa, S. Nakashima and H. Okamura, *e-J. Surf. Sci. Nanotech.*, 2011, **9**, 63.
84 E. Bründermann, D. A. Schmidt, I. Kopf and M. Havenith, *AIP Conference Proceedings*, 2010, **1214**, 7–9.

Metal-based molecular switches generated from dithienyl ethene (DTE)

Gearóid M. Ó Máille and Sylvia M. Draper*
DOI: 10.1039/9781849734899-00166

This report reviews the current literature on metal complexes whose active component is a dithienylethene (DTE)-derived molecular switch. It is a comprehensive account of the rationale for and the variations in the ligand design. In detailing and tabulating the consequent photophysical properties of the complexes and materials produced, it provides an overview of the advances made and directions undertaken in this burgeoning research field.

1 Introduction

Photoresponsive molecules have become lead targets in the drive to miniaturise opto-electronic devices. They can respond almost instantaneously to external stimuli (*e.g.* light) by registering the event at a molecular level. Once excited, these molecules can undergo a range of processes, from luminescence and energy transfer to bond forming/breaking. As a result they offer an exciting prospect for high-density data storage and fast data processing. Taking diffraction-limited spot size into consideration, an estimated 2×10^9 data bits cm^{-2} can be realised.[1] Photophysical processes on the subpicosecond timescale can be exploited, leaving the development of instrumentation with suitable response times as the only limiting factor. The use of fast, high quantum yield processes can also potentially reduce the heat load of computing devices, and provide us with a route to more energy efficient systems.[2] In all such future technologies the molecular switch is fundamental.

The use of molecular switches in integrated molecular devices is still in its infancy but nature presents us with some effective and elegant examples of their application. Vision is essentially the 'readout' of a nerve impulse, triggered as a response to the photoisomerisation of the polyenyl chromophore *retinal*, and its conformational consequences on the protein to which it is bound. For synthetic molecular switches the potential readout responses include light-induced changes in refractive index, chirality and geometric/structural transformations. Amongst these however, it is the photoisomerisation of photochromic materials that has become a particularly promising and deliberate line of enquiry.

Photochromism is the reversible, light-driven conversion of a molecule between two forms with differing absorption properties. Molecular examples are numerous and include fulgides, dihydropyridazines, spiropyrans, spirooxazines and diarylethenes. The reorganisation of the electrons involved in the transformation means that photochromism is often accompanied by the modulation of other attributes such as fluorescence intensity, redox properties, magnetic interactions and non-linear optical

University of Dublin, Trinity College, D2, Ireland; E-mail: smdraper@tcd.ie, omailleg@tcd.ie

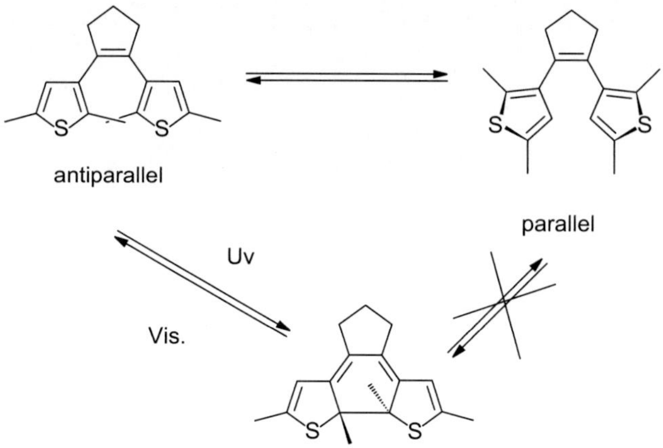

Scheme 1 Photochromic behaviour of a typical DTE.

(NLO) responses. Such changes are indicative of the switching state and can be exploited as the readout.

As outlined by Irie in his seminal review,[3] an ideal switching system requires certain physical properties: namely, thermal stability of both isomers, high fatigue resistance, high sensitivity and rapid response. Thermal irreversibility is also an important property in the application of photochromic compounds but it can be difficult to achieve. Theoretical studies carried out by Nakamura and Irie[4] on *cis*-diarylethenes have demonstrated that, in this class of molecule, the (Woodward-Hoffmann allowed) thermal disrotatory cyclisation is prohibited by a large energy barrier, but that no such barrier exists in the photoexcited state. As a result photocylisation (the fusion of the aryl components), can occur readily (as illustrated in Scheme 1). The same work established that the height of the thermal energy barrier required to undergo cycloreversion is inversely related to the difference in the aromatic stabilisation energy of the aryl groups in the open and closed form. The greater the energy difference, the lower the barrier. In a series of comparative semi-empirical MNDO calculations of a number of aryl substituents, thienyl substituents came out on top, identifying dithienyl ethenes (DTE) as the most thermally stable systems (an energy difference of 4.7 kcal/mol, compared to pyrrolyl at 13.8 kcal/mol).

This theory has been borne out by countless experimental studies. As an example the ring closed photogenerated isomer of 1,2-bis(2-cyano-1,5-dimethyl-4-pyrrolyl)perfluorocyclopentene has a half-life of 37 s at 25 °C, whereas its thiophene counterpart (1,2-bis(2-methyl-5-phenylthiophen-3-yl)perfluorocyclopentene) has a half-life of up to 1900 years at 30 °C.

While exceeding many other switching systems in terms of their thermal irreversibility, diarylethenes also often excel in addressing many of the other requirements. DTEs in particular have held a position at the forefront of molecular switch research for the last 20 years with the best performing systems offering the following characteristics:[5]

1. *Both isomers are thermally stable: a well-designed derivative can have a half-life at room temperature of up to 470 000 years*
2. *Colouration/decolouration cycles can be repeated more than 10^5 times*

3. *The quantum yield of colouration is close to 1 (100%)*
4. *Response times of both colouration and decolouration are less than 10 ps*
5. *Photochromic reactions occur even in the single-crystalline phase.*

Another particularly important property that sets DTEs apart from other systems is their exemplary fatigue resistance. This parameter is important in the construction of an efficient and cost-effective molecular device. To give an example, in a notional switch which cleanly interconverts between open and closed states, but where there is a competing side product generated from the open form with a quantum yield of 0.001; the initial concentration of the open isomer will have degraded by 63% after 1000 cycles.[3]

Most DTE derivatives are based on the 1,2-diarylperfluorocyclopentene motif and are synthesised by reacting lithiated thiophene fragments with octafluorocyclopentene, thereby generating a *cis*-configuration about the double bond. The non-fluorinated cyclopentene analogues can be synthesised *via* McMurry methodology and are also encountered in the literature despite the fact that their fatigue resistance and the degree of spectral difference between the open and closed forms are generally inferior.[3,6]

A difficulty encountered with many DTE derivatives is that due to steric considerations, a mixture of the parallel and anti-parallel conformers co-exist (Scheme 1). As only the antiparallel configuration is photoactive, the quantum yield of the system is determined by the proportion of just this conformer present. This is generally 50%, though quantum yields are often corrected in the literature, making the values artificially high. Rational synthetic designs are often developed to increase the quantum yields by increasing the preponderance of the antiparallel form. Such strategies include increasing the steric bulk at the 5-thienyl position,[7] encasing the DTE in a cyclodextrin,[8] or even tethering the thiophene rings.[9] In the latter case the quantum yield of photocyclisation was increased from 0.40 (untethered) to 0.67 (tethered).

Despite the wide-range of DTE-derived systems that exist, it is possible to describe their fundamental spectroscopic behaviour in solution in general terms – a typical absorption spectrum is shown in (Fig. 1).

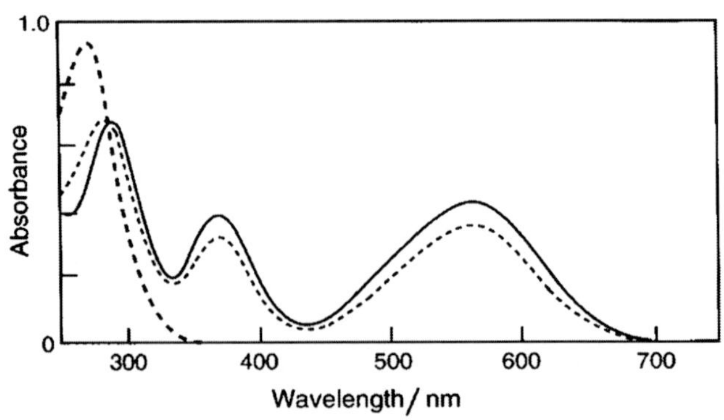

Fig. 1 Absorption spectrum of 1,2-bis(2,4-dimethyl-5-phenylthiophene-3-yl)perfluorocyclopentene: open (- - -) and closed- ring (—) isomers and at the photostationary state (---) by irradiation with Uv light. Reprinted (adapted) with permission from (M. Irie, *Chemical Reviews*, 2000, **100**, 1685–1716). Copyright (2000) American Chemical Society.

Common features of the absorption spectrum of the ring open form are the thiophene n–π* and π–π* bands at wavelengths less than *ca.* 300 nm. Irradiation into these states using Uv light results in photocyclisation. This eventually leads to a photostationary state (PSS) whose composition chiefly consists of the closed isomer. During this process the most prominent change is the colouration of the solution due to the growth of the characteristic highly red-shifted $S_0 \rightarrow S_1$ absorption band of the thiophene moiety in the visible region.[6] More subtle changes involve the bands becoming more structured with the increased rigidity of the system on closure, and a red shift in some bands due to the increased conjugation arising from the cyclisation. Irradiation into the visible region of the PPS absorption spectrum results in decolouration or bleaching of the solution due to photocycloreversion. Photocycloreversion often has a lower quantum yield than the photocyclisation process, but the solution is generally bleached within a few minutes of continuous irradiation. A significant challenge is that the readout must be non destructive – *i.e.* the probing light must not induce a photochromic response while the data is being read. This requires that both the open and closed isomers have zero absorbance at the readout wavelength.

The concept of introducing a metal centre into the DTE motif is relatively recent. Our intention in this report is to expose the rationale behind this fascinating new direction and to provide a particular photophysical focus. The marriage of the optical, redox and magnetic properties of metal complexes and the desirable characteristics of DTE-based molecular switches have numerous advantages. Many of the articles reviewed, reveal how metal-coordination has modulated the electron transfer, fluorescence and NLO responses of the resulting switch. In some cases the inclusion of metal complexes has been used as an effective means to lock a certain organic conformation in place. In others there is a clear intent to tap into the manifold of states arising from the metal to facilitate cyclisation *via* a different pathway than that adopted by the purely organic analogue.

The report is divided into sections that reflect similarities in DTE-ligand design. Systems with common pendant moieties or coordination modes are discussed together. The abbreviations used are based on a compound number given to the ligand and the element symbols of the metal centres present in any resulting complexes.

1.1 Pendant 4-pyridyl or 2-pyridyl coordination sites

Seminal work in 1999 by Lehn *et al.*[10] established the effect of incorporating metal fragments onto the periphery of perfluorinated DTE and opened the door to a number of later investigations. This early work used either one pendant 4-pyridyl moiety on the switching thiophene **1a(o)**–**1b(o)** to generate, mononuclear tungsten and rhenium complexes (Scheme 2) or two 4-pyridyl moieties **2(o)** to generate a dinuclear Re, W and Ru complex (Scheme 3). Photocycloreversion was readily achieved by irradiation with light of $\lambda > 600$ nm and the mononuclear complexes displayed efficient photochromic behaviour, with conversions of up to 90% being achieved in 15 mins. Despite the fact that the **RuRu2(o)** complex was found to degrade under Uv irradiation, the authors were able to postulate that the treatment of the free ligand with Ru^{II} might lock-in the switching state. The

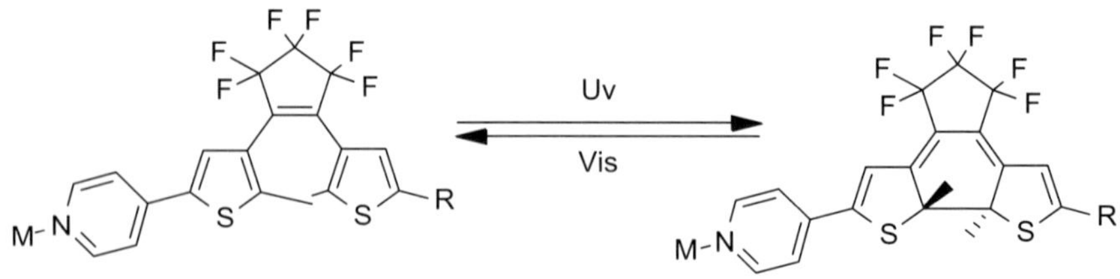

1a(o)R = *p*-phenol
1b(o)R = *p*-methoxyphenyl
1c(o)R = Me
1d(o)R = Ph
2(o)R = 4-Pyridyl
WW2(o)M = $W(CO)_5$, R = *N*-$[W(CO)_5]$pyridyl
W1a(o)M = $W(CO)_5$, R = *p*-phenol
ReRe2(o)M = $[Re(bpy)(CO)_3](CF_3SO_3)$, R = *N*-$[Re(bpy)(CO)_3CF_3SO_3]$pyridyl
Re1b(o)M = $[Re(bpy)(CO)_3](CF_3SO_3)$, R = *p*-methoxyphenyl

Scheme 2 Photochemical interconversion between the open (o) and closed (c) forms of **1** and **2**.

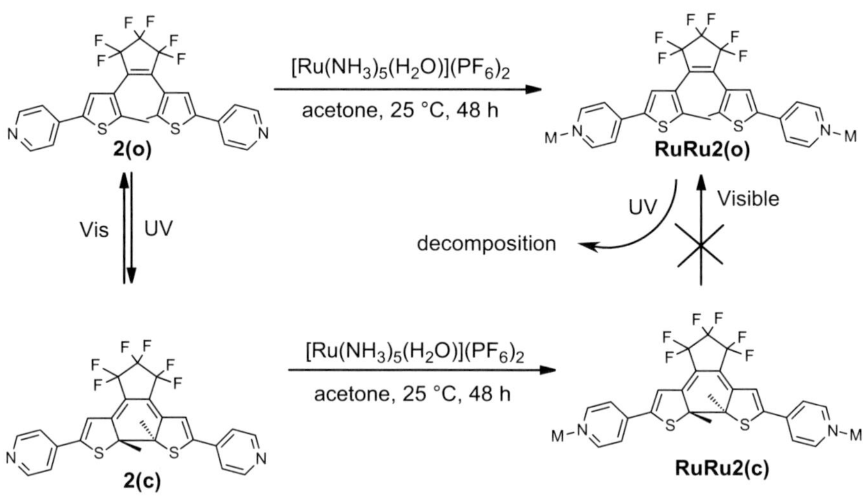

Scheme 3 Synthesis and photochromic processes of the dithienylethene photochromic complex **RuRu2(o)**, (M = $[Ru(NH_3)_5](PF_6)_2$).

mononuclear complexes exhibited a higher degree of luminescence in their closed as opposed to open forms and this demonstrated a potential mechanism for a responsive readout in a device. Although tungsten carbonyls are generally not luminescent as a result of photodissociation and efficient non-radiative decay processes, the addition of the pyridyl-coordinated DTE gave rise to MLCT-derived emission in both **WW2(o)** and **W1b(o)**.

A similar monopendant 2-pyridyl perfluorinated DTE (**1c(o)**, in Scheme 2) was used by Cui *et al.*[11] These authors examined its selective binding to Cu(II) and revealed that both protonated and non-protonated versions of this ligand undergo photoswitching.

Both monopyridyl and bipyridyl perfluorinated diarylethene ligands (**1d(o)** and **2(o)**) have been used more recently to prepare metal complexes

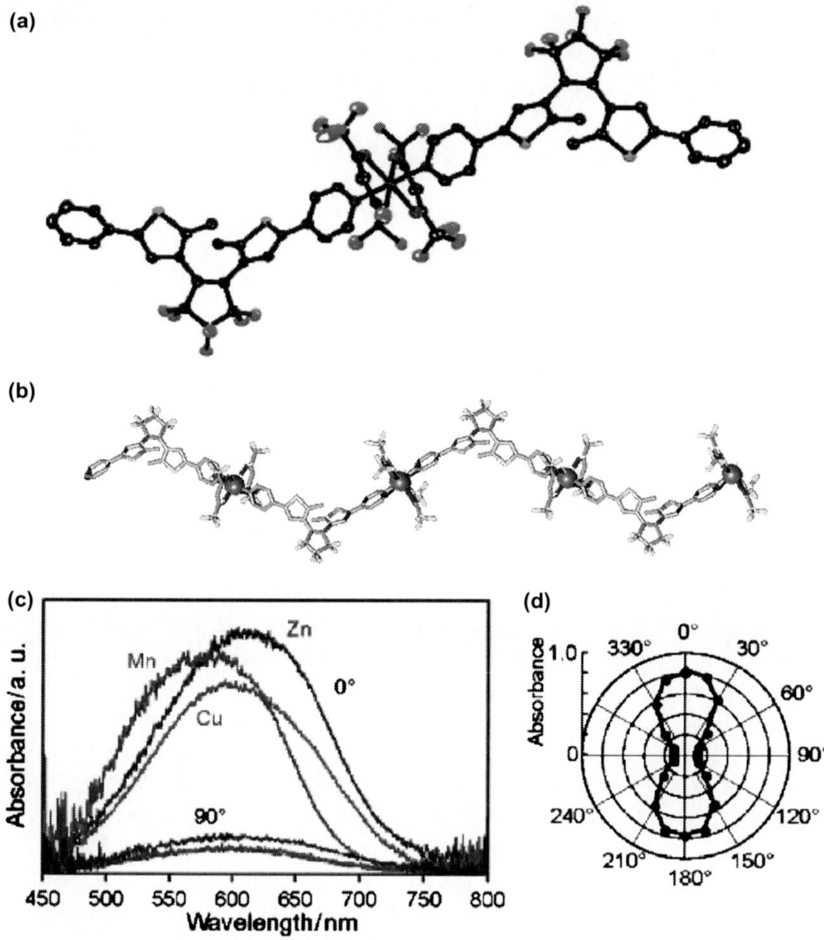

Fig. 2 X-ray structures of (a) discrete Zn:**2(1d(o))** complexes and (b) Cu:**2(o)** linear coordination polymers. (c) Polarised absorption spectra of M:**2(o)**. (d) Polar plot of polarized absorption of Zn:**2(o)** measured at 620 nm. Reprinted (adapted) with permission from (K. Matsuda, K. Takayama and M. Irie, *Inorganic Chemistry*, 2003, **43**, 482–489). Copyright (2003) American Chemical Society.

from $M(hfac)_2 \cdot H_2O$ (M = Zn(II), Mn(II), and Cu(II)).[12] The absorption maxima of the ligands were $\lambda = 586$ nm (**1d(o)**) and $\lambda = 589$ nm (**2(o)**) and both formed their respective closed isomers in over 95% upon irradiation at $\lambda = 333$ nm. On metal complexation, both discrete M:2(**1d(o)**) complexes (Fig. 2(a)) and M:**2(o)** linear coordination polymers (Fig. 2(b)) were obtained. In each case the diarylethene unit adopted an antiparallel conformation and the distance between the 2-methyl carbon atoms were in the range 3.53–3.77 Å *i.e.* within the region (< 4 Å) for photocyclisation to occur in the single-crystalline phase. The Cu(II)complexes of the closed form of the ligands were discrete 3Cu:2(**2(c)**) and Cu:**1d(c)** complexes and although Jahn Teller distorted, showed differences in axial Cu–N bond lengths and therefore the metal coordination sphere compared to their open forms. These differences were explored further using ESR spectroscopy.

The complexes underwent reversible photochromic reactions by alternate irradiation with UV ($\lambda = 313$ nm) and visible light ($\lambda = 578$ nm) in solution as well as in the single-crystalline phase. Like the free ligands, the solutions/crystals of the complexes reversibly changed (from colourless Zn(II), yellow Mn(II) or pale green Cu(II)) to intense blue (λ_{abs} 620 nm, 580 nm and 600 nm respectively) upon photoirradiation. Polarised absorption spectra of face-indexed single crystals (Fig. 2(c) and (d)) showed that the intensity of the blue colour changed on crystal rotation, indicative of an ordered orientation of the photoactivated dithienyl units in the crystals. ESR spectra of the linear chain Cu:**2(o)** changed on photoisomerisation of the ligand. While the open form gave an axial-type spectrum consistent with elongated tetragonal octahedral metal geometry, the photogenerated ring-closed form showed a rhombic type spectrum. It was felt this might reflect the conformational changes induced in the complexes by the open and closed forms of the ligands (complexes of monodentate **1d(o)** being able to undergo shrinkage and axial elongation on ring-closure and complexes of **2** unable to undergo this distortion such that the ESR spectra remain unaffected).

Garcia *et al.*[13] have also used **2(o)** as a bidentate bridge in the formation of an Fe(II) coordination polymer Fe(**2(o)**)$_2$(NCS)$_2$.2MeOH (**Fe2(o)**). IR spectroscopy indicated Fe coordination of the pendant pyridines and the *trans* arrangement of the NCS^- ligands which are N-coordinated to the Fe centres. On Uv irradiation ($\lambda = 365$ nm) the orange, open form of the polymer becomes blue with the emergence of two new bands ($\lambda = 386$, 600 nm). Attempts to record the absorption spectra in MeOH failed due to decomplexation. The ligand and the complex were found to be emissive at $\lambda = 418$, 436 nm (**2(c)**) and $\lambda = 374$ nm ((**Fe2(c)**) respectively. The photocyclisation process was followed by diffuse reflectance spectroscopy and was found to survive several cycles. The sample showed improvements in crystallinity and lattice rigidity according to powder X-ray diffraction data and ^{57}Fe Mossbauer spectroscopy. The latter and magnetic susceptibility measurements were carried out between 4-293 K and 1 bar–2 Kbar. Three resonance signals were observed at RT for Fe**2(o)** in the ^{57}Fe Mossbauer spectrum. The outer lines were found to correspond to high-spin Fe(II) with isomer shift and quadruple splitting values indicative of a non-cubic environment. At low temperature no magnetic order was observed and therefore no super-exchange was occurring through a Fe-NCS-FE magnetic pathway. On irradiation ($\lambda = 365$ nm) and grinding of the resultant blue material to produce a homogeneous sample, Mossbauer revealed three resonances with clear identification of Fe(II) and Fe(III) ions. Further careful experimentation ruled out photooxidation as the cause and a modification of the $Fe^{II} - (L)/Fe^{III} - (L^{\bullet})$ equilibrium on photocyclisation was proposed. The *tHS*/*tLS* ratios before and after irradiation were similar suggesting negligible spin state change, however the magnetic susceptibility doubled at room temperature. The good fatigue resistance, low addressing power (mW) and other properties of this solid were evidence to the authors of its potential in a multi-functional readout/memory device.

The non-fluorinated version of this bidentate ligand (**4(o)**) was used in a study by Tian *et al.* to bind in a 1:1 stoichimetric ratio with Zn(II).[14] The

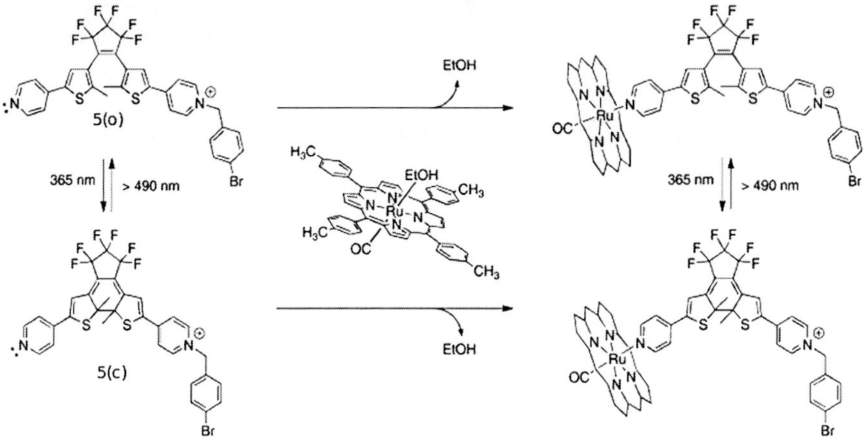

Scheme 4 Photochromic behaviour of **5(o)** and [Ru(**5(o)**)(porphyrin)(CO)].

resulting complexes (**Zn4(o)** and **Zn4(c)**) showed an almost 3 fold increase in their photocyclisation quantum yield compared to the free ligand.

With a different investigation in mind, Branda and co-workers modified **2(o)** further by making it undergo a one-step alkylation with 4-bromobenzyl bromide, to yield (**5(o)**).[15] The aim of this investigation was to use the differences in the π-conjugation of the open and closed form of this perfluorinated DTE switch (Scheme 4) to alter the electronic communication across the molecule and thereby monitor a supramolecular event. On irradiation (**5(o)** $\lambda = 364$ nm, **5(c)** $\lambda = 490$ nm), this DTE ligand reversibly switches without degradation over 10 cycles. After prolonged irradiation ($\lambda = 365$ nm) a 50:50 mixture of the open:closed forms could be obtained to which a slightly less than 1 molar equivalent of Ru(porphyrin)(CO)(EtOH) was added. *Via* ^{1}H NMR spectroscopy of the resulting mixture of the open and closed axially coordinated complexes, and *via* competitive coordination studies using 3-bromo-2-methyl-5-pyridylthiophene, the authors were able to establish that the twisted ring-open form of **5(o)** in [Ru(**5(o)**)(porphyrin)(CO)] was the more effective coordinating ligand, providing the stronger Lewis basic nitrogen. In a series of IR experiments involving the prolonged irradiation of a KBr pellet of the complex, the electron donating ability of the DTE pyridine could be monitored *via* the change of the ν(CO) stretching frequency of the axial CO. When the closed form of the complex was irradiated ($\lambda > 490$ nm, 1 h) the ν-CO changed from 1954 cm^{-1} to 1949 cm^{-1} corresponding to the increased Ru-CO back bonding in the open form of the complex. In the closed form the coordinated Ru porphryin was clearly electronically responsive to the electron withdrawing pyridinium located on the other side of the DTE switch (Scheme 5).

In an interesting twist on the pendant pyridyl design Tian *et al.* reported the first photochromic Iridium(III) diarylethene in which the metal fragment is directly coordinated to the photochromic core.[16] The free ligand, bis-2-pyridyl DTE (**6(o)**), exhibits a λ_{max} absorption at 320 nm, ascribed to singlet π–π* and n–π* transitions centred on the pyridine and thiophene rings. It is weakly fluorescent ($\lambda_{em} = 440$ nm). Irradiation of **6(o)** in THF

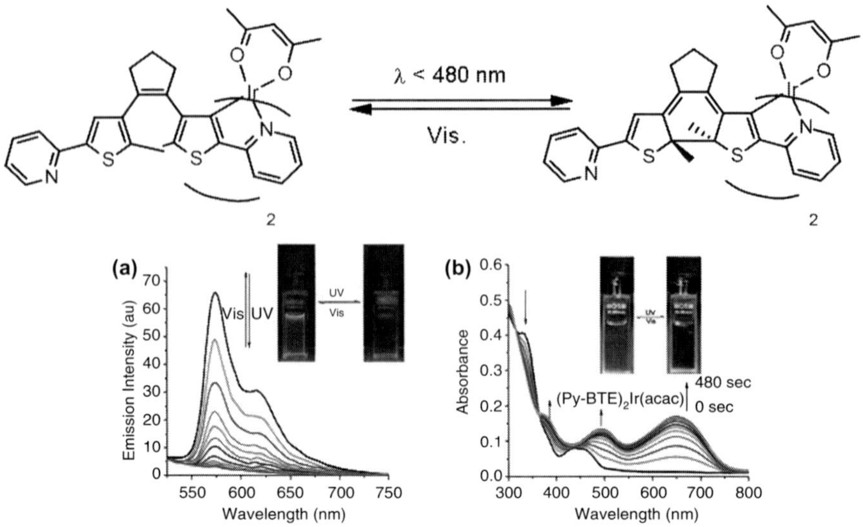

Scheme 5 Top: Schematic representation of the photochromic reaction of **[Ir(acac)(6(o))2]**. Below: Effects of cyclisation on the (a) emission and (b) absorption profiles of **[Ir(acac)(6(o))2]**. Reprinted (adapted) with permission from (W. Tan, Q. Zhang, J. Zhang and H. Tian, Organic Letters, 2008, **11**, 161–164). Copyright (2008) American Chemical Society.

solution ($\lambda = 313$ nm) results in the clear solution becoming purple, and the emergence of a new absorption maximum band at 560 nm ($\Phi_{OC\ 330\,nm} = 0.30$). Cycloreversion can also be achieved, but at a lower efficiency ($\varphi_{OC\ 550\,nm} = 0.016$). The complex [Ir(acac)(**6(o)**)$_2$] has a similar absorption profile but with an MLCT band appearing at λ 450 nm (incorporating some triplet π–π* and metal-perturbed intraligand π–π* character). (Scheme 5). Irradiation ($\lambda \leq 330$ nm), or excitation into the MLCT bands brings about photocyclisation, and a new band at $\lambda_{max} = 647$ nm. Both open and closed forms are stable in the dark, even on heating to 100 °C for seven days. The complex emits *via* ^{3}MLCT derived phosphorescence which is efficiently quenched upon photocyclisation by intramolecular quenching as the absorption overlaps with the emission spectra of the closed form.

1.2 Pendant bidentate bipyridyl coordination sites

In one of the first rigorous investigations of the triplet pathway in metal-containing DTEs, De Cola *et al.*[17] incorporated a bipyridyl coordination site onto the perfluorinated DTE moiety. Metal fragments were introduced in order to promote the formation of the photoactive antiparallel conformation and to induce triplet sensitised photochromism.

The ligand **7(o)** was synthesised using typical Suzuki methodology and subsequently treated with suitable metal precursors to form the homonuclear complexes **RuRu7(o)** and **OsOs7(o)** as PF_6 salts (Fig. 3 (top)). **RuRu7(o)** is yellow in acetonitrile solution, with typical metal-perturbed ligand-localised π–π* bands from *ca.* 245 nm to 345 nm and an MLCT band at 458 nm. This MLCT band is red shifted relative to $[Ru(bpy)_3]^{2+}$, indicating that electronic charge is localised mainly on the more conjugated DTE-containing ligand. Irradiation with UV light results in almost

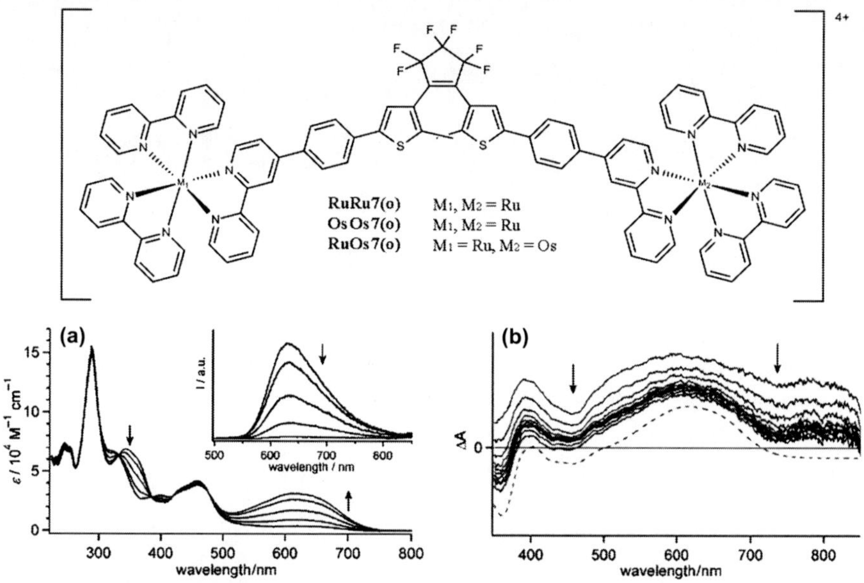

Fig. 3 (top) Representation of the complexes **RuRu7(o)**, **OsOs7(o)**, **RuOs7(o)**. (below) Transient absorption difference spectrum of **RuRu7(o)**, (b), employing a step size of 50 ns between frames. Reprinted (adapted) with permission from (R. T. F. Jukes, V. Adamo, F. Hartl, P. Belser and L. De Cola, *Inorganic Chemistry*, 2004, **43**, 2779–2792). Copyright (2004) American Chemical Society.

quantitative conversion to the closed form (Fig. 3(a)) (PSS composition 97% as determined by NMR studies). Irradiation into the broad ^{1}IL band (600–740 nm) of **RuRu7(c)** results in slow photocycloreversion to the open form (Fig. 3(a) inset).

RuRu7(c) is non-emissive, and its PSS emission intensity is 3% that of pure **RuRu7(o)**. **RuRu7(o)** is weakly emissive, with emission lifetimes and quantum yields up to 8 times less than those of reported non-photochromic model compounds. These quantum yields are highly oxygen sensitive (0.88 in deaerated solution and 0.37 in aearated solution) and differ only slightly on irradiation into either the IL or MLCT bands. In combination, these results suggest that the emissive ^{3}MLCT state is being quenched by energy transfer into the ^{3}IL state and that these two states are comparable in energy. As the free ligand is not phosphorescent at low temperatures or when coordinated to an innocent metal ion, a determination of the energy of the ^{3}IL was not possible, however the authors used transient absorption studies to explore the relative properties of the excited states more closely.

Using a flow through cell to avoid build-up of the closed photoproduct, nanosecond transient spectra were collected of air equilibrated acetonitrile solutions of **RuRu7(o)**. Following excitation into the IL (334 nm) and the MLCT (450 nm) states the same spectra where obtained, but the absence of an MLCT bleach at 460 nm indicated that the transient species was not of CT origin. It was therefore assigned as a ^{3}IL species (Fig. 3(b)). The photocycloreversion process proved far less efficient, with quantum yields of less than 0.1. The slow nature of this process, meant that picosecond transient

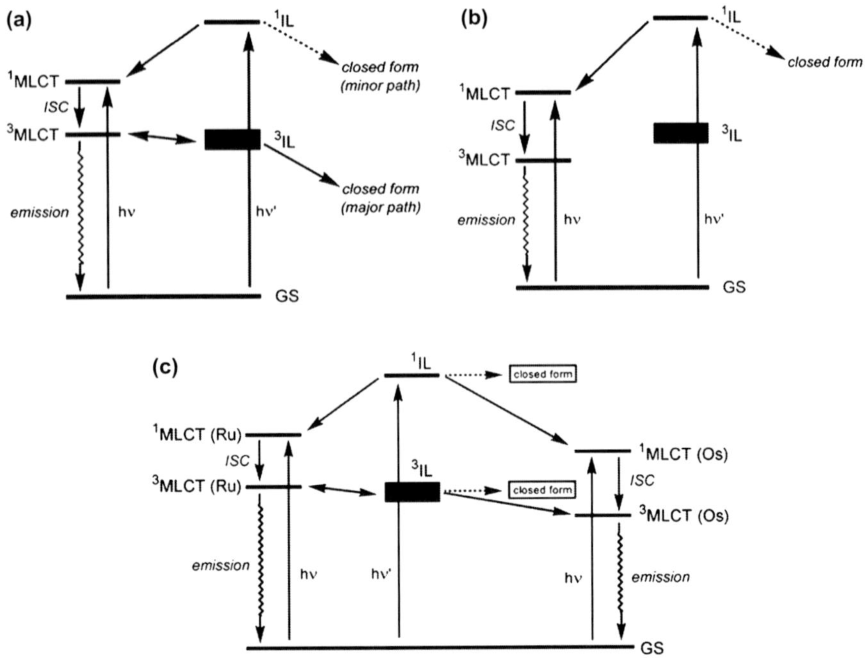

Scheme 6 Qualitative energetic schemes for (a) **RuRu7** (b) **OsOs7** (c) **RuOs7**. (a), (b) Reprinted (adapted) with permission from (R. T. F. Jukes, V. Adamo, F. Hartl, P. Belser and L. De Cola, *Inorganic Chemistry*, 2004, **43**, 2779–2792). Copyright (2004) American Chemical Society. (c) From R. T. F. Jukes, V. Adamo, F. Hartl, P. Belser and L. De Cola, *Coordination Chemistry Reviews*, 2005, **249**, 1327. Copyright © 2005 by John Wiley & Sons, Inc. Reprinted by permission of John Wiley & Sons, Inc.

spectra could be recorded of the closed isomer. These showed a bleach of the band at 550 nm–670 nm together with the appearance of structureless bands above 550 nm and below 670 nm indicative of ^{1}IL species. Excitation into the MLCT state of **RuRu7(c)** resulted in similar transient spectra, except that a new band at 525 nm appeared within 5 ps, that overlapped with a bleaching of the S_0–S_1 between 550 nm and 670 nm. The long lifetime of this transient species (12 nm) and the absence of MLCT bleaching features led the authors to associate it with ^{3}IL state of the closed species, and the initial absorption increase at 525 nm was ascribed to energy transfer from the ^{3}MLCT to the ^{3}IL (Scheme 6(a)).

The dinuclear **OsOs7(o)** gave similar absorption spectra, apart from ^{3}MLCT absorptions between 500 nm and 700 nm which give rise to its green colour. Irradiation into the high energy states resulted in the IL of the closed form appearing at $\lambda = 550$ nm–750 nm, and a darkening of the green colour. The quantum yield in this case was found to be independent of O_2 and very low, (0.0045 whereas the free ligand quantum yield is 0.72). The complex degraded on prolonged irradiation, preventing study of its cycloreversion. **OsOs7(o)** gave similar emission quantum yields and lifetimes to non-switching analogues, *i.e.* $[Os(bpy)_3]^{2+}$, and irradiation into the ^{1}MLCT state did not result in photocyclisation, suggesting the ^{3}MLCT state lies below the ^{3}IL level, and that cyclisation only occurs through the ^{1}IL state. Photocyclisation quenches the emission – the PSS (73% from NMR) composite emission is 25% that of the open form, and is due to residual **OsOs7(o)**. Scheme 6(b) shows qualitative energetic diagrams, summarising these results.

The synthesis of a mixed metal Ru/Os system was undertaken in order to study the energy transfer processes between the two metal centres depending on the switching state of the perfluorinated DTE core.[18] The heterodinuclear complex was synthesised by reacting the ligand **7(o)** with one equivalent of [$Os(bpy)_2Cl_2$] and isolating the PF_6^- salt (**Os7(o)**). This complex was subsequently treated with [$Ru(bpy)_2Cl_2$], resulting in **RuOs7(o)** in 12% overall yield. The UV-vis spectrum of the system was an average of **RuRu7(o)** and **OsOs7(o)** indicating a low level of interaction between the metal-containing moieties. Irradiation with light of $\lambda = 334$ nm leads to the initially green solution becoming a darker green, with a PSS composition of 90% (as verified by NMR spectroscopy). The emission spectrum of **RuOs7(o)** was obtained by excitation into an isoabsorptive point of the Ru-bpy and Os-bpy fragments. Both Ru ($\lambda_{em} = 630$ nm) and Os ($\lambda_{em} = 730$ nm) derived emissions were observed. The Ru-based emission was much less intense relative to the Os-based emission, indicating efficient energy transfer from the Ru to the Os moiety. The lifetime of the Ru-based emission was also significantly reduced relative to **RuRu7(o)**, while the Os emission lifetime remained relatively unchanged.

Photocyclisation could be achieved by irradiation into the MLCT band at $\lambda = 450$ nm, but the quantum yield showed a 2-fold increase when the system was irradiated into the IL states at higher energy. It was postulated that this occurred as a result of the populated IL state transferring energy to both the Ru and Os-derived ^{1}MLCT states, while also leading to cyclisation (Scheme 6(c)). Following intersystem crossing from the ^{1}MLCT, the transfer of energy from the Ru ^{3}MLCT to the Os ^{3}MLCT is sufficiently slow to allow some ^{3}IL population and hence triplet sensitised cyclisation. Excitation into the ^{1}MLCT band results in a lower cyclisation yield as the ^{1}IL state is not populated, and only the triplet sensitisation pathway, which competes with metal-metal energy transfer, yields cyclised product. Relative to **RuRu7(o)**, the photocyclisation efficiency is reduced by 90 times, indicating that photocyclisation competes with efficient Ru to Os energy transfer. Emission studies gave a lifetime of 0.56 ns. This can be considered as being the total rate of energy transfer from both the Ru ^{3}MLCT and the ^{3}IL to the Os ^{3}MLCT level, and equates to a rate constant of $k_{ET} = 1.8 \times 10^9\,s^{-1}$.

Nanosecond transient absorption spectra obtained by excitation at $\lambda = 334$ nm resulted in a bleach of the MLCT at $\lambda = 460$ nm. The transient species has the same excited state lifetime as the Os emission, indicating an Os localised ^{3}MLCT species with the remaining absorptions ascribed to the cyclised isomer. Excitation into the isoabsorptive point at $\lambda = 450$ nm results in a lower photoproduct absorption, and excitation directly into the Os MLCT at $\lambda = 650$ nm results in the ^{3}MLCT decaying to the ground state but no photoproduct is observed. Transient spectra of the closed isomer **RuOs7(o)** results in similar spectra to those obtained for the closed forms of the homodinuclear forms, and are by analogy ascribed to a ^{3}IL state. After this state decays, a persistent **RuOs7(o)** absorption is observed, indicating the occurrence of photocycloreversion. The authors indicate this process could happen from a singlet or triplet IL state. On the picosecond timescale, an initial bleach can be observed at $\lambda = 475$ nm due to population of both MLCTs from the ^{1}IL state. As this feature disappears, a band grows in at $\lambda = 540$ nm, accompanied by a further bleach at $\lambda = 660$ nm. The resulting spectrum is attributed to arising from a ^{3}IL state. The authors associate the calculated short time constant of 4 ps to ^{3}MLCT to ^{3}IL energy transfer, giving a rate constant of $k_{ET} = 2.5 \times 10^{11}\,s^{-1}$ which is in excellent agreement with that observed in **RuRu7(o)**-$k_{ET} = 2 \times 10^{11}\,s^{-1}$. The closed form is also non-emissive as the DTE energy level is below the energy levels of the two metal centres, quenching the emission. The closed switching unit can then trap the excited states of both metals, in contrast with its open isomer, which acts as a bridge, allowing efficient energy transfer from the Ru to the Os centre.

In a rare example in which both NLO investigations and photoresponsive DTE-switches were combined, a cationic Ir(III) substituted complex incorporating bipyridyl vinyl perfluorinated-dithienylethene was generated as part of a series of [Ir(C^N-ppy-R′)$_2$(N^N-bpy-CH=CH-C$_6$H$_4$R)][X] with varying electron-donor and acceptor substituents (R, R′).[19] The DTE-based ppy ligand (**8(o)**) (Fig. 4) absorbs at $\lambda = 348$ nm, closes on irradiation at 400 nm (PSS 669 nm, conversion 95%) and is non-emissive even in frozen EPA glass at 77 K. In comparison, irradiation of the complex (**Ir8(o)** ($\lambda = 350$ nm, CH_2Cl_2) results in a new broad band at $\lambda_{max} = 715$ nm corresponding to the shifted S_0–S_1 transition of the closed-perfluorinated DTE unit. The photocyclisation of the DTE unit can be triggered by using either UV ($\lambda = 350$ nm) or visible light ($\lambda = 450$ nm) (79% conversion, determined by ^{1}H NMR spectroscopy), the latter suggesting intersystem crossing from ^{1}MLCT to ^{3}MLCT and efficient energy transfer into the lowest-lying ^{3}IL state. The complex (o) emits (650, 717 nm, τ 23 μs) from what is thought to

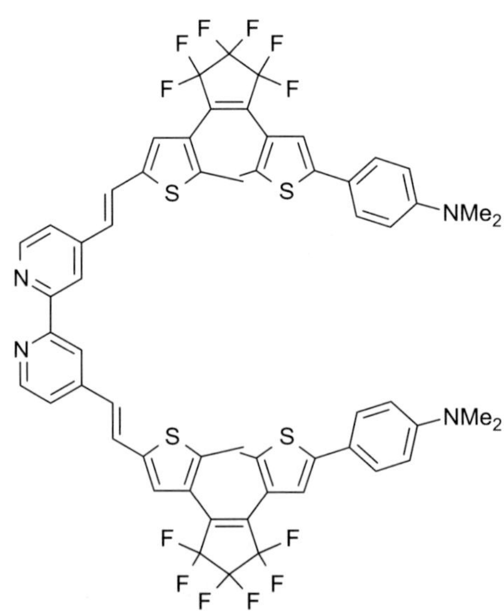

Fig. 4 Schematic of ligand **8(o)**.

be a ^{3}IL state, localised on the bpy-C=C-DTE. On conversion (79%) to the PSS, efficient quenching of this 77 K luminescence occurs but it can be restored upon irradiation at $\lambda = 715$ nm.

Unlike the dramatic change in NLO activity of the Zn(II) complex of the same ligand on photocyclisation, no significant modification of the EFISH value $\mu\beta_{1.907}$ (**Ir8(o)** -2190×10^{-48} esu, **Ir8(c)** -2000×10^{-48} esu) is observed upon photocyclization. The authors suggest that this is evidence that the quadratic NLO response is dominated by the MLCT/L'LCT processes, rather than by the intraligand excited states localized on the substituted bipyridine ligand.

1.3 Pendant bidentate terpyridyl coordination sites

Abruña *et al.*[20] have incorporated a terpyridyl coordinating site into the DTE moiety in order to exploit the appealing photophysical and electrochemical properties associated with the well-studied motif. Although possessing photophysical properties inferior to related metal bipyridyl complexes, terpyridyl complexes are far better suited to single molecular device applications due to their well-defined geometries and their ability to form strong connections between electrodes.

The systems studied were synthesised *via* Suzuki methodology by reacting the *bis* boronic acid derivative of the DTE core with either the preformed Ru or Os terpyridyl units, to give **RuRu9(o)** and **OsOs9(o)** respectively (Fig. 5). Due to their higher rates of ligand exchange, Fe and Co analogues were formed by first synthesising the free ligand **9(o)** and subsequently reacting it with the terpyridyl–coordinated metal fragment. The electronic spectra of all the complexes showed mainly tpy based absorptions below 340 nm, with a clear shoulder at $\lambda = 370$ nm indicating the DTE-localised $S_0 \rightarrow S_1$

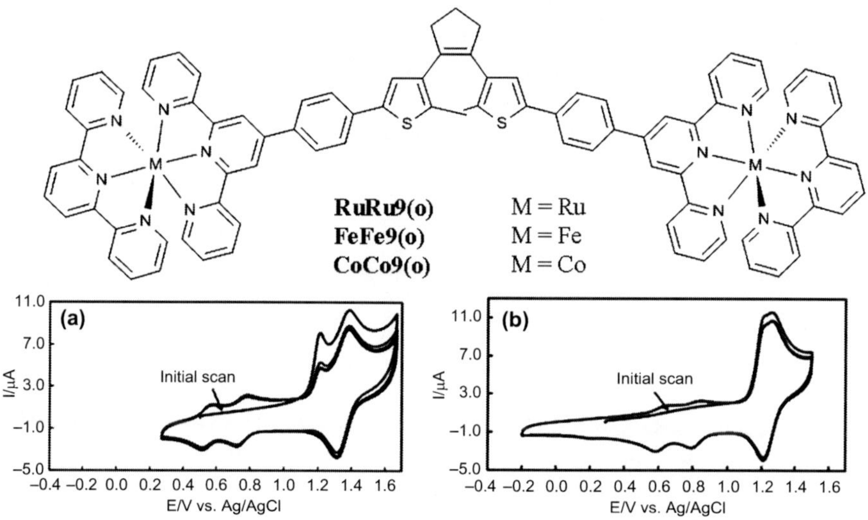

Fig. 5 (top) Representation of the complexes **RuRu9(o)**, **FeFe9(o)**, **CoCo9(o)**, (below) CV of (a) **RuRu9(o)** and (b) **FeFe9(o)** (0.3 mM in acetonitrile) at a scan rate of 100 mV/s. Reprinted (adapted) with permission from (Y. W. Zhong, N. Vila, J. C. Henderson, S. Flores-Torres and H. D. Abruña, *Inorganic Chemistry*, 2007, **46**, 10470–10472). Copyright (2007) American Chemical Society.

transition. Broad ^{1}MLCT bands were also observed in the visible region (from 489 nm–520 nm) in complexes **RuRu9(o)**, **OsOs9(o)** and **FeFe9(o).** Irradiation into any of the bands of **RuRu9(o)**, **OsOs9(o)** and **FeFe9(o)** failed to elicit ring closure. The authors rationalised this as the excited S_1 state of the central DTE core being quenched through the complex's MLCT state. The **CoCo9(o)** did not display any significant absorption above 450 nm, and irradiation at 350 nm did indeed result in the formation of the closed form, though the paramagnetism of the Co^{II} centre made PSS composition difficult to establish by NMR. Photocycloreversion did not take place. Electrocyclisation could however be carried out on the systems. An initial anodic sweep resulted in the irreversible oxidation of the thienyl groups at *ca.* 1.22 V *vs.* Ag/Ag, followed by the oxidation of the metal centres at higher potential (Fig. 5 (a) and (b)). Subsequent scans showed two reversible waves at + 0.54 V and + 0.75 V, ascribed to the redox processes of the fused thienyl system. The Co and Fe systems did not display electrocyclisation due to the oxidation of the metal occurring at lower potentials than the DTE core.

In a modification of their previous **MM9(o)** systems, Abruña *et al.*[21] synthesised a series of **M(10(o)R)$_2$** complexes using Fe^{II} and Co^{II} metal centres. The new design circumvented ligand exchange issues while allowing terminal groups to be attached to the non-coordinated thienyl ring in order to modify the physical properties (Fig. 6). **Fe(10(o)Cl)$_2$** showed no switching behaviour on irradiation, though electrochemical cyclisation did occur in separate studies. Irradiation of **Fe(10(o)Ph)$_2$** did show some formation of the closed isomer, though very slowly (10 h). This and its overlapping NMR spectra hampered further study. The cobalt containing compounds had better photochromic properties. Irradiation of **Co(10(o)Cl)$_2$** at $\lambda = 350$ nm for 3 h resulted in the "closed" IL peak at $\lambda = 534$ nm, indicative of the formation of the closed isomer. Despite its paramagnetic metal centre, ^{1}H NMR study was possible, giving a PSS composition of 40% photoconversion. Its diphenyl-terminated counterpart, **Co(10(o)Ph)$_2$**, gave quantitative conversion when irradiated at $\lambda = 380$ nm for 4 h, and cycloreversion could be achieved by irradiating into the low energy band at 580 nm of the closed form. The low energy bands of both **Co(10(o)Cl)$_2$** and **Co(10(o)Ph)$_2$** are significantly red shifted (*ca.* 30 nm) relative to their closed free ligands due to the electron withdrawing character of the conjugated metal centres. Despite its favourable photochromic activity, **Co(10(o)Cl)$_2$** proved to be electrochromically inactive. **Co(10(o)Ph)$_2$** did cyclise under electrochemical conditions however, showing the typical two reversible fused thienyl

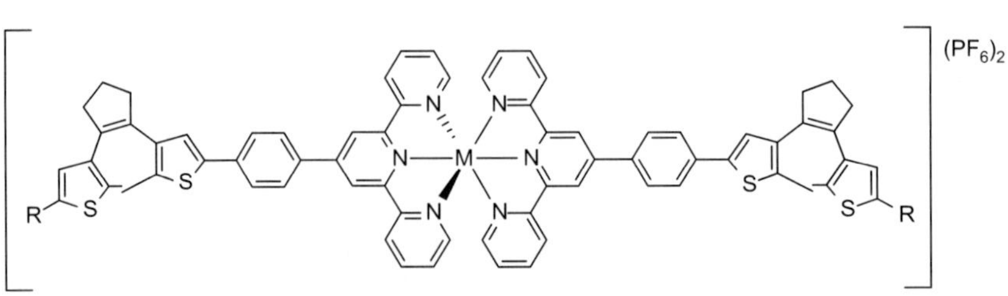

Fig. 6 Representation of the general **M(10(o)R)$_2$** system. R = –Cl, Ph.

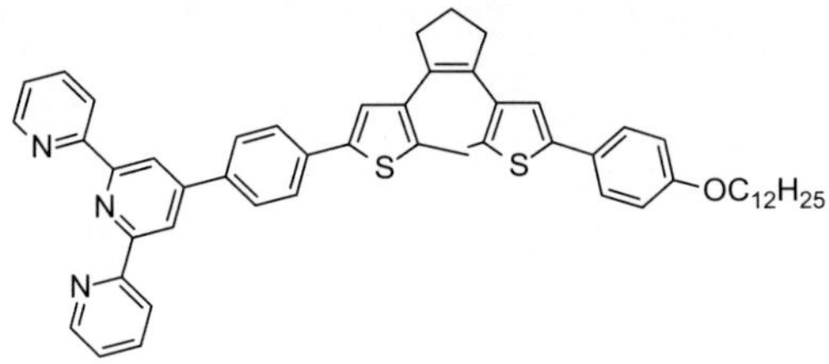

Fig. 7 Representation of **11(o)**.

oxidation processes (at +0.60 and +0.83 V) after the initial scan which oxidised the thienyl moiety.

Yi *et al.* have used a modified version of this DTE-based ligand **11(o)** (Fig. 7), replacing one of the pendant phenylterpyridine units with a non-coordinating solubilising moiety, $C_6H_4OC_{12}H_{25}$.[22] Over 5 cycles of irradiation in the Uv/visible region (λ = 365 nm, Φoc 23%, λ = 549 nm, Φco 22%, λ = 650 nm, Φco 2%) of the ligand showed good fatigue resistance, switching from the colourless, blue fluorescent open form (λ = 440 nm, $\Phi_F = 0.041$) to its non-emissive and slightly purple closed form (95% quenching in PSS). **11(o)** is found to coordinate to Zn(II) and Cu(II) with some decrease in fluorescent intensity, to be unaffected by 3.5 fold excess of alkali and alkaline earth metal ions and weakly responsive to the presence of Fe(II), Cd(II) and Hg(II).

When a THF solution of **11(o)** was titrated with an aqueous solution of $Zn(NO_3)_2$ a 1:1 complex was formed ($K = 1.76 \times 10^6\,M^{-1}$) with quenching of the fluorescence. The fluorescence could be switched back on by the addition of the chelate EDTA ($K = 3.16 \times 10^{16}\,M^{-1}$). This effect was examined in KB (human nasopharyngeal epidermal carcinoma) cells after incubation with **11(o)** and studied using confocal laser scanning microscopy. **11(o)** was found to have low toxicity (cellular viability 85% after 24 h exposure to 100 μM) and to undergo photoswitching within the cells; on the addition of Zn(II) solutions the fluorescence of **11(o)** was quenched and restored on exposure to EDTA. The ligand was presented as a potential fluorescent probe for metal ion transport through the cellular membrane.

1.4 Unusual N,O and P coordinating substituents

Park *et al.*[23] have synthesised a series of bis-cyclometalated Ir^{III} systems linked to a distal perfluorinated DTE moiety *via* a picolinate-derived ancillary ligand **12(o)** (Fig. 8). Ether linkages were used to electronically isolate the separate components of the system, as delocalisation of the excited states can interfere with the photochromic reaction. The open forms of each **Ir12(o)** compound were moderately luminescent, emitting between 483 and 556 nm, with quantum yields in the range 0.03–0.08. Upon photocyclisation with 355 nm light the emission was quenched to some extent due to spectral overlap, despite the differing multiplicities of the emission (being chiefly triplet in origin) and the singlet HOMO-LUMO transition of the switching moiety. This did not occur in a simple mixture of the untethered DTE switch and free Ir^{III} complex.

When **13** was synthesised with two DTE moieties an enhanced change in emission intensity was observed on closure, with a $\Delta\Phi_P = 10$, compared with

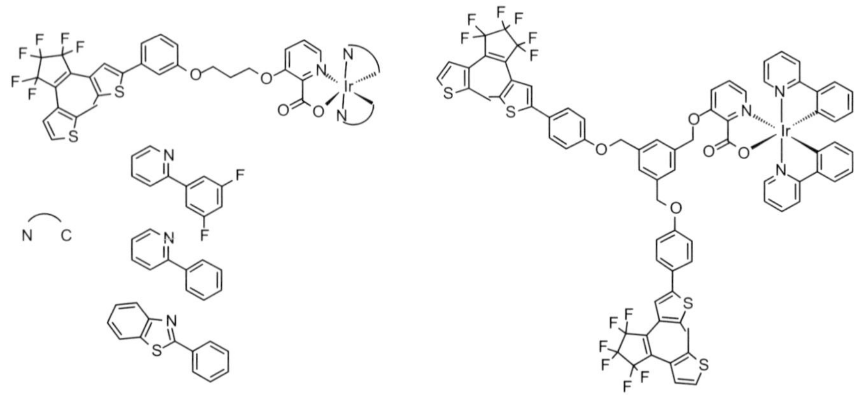

Fig. 8 Representations of **12(o)** (top left) and **13(o)** (right).

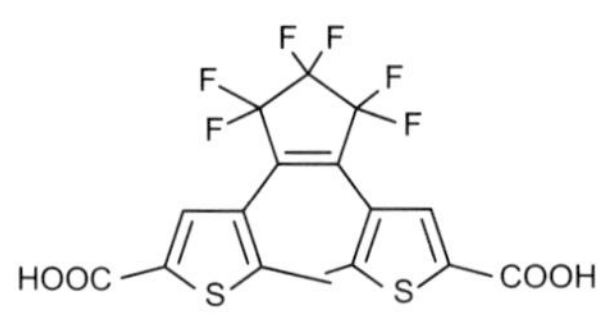

Fig. 9 Representations of **14(o)**

3–4 for the previous complexes. At PSS however, only 3% is bi-closed, with 87% having just one switching unit closed. All the systems were fully reversible. Further improvement in $\Delta\Phi_P$ was achieved by doping a polymer matrix with a high concentration of the complexes whereupon intermolecular energy transfer increases it to >50.

1.5 Pendant carboxylic functionalities

A dicarboxy perfluorylated DTE ligand, 1,2-Bis(2′-methyl-5′-(carboxylic acid)-3′-thienyl)perfluorocyclopentene **14(o)** was used as a bis-monodentate ligand (Fig. 9).[24] Binding through one carboxylic oxygen on each acid moiety it bridges in three separate experiments to distorted square based pyramidyl Cu(II) and Zn(II), and octahedral Co(II) metal centres to generate a set of 1D polymers. The single crystal X-ray structures of these reveal that in each case the two thienyl rings are antiparallel, with sufficiently short distances between reactive carbon centres for photochromism to occur. All three systems behave as photoreversible switches in the crystalline phase; one, [Co**14(o)**$(py)_2(MeOH)_2$], shows solvation/desolvation dependent properties.

Colourless **14(o)** closes as a solid on irradiation into its intense intraligand absorption band (305 nm), producing its purple, closed form with new absorptions at 390 and 610 nm. Its blue Cu(II) and colourless Zn(II) complexes become purple on irradiation at 278 and 300 nm respectively, to give the characteristic absorption bands of the closed form at 575 and 589 nm, however the Zn(II) complex shows less efficient conversion due to considerable H-bonding and the steric constraints of its auxiliary phenanthroline. On heating, the solid, red, solvated Co(II) complex loses both

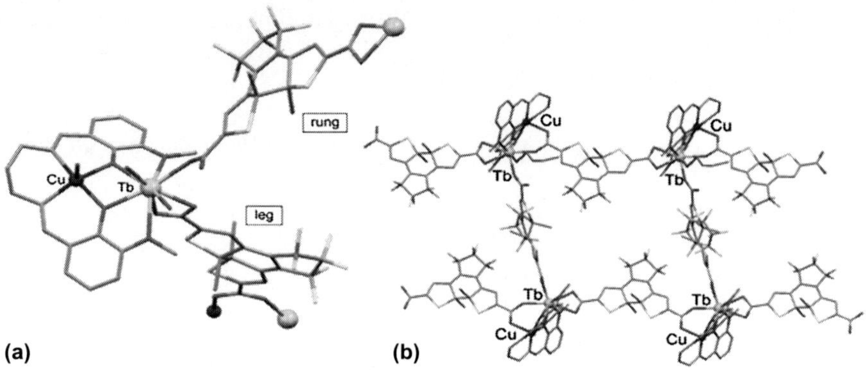

Fig. 10 Molecular structure of asymmetric unit of **2[CuTb]:2(14)(c)**, solvent and disorder removed for clarity. (b) Packing of **3[CuTb]:2(14)(c)** along the *a* axis.

coordinated MeOH molecules to give a pale purple desolvated form from which the solvated form can be restored by dissolving in MeOH. Both forms undergo photoswitching (MeOH solvated, **Co14(o)** absorbs at 306 nm, **Co14(c)** 575 nm, MeOH desolvated, **Cu14(o)** absorbs at 295 nm, **Co14(c)** 551 nm). The room temperature $\chi_M T$ of the solvated and desolvated open forms were different (3.55 compared to 2.89 $cm^3 K mol^{-1}$) as were the ESR spectra. The latter changed reversibly with photocyclisation of the ligand. Loss of solvent appears to cause changes in the environment of the metal centre but not in the interaction between metal centres.

The same ligand was used to assemble two single molecular magnets containing [Cu(II)Tb(III)] units.[25] The first (Fig. 10(a)) is a green, tetranuclear, ring-like cluster, with two [CuTb] units and two **14(o)** bound through O to both metal centres. The other is a purple 1D ladder-like structure involving a **3[CuTb]:2(14)(c)** metal:ligand ratio (Fig. 10(b)). Both demonstrate ferromagnetic interactions between the Cu(II) and Tb(III) ions due to the slow magnetic relaxation of the [CuTb] single molecule magnetic units and frequency dependence of their magnetic susceptibilities. The Uv/vis spectrum of the **CuTb14(o)** complex before irradiation with Uv light (323 nm) was very similar to that of **CuTb14(c)** after irradiation with visible light (> 480 nm) with the appearance of a band due to the closed DTE ligand at 600 nm. This corresponded to the colour change observed in powdered samples immersed in a $BaSO_4$ matrix. IR spectroscopy monitoring of the sample before and after irradiation revealed changes in the DTE stretching frequencies. Only the magnetic behaviour of the 3:2 ladder material was affected on irradiation as a result of photo-induced changes in its intermolecular (most probably inter-ladder) interactions.

1.6 Pendant phenanthroline coordinating units

A novel symmetric diarylethene **15(o)** bearing two 1,10-phenathroline moieties (1,2-*bis*[2-methyl-5-(1,10-phenanthrolin-2-yl)-3-thienyl]hexafluorocyclopentene) was found to reversibly switch between open and closed forms (Φoc 0.75, Φco 0.0066 (366 nm), PSS 99.5% conversion, Φco 0.0035 (578 nm)).[26] Complexation of this ligand with Fe(II/III), Cu(II), Ni(II), Zn(II) was monitored on the addition of equimolar, 2 and 10-fold amounts of metal salts to a fixed concentration of ligand and irradiating the solutions with visible light. Only when $[Cu(CH_3CN)_4]PF_6$ was added was any significant (bathochromic) shift observed on irradiation which could be indicative of complexation. Job plots indicated a 1:1 metal:ligand ratio and PM3 semi-empirical MO calculations suggested a 2:2 cyclic and C2-symmetric **CuCu(15(o))₂** complex. The complex underwent photocyclisation (Φoc 0.026, Φco 0.0028 (366 nm), PSS 92.1% conversion, Φco 0.0035 (578 nm)) less efficiently than the free ligand due to restraints (long CC-distances) in the complex. **15(c)** also forms a stable, predicated to be 2:2 symmetric D_2 complex (λ_{max}(abs) = 643 nm) which on irradiation with visible light regenerates the colourless **CuCu(15(o))₂** complex.

Scandola *et al.*[27] have presented detailed photophysical studies of dyads comprising a standard Ru(II) polypyridine chromophore attached directly or through a methylene spacer to a dithienylmaleimide (DAE) moiety (**Ru17**). For comparison purposes, analogues of both components of the dyad were also studied, including DAE (Fig. 11), and $[Ru(bpy)_2(phen)]^{2+}$. The Uv-vis spectrum of the complex **Ru17** is almost identical to the sum of the model compounds' spectra – indicating that the system is truly *supramolecular i.e.* consisting of separate photochromic and photosensitising moieties which are electronically decoupled in the ground state. The visible portion of the spectrum is dominated by the broad MLCT band of the Ru(II) complex unit, centred at approximately λ 470 nm. Irradiation at this wavelength results in the observation of the typical closed-form bands at λ 360 nm and 510 nm, with a PSS composition corresponding to 90% photocyclisation, as was confirmed by NMR. Photoreversion is achieved using λ > 550 nm irradiation and behaves similarly to DAE-Me, indicating that the Ru centre minimally perturbs the singlet pathway of cycloreversion. Irradiation into the isosbestic point (λ 470 nm) of the PSS still yields a composition of *ca.* 90% - *i.e.* irradiation into the MLCT state of the closed form is not an efficient means of effecting ring-opening (while being a very

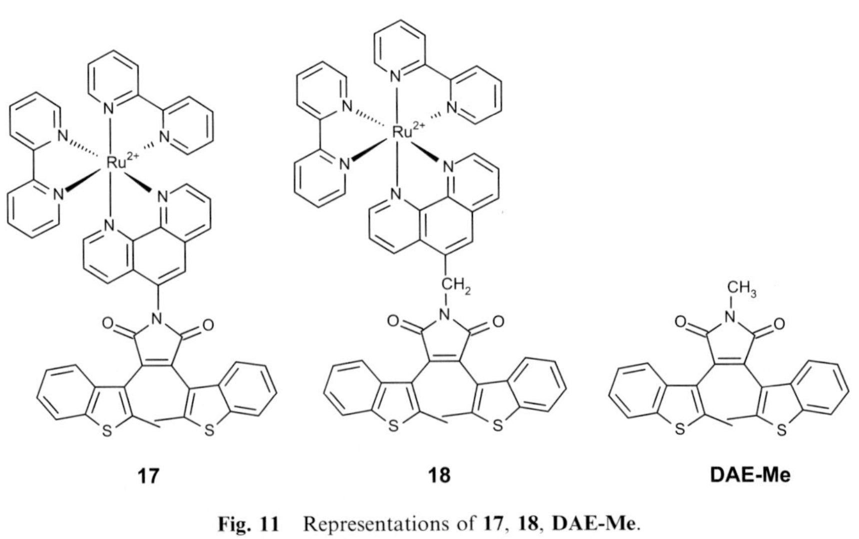

Fig. 11 Representations of **17**, **18**, **DAE-Me**.

efficient means of closing the open-form). DFT calculations show that this is due to the barrier to ring opening along the triplet potential energy surface being slightly higher (by *ca.* 5 kcal/mol) than for ring-closure.

Ru17 is weakly luminescent in CH_3CN solution, the emission ($\lambda_{max} = 600$ nm) assigned to ^{3}MLCT phosphorescence, with a slight shoulder visible at λ 570 nm arising from the DAE-like moiety's fluorescence. The intensity of both emissions is significantly quenched (about 100 fold) relative to optically matched solutions of the respective model compounds. In ultrafast spectroscopy studies (MeCN, λ_{exc} 400 nm) (see Fig. 12(a)) *ca.* 75% of the exciting light is absorbed by the metal complex and 25% by the DAE-derived chromophore. Typical signatures of both chromophores are visible in the initial spectra (1 ps after excitation): the bleaching of the MLCT band and the flat absorption of the reduced ligand, and a persistent absorption at λ 520 nm which is indicative of the formation of the closed-form. In the longer timescale spectra, a build-up of closed-form bands are observed. Kinetic analysis of the biphasic spectral changes (Fig. 12(b)) show the initial bleaching of the ^{1}MLCT, which from $t > 80$ ps, is replaced by an increase in the differential absorbance which is ascribed to the formation of the DAE-localised triplet. From these results it is evident that the excited singlet Ru(II) unit (whether populated directly or by singlet energy transfer from the excited DAE moiety) quickly transfers its energy to the triplet state Ru(II) unit. Intramolecular energy transfer then results in the sensitisation of the triplet DAE which results in cyclisation. This is

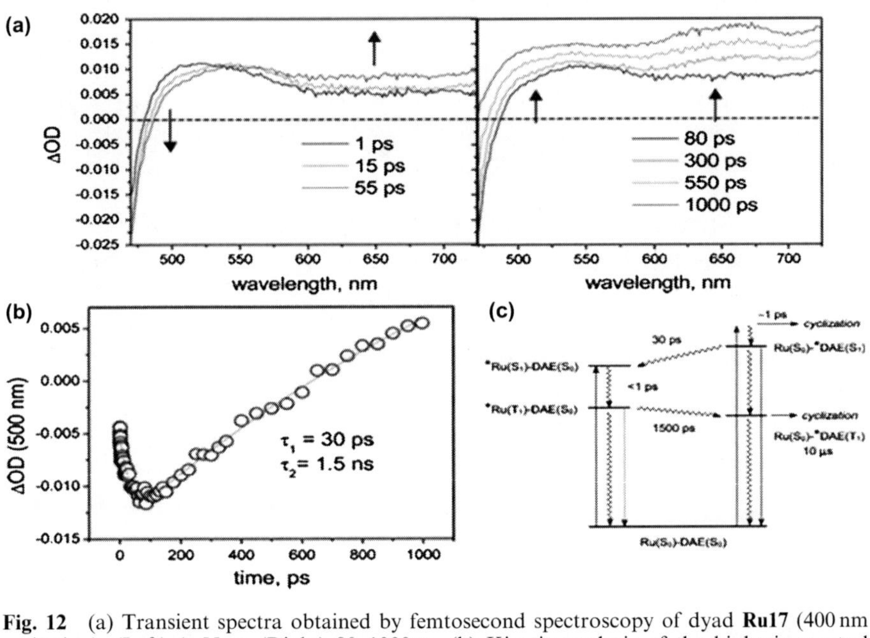

Fig. 12 (a) Transient spectra obtained by femtosecond spectroscopy of dyad **Ru17** (400 nm excitation). (Left) 1–55 ps; (Right) 80–1000 ps. (b) Kinetic analysis of the biphasic spectral changes observed in (a). (c) Qualitative energy level diagram for **Ru17**. Reprinted (adapted) with permission from (M. T. Indelli, S. Carli, M. Ghirotti, C. Chiorboli, M. Ravaglia, M. Garavelli and F. Scandola, *Journal of the American Chemical Society*, 2008, **130**, 7286–7299). Copyright (2008) American Chemical Society.

summarised in Fig. 12(c). An odd result arising from these measurements indicates that the photocyclisation quantum yield is oxygen quenched by a factor of 2, whereas the triplet lifetime is oxygen quenched by a factor of ten. The authors ascribe this anomaly to the impossibility of discriminating the excited states of the parallel and antiparallel conformers.

Ru18 behaviour is broadly very similar, though the singlet energy transfer time constant has slowed from 30 ps for **Ru17** to 150 ps. This, however, facilitated the study of the system using nanosecond flash photolysis. The spectral changes observed show the features of the Ru(II) triplet state: bleaching at 450 nm and emission at 610 nm, evolving into the typical spectrum of the DAE triplet (with a time constant of 40 ns, compared to 1.5 ns in **Ru17**). The authors attribute the slower processes arising from insertion of the methylene spacer to the difference in energy transfer processes – Förster for the singlet singlet processes, Dexter for the triplet energy transfer processes.

1.7 Inherent phenanthroline units directly incorporated into the DTE

By attaching the switching moiety directly to a phenanthroline unit using Suzuki methodology, Yam *et al.* suceeded in incorporating the ligand itself into part of the overall DTE unit. Treatment of the ligand with [Re(CO)$_5$Cl] resulted in the rhenium(I) diimine system **Re19(o)** (Scheme 7(a). In collaboration with Phillips,[28] an in-depth study of the system's photocyclisation pathways was carried out through spectroscopic studies. The electronic spectrum of **Re19(o)** shows an intense IL (π–π*) band at 300 nm, with a shoulder at 400 nm ascribed to [dπ(Re)→π*(phen)] MLCT derived transition. The complex is emissive in solution, with a relatively broad emission centred at λ_{max} = 595 nm, and as a glass at 77 K, with the emission blue shifted by 60 nm due to the rigidochromic effect. Irradiation into the IL or MLCT states results in the formation of **Re19(c)** ($\Phi_{photocyc\ (MLCT)}$ = 0.648), evidenced by the appearance of the typical closed-switch transitions as a broad band at *ca.* 580 nm, accompanied by a very intense band at 386 nm. Both bands were also observed in the photocyclised free ligand. The emission spectrum of the closed form, **Re19(c)**, is red-shifted to *ca.* 640 nm and shows vibrational progressions typical of ν(C-S). Both observations are indicative of the emissive state now being chiefly of IL origin (due to the increased conjugation of **Re19(c)** resulting in the lowering the energy of the

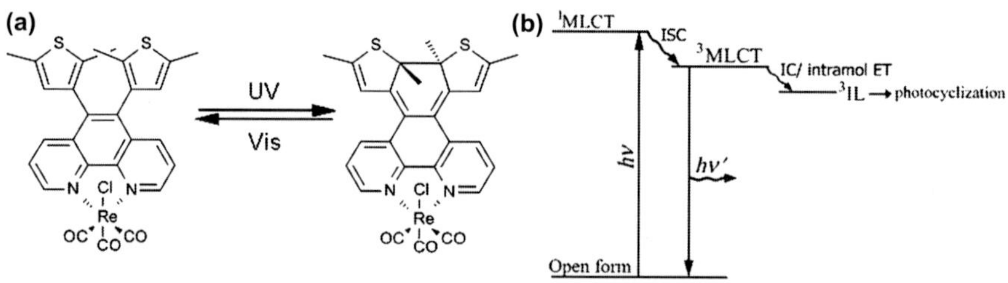

Scheme 7 (a) Schematic of the photochromic process of **Re19(o)** (b) Qualitative energy level diagram for the triplet-sensitised photocyclisation of **Re19(o)**. Reprinted (adapted) with permission from (V. W.-W. Yam, C.-C. Ko and N. Zhu, *Journal of the American Chemical Society*, 2004, **126**, 12734–12735). Copyright (2004) American Chemical Society.

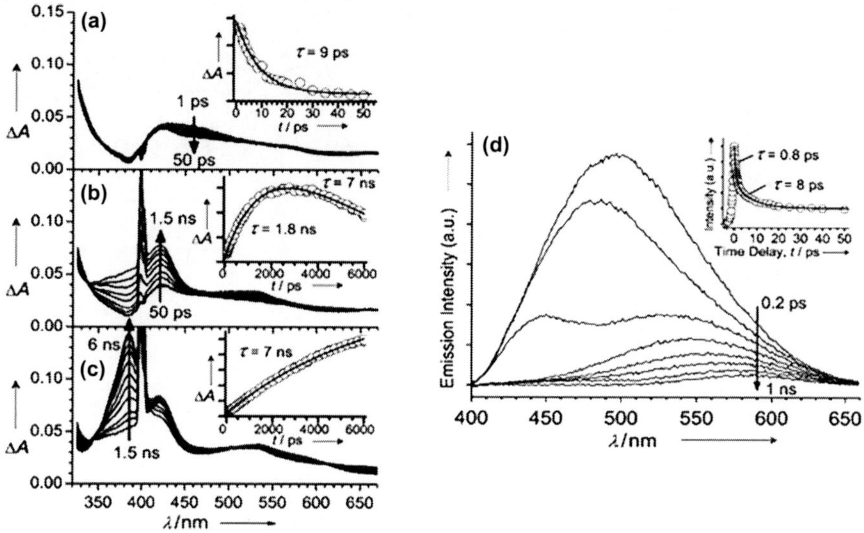

Fig. 13 Transient absorption spectra 400 nm excitation at a series of pump–probe delays: a) from 1 to 50 ps; b) from 50 to 1500 ps; c) from 1.5 to 6 ns. The insets show absorption–time profiles at a) 460 nm, b) 420 nm, and c) 386 nm. The sharp features around 400 nm are due to artefacts from the pump pulses. (d) Kerr gated time-resolved emission. The time delays from top to bottom are 0.2, 0.5, 1, 2, 5, 10, 30, and 1000 ps. The inset shows the emission time profile at 590 nm. From C. C. Ko, W. M. Kwok, V. W. W. Yam and D. L. Phillips, *Chemistry – A European Journal*, 2006, **12**, 5840–5848. Copyright © 2006 by John Wiley & Sons, Inc. Reprinted by permission of John Wiley & Sons, Inc.

IL state below that of the MLCT state). Irradiation with light of $\lambda = 510$ nm resulted in slow cycloreversion, with a quantum yield of $\Phi_{cyclorev} = 0.009$. Transient absorption spectra of **Re19** (Fig. 13(a)) showed a fast decay ($\tau = 9$ ps) at *ca.* 460 nm soon after excitation ($\lambda_{exc} = 400$ nm) and was ascribed to vibrational relaxation of the hot ^{3}MLCT, implying that ISC from the ^{1}MLCT occurred quickly and far from the equilibrium ^{3}MLCT geometry. This is followed by the growth ($\tau = 1.8$ ns) and slower decay ($\tau = 7$ ns) of an absorbance feature at 460 nm, indicating the internal conversion of the ^{3}MLCT to the ^{3}IL state, which slowly decays giving rise to bands also seen in the steady state spectra of the closed form. The increase in the most intense of these, at 386 nm, was monitored, giving rise to a time constant of $\tau = 7$ ps. The fact that the time constant for the decay of the ^{3}IL state matches that of the formation of the closed form, and the observation of two isosbestic points in the transient spectra, indicate the clean conversion of the excited IL triplet to the closed form. Kerr gated time-resolved emission studies (Fig. 13(d)) showed an initial ^{1}MLCT emission at 500 nm, lasting for 2 ps after excitation with a 300 nm pulse. This transient band was soon replaced by emission arising from the vibrationally hot ^{3}MLCT at *ca.* 550 nm. This emission gradually red-shifts as the system relaxes to the equilibrated ^{3}MLCT state, resulting in emission at 590 nm. Biexponential fitting of the decay of this peak results in fast ($\tau = 0.8$ ps) and slow ($\tau = 8$ ps) time constant components. These are ascribed to ISC from the ^{1}MLCT to the ^{3}MLCT, and the relaxation of the hot ^{3}MLCT respectively. The processes are summarised in Scheme 7(b).

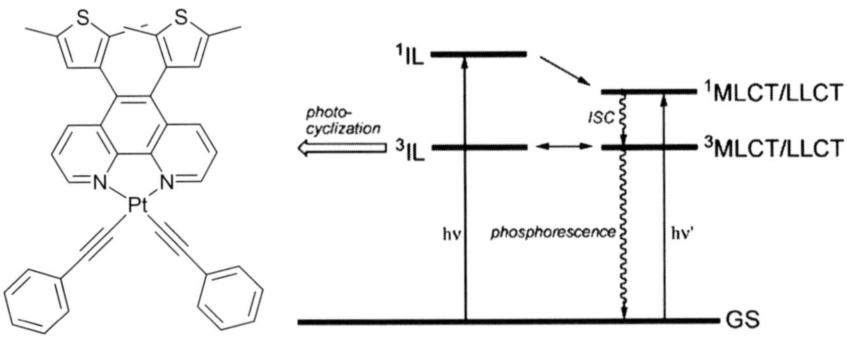

Scheme 8 (left) Schematic of Re**19(o)**. (right) Qualitative energy diagram for the photocyclisation process. Reprinted (adapted) with permission from (J. K. W. Lee, C. C. Ko, K. M. C. Wong, N. Zhu and V. W. W. Yam, *Organometallics*, 2007, **26**, 12–15). Copyright (2007) American Chemical Society.

1.8 DTE-based N-coordination to Pt(II) acetylide

The complex **[Pt(19(o))(C_2Ph)$_2$]** (Scheme 8). was synthesised from **[Pt(19(o))Cl_2]** and phenyl acetylene using Sonogashira methodology.[29] As previously observed with this ligand, evidence for parallel and antiparallel conformers is observed in the NMR spectra, and X-ray crystal structures of both parallel and antiparallel conformers were resolved from a co-crystal of the complex. The Uv-vis spectrum of the complex has a MLCT band at 390 nm which probably incorporates some LLCT character. Irradiation of this ($\lambda_{exc} \geq 480$ nm) or the IL band (λ_{exc} 313 nm) results in photocyclisation, generating the expected "closed" band at 560 nm, and the hyperchromic shift of a band at 360 nm. The photocyclisation quantum yields obtained for MLCT and IL irradiation are similar, indicating efficient intersystem crossing into the ^{3}MLCT, in turn sensitizing the ^{3}IL state and leading to cyclisation (described in Scheme 8 (right)). Photocycloreversion can also be achieved using light of $\lambda_{exc} \geq 500$ nm, again with relatively low quantum yields. The complex is luminescent in its open form ($\lambda_{max} = 405$ nm) possibly arising from a phenanthroline-based IL fluorescence.

2 Pendant Pt(II) acetylides

Although several authors have reported the inter-mixing of MLCT and IL states in a perfluorinated DTE metal complex, the recent work of Raithby *et al.*[30] deftly uses single crystal analysis to explore the photochromism of an organometallic DTE complex bearing two symmetrically placed platinum acetylide units (Scheme 9). The striking feature of this complex is that the switching process is highly efficient and driven by visible light.

The ground state UV-vis spectra are shown in Fig. 14(A). The introduction of the {Pt(PEt_3)Ph} units into the uncomplexed ligand results in a new band at 335 nm ($\varepsilon = 5.4 \times 10^4 M^{-1} cm^{-1}$) which DFT calculations suggest is a HOMO – LUMO transition, incorporating partial charge transfer character from electron density in the Pt/alkynyl/thiophene region shifting onto the perfluorocyclopentene. In addition the spectrum of **PtPt20(o)** has a broad absorption that extends into the lower energy region and slightly beyond λ 400 nm. This is assigned to a disallowed transition

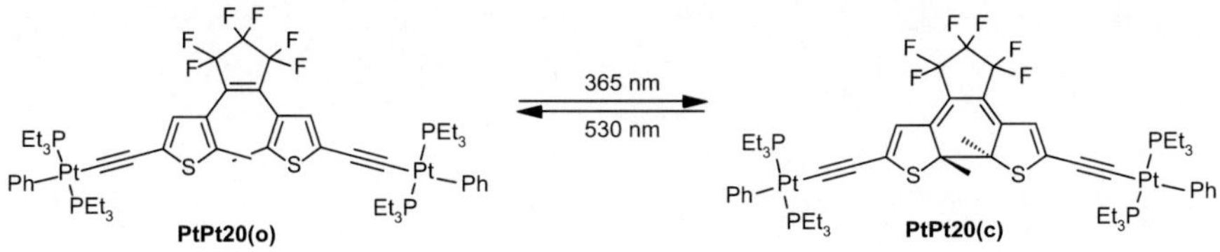

Scheme 9 Photochromic reaction of **PtPt20**.

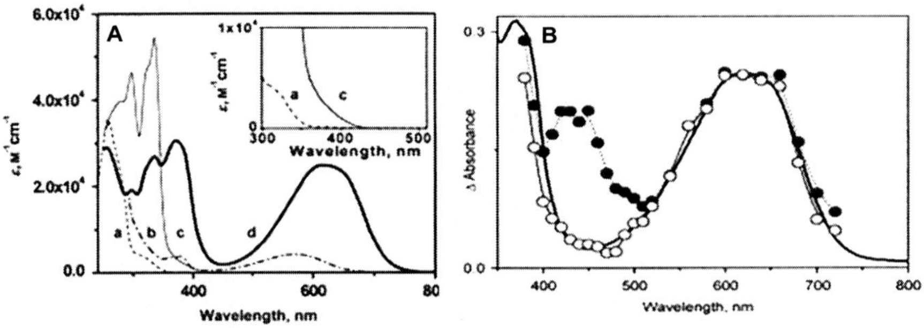

Fig. 14 (A) Absorption spectra of **20(o)** (a), **20(c)** (b), **PtPt20(o)** (c) and **PtPt20(c)** (d, photostationary state) in CH_2Cl_2 at RT. Inset: An expanded low energy region of the absorption spectrum of **20(o)** and **PtPt20(o)**. (B) Transient absorption spectra obtained following a 7 ns laser pulse: 270 ns after the laser pulse (black circles); 18 ms after the laser pulse (unfilled circles). The scaled absorption spectrum of **PtPt20(c)** is shown for comparison (solid line). From S. K. Brayshaw, S. Schiffers, A. J. Stevenson, S. J. Teat, M. R. Warren, R. D. Bennett, I. V. Sazanovich, A. R. Buckley, J. A. Weinstein and P. R. Raithby, *Chemistry – A European Journal*, 2011, **17**, 4385–4395, 4th Edition. Copyright © 2011 by John Wiley & Sons, Inc. Reprinted by permission of John Wiley & Sons, Inc.

facilitated by the heavy metal centres into triplet state(s), as is observed in other similar systems.[31] On irradiation with visible light (λ_{exc} 420 nm, corresponding to an absorption of $\varepsilon = 240\,M^{-1}cm^{-1}$) a typical broad, intense absorption band at $\lambda \sim 635$ nm emerges, characteristic of the ring-closed DTE species (the same absorption spectrum is also achieved on irradiation with UV light). If the assignments are correct the implication is that ring closure in the complex has been achieved through *direct* irradiation into a state of triplet multiplicity. By irradiating into the intense absorption of **PtPt20(c)**, an almost quantitative reversion to **PtPt20(o)** is achieved, albeit at a far slower rate – the reaction takes several hours to complete.

Fig. 14(B) shows transient spectra of **PtPt20(o)** (in acetonitrile) generated from a 7 ns laser pulse and acquired after 270 ns, and 18 μs, delays. The spectrum after the long delay is attributed to the closed form of the complex, as it is identical to the steady state Uv-vis spectrum of **PtPt20(c)**. The spectrum obtained at the shorter delay time, however, shows a transient species absorbing at $\lambda = 450$ nm with a lifetime of 32 ± 4 μs. The intensity of this band is proportional to the concentration of the open isomer, and therefore must result from it. This behaviour along with its long lifetime and independent kinetics (relative to the other persistent band at λ 635 nm) provides strong evidence for its assignment as an absorption arising from a twisted triplet excited state. The band is absent in similar studies of the cycloreversion of **PtPt20(c)** to **PtPt20(o)**, which is a purely "singlet" phenomenon. Although not without precedent, the band is not significantly quenched by oxygen as is more usual for a species attributed to a triplet state.[32] The absence of any charge transfer character was established from the lack of solvatochromism in the band on repeating the experiment in toluene. In this system Resonance Raman spectroscopy proved a potentially useful method for data readout. The acetylide $\nu(C{\equiv}C)$ stretch is a useful marker for monitoring the ring closing/opening reaction as excitation wavelengths (λ 632.8/830 nm) are suitable, and do not initiate photochromic behaviour. The vibration also exists in a region free from other vibrations, and exhibits a dramatic increase in intensity on going from **PtPt20(o)** to **PtPt20(c)**. Uv irradiation of crystals of **PtPt20(o)** (over 20 mins) results in a "single crystal to single crystal" transformation to the ring closed form with an impressive (though insufficient for many applications) conversion of 80%. Unfortunately, irradiation with longer wavelength light does not result in cycloreversion, probably due to the low Φ_{OC} or constraints within the lattice. The powder form of the compound also demonstrates photocyclisation (60% conversion as determined by NMR spectroscopy).

A perfluorinated DTE ligand comprising an acetylene coupled Pt(II) terpyridine was investigated in order to observe how photoswitching would perturb the metal-based excited state and thereby its application as a photosensitiser for water reduction and in the photoinduced oxidation of alkenes.[33] **Pt21(o)tpy** exhibits both intraligand and mixed MLCT/LLCT states (λ_{max} 285 nm (ε 52,800 $dm^3 mol^{-1} cm^{-1}$), 465 nm (ε 6,700 $L mol^{-1} cm^{-1}$) and photocloses on irradiation into either the DTE chromophore (λ_{exc} 302 nm) or the MLCT/LLCT absorption bands (λ_{exc} 465 nm) to produce **Pt21(c)tpy**. The latter has a new band at 600 nm which is a combination of the red shifted MLCT and the π–π* transition of the ring-closed DTE unit. The red-shift is induced by the increased π-conjugation in **Pt21(o)tpy** and more particularly the higher energy of the HOMO of the ring-closed isomer, which DFT indicates has a lesser contribution from the metal. Indirect excitation produces a lower amount of **Pt21(c)tpy** at PSS however electronic coupling allows energy transfer between adjacent ^{3}MLCT/LLCT and ^{3}IL states. DFT calculates these as ≥ 2.0 eV and 1.9 eV respectively.

Building on the work of Castellano who demonstrated that DABCO is an efficient reducing agent for the transient hole localized on Pt-acetylides in charge separated excited states,[34] Wolf *et al.* followed up their earlier work by performing quenching studies on **Pt21(c)tpy** (Fig. 15) in the presence and absence of DABCO.[35] At high concentrations of DABCO the cyclisation

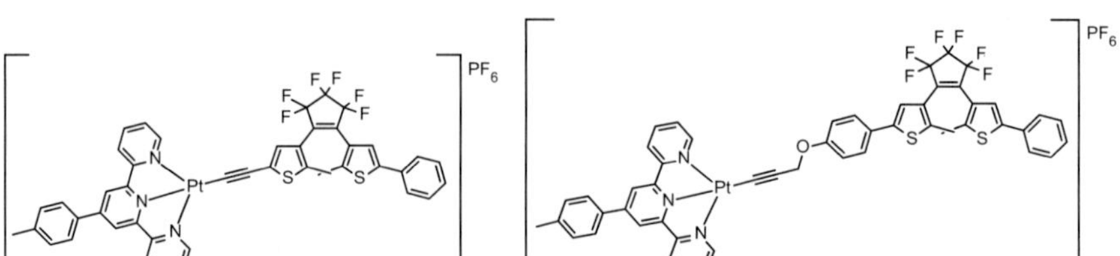

Fig. 15 Representations of **Pt21(o)tpy** and **Pt22(o)tpy**.

process in **Pt21(c)tpy** is inhibited when the CT state is irradiated with visible light, but relatively unaffected when the DTE-localised IL states are directly irradiated using UV light. The metal sensitized ring closure is a slower process (ns range), and therefore more susceptible to intermolecular quenching, unlike the faster IL route (ps timescale). The rate of energy transfer was found to reduce on cooling: a EtOH:MeOH (4:1) solution of **Pt21(o)tpy**, frozen to a glass at 85 K did not undergo cyclisation under visible irradiation and only began to increase at the melting point of the glass (130–150 K). Photocyclisation was found to occur down to 90 K when UV light was used. DFT calculations indicated a 0.02 eV barrier to energy transfer from the ^{3}CT to the ^{3}IL states but below temperatures of *ca.* 140 K there was insufficient thermal energy to surmount this barrier. At low temperatures phosphorescence also competes with photocyclisation, and a strong orange emission is observed at λ_{max} *ca.* 550 nm at 85 K, disappearing at temperatures above 125 K.

The transient spectrum of **Pt21(o)tpy**, (Fig. 16) was measured at 800 ps following a 355 nm pulse and assigned to the ^{3}IL state of the DTE, which is formed faster than the time resolution of the instrument. Excitation of, **Pt21(c)tpy**, resulted in the intense bleaching of the band due to the closed form (560 nm–700 nm), indicative of photocycloreversion. The addition of the ether linkage in **Pt22(o)tpy** (Fig. 15) results in limited ground-state interaction between the ^{3}CT and ^{3}IL state. Although the excited states were proven to be similar to those of **Pt21(o)tpy**, the processes involved in photochromic behaviour were slower due to poorer orbital overlap. Irradiation of the CT state of **Pt22(o)tpy** induces photocyclisation, but at a slower rate. The quantum yields of cyclisation are about four times lower for **Pt22(o)tpy**. The transient absorption spectra of **Pt22(o)tpy** obtained at 800 ps following 355 nm excitation is again ascribed to the ^{3}IL state due to comparison with the spectrum of a similar complex not containing the photochromic unit. The transient spectra of **Pt21(c)tpy** and **Pt22(c)tpy** were essentially identical, and were ascribed to the DTE-delocalised ^{3}IL state. This state is characterised by bleaching of the "closed" π–π* band, accompanied by a new band at higher energy. Considering that the ring closed form of the DTE exhibits strong absorption π–π* around 355 nm the

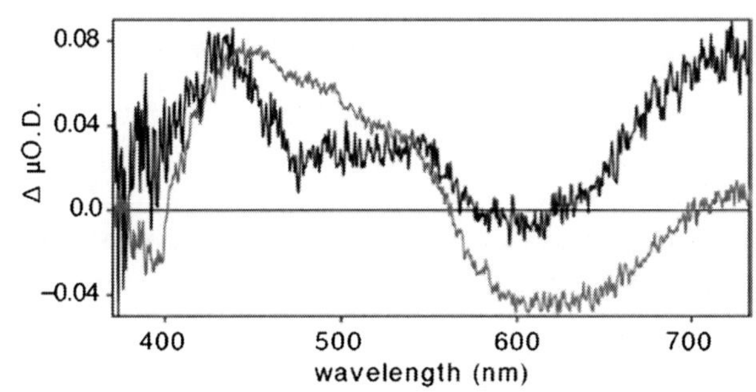

Fig. 16 Transient absorption difference spectrum of **Pt21(o)tpy** (black) and **Pt21(o)tpy** (grey) measured at 800 ps following a 355 nm laser pulse. Reprinted (adapted) with permission from (M. N. Roberts, J. K. Nagle, J. G. Finden, N. R. Branda and M. O. Wolf, *Inorganic Chemistry*, 2008, **48**, 19–21). Copyright (2008) American Chemical Society.

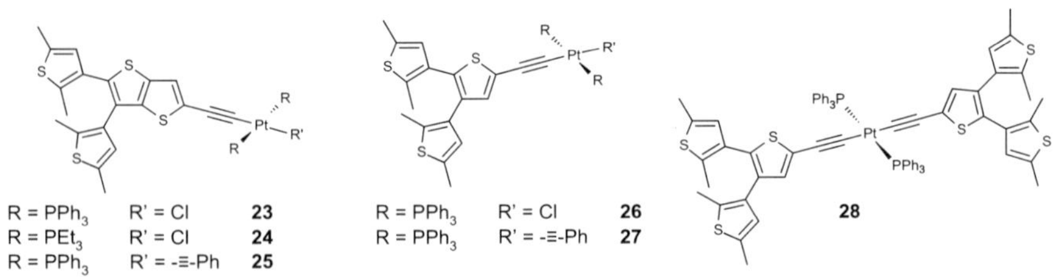

Fig. 17 Representations of compounds **23**–**28**.

authors considered that direct excitation of the ^{1}IL state is followed by rapid intersystem crossing to the ^{3}IL state, facilitated by the metal, but they also did not discount population of the ^{3}CT followed by population of the ^{3}IL state (Fig. 16).

Yam *et al.*[36] have synthesised a range of Pt acetylides containing a DTE moiety attached to either a thienyl (**26**, **27** and **28**) or a fused bis-thienyl core (**23**–**25**) (see Fig. 17).The complexes dissolve in benzene, giving clear yellow solutions. Their intense absorptions in the $\lambda = 300$–350 nm region are ascribed to intraligand transitions centred on the phosphine and thiophene portions of the complexes, while lower energy bands centred at 370–390 nm are ascribed to MLCT transitions [dπ(Pt) - π*(thiophene-containing alkynyl)] due to their sensitivity to both the auxiliary ligands on the Pt(II), and the nature of the thiophene core. The large ε of this low energy band (in the order of $10^4\,dm^3\,mol^{-1}\,cm^{-1}$) also points toward a significant IL [π–π*(thiophene- containing alkynyl)] contribution. Excitation ($\lambda_{exc} = 360$ nm) results in a red emission (λ_{em} 630–670 nm). The long lifetimes (μs) and the large Stokes shift both indicate phosphorescence, and hence the emission is tenuously assigned to comprise mainly ^{3}IL facilitated by the heavy metal. An assignment of the emission as arising from a ^{3}MLCT state was deemed improbable due to the lack of any obvious trend in the luminescence of the compounds relative to each another. Low temperature studies showed well-defined vibrational progressions, characteristic of the ligands and confirming the ^{3}IL nature of the assignment, but the authors hint at a possible ^{3}MLCT contribution due to small changes observed on changing the ancillary ligand.

Exciting the complexes ($\lambda_{exc} = 370$ nm) resulted in the purple colouration of the solutions and the emergence of the typical low energy band at *ca.* 560 nm of the closed form. This band is only slightly perturbed by the Pt centre and ancillary ligands and is indicative of an allowed transition in the ring-closed *8a,8b*-dimethyl-1,8-thia-*as*-indacene moiety. As evidenced by the appearance of clear isosbestic points in its Uv-vis spectrum during irradiation, **28** does not undergo dual cyclisation due to one adjacent closed form quenching the photocyclisation step of the second. Photocyclisation quantum yields are relatively high (between 0.226 and 0.117). The lower of these was associated with complexes bearing weak field chloro ligands, as these give rise to a low lying ligand field excited state which undergoes facile deactivation and competes with cyclisation and luminescence. The thermal stability of the closed form of **28** was determined by Arrhenius methods in argon flushed toluene, and was found to be 1270 minutes at 298 K, falling to 257 minutes at 313 K (Fig. 18).

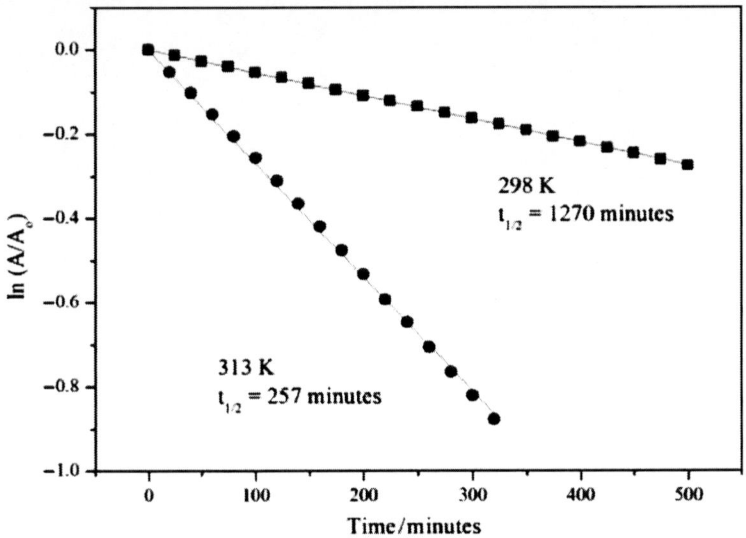

Fig. 18 Thermal stability studies of **28(c)**. Reprinted (adapted) with permission from (H. L. Wong, C. H. Tao, N. Zhu and V. W. W. Yam, *Inorganic Chemistry*, 2010, **50**, 471–481). Copyright (2010) American Chemical Society.

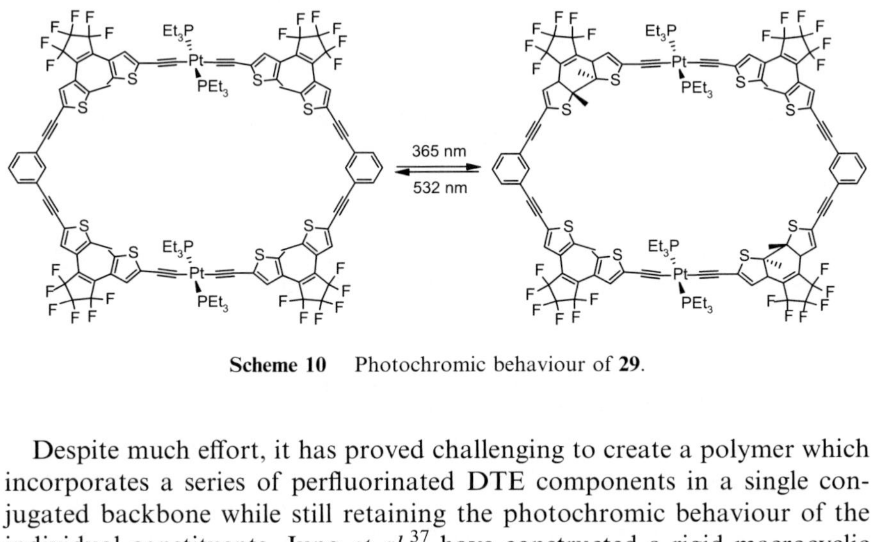

Scheme 10 Photochromic behaviour of **29**.

Despite much effort, it has proved challenging to create a polymer which incorporates a series of perfluorinated DTE components in a single conjugated backbone while still retaining the photochromic behaviour of the individual constituents. Jung *et al.*[37] have constructed a rigid macrocyclic system consisting of four DTE moieties bridged by two Pt centres, **29** (Scheme 10).

Irradiation in chloroform solution with light of λ 325 nm results in a new band at 628 nm, giving rise to a dark blue colouration to the solution. The blue compound was isolated using HPLC techniques and NMR characterisation showed the formation of a closed-open-closed-open (COCO) isomer. The fact that no further cyclisation would occur is thought to be due to the excited states of the open form being efficiently quenched by energy transfer to the adjacent closed form. Exposure to light of λ > 532 nm for 3 mins, results in the solution becoming completely bleached to the open

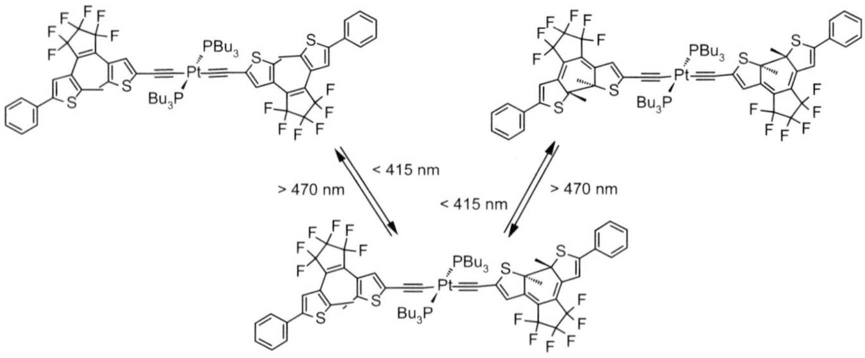

Scheme 11 Schematic of the photochromic processes in Pt**21**(o)**21**(o).

form. The quantum yield from OOOO to COCO was determined to be 0.64, the high value being attributed to the photoactive antiparallel conformation being enforced by the restrictions of the ring. To circumvent this, Wolf *et al.* have designed a system consisting of two acetylene-terminated DTEs coupled in a *trans* geometry to a Pt–phosphine centre (**Pt21(o)21(o)**). Pt acetylides are ideal for this purpose as calculations show that the singlet state is generally located on one conjugated ligand, not on the entire oligomer. This means that successive cyclisations of adjacent DTEs can be triggered through the triplet states localized on the ligand (*via* irradiation with visible light), without the open form having the possibility of transferring its energy to an adjacent closed form (Scheme 11).

In this example[38] the sequential manner in which the photochromic process takes place is evidenced by the observation of isosbestic points which arise upon irradiating the system with UV light (< 415 nm), and mark the presence of two species (**Pt21(o)21(o)** and **Pt21(o)21(c)**) in equilibrium. The isosbestic points disappear on further irradiation as the formation of the third component **Pt21(c)21(c)** establishes a more complex equilibrium. The relative concentration of each species was monitored using ^{31}P NMR spectroscopy. This revealed that significant amounts of **Pt21(o)21(c)** were formed before **Pt21(c)21(c)** (PSS composition 80% **Pt21(c)21(c)** and 20% **Pt21(o)21(c)**). A significant red shift (20 nm) is observed when **Pt21(o)21(o)** is irradiated to the PSS. This is ascribed to the singlet state becoming delocalized over the entire conjugated system upon closure, a view that was further supported by DFT investigations. Differential pulse voltammetry of **Pt21(o)21(o)** shows a single oxidation at 1.00 V (*vs.* SCE), corresponding to oxidation of the DTE portion of the complex (Fig. 19(a)). Two additional oxidations appear on conversion to the **Pt21(o)21(c)** form (0.65, 0.90 V) and a single oxidation process is observed for **Pt21(c)21(c)** (0.74 V). The cathodic shift in the DTE oxidation of **Pt21(c)21(c)** relative to the others is a result of the increased electron-accepting character of the conjugated fully-closed form, with increasing backbonding to the Pt centre.

Ground state electronic communication in the fully-closed **Pt21(c)21(c)** was investigated by studying the vis-NIR spectral properties of the chemically oxidised PSS solution. Oxidation with $[SbCl_6]^-$ generates several new

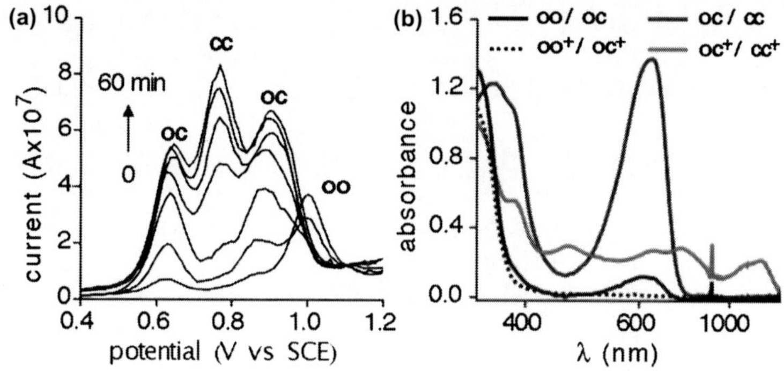

Fig. 19 (a) Changes in the differential pulse voltammograms of **Pt21(o)21(o)** as it is irradiated with 365 nm light. (b) vis-NIR absorption spectra of **Pt21(o)21(o)/Pt21(o)21(c)** (black) and **Pt21(o)21(c)/Pt21(c)21(c)** (blue) before oxidation and after 1 equivalent of oxidant is added to generate **Pt21(o)21(o)**$^{+}$/**Pt21(o)21 (c)**$^{+}$ (dash) and **Pt21(o)21(c)**$^{+}$/**Pt21(c)21(c)**$^{+}$(grey). Colour representation available in the online version of this manuscript. Reprinted (adapted) with permission from (M. N. Roberts, J. K. Nagle, J. G. Finden, N. R. Branda and M. O. Wolf, *Inorganic Chemistry*, 2008, **48**, 19–21). Copyright (2008) American Chemical Society.

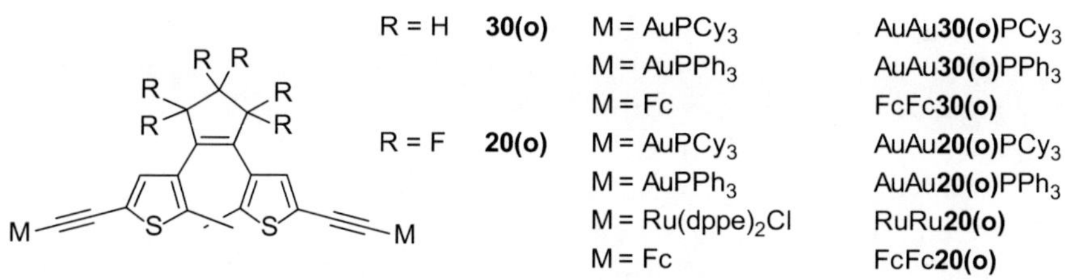

Fig. 20 Various acetylide/ethynyl-ferrocene complexes based on ligands **20(o)** and **30(o)**.

bands not observed in solutions containing a mixture of **Pt21(o)21(o)** and **Pt21(o)21(c)**. The most notable of these at $\lambda = 1301\,cm^{-1}$ was assigned to Intervalence Charge Transfer (IVCT) and arises from the optically-induced exchange of charge between the DTE moieties through the Pt bridge of **Pt21(c)21(c)**$^{+}$ (Fig. 19(b)). This is clear evidence for a large degree of electronic coupling between the linked DTEs in the delocalised singlet state.

3 Alternative metal acetylides (Au(I), Ru(II), CpFe)

Alkynyl gold(I) complexes incorporating tertiary phosphine ligands have received considerable attention due to their stability, ease of preparation and rich photophysical properties. Liu *et al.*[39] have incorporated this motif into the typical perfluorinated and hexahydro DTE design through the reaction of PCy_3 and PPh_3 chlorogold(I) complexes with ligand perfluorinated and perhydro versions of the bis-alkynyl ligands, **20(o)** and **30(o)** (Fig. 20).

The Uv spectra of the complexes in CH_2Cl_2 are unsurprising and consist of structured bands at low energy (between 283 nm–320 nm) ascribed to metal perturbed intraligand transitions with some MLCT character. The complexes exhibit room temperature luminescence in CH_2Cl_2 solution and all undergo photocyclisation on UV irradiation (λ_{exc} 302 nm); the hexahydro complexes turning red and the hexafluoro derivatives blue. Altering the nature of the phosphine ligands has no significant effect on the absorption spectra of either form.

Visible light ($\lambda > 420$ nm) bleaches the solutions, converting them back to their colourless open forms. The reversibility of the reaction was demonstrated for **AuAu30(o)PCy$_3$** by carrying out ten cycles at room temperature with no appreciable loss in absorption. On comparison with the free ligands metallation leads to significant increases in cyclisation quantum yields (*e.g.* 0.49 to 0.61 in **AuAu30(o)PCy$_3$**). PSS compositions were also increased for the metal containing species. Although photocycloreversion quantum yields showed no such consistent pattern, the metal complexes reverted to their open form at a slower rate than the metal-free species. The authors suggest that this is due to the metal centres affecting the stability of the ring closed isomers.

Rigaut *et al.*[7] have synthesised a novel dinuclear ruthenium acetylide switch **RuRu20(o)** by the reaction of an ethynyl perfluorinated DTE with $[Cl(dppe)_2Ru]^+$ (Fig. 20). ^1H and ^{31}P NMR spectroscopy were used to characterise the product and complimented the molecular modelling. These showed that the antiparallel form is highly favoured over its photochromically inactive isomer. In toluene solution the colourless complex **RuRu20(o)** displays an intense absorption at λ_{max} 346 nm which is ascribed to charge transfer from π orbitals consisting of DTE and "Cl-Ru-C2" character to DTE and "Ru-P" derived π* orbitals. Irradiation of this CT state results in a green colour change due to the appearance of a new shouldered ^{1}IL band at λ 722 nm. The unusual shape of the band is thought to be due to rotamers of **RuRu20(c)** which have differing conformations of the metal units and overlapping absorption bands (Fig. 21 (a)). Calculations indicate that the new band is chiefly an allowed HOMO [Mdπ – C_2DTEπ] – LUMO [Mdπ – C_2DTEπ*]* transition. Irradiation with visible light (λ 750 nm) results in the solution becoming bleached to its colourless form. NMR studies indicate complete conversion.

The cyclic voltammogram of **RuRu20(o)** in CH_2Cl_2 (scan rate 50 Vs^{-1}, Fig. 21(b) inset) shows two almost coincident waves at 0.49 V *vs.* SCE, attributed to the one-electron oxidations of the two non-communicating Ru centres. At 0.2 Vs^{-1} this wave becomes irreversible, and during the reverse scans two well separated new waves at less positive potentials ($E^\circ_{c1} = 0.009$ V and $E^\circ_{c2} = 0.130$ V) appear (Fig. 21(b)). These oxidations arise from the closed species, as verified by NMR. Uv-Vis-NIR spectroelectrochemistry in CH_2Cl_2 (Fig. 21(a)) reveals that after two electron oxidation of **RuRu20(o),** two absorptions associated with **RuRu20(c)**$^{2+}$ appear ($\lambda_{max} = 452$ nm, $\varepsilon = 13{,}200$ $dm^{-3} mol^{-1} cm^{-1}$ and $\lambda_{max} = 652$ nm, $\varepsilon = 20{,}200$ $dm^{-3} mol^{-1} cm^{-1}$) as the absorption of **RuRu20(o)** (λ 344 nm in CH_2Cl_2) decreases in intensity. Electrochemical reduction of this species then leads initially to **RuRu20(c)**$^+$, and ultimately yields the characteristic spectrum of **RuRu20(c).** The spectrum of **RuRu20(c)**$^+$ shows an extremely broad absorption in the NIR from λ 715 nm to 1600 nm with $\lambda_{max} = 1100$ nm ($\varepsilon = 13{,}000$ $dm^{-3} mol^{-1} cm^{-1}$). This is ascribed to a metal perturbed π–π* transition in the extended conjugated system. The results obtained, supported by TD-DFT calculations, suggest ring closing *via* **RuRu20(o)**$^{2+}$ with a rate constant of 15 s^{-1}, possibly indicating kinetic

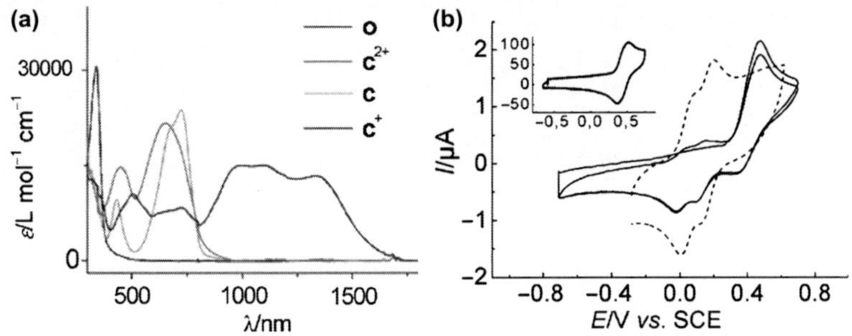

Fig. 21 (a) UV-Vis-NIR absorption spectra obtained during spectroelectrochemical analysis of **RuRu20**, (o = open, c = closed), (b) CV of **RuRu20(o)** at 0.2 Vs^{-1} with two consecutive scans (solid line), and scan of **RuRu20(o)** obtained after irradiation at 350 nm (dotted line). The inset shows a scan of **RuRu20(o)** at 50 Vs^{-1}.

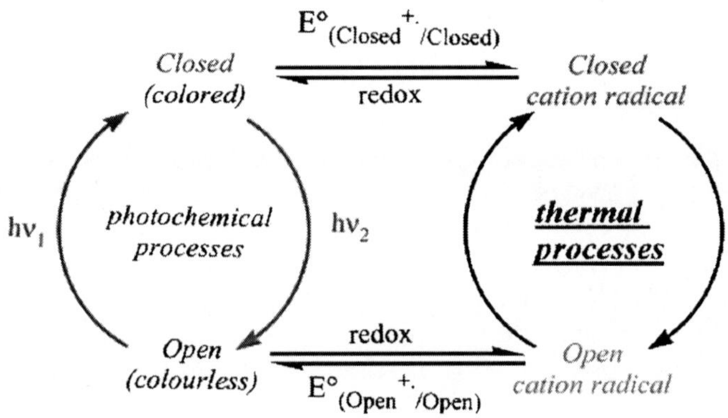

Scheme 12 Unlocking further switching states through electrochemical means. Reprinted (adapted) with permission from (G. Guirado, C. Coudret and J.-P. Launay, *The Journal of Physical Chemistry C*, 2007, **111**, 2770–2776). Copyright (2007) American Chemical Society.

control. The closing process occurs at low potential, far from the *ca.* 1 V generally required to bring about electrocyclisation. The authors attribute this to the unique electronic structure afforded by the Ru metal centre, which facilitates radicals to delocalise into the DTE ligand and promotes the radical coupling which forms the more favoured **RuRu20(c)**$^{2+}$ form.

Launay *et al.*[40] have shown that following electrochemical oxidation, many DTE systems undergo a thermal isomerisation from one photoisomer to the cation radical of the other isomeric form. This opens up the possibility of "electrochromism with memory" *via* thermal transformations of the oxidised switch, as a reduction step does not return the molecule to its initial state but to its photoisomer. The electronic properties of 5-thienyl susbstituents on the DTE switch does exert a degree of control over these characteristics because these groups can modify the oxidation potential of the dithienyl moiety and therefore provide the trigger required to unlock the thermally-derived forms (Scheme 12).

In order to exert further control on this process, it was considered desirable to introduce a substituent which could be converted from electron donating to electron withdrawing. To this end, the authors included an ethynyl ferrocene moiety onto both the perhydro and perfluoro DTE motifs, creating the compounds **Fc30(o)Fc** and **Fc20(o)Fc** respectively (Fig. 20).[41] Oxidation of ferrocene (which occurs at lower potentials relative to typical DTE oxidation) results in a large umpolung, converting Fc from being a mild donor in the Fe(II) state to an acceptor in the Fe(III) state.

Electrochemical studies show that the open forms undergoes two reversible oxidation processes, the first at *ca.* 0.5 V *vs.* SCE is a two electron wave, attributed to the oxidation of both Fc moieties. The second wave is at higher potential, and is ascribed to oxidation of the open photochromic unit. Further study revealed that oxidation of the lateral ferrocene efficiently destabilises the ring-closed perfluorinated form. As expected, the presence of two ferrocenium acceptor moieties, greatly enhances the reactivity of the photochromic unit's closed cation radical. Attempts to generate the twice oxidized intermediate **Fc^{+}20(c)Fc^{+}** by controlled potential electrolysis at 0.9 V resulted in the formation of the open isomer, showing that the ferrocenium substituents also destabilize the neutral photochromic closed form, hence lowering the global activation energy for the thermal ring opening. The authors postulate a possible, though energetically unfavourable, **Fc20(c)$^{+}$Fc^{+}** intermediate species resulting from a charge transfer process. The ring does stay open on back reduction, with a new colour arising from the absorption properties of oxidized redox sites. Both open and closed forms of the perhydro system behaved similarly in CV experiments. Again, the ring opening reaction occurred readily following oxidation, and spectrochemical analysis showed that ring closing was catalytic in electrons. Further oxidation leads to the fully oxidised **Fc^{+}30(o)Fc^{+}**, evidenced by new absorptions at 583 nm and 898 nm. As electron donating groups pull the oxidation potential of the DTE into the same region as ferrocene itself, it is impossible to say if the ferrocene has indeed oxidised first in this case. If not, however an oxidized **30(c)** core surrounded by electron donating ferrocene groups would not be expected to give rise to ring opening.

Photoswitchable molecular wires have been synthesised by Akita *et al.* through bridging {Ru/FeCp*(dppe)} fragments to DTE through an acetylide linker *via* the deprotonation of the vinylidene precursor, yielding **31** (Fig. 22).[42] Uv irradiation (< 360 nm) caused the systems to change colour from red to green (**FeFe31**) and from pale yellow to green **(RuRu31)** due to

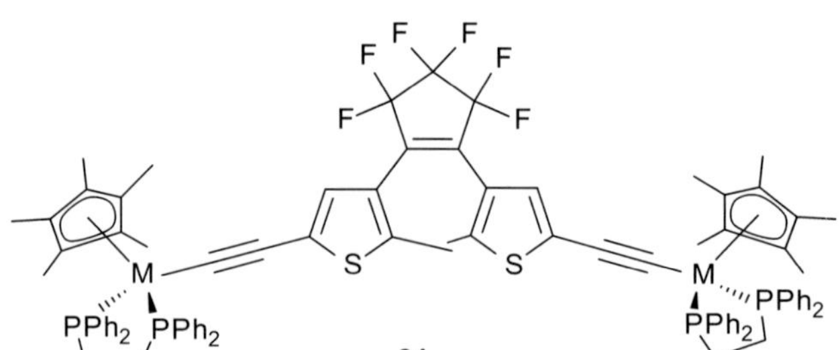

31
M = Ru or Fe

Fig. 22 Representation of **31**.

the appearance of the closed species' S_0–S_1 absorption at 700–800 nm. These changes were fully reversible on exposure of the solutions to visible light >560 nm. The PSS composition following Uv irradiation was quantitative for **RuRu31**, though in the case of **FeFe31** the PSS was found to be solvent dependent, ranging between 91% (benzene) – 37% (acetone), and the cycloreversion reaction took longer and did not go to completion for **FeFe31(o)**.

Wire-like performance was chiefly evaluated from the electronic coupling value (V_{ab}), which can be assessed from the IVCT bands of the chemically-generated monocationic species (see Fig. 23). **RuRu31(o)**$^+$ could not be produced due to disproportionation to **RuRu31(o)** and **RuRu31(o)**$^{2+}$ which quickly cyclised to **RuRu31(c)**$^{2+}$ preventing its wire-like properties from being thoroughly appraised. The system has also been found to

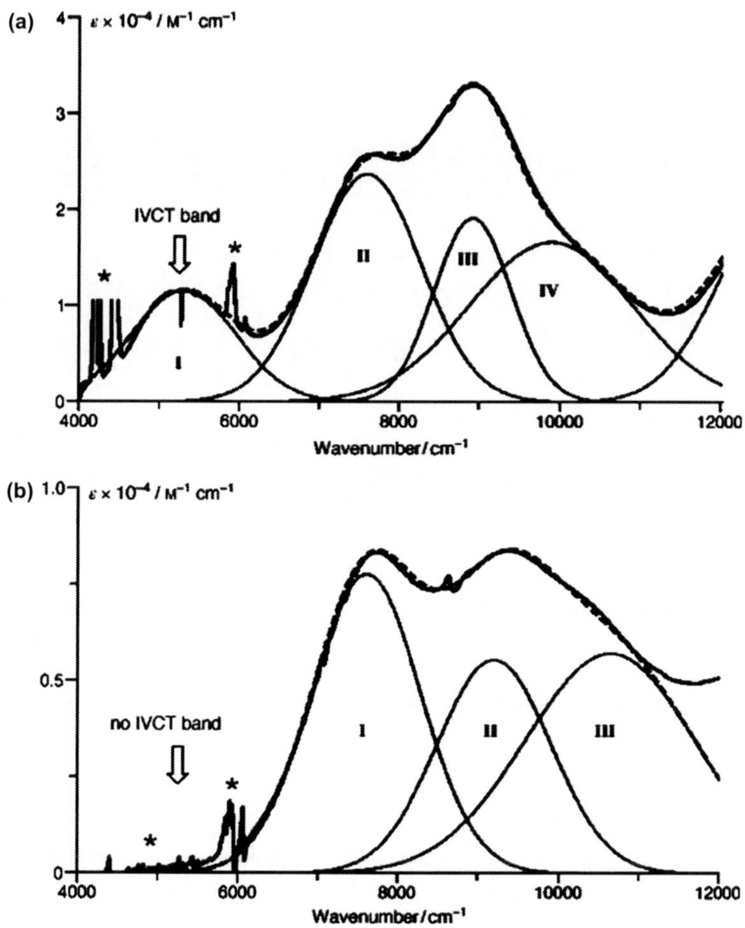

Fig. 23 NIR spectra for (a) **FeFe31(c)**$^+$ and (b) **RuRu31(c)**$^+$ recorded in CH_2Cl_2 (thick solid line = observed spectra, thin solid line = deconvoluted Gaussian curves, dashed line = the sum of the deconvoluted curves). The signals indicated by * are due to the solvent. From Y. Tanaka, T. Ishisaka, A. Inagaki, T. Koike, C. Lapinte and M. Akita, *Chemistry – A European Journal*, 2010, **16**, 4762–4776. Copyright © 2010 by John Wiley & Sons, Inc. Reprinted by permission of John Wiley & Sons, Inc.

electrocyclise. On the basis of the IVCT band of **FeFe31(c)**$^+$, its V_{ab} value was determined to be 0.0047 eV, assuming a Class II Robin-Day compound (intermediate charge delocalisation). Other NIR bands observed are ascribed to the closed DTE functionality. The absence of an IVCT band in the open form **FeFe31(o)**$^+$ indicated a value of $V_{ab}=0$. Electrochemically-derived comproportionation constants (K_C) (eqn. 1) were also calculated for both the iron and ruthenium systems, and were used to calculate the switching factor (SF) (eqn. 2) which were quite high, at 39 for **FeFe31** and 4.2 for **RuRu31**. These results indicate that the DTE moiety can be considered as being an efficient photoresponsive switch embedded in a molecular wire.

$$K_C = \exp\left(\frac{\Delta E \times F}{R \times T}\right) \quad \text{(eqn. 1)}$$

$$SF = \frac{K_{C(open)}}{K_{C(closed)}} \quad \text{(eqn. 2)}$$

4 Coordinating acac-derived metal complexes

Kawai *et al.*[43] have incorporated a coordination site into the core of a DTE- system comprising three linked thiazole groups (Scheme 13). Ligand **32(o)** was synthesised *via* the cross coupling of 4,5-dibromo-2-phenylthiazole and 5-methyl-2-phenyl-4-(4,4,5,5-tetramethyl-1,3,2-dioxaborolane-2-yl)thiazole. Colourless solutions of **32(o)** turned blue upon irradiation with UV light ($\lambda = 365$ nm) with the appearance of the closed-form (low energy absorption at λ 591 nm). Cycloreversion was achieved by irradiation with light of $\lambda > 440$ nm. **32(o)** was treated with $[Eu^{III}(hfac)_3(H_2O)_2]$ to give **Eu(hfac)$_3$(32(o))$_2$**, which had similar photochromic UV-vis properties to the free ligand (ligand dissociation was discounted as an explanation from NMR spectroscopic studies). The authors suggest that the PSS consists of a mixture of **[Eu(hfac)$_3$32(c)32(c)]** and **[Eu(hfac)$_3$32(o)32(c)]**, although an isosbestic point was observed at 330 nm. When irradiated with light (465 nm, low absorption for the ligand) emission ascribed to the f-f transitions $^5D_0$–$^7F_0$, $^5D_0$–$^7F_1$, $^5D_0$–$^7F_2$ were observed at 579, 592, and 614 nm respectively. As this overlaps with the absorption of **[Eu(hfac)$_3$32(c)32(c)]**, photocyclisation resulted in the quenching of the Eu^{III} luminescence. The emission from **[Eu(hfac)$_3$(32(o)32(o)]** was found to drop to 30% in the PSS ($\lambda_{em} = 614$ nm) partly due to changes in the symmetry about the Eu^{III} centre.

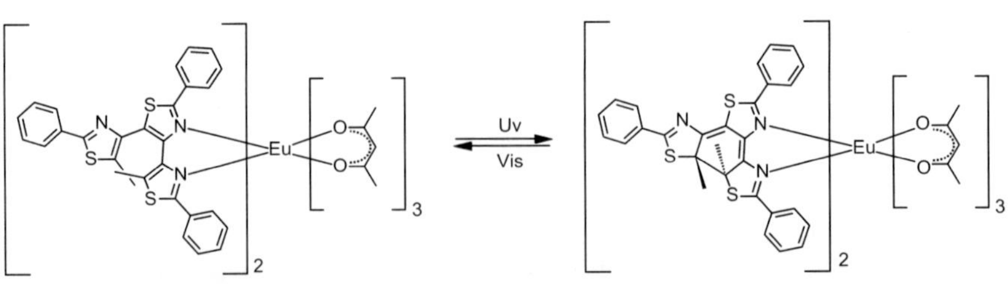

Scheme 13 Photochromic process of **[Eu(hfac)$_3$32(o)32(o)]**.

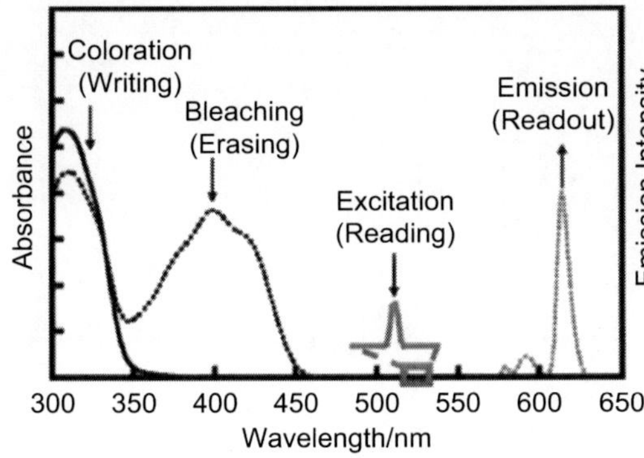

Fig. 24 Representation of the ideal non-destructive reading system.[44]

Emission intensity modulation is an ideal indicator of the switching state in molecular switches due to the high contrast, sensitivity, resolution and fast response times associated with luminescence sensing. However for luminescence measurements to be useful, a readout excitation wavelength must exist at which the absorption of both switch states is very low, thus ensuring that no photoreaction occurs during luminescence reading. Modulation intensity of fluorescence is often achieved by FRET, where the absorption profile of the closed form of the DTE overlaps with the emission of the luminescent centre, resulting in the quenching of the luminescence. This energy transfer to the closed form often efficiently quenches the luminescence – but can also result in photocycloreversion. Although uncommon in the literature, the utilisation of "readout mechanisms" outside of FRET are of importance as they allow for the separation of the "bleaching" and emission wavelengths, facilitating non-destructive reading (excitation) (Fig. 24).

Kawai *et al.*[44] have attempted to separate the absorption and emission wavelengths of the ligand-sensitive luminescence of a Eu^{III} centre using **33(o)** (Scheme 14) a ligand which was initially synthesised by Kim *et al.*[45] On reaction with **33(o)**, the coordinated water molecules of $[Eu(hfac)_3(H_2O)_2]$ are displaced by the coordinating sulfone groups to give **[Eu(hfac)$_3$33(o)]**. Ethyl acetate solutions are reversibly photochromic upon irradiation with 313 nm light, with the colourless solutions turning yellow ($\lambda_{abs} = 440$ nm in the closed form of the complex). The emission spectrum of **Eu(hfac)$_3$33(o)]** shows the expected $^5D_0$–^7F_J ($J = 0$–4) transitions of which the most intense (615 nm) decreases in intensity on irradiation, as the typical S_0–S_1 absorption increases (Scheme 14 (bottom)). Surprisingly the kinetic studies showed that the rate constants for the non-radiative processes of both open and closed forms were the same, but that the radiative rate constant was 1.5 times larger for the open form. Readout was also achieved using 526 nm light, with no change observed in the absorption properties of the system.

Yam *et al.* have combined a strongly σ-donating NHC ligand and an η^6–arene Ru complex (Fig. 25, compounds **Ru34–Ru40**).[46] The ligands were synthesised by attaching a co-ordinating heterocyclic substituent to an imidazole core *via* copper(I) catalysed cross coupling reactions, followed by selective methylation. An initially formed Ag(I) carbene complex was then transmetallated using $[Ru(mesitylene)Cl_2]_2$. The Uv-vis spectra of the

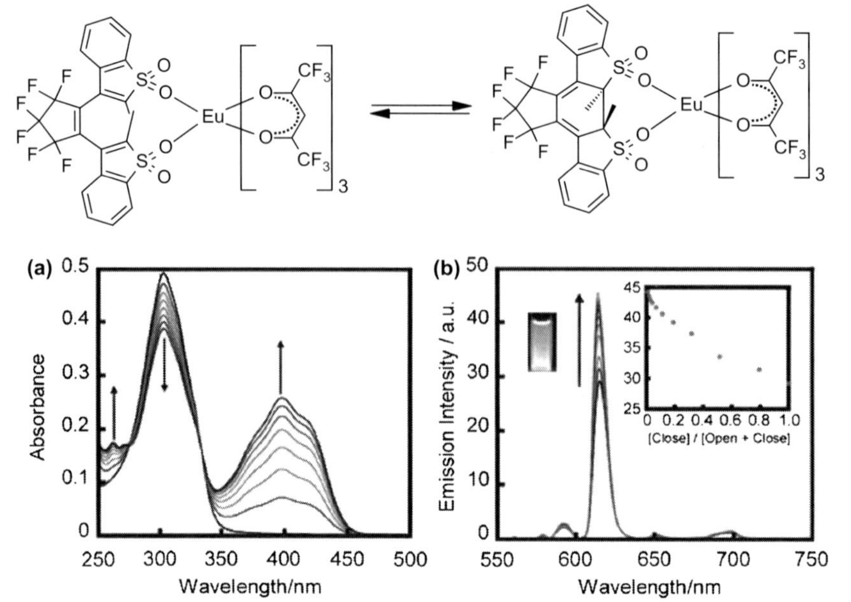

Scheme 14 (top) Schematic of the photochromic process of [**Eu(hfac)333(o)**]. (bottom) (a) Uv-vis spectrum of [**Eu(hfac)333(o)**] under Uv irradiation. (b) Emission spectrum showing a decreased intensity on irradiation with Uv light. (inset) This decrease is shown as a function of increasing [closed].

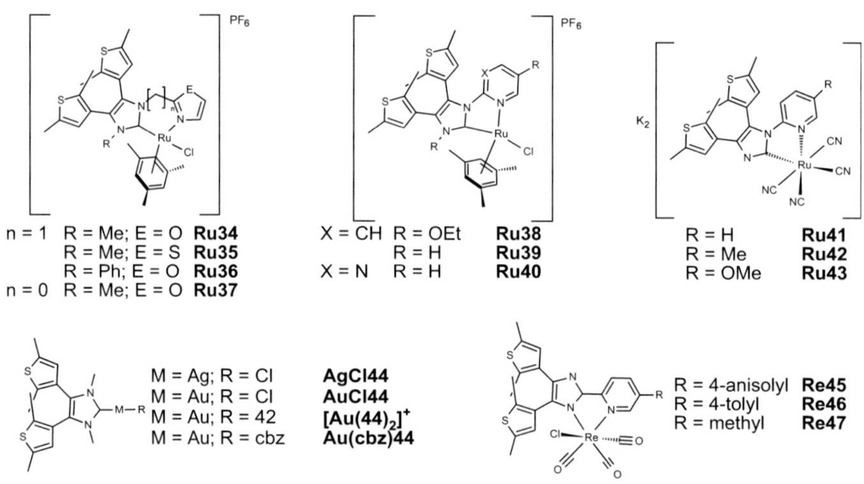

Fig. 25 A series of imidazole-derived DTEs synthesised by Yam *et al.* (cpz = the *N* – carbazole ligand).

complexes are typical, showing intense thiophene derived π–π* transitions in the high energy region and a less intense band between 300–350 nm which progressively red shifts throughout the series as the π* orbital of the chelating ligand is lowered in energy, indicating a dπ(Ru) - π*(NHC-L) MLCT transition. A weak band at 400 nm of $\varepsilon = 300–800\,\mathrm{nm\,dm^3\,mol^{-1}\,cm^{-1}}$ is ascribed to a metal centred transition, and rationalises the lack of room temperature emission from the complexes due to deactivation through the low lying d–d excited state.

Photochromic behaviour was observed in all the complexes in methanol and acetonitrile solution. The typically intense IL band associated with the closed form appears at *ca.* 550 nm. The quantum yields of photocyclisation are low, and decrease dramatically down the series, *e.g.* **Ru34** has a Φ_{cyc} of 0.044, while **Ru40** has a Φ_{cyc} of 0.0037. These low values suggest the efficient quenching of the excited state through the d–d energy levels. Complexes **38–40** can be cyclised by light of λ 340 nm, corresponding mainly to the MLCT band. Although triplet sensitisation could be implicated, the substantial overlap of the MLCT band with transitions of IL character suggest that the cyclisation is more than likely proceeding through a singlet pathway, particularly as the quantum yields were unaffected by the presence of oxygen.

The same authors have synthesised a similar series of photochromic carbene-ligated Ru(II) complexes (Fig. 25, compounds **Ru41–Ru43)** by treating an *N*-methylated 1-("substituted"pyridyl)-4,5-bis(2,5-dimethyl-3-thienyl)-imidazolium derivative with $K_4[Ru(CN)_6].3H_2O$ *via* an acid catalysed ligand exchange reaction.[47] The complexes have intense intraligand absorptions between 230 nm–300 nm, possibly bearing some intraligand charge transfer character [π(im) – π*(py)]. The lower energy bands observed at 300–450 nm and has been ascribed to MLCT transitions, and undergo a large bathochromic shift in acidic medium as is characteristic of the MLCT bands of similar compounds. The band is also sensitive to the nature of the pyridyl substituent, and follows the expected $CF_3 < H < 4\text{-OMe}$ trend in absorption energy. All the complexes are luminescent, with λ_{max} ranging from 500 to 600 nm and lifetimes in the 0.5–2 μs range in room temperature solution. The emission is IL derived. Excitation into the either IL or MLCT bands results in photocyclisation. The expected low-energy "closed" band appears at 500 nm, accompanied by a large increase in the band at 370 nm. Excitation into the lowest energy band (*ca.* 500 nm) results in photocycloreversion, with a quantum yield of approximately $\Phi = 0.092$. Excited state lifetimes and luminescence quantum yields are found to be consistently smaller than in similar complexes not bearing the photochromic unit, indicating a deactivation process competing with the phosphorescence of ^{3}MLCT: the DTE systems – the triplet sensitised pathway for photocyclisation.

Using the methyl derivative of the imidazolium precursor, novel Au(I) and Ag(I) complexes were synthesised as outlined in Fig. 25 (complexes of ligand **42**).[48] The systems were found to be photochromic, though not very thermally stable - conversion to their closed forms is observed even at 25 °C (half lives: **AgCl44** 11.7; **AuCl44,** 25.7; **[Au(44)$_2$]$^+$** 119; and **Au(cbz)44** 46.7 mins), making them unsuitable for most switching applications.

In order to enhance the semiconductor laser susceptibility for applications in optical memory storage it is desirable to design systems displaying NIR absorption. To this end, a similar highly conjugated imidazole-derived DTE ligand was synthesised in which the wavelength of the closed-isomer's IL band has been shifted into the NIR.[49] This was achieved by enforcing the closed form's planarity through incorporating a metal centre. The complexes **Re45(o)** and **Re46(o)** (Fig. 25) all show IL absorptions at *ca.* 352 nm in chloroform solution, with a shoulder at *ca.* 425 nm, primarily ascribed to an MLCT transition. Excitation into either of these bands results in the closed form IL absorption, appearing at *ca.* 715 nm, significantly red-shifted from the free ligand's band (580 nm) due to the coordinating pyridyl ring being held in the same plane as the closed DTE-imidazole moiety by the metal centre. Photocyclization quantum yields were approximately 0.40, though cycloreversion was lower (*ca.* 0.004). The compounds are also luminescent, excitation into the IL or MLCT states resulting in ^{3}MLCT-derived phosphorescence at 570 nm, the intensity of which decreases steadily on photocyclisation.

In an unusual macrocyclic switch Zhang *et al.* have included three photochromic units into the periphery of a Zn-coordinated pthalocyanin[50] (**Zn48**) (Fig. 26). Irradiation of the solution at 254 nm resulted in a dramatic decrease in the Q-band absorption intensity and a small red shift of 6 nm. Keeping the solution in darkness for 1 day meant that the original spectra could be regenerated. This was ascribed to enhanced aggregation of the ring-closed form, as other routes to such spectral changes, such as photo-oxidation of the Pc ring, protonation due to chloroform photolysis or isolated photochromic species were ruled out by either the reversibility of the changes or by ^{1}H NMR studies. This solvent-dependent photocontrolled J-aggregation was thought to be due to planarization of the Pc ring on photocyclisation, as indicated by DFT studies. This was a potentially useful result as head to tail J-aggregated systems are more suitable for NLO, optical storage and spectral sensitisation than their H-aggregated counterparts, however it was impossible to ascertain the number of units closing on the molecule, as any attempt to disrupt the aggregates resulted in ring opening.

5 Future directions

There are anticipated bottle-necks in the development of nanoscale devices to overcome and future socio-economic challenges to face. Within this futuristic backdrop, the fabrication of new molecular systems capable of efficiently processing data in organic systems, or inter-converting optical-electronic information is essential. There is no doubt that DTE-based systems have a role to play in responding to the need for easy information access and exchange and the demand for fast, efficient, high-density data

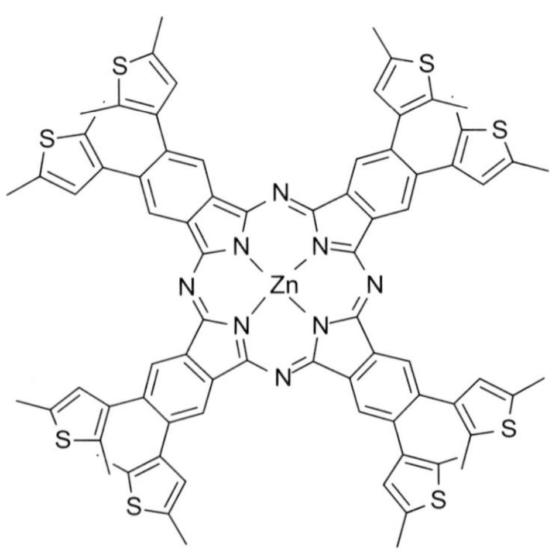

Fig. 26 Representation of **Zn48**.

storage and retrieval systems. In fact, they are favoured candidates in the synthetic hunt for photochromic systems which exhibit good fatigue resistance, high quantum yields, short response times, and typically large and measurable optically-induced changes.

Contributors to this review have demonstrated that it is both feasible and desirable to co-ordinate a photoresponsive metal centre into a DTE moiety. In general, it broadens the spectrum of activity of the switch by substantially differentiating the properties of the open and closed forms. Ring-closure can be brought about by light excitation into the MLCT state of the molecule and it opens up metal-facilitated, triplet energy transfer to the dithienylethene (DTE) moiety. However there are so many potentially advantageous synthetic modifications that can be made (including systems based on 2,3-bis(thienyl) quinones and bis(thienyl)ethene-based tetraazaporphyrin and phthalocyanine hybrids in which the ethene linkage is incorporated into a backbone with extended conjugation capabilities) that in some ways it is difficult to know where to begin or where this synergy of synthesis and spectroscopy might lead. Behind it all is the lurking and unspoken hurdle of incomplete conversion, courtesy of the optically inactive parallel form of all DTE-based switches.

Despite these reservations there are several systems that are actively pushing out current boundaries to give a glimpse of how these materials might be exploited in the future. One example comes in the form of a DTE complex based on ligand **20**, containing pendant alkynyl Ru(II) bisphenylphosphinochlorides. This has been shown to afford six stable and switchable states each with distinct NLO properties.[51] The complex exhibits orthogonal responses to protic (alkynyl-vinyl ligand), electrochemical (Ru(II)/(III)) and optical (L(o)-L(c)) stimuli. The individual states, connected by seven pathways (electrochemical oxidation of the metal centre in the vinylidene complex is irreversible), all give rise to cubic nonlinearity in specific regions of the NLO spectra as determined by femtosecond Z scan techniques (600–1600 nm).

The scientific community is looking beyond the individual metal-centre or ligand-metal bond. Inorganic nanorods and nanowires are garnering interest due to their anisotropic shape and fascinating properties. In their work, Fages *et al.*[52] aim to construct building blocks for logic gates and memory circuit elements by marrying the properties of nanoscopic science with the fundamental molecular behaviour of a DTE switch. To this end, they have synthesised carboxylic acid terminated DTEs and grafted them to ZnO nanorods, achieving approximately 40% surface coverage. The nanorods do not efficiently quench the excited state of the switch, therefore ring closing and opening could be observed in the Uv-vis spectra.

The distinct but complementary area of plasmonic nanoparticles is also well developed and to investigate the enhancement of photochemistry on the surface of such particles, Kobatake *et al.*[53] have synthesised a thiol-terminated polystyrene-derived polymer containing DTE side chains (Scheme 15). Gold nanoparticles coated with this polymer were grown by the seeding method, followed by ligand exchange. Uv-vis spectra were collected from samples cast on quartz slides, and showed a strong localised surface plasmon resonance (LSPR) band at 620 nm, shifting to 659 nm upon photocyclisation of the DTEs. This shift was reversed by visible light irradiation, although studies of the difference spectra showed strong wavelength

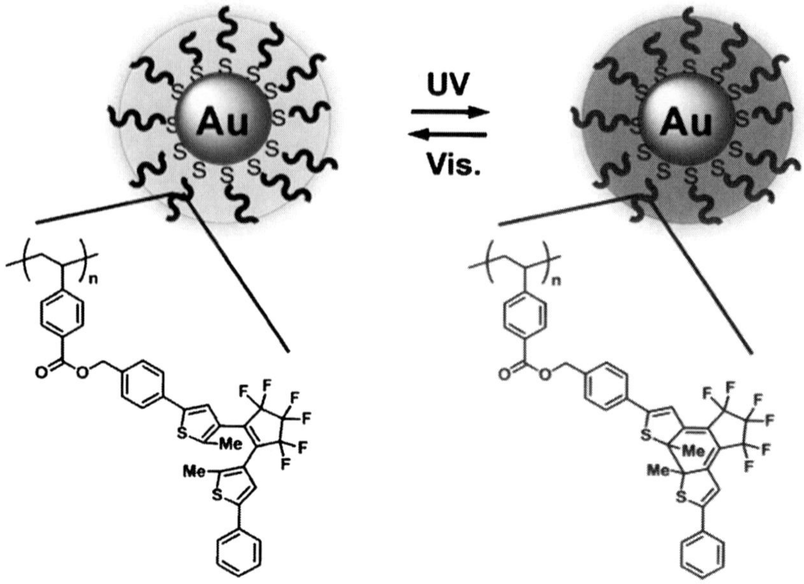

Scheme 15 Schematic of the gold nanoparticle-bound switch. Journal of Photochemistry and Photobiology A: Chemistry, Vol. 221, H. Nishi, T. Asahi and S. Kobatake. Enhanced photocycloreversion reaction of diarylethene polymers attached to gold nanoparticles in the solid state, pages no. 256–260. Copyright (2011), with permission from Elsevier.

dependence (irradiation at 500 nm and 600 nm gives rise to two isosbestic points, whereas irradiation at 650 nm and 700 nm showed none). The observation of isosbestic points indicated clean photocycloreversion of the entire poly(DE) shell, whereas their absence, indicated differing rates of cycloreversion - the process was enhanced nearer the surface of the nanoparticle on irradiation with longer wavelength visible light.

Castellano *et al.*[54] have adroitly tapped into the intermolecular Förster-type resonance energy transfer (FRET) between a mixture of benzothienyl DTE **BTF6** and $[Ru(dpp)_3]^+$ in solution. The read (390 nm), write ($\lambda < 360$ nm), and erase ($\lambda > 500$ nm) wavelengths are well-separated, and the physical coupling of the chromophore to the switching moiety is avoided. The complex $[Ru(dpp)_3]^+$ possesses exceptional photostability, solubility, high quantum yield (0.37 ± 0.04) and a long excited state lifetime (6.0 ± 0.30 µs). Its bright orange emission ($\lambda_{max} = 610$ nm) was almost completely unaffected by the solution addition of **BTF6** (which absorbs little from $\lambda > 380$ nm). Ring-closing of the DTE using light of $\lambda < 360$ nm was achieved, and resulted in a red non-emissive solution. The UV spectrum of the photostationary state of the mixture revealed that the absorption of the closed form (at λ 525 nm) overlapped significantly with the $[Ru(dpp)_3]^{2+}$ emission band, thus facilitating its quenching, which Stern – Volmer plots indicated was dynamic only (quenching constant $3.23 \times 10^9 M^{-1} s^{-1}$). The authors also demonstrated the first use of luminescence lifetimes to transduce the photochromic state of the quencher, instead of the typical intensity-based method (Fig. 27).

Although many hurdles remain in the quest for molecular-switch containing devices, we hope this comprehensive review shows that the future is bright whatever side of the switch one is on (Table 1).

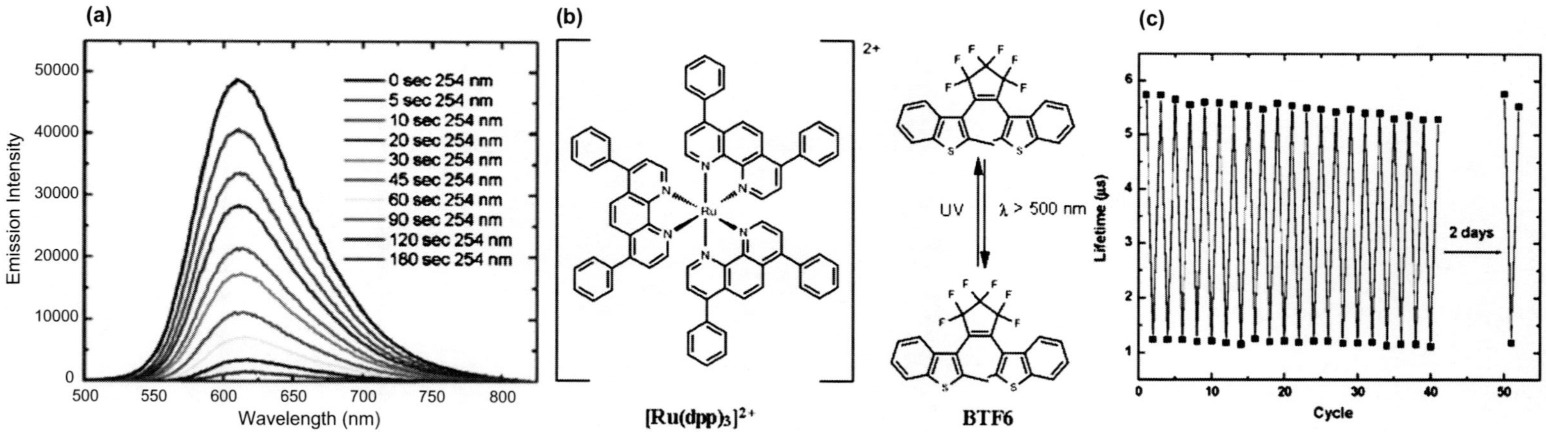

Fig. 27 (a) Luminescence of $[Ru(dpp)_3]^{2+}$ by the closed form of BTF6 in degassed CH_3CN solution. (b) Schematic of the systems involved in the photochromic process. (c) Demonstration of fatigue resistance, using luminescence lifetime as readout. Reprinted (adapted) with permission from (D. V. Kozlov and F. N. Castellano, *The Journal of Physical Chemistry A*, 2004, **108**, 10619–10622). Copyright (2004) American Chemical Society.

Table 1 Selected available photophysical data from the bibliography (see individual references for additional data).

	L photocyc/mlct (ε/dm^3 mol^{-1} cm^{-1})	Pss composition %	QY photchromic (λ_{exc}/nm)	Medium (T/K)	Emission λ_{max}/nm (τ/μs)	QY lum	Ref.
WW**2**(o)	209 (50000)			MeOH: H_2O [5:1] (298)	382 (-)	0.03 (240); 0.04 (311)	10
WW**2**(c)	615 (4000)				382 (-)	0.08 (240), 0.12 (311)	10
W**1a**(o)	233 (10000)				400 (-)	0.03 (240); 0.01 (394)	10
W**1a**(c)	632 (4000)				400 (-)	0.15 (240); 0.07 (394)	10
ReRe**2**(o)	319 (18000)			MeOH (298)	372 (-)	0.01 (270), 0.04 (320)	10
ReRe**2**(c)	621 (6000)				372 (-)	0.004 (270), 0.02 (320)	10
Re**1b**(o)	239 (8000)				368 (-)	0.03 (270); 0.18 (315)	10
Re**1b**(c)	604 (4000)				368 (-)	0.01 (270); 0.11 (315)	10
[Ir(acac)(**6(o)**)$_2$]	480 (-)		0.20 (330)	THF (298)	570 (0.1)	0.06 (440)	16
[Ir(acac)(**6(c)**)$_2$]	550 (-)		0.01 (550)	THF (298)			16
RuRu**7(o)**	458 (42000)		0.037 (334)a, 0.88 (334)d; 0.38 (450)a, 0.80 (450)d	Acetonitrile (RT)	630 (0.140a, 0.310d)	0.009 (334)a, 0.014 (334)d; 0.007 (450)a, 0.014 (450)d	17
RuRu**7(c)**	614 (32000)	97%					17
OsOs**7(o)**	492 (34000, ^{1}MLCT); 590 (9000, ^{3}MLCT)		0.0045 (334)a; 0.0041 (334)d	Acetonitrile (RT)	759 (0.044a, 0.062d)	0.004 (334)a, 0.005 (334)d; 0.004 (450)a, 0.006 (450)d	17
OsOs**7(c)**	618 (37000)	73%					17
RuOs**7(o)**	590 (5600 acetonitrile)		0.010 (334)a,d; 0.0042 (450)a, 0.0045 (450)d	MeCN (293 K)	630 (7×10^{-5}, 61×10^{-5} Ru)a 758 (5.6×10^{-5}, 46×10^{-5} Os)a	2×10^{-4} (471, Ru) 0.0035 (471, Os)	18

RuOs**7(c)**	617 (33000 acetonitrile)	90%					18
Ru**17(o)**	470 (-)		0.22 (470)a; 0.44 (470)d	Acetonitrile (298)	600 (5×10^{-4})	See text	27
Ru**17(c)**	540 (-)	90%	0.50 (500)a				27
Re**19(o)**	400 (4690)		0.552 (313)	Benzene (298 K)	595 (0.26)	0.02 (417)	55
				Solid (298 K)	558 (0.1)		
				Solid (77 K)	562 (1.9)		
				Glass (4:1 Meoh:EtOH) (77 K)	535 (7.2)		
Re**19(o)**	580 (5050)	Up to 98	0.648 (440), 0.009 (510)	Benzene (298 K)	626 (0.1)		55
[Pt(**19(o)**)($C_2Ph)_2$]	396 (7900)		0.035	CH_2Cl_2 (298)	580 (1.6)	0.18	29
[Pt(**19(c)**)($C_2Ph)_2$]	560 (-)		0.012	CH_2Cl_2 (298)			29
Pt**21(o)**tpy	465 (-)		0.14 (425), 0.41(302)	CH_2Cl_2 (298)	EtOH:MeOH glass (90 K)	568, 608 (-)	35
Pt**21(c)**tpy	602 (-)			CH_3CN (298)			35
Pt**22(o)**tpy	417 (-)		0.03 (425)				35
Pt**22(c)**tpy	592 (-)						35
23(o)	378 (35600)		0.118 (370)	Benzene (298 K)	640 (2.0)	0.0095	36
				Solid (298 K)	646 (9.8)		
				Solid (77 K)	641 (59.1)		
				Glass (77 K)	574, 625 (135.3)		
23(c)	566 (7830)	48 (500)	0.026 (500)	Benzene (298 K)			36
				Solid (298 K)			
				Solid (77 K)			
				Glass (77 K)			
24(o)	386 (54830)		0.243 (370)	Benzene (298 K)	630 (16.6)	0.0222	36
				Solid (298 K)	573, 612 (21.8) 573, 614 (65.4) 572, 620 (190.2)		
				Solid (77 K)			
				Glass (77 K)			

Table 1 (*Continued*)

	L photocyc/mlct (ε/dm^3 mol^{-1} cm^{-1})	Pss composition %	QY photchromic (λ_{exc}/nm)	Medium (T/K)	Emission λ_{max}/nm (τ/μs)	QY lum	Ref.
24(c)	564 (8350)	>95 (500)	0.015 (500)	Benzene (298 K)			36
				Solid (298 K)			
				Solid (77 K)			
				Glass (77 K)			
25(o)	380 (66060)		0.117 (370)	Benzene (298 K)	632 (17.8)	0.0162	36
				Solid (298 K)			
				Solid (77 K)	635, 670 (6.4) 578, 632 (145.3)		
				Glass (77 K)			
25(c)	570 (15780)	44 (500)	0.010 (500)	Benzene (298 K)			36
				Solid (298 K)			
				Solid (77 K)			
				Glass (77 K)			
26(o)	368 sh (15570)		0.063 (370)	Benzene (298 K)	647 (0.1)	0.0027	36
				Solid (298 K)	570 (8.5)		
				Solid (77 K)	568 (23.8)		
				Glass (77 K)	577 (99.2)		
26(c)	562 (8400)	22 (500)	0.025 (500)	Benzene (298 K)			36
				Solid (298 K)			
				Solid (77 K)			
				Glass (77 K)			
27(o)	376 (44500)		0.170 (370)	Benzene (298 K)	640 (2.4)	0.0276	36
				Solid (298 K)	567 (5.3)		
				Solid (77 K)	562 (93.9)		
				Glass (77 K)	567 (122.4)		

27(c)	562 (10400)	94 (500)	0.0025 (500)	Benzene (298 K)			36
				Solid (298 K)			
				Solid (77 K)			
				Glass (77 K)			
28(o)	366 (53040)		0.242 (370)	Benzene (298 K)	677 (2.0)	0.0077	36
				Solid (298 K)			
				Solid (77 K)	562 (58.2)		
				Glass (77 K)	547, 581 (226.2)		
28(c)	556 (11770)	>95 (500, one closure)	0.029 (500)	Benzene (298 K)			36
				Solid (298 K)			
				Solid (77 K)			
				Glass (77 K)			
29 oooo	*ca.* 325 (-)		0.64				37
29 coco	628 (-)	88	0.0094				37
AuAu**30(o)**PCy_3	320 (29500)		0.61 (302)	CH_2Cl_2 (298 K)	416, 433 (-)		39
AuAu**30(c)**PCy_3	530 (16300)	92	0.0063 (>420)	CH_2Cl_2 (298 K)			39
AuAu**30(o)**PPh_3	312 (38800)		0.78 (302)	CH_2Cl_2 (298 K)	395, 422 (-)		39
AuAu**30(c)**PPh_3	536 (22500)	85	0.040 (>420)	CH_2Cl_2 (298 K)			39
AuAu**20(o)**PCy_3	312 (48500)		0.41 (302)	CH_2Cl_2 (298 K)	436, 460 (-)		39
AuAu**20(c)**PCy_3	601 (21000)	95	0.0096 (>420)	CH_2Cl_2 (298 K)			39
AuAu**20(o)**PPh_3	316 (47200)		0.72 (302)	CH_2Cl_2 (298 K)	398, 422 (-)		39
AuAu**20(c)**PPh_3	605 (19900)	>95	0.011 (>420)	CH_2Cl_2 (298 K)			39
FeFe**31(o)**	*Ca.* 355 (*ca.* 18000) thf		0.0021 (366)				42
FeFe**31(c)**	*Ca.* 800 (28000) thf	91 (benzene); 63 (THF); 37 (acetone)	0.00018 (768)				42
RuRu**31(o)**	*Ca.* 350 (*ca.* 50000)		0.38 (366)				42
RuRu**31(c)**	*Ca.* 700 (*ca.* 55000)	Quant	0.00044 (710)				42

Table 1 (*Continued*)

	L photocyc/mlct (ε/dm^3 mol^{-1} cm^{-1})	Pss composition %	QY photchromic (λ_{exc}/nm)	Medium (T/K)	Emission λ_{max}/nm (τ/μs)	QY lum	Ref.
Eu(hfac)$_3$**33(o)**	300 (-)			Ethyl acetate (298)	579		44
					592		
					615 (710)	0.16	
					650		
					698		
Eu(hfac)$_3$**33(c)**	400 (-)			Ethyl acetate (288)	579		44
					592		
					615 (680)	0.10	
					650		
					698		
34(o)	424 (300)		0.044 (280)	MeOH			46
34(c)	548 (10400)		0.094 (500)	MeOH			46
35(o)	426 (300)		0.021 (280)	MeOH			46
35(c)	546 (11100)		0.085 (500)	MeOH			46
36(o)	426 (400)		0.029 (280)	MeOH			46
36(c)	550 (10300)		0.092 (500)	MeOH			46
37(o)	400 (400)		0.0078 (280)	MeOH			46
37(c)	538 (9810)		0.061 (500)	MeOH			46
38(o)	406 (700)		0.0079 (280)	MeOH			46

38(c)	534 (4500)	0.16 (500)	MeOH			46
39(o)	400 (700)	0.0043 (280)	MeOH			46
39(c)	532 (5100)	0.08 (500)	MeOH			46
40(o)	398 sh (800)	0.0037 (280)	MeOH			46
40(c)		0.4 (500)	MeOH			46
	526 (7000)					
41(o)	366 (5780)	0.58 (360)d	MeOH (RT)	550 (0.15)	0.0023	47
			4:1 EtOH:MeOH (77)	484 (250)		
41(c)	502 (7830)	0.0096 (500)d	MeOH (RT)	585 (-)		47
			4:1 EtOH:MeOH (77)	568 (<0.1)		
42(o)	362 (6650)	0.47 (360)d	MeOH (RT)	568 (0.13)	0.0021	47
		0.40 (360)a	4:1 EtOH:MeOH (77)	478 (26.6)		
42(c)	502 (8190)	0.0092 (500) a,d	MeOH (RT)	591 (-)	0.0003	47
			4:1 EtOH:MeOH (77)	565 (<0.1)		
43(o)	366 (5950)	0.42 (360)d	MeOH (RT)	572 (0.13)	0.0023	47
		0.32 (360)a	4:1 EtOH:MeOH (77)	484 (36.9)		
43(c)	508 (8610)	0.089 (500)d	MeOH (RT)	591 (-)	0.0005	47
			4:1 EtOH:MeOH (77)	565 (<0.1)		

d = degassed solution, a = aerated solution

References

1 D. A. Parthenopoulos and P. M. Rentzepis, *Science*, 1989, **245**, 843–845.
2 B. L. Feringa, *Molecular Switches*, Wiley-VCH, Weinheim, Germany, 2001.
3 M. Irie, *Chemical Reviews*, 2000, **100**, 1685–1716.
4 S. Nakamura and M. Irie, *The Journal of Organic Chemistry*, 1988, **53**, 6136–6138.
5 M. Irie, *Photochemical & Photobiological Sciences*, 2010, **9**, 1535–1542.
6 L. N. Lucas, J. van Esch, R. M. Kellogg and B. L. Feringa, *Chemical Communications*, 1998, 2313–2314.
7 Y. Liu, C. Lagrost, K. Costuas, N. Tchouar, H. L. Bozec and S. Rigaut, *Chemical Communications*, 2008, 6117–6119.
8 M. Takeshita, N. Kato, S. Kawauchi, T. Imase, J. Watanabe and M. Irie, *The Journal of Organic Chemistry*, 1998, **63**, 9306–9313.
9 M. Takeshita, M. Nagai and T. Yamato, *Chemical Communications*, 2003, 1496–1497.
10 A. Fernandez-Acebes and J. M. Lehn, *Chemistry – A European Journal*, 1999, **5**, 3285.
11 S. Cui, S. Pu, W. Liu and G. Liu, *Dyes and Pigments*, 2011, **91**, 435–441.
12 K. Matsuda, K. Takayama and M. Irie, *Inorganic Chemistry*, 2003, **43**, 482–489.
13 Y. Garcia, V. Ksenofontov, R. Lapouyade, A. D. Naik, F. Robert and P. Gütlich, *Optical Materials*, 2011, **33**, 942–948.
14 B. Qin, R. Yao, X. Zhao and H. Tian, *Org. Biomol. Chem.*, 2003, **1**, 2187.
15 H. D. Samachetty and N. R. Branda, *Chemical Communications*, 2005, 2840–2842.
16 W. Tan, Q. Zhang, J. Zhang and H. Tian, *Organic Letters*, 2008, **11**, 161–164.
17 R. T. F. Jukes, V. Adamo, F. Hartl, P. Belser and L. De Cola, *Inorganic Chemistry*, 2004, **43**, 2779–2792.
18 R. T. F. Jukes, V. Adamo, F. Hartl, P. Belser and L. De Cola, *Coordination Chemistry Reviews*, 2005, **249**, 1327.
19 V. Aubert, L. Ordronneau, M. Escadeillas, J. A. G. Williams, A. Boucekkine, E. Coulaud, C. Dragonetti, S. Righetto, D. Roberto, R. Ugo, A. Valore, A. Singh, J. Zyss, I. Ledoux-Rak, H. Le Bozec and V. r. Guerchais, *Inorganic Chemistry*, 2011, **50**, 5027–5038.
20 Y.-W. Zhong, N. Vila, J. C. Henderson, S. Flores-Torres and H. D. Abruña, *Inorganic Chemistry*, 2007, **46**, 10470–10472.
21 Y.-W. Zhong, N. Vila, J. C. Henderson and H. c. D. Abruña, *Inorganic Chemistry*, 2008, **48**, 991–999.
22 X. Piao, Y. Zou, J. Wu, C. Li and T. Yi, *Organic Letters*, 2009, **11**, 3818–3821.
23 I. Lee, Y. You, S.-J. Lim and S. Y. Park, *Chemistry Letters*, 2007, **36**, 888–889.
24 J. Han, M. Maekawa, Y. Suenaga, H. Ebisu, A. Nabei, T. Kuroda-Sowa and M. Munakata, *Inorganic Chemistry*, 2007, **46**, 3313–3321.
25 T. Shiga, H. Miyasaka, M. Yamashita, M. Morimoto and M. Irie, *Dalton Transactions*, 2011, **40**, 2275–2282.
26 S. Uehara, Y. Hiromoto, S. Minkovska, K. Suzuki, T. Ubukata and Y. Yokoyama, *Dyes and Pigments*, 2012, **92**, 861–867.
27 M. T. Indelli, S. Carli, M. Ghirotti, C. Chiorboli, M. Ravaglia, M. Garavelli and F. Scandola, *Journal of the American Chemical Society*, 2008, **130**, 7286–7299.
28 V. W.-W. Yam, C.-C. Ko and N. Zhu, *Journal of the American Chemical Society*, 2004, **126**, 12734–12735.
29 J. K. W. Lee, C. C. Ko, K. M. C. Wong, N. Zhu and V. W. W. Yam, *Organometallics*, 2007, **26**, 12–15.

30 S. K. Brayshaw, S. Schiffers, A. J. Stevenson, S. J. Teat, M. R. Warren, R. D. Bennett, I. V. Sazanovich, A. R. Buckley, J. A. Weinstein and P. R. Raithby, *Chemistry – A European Journal*, 2011, **17**, 4385–4395.
31 C. J. Adams, N. Fey, Z. A. Harrison, I. V. Sazanovich, M. Towrie and J. A. Weinstein, *Inorganic Chemistry*, 2008, **47**, 8242–8257.
32 T. Fukaminato, T. Doi, M. Tanaka and M. Irie, *The Journal of Physical Chemistry C*, 2009, **113**, 11623–11627.
33 M. N. Roberts, J. K. Nagle, J. G. Finden, N. R. Branda and M. O. Wolf, *Inorganic Chemistry*, 2008, **48**, 19–21.
34 X. Wang, S. b. Goeb, Z. Ji, N. A. Pogulaichenko and F. N. Castellano, *Inorganic Chemistry*, 2011, **50**, 705–707.
35 M. N. Roberts, J. K. Nagle, M. B. Majewski, J. G. Finden, N. R. Branda and M. O. Wolf, *Inorganic Chemistry*, 2011, **50**, 4956–4966.
36 H.-L. Wong, C.-H. Tao, N. Zhu and V. W.-W. Yam, *Inorganic Chemistry*, 2010, **50**, 471–481.
37 I. Jung, H. Choi, E. Kim, C.-H. Lee, S. O. Kang and J. Ko, *Tetrahedron*, 2005, **61**, 12256–12263.
38 M. N. Roberts, C.-J. Carling, J. K. Nagle, N. R. Branda and M. O. Wolf, *Journal of the American Chemical Society*, 2009, **131**, 16644–16645.
39 Y. Lin, J. Yin, J. Yuan, M. Hu, Z. Li, G.-A. Yu and S. H. Liu, *Organometallics*, 2010, **29**, 2808–2814.
40 G. Guirado, C. Coudret, M. Hliwa and J. P. Launay, *The Journal of Physical Chemistry B*, 2005, **109**, 1744.
41 G. Guirado, C. Coudret and J.-P. Launay, *The Journal of Physical Chemistry C*, 2007, **111**, 2770–2776.
42 Y. Tanaka, T. Ishisaka, A. Inagaki, T. Koike, C. Lapinte and M. Akita, *Chemistry – A European Journal*, 2010, **16**, 4762–4776.
43 T. Nakagawa, K. Atsumi, T. Nakashima, Y. Hasegawa and T. Kawai, *Chemistry Letters*, 2007, **36**, 372–373.
44 T. Nakagawa, Y. Hasegawa and T. Kawai, *Chemical Communications*, 2009, 5630–5632.
45 Y.-C. Jeong, S. I. Yang, K.-H. Ahn and E. Kim, *Chemical Communications*, 2005, 2503–2505.
46 G. Duan, W.-T. Wong and V. W.-W. Yam, *New Journal of Chemistry*, 2011, **35**, 2267–2278.
47 G. Duan and V. W.-W. Yam, *Chemistry – A European Journal*, 2010, **16**, 12642–12649.
48 V. W.-W. Yam, J. K.-W. Lee, C.-C. Ko and N. Zhu, *Journal of the American Chemical Society*, 2009, **131**, 912–913.
49 P. H.-M. Lee, C.-C. Ko, N. Zhu and V. W.-W. Yam, *Journal of the American Chemical Society*, 2007, **129**, 6058–6059.
50 J. Yi, Z. Chen, J. Xiang and F. Zhang, *Langmuir*, 2011, **27**, 8061–8066.
51 K. A. Green, M. P. Cifuentes, T. C. Corkery, M. Samoc and M. G. Humphrey, *Angewandte Chemie International Edition*, 2009, **48**, 7867–7870.
52 S. Remy, S. M. Shah, C. Martini, G. Poize, O. Margeat, A. Heynderickx, J. Ackermann and F. Fages, *Dyes and Pigments*, 2011, **89**, 266–270.
53 H. Nishi, T. Asahi and S. Kobatake, *Journal of Photochemistry and Photobiology A: Chemistry*, 2011, **221**, 256–260.
54 D. V. Kozlov and F. N. Castellano, *The Journal of Physical Chemistry A*, 2004, **108**, 10619–10622.
55 C.-C. Ko, W.-M. Kwok, V. W.-W. Yam and D. L. Phillips, *Chemistry – A European Journal*, 2006, **12**, 5840–5848.

Nuclear quadrupole resonance spectroscopy

Keith B. Dillon
DOI: 10.1039/9781849734899-00216

This report takes a critical look at new publications in pure nuclear quadrupole resonance (NQR) spectroscopy for inorganic and organometallic species up to the end of 2010.

1 Introduction

This chapter reports and considers new publications in pure nuclear quadrupole resonance (NQR) spectroscopy of inorganic and organometallic compounds, which have appeared since volume 42 in this series,[1] up to the end of 2010. The number of new articles has remained virtually constant, but with a change of emphasis. There are more papers this time on Group 17 nuclei (chlorine, bromine and iodine), but fewer on transition metals, and once again none on lanthanides. Nuclei which did not feature in the previous volume are ^{73}Ge (Group 14), ^{17}O (Group 16), and the transition metals ^{105}Pd and ^{185,187}Re. Only one specialised review has appeared in 2010, about ^{59}Co NQR experiments on the ferromagnetic superconductor UCoGe; a "self-induced vortex state" was suggested in its superconducting (SC) state.[2]

The normal format is followed in the main body of this review, with results for main group elements followed by those for transition metals.

2 Main group elements

2.1 Group 13 (Boron-11, Aluminium-27, Gallium-69 and -71, and Indium-115)

Density functional theory (DFT) calculations using the BLYP method and 6-31G* standard basis set have been performed for perfect and defective boron nitride nanotubes.[3] The NQR parameters for ^{11}B and ^{14}N nuclei were evaluated, to study the influence of defects on the electronic structure properties, for single-walled nanotubes of 36 boron, 36 nitrogen and 12 hydrogen atoms, and for those of 36 boron, 36 nitrogen and 16 hydrogen atoms. The results indicated that, due to the formation of non-hexagonal rings in the defective model because of the removal of a B-N bond, the NQR parameters at the site of first neighbouring nuclei were significantly influenced, while those further away remained almost unchanged. It seems a great pity that this research group does not carry out some experimental NQR work to test these interesting predictions, or alternatively arrange a collaboration with an experimental NQR group if they do not have the facilities themselves. (They have previously published several theoretical

Department of Chemistry, Durham University, South Road, Durham, DH1 3LE, UK; E-mail: k.b.dillon@durham.ac.uk

studies of NQR parameters in organic materials, which lie outside the scope of the present series of reports.)

The novel phase transition observed at $T_0 = 27\,K$ in $CeRu_2Al_{10}$ has been investigated theoretically as a spin-Peierls transition on the one-dimensional zigzag chains formed by nearest-neighbour Ce ions along the *c*-axis.[4] A lattice distortion was proposed to stabilise the spin-Peierls state, compatible with the ^{27}Al NQR spectra and neutron diffraction patterns. ^{27}Al NQR (and NMR) data for $CeFe_2Al_{10}$, relevant to the novel phase below $T_0 = 27\,K$ of isostructural $CeRu_2Al_{10}$, have enabled the signals to be assigned to five crystallographically inequivalent Al sites.[5] No splitting of the signals was observed down to 1.5 K, confirming the lack of a phase transition in this compound. A gaplike decrease in the spin-lattice relaxation rate (SLR) above *ca.* 20 K, with a Korringa law dependence below 20 K, indicated a pseudogap opening near the Fermi level. A gap magnitude of 70 K with a mid-gap state of 42 % was estimated from the results. This gap magnitude was much larger than that of 15 K evaluated previously from electrical resistivity, suggesting that the compound is a Kondo semiconductor, with a Kondo temperature much higher than 300 K.

$^{69,71}Ga$ NQR measurements, including the SLR, together with muon spin relaxation, have been carried out for the kagome compound $Pr_3Ga_5SiO_{14}$.[6] A very weak quasistatic magnetism was found to develop below 40 K, persisting down to the base temperature (T) of 21 mK. This magnetism was deduced to originate from hyperfine-enhanced ^{141}Pr nuclear magnetism, unambiguously assigning the compound as a Van Vleck paramagnet. A wide distribution of local environments was found, leading to broad $^{69,71}Ga$ NQR signals for both $Pr_3Ga_5SiO_{14}$ and its Nd analogue. A broad maximum in the relaxation rate at $\approx 10\,K$ was attributed to population of the first excited magnetic crystal field level.

A phase diagram has been presented for antiferromagnetism and superconductivity in Ce_2RhIn_8, as probed by ^{115}In NQR under pressure (P).[7] The quasi-2D character of antiferromagnetic (AF) spin fluctuations in the paramagnetic state at $P = 0$ evolved into a 3D character because of the suppression of AF order for $P > \sim 1.36\,GPa$. SLR measurements revealed superconducting order only in the narrow P range 1.36–1.84 GPa, with a maximum $T_C \sim 0.9\,K$. The phase diagram was considered to differ from those published previously because of contamination in the latter by impurity phases, such as $CeRhIn_5$. ^{115}In NQR data, including the SLR, have been obtained under pressure for $CeIr(In_{0.925}Cd_{0.075})_5$, mainly for the In(1) site located in the CeIn layer, and for $CeRh_{0.4}Ir_{0.6}In_5$.[8,9] In the former material, Cd substitution induced AF order with $T_N \sim 2.5\,K$ at ambient P, but the SLR above T_N was similar to that in $CeIrIn_5$. The results were consistent with the droplet model proposed for Cd-doped $CeCoIn_5$. In the Rh-substituted compound a large enhancement of the SLR on cooling down to $T_C = 0.9\,K$ was observed, due to the existence of AF spin fluctuations. In $CeCoIn_5$ and $CeRhIn_5$ the occurrence of superconductivity was related with AF spin fluctuations, originating from the AF quantum-critical point. In $CeIrIn_5$, however, not near the AF critical point, superconductivity occurred without AF spin fluctations, and a quite small value of T_C ($\sim 1\,K$) was seen around $P = 3\,GPa$. The suggested mechanism of superconduction could be different in $CeIrIn_5$ (SC2) from that in the Co and Rh compounds (SC1).

An analysis of ^{115}In NQR SLR data under P for $CeCoIn_5$ has revealed the presence of 2D magnetic quantum critical fluctuations in the heavy electron component, providing a promising candidate for the pairing mechanism in this material.[10] The results were compatible with an AF quantum critical point located at slightly negative P in this compound, and suggested significant similarities between heavy electron materials and high-T_C cuprates. Three sets of ^{115}In NQR signals have been observed for $CePt_2In_7$ at 10 K, which could be assigned to distinct sites.[11] The compound proved to be AF below $T_N = 5.2$ K, with staggered AF ordering in-plane, and collinear AF correlations between adjacent planes; it exhibited commensurate antiferromagnetism above T_N. The T-dependence of the SLR revealed enhanced fluctuations at T up to nearly 4 T_N, different from the $CeMIn_5$ systems (M = Co, Rh or Ir).

2.2 Group 14 (Germanium-73)

NQR data have been reported for ^{73}Ge nuclei in the filled skutterudites $LaPt_4Ge_{12}$ (T_C 8.3 K) and $PrPt_4Ge_{12}$ (T_C 7.9 K).[12,13] The SLR in the Pr compound revealed a coherence peak just below T_C, indicating that SC could be explained in terms of a conventional BCS regime. The material was classified as an anisotropic *s*-wave superconductor with a point node gap when other physical measurements were taken into account. In the conventional superconductor $LaPt_4Ge_{12}$, the coherence peak was markedly suppressed by a large damping effect of quasiparticles, possibly related to "rattling" of the La ions. The higher T_C values for these compounds than for other RPt_4Ge_{12} systems were attributed to a larger density of states, rather than to local anharmonic phonon modes derived from possible rattling of R ions.

2.3 Group 15 (Nitrogen-14, Arsenic-75, Antimony-121 and -123, and Bismuth-209)

As mentioned in previous reports in this series, both $NaNO_2$ and NH_4NO_3 are frequently used as test compounds for new equipment and/or processes in the recording of ^{14}N NQR spectra. The next three papers fall into this category.[14–16] Nonuniform line broadening in quadrupolar spin systems with integral spins has been analysed theoretically, and shown to be of the tensor type.[14] This formed the basis of a two-frequency method for analysing the distribution of local inhomogeneities in a crystal lattice, which was verified experimentally for the ^{14}N nuclei in a sample of $NaNO_2$ at room temperature (RT). Singular value decomposition (SVD) based on a Hankel matrix has been proposed for enhanced detection of ^{14}N NQR signals from materials such as explosives.[15] The method was tested using the signals from NH_4NO_3 (250 g sample). Frequency domain analysis of the ^{14}N signals of NH_4NO_3 *via* Fourier transform lacks time domain information.[16] A threshold de-noising method, based on wavelet transform, for analysing time-domain NQR signals of this compound in the presence of strong background noise, has been proposed. The correlation coefficients between the processed signals and those of a standard sample of NH_4NO_3

were calculated. They were found to increase with the amount of NH_4NO_3 used. Other references to ^{14}N NQR spectroscopy have been described in the sub-section on Group 13.[3]

^{75}As NQR (and NMR) measurements as a function of T have been reported for LiFeAs ($T_C \sim 17$ K).[17] The SLR decreased below T_C with no coherence peak, and could be fitted by gaps with $s^{\pm}$ wave symmetry in the presence of impurity scattering. In the normal state, both $1/T_1T$ and the Knight shift decreased with decreasing T, but became constant below $T \leq 50$ K. Estimates of the Korringa ratio showed that the electron correlations were weak in this material, which might account for a lower T_C than in other families of iron arsenides. A new approach using a self-flux technique has been used to grow large and high quality single crystals of LiFeAs.[18] The ^{75}As NQR at RT showed a very sharp resonance, confirming the quality of the crystal, which was also demonstrated by other physical techniques. ^{75}As NQR (and NMR) spectra have been recorded for $CaFe_2As_2$ under P.[19] At $P = 4.7$ and 10.8 kbar, the T-dependence of the SLR in the tetragonal phase showed no coherence peak just below T_C, and decreased with decreasing T. The superconductivity was gapless at $P = 4.7$ kbar, but evolved to a system with multiple gaps at $P = 10.8$ kbar. Superconductivity was found to appear near to a quantum critical point in this P range. A systematic study under P indicated that electron correlations played a vital role in forming Cooper pairs. Both electron correlation and superconductivity disappeared in the collapsed tetragonal phase.

^{75}As NQR measurements, including the SLR, have been described for heavily electron-doped $LaFeAsO_{1\text{-}x}F_x$ ($T_C = 5$ K).[20] The Hebel-Slichter (HS) peak partially recovered in the heavily electron-overdoped regime, where the nesting condition of hole and electron Fermi surfaces became significantly worse. The results contrasted with those for the optimally doped material ($T_C = 28$ K), which showed a lack of the HS peak, and an explanation for these differences was proposed. The F-doping dependence of the ^{75}As NQR frequency in the series $LaFeAsO_{1\text{-}x}F_x$ has been established for x = 0, 0.07, 0.08, 0.11 and 0.14.[21] The frequency was found to increase with increasing x. From point charge calculations, this effect could not be explained in terms of the effect of surrounding ions. The on-site charge distribution at the ^{75}As site modified by hybridisation between As-4p and Fe-3d orbitals was considered to give the dominant contribution to the ^{75}As frequency. ^{75}As NQR and ^{139}La NMR data have been recorded for $LaNiAsO_{1\text{-}x}F_x$ (x = 0, 0.06, 0.10 or 0.12).[22] The SLR decreased below T_C with a well-defined coherence peak, and followed an exponential decay at low T. The results indicated that the superconducting gap was fully opened (different from iron-pnictide analogues). The difference was ascribed to a different topology of the Fermi surface for these types of compound. In the normal state, $1/T_1T$ was constant between T_C and ~ 10 K for all compounds, and up to 250 K for x = 0 and 0.06, indicating weak electron correlations. A possible interpretation was suggested for the "pseudogap-like" behaviour in the normal state. ^{75}As NQR spectroscopy has been used to probe the charge distribution in RFeAs$O_{1\text{-}x}F_x$ (R = La or Sm).[23] The undoped, optimally doped or overdoped compounds featured a single charge environment, but two charge environments were detected in the underdoped region. SLR measurements showed their coexistence at the nanoscale. These were ascribed to a local electronic order in the iron layers, where low- and high-doping-like regions could coexist. A phase diagram was proposed, in which the local order served as an electronic background for the interplay of static magnetism and superconductivity. ^{75}As NQR (and NMR) data, including the T-dependence of the SLR, have been obtained for superconducting samples of $LaFeAsO_{1-x}F_x$ ($0.11 \leq x \leq 0.25$).[24] The smallest linewidth was found for an optimally doped sample with $T_C \sim 28$ K. A simple relationship SLR $\sim T^n$ was used, which held quite well in the region $0.3\ T_C < T \leq T_C$, with n varying from 2.5–6 with variation in x. The maximum value occurred at optimal x. The n value for optimally doped samples was considered to be intrinsic to this system, and the x dependence of n to originate from the spatial distribution of the superconducting order parameter Δ.

In an extension of previous ^{75}As NQR studies by this group on Ge-As-Se glasses, high magnetic fields have been used to perform ^{75}As NMR relaxation measurements.[25] The results in conjunction with the NQR data revealed the presence of indirect dipolar coupling of the As atoms in $Ge_2As_2Se_5$, indicative of increased clustering effects in the glass. The acquisition of NQR spectra with wideband uniform rate and smooth truncation (WURST) pulses has been investigated.[26] Results were obtained for ^{75}As (As_2O_3), ^{35}Cl (an organic compound) and ^{63,65}Cu (CuCN) nuclei. The results for As and Cl were compared with those acquired using standard Hahn-echo sequences, or echo sequences employing composite refocusing pulses.The WURST sequence was demonstrated to possess superior excitation bandwidths, which allowed for differences in the characteristic impedance of the receiving and excitation circuits of the spectrometer to be detected. The use of the Carr-Purcell-Meiboom-Gill (CPMG) pulse sequence for signal enhancement of NQR spectra acquired with WURST and conventional pulses was also investigated, and was found to be useful for samples with a broad distribution of NQR frequencies, such as CuCN. Advantages and disadvantages of the method were discussed.

The parameters of the Voronoi-Derichlet (VD) polyhedra characterising the asymmetry of the crystal field in the vicinity of Sb nuclei, *i.e.* the displacement of the Sb nucleus from the centroid of its polyhedron (D_A) and the noncentrosymmetry parameter of the atomic environment (A_{ns}), have been determined for 53 compounds containing 73 crystallographic types of SbX_n coordination polyhedra (X is a halogen and/or chalcogen).[27] A linear correlation was found between the experimental ^{121}Sb quadrupole coupling constants from NQR spectra and the D_A and A_{ns} values. This could provide a method for assigning quadrupolar resonances in compounds with several crystallographically non-equivalent positions of the quadrupolar nuclei.

^{209}Bi NQR spin-lattice relaxation as a function of T has been studied for $Bi_4Ge_3O_{12}$ single crystals doped with minor amounts of paramagnetic atoms of Cr, Nd or Gd.[28] A dramatic increase (up to eight-fold) in effective spin-spin relaxation time T^*_2 had been found previously at RT on doping. Unlike T^*_2, the effective spin-lattice relaxation time T^*_1 at RT differed insignificantly for doped and pure samples. At lower T, however, different T_1 behaviour was observed, caused by different contributions to the relaxation process by the dopant paramagnetic ions. A distinctive maximum in the T-dependence of T_1 for the Nd-doped crystal was shown to result from crystal electric field effects.

2.4 Group 16 (Oxygen-17)

A modification of the NQ double resonance with coupled multiplets method has been proposed, which can be used for the measurement of the dipolar structure of ^{17}O NQR lines in the case of a strong ^{1}H–^{17}O dipolar interaction.[29] The technique was based on magnetic field cycling between a high magnetic field and zero magnetic field, and a simultaneous application of three radiofrequency (rf) magnetic fields with frequencies close to the three ^{17}O NQR frequencies during the time spent in zero magnetic field. When the sum of the two lower frequencies was not equal to the highest frequency, the method produced an energy flow to the proton spin system, and the proton NMR signal at the end of the magnetic field cycle dropped to a lower value. The new technique was analysed theoretically, and compared with single-frequency and two-frequency irradiation techniques. Its sensitivity was found to exceed that of two-frequency irradiation. As a test, the shape was measured of the highest-frequency ^{17}O resonance from paraelectric KH_2PO_4 at 160 K. On the assumption of a spin-triplet superconducting state in Sr_2RuO_4, the spin-orbit coupling associated with relative motion in the Cooper pairs has been calculated, by extending the method for dipole-dipole coupling given by Leggett for superfluid ^{3}He.[30] A formula was obtained, giving rise to the internal Josephson oscillations of the d-vector in the ab-plane if the stable direction of this vector was in the ab-plane. This formula resolved the problem of the anomalous T-dependence of the SLR for ^{17}O nuclei in this compound.

2.5 Group 17 (Chlorine-35 and -37, Bromine-79 and 81, and Iodine-127)

Quantum-chemical calculations by the RHF/6-31G(d) and MP2/6-31G(d) methods have been performed for Ph_2PCl_3, with complete or partial optimisation of its geometry.[31] The results were used to evaluate the ^{35}Cl NQR frequencies and asymmetry parameters (η) of the electric field gradient (efg). Bond lengths were found from the calculations with partial optimisation of the geometry and the experimental ^{35}Cl NQR data. The ^{35}Cl frequencies of the axial and equatorial Cl atoms in the chlorophosphorane were shown to decrease with a decrease in the bond length to P. Similar calculations have been carried out for PCl_5, $PhPCl_4$ and $PhCH{=}CHPCl_4$.[32] A substantial change in the NQR frequencies of the axial Cl atoms and a weak change in the frequencies of the equatorial Cl atoms with a change in the number of the latter were explained. Parallel calculations on $PFCl_4$, $P(CCl_3)Cl_4$, PF_2Cl_3 and $P(CCl_3)_2Cl_3$ have revealed the mutual influence of the axial and equatorial substituents.[33] A decrease in the P-Cl bond length led to a decrease in the NQR frequency for both axial and equatorial chlorines. The results of quantum-chemical calculations by the RHF/3-21G* and MR2/3-21G* methods have been given for $Cl(CH_2)_3SnCl_3$ and $ClCH_2SnCl_3$, with full optimisation of geometry.[34] The ^{35}Cl NQR frequencies and η values were estimated, and compared with experimental data. The experimental ^{35}Cl signal at 77 K for the chloropropyl group occurred at an anomalously low frequency of 31.750 MHz. This was attributed to polarisation of this compound under the influence of a significant positive charge on the Sn atom directly through the field. The polarisation of the C-Cl bond was deduced to be in the opposite direction to that of the related bond in $ClCH_2SnCl_3$, because of peculiarities in their geometry. The overall results indicated that the RHF/3-21G* method was better for organotin molecules than the MP2/3-21G* method.

Quantum-chemical calculations *via* the MP2/6-31G* method have been applied to isolated molecules of $Cl_3P{=}NCOCX_3$, where X = F or Cl; the barriers to intramolecular reorientation of the PCl_3 group relative to the P=N bond, and of the CX_3 groups relative to the C–C bond, were estimated.[35] The values of the potential barriers to internal rotation were compared with data obtained from NQR and NMR spectra in the crystalline state. Structural and dynamic factors were discussed. Intermolecular interactions contributed to impeding PCl_3 and CX_3 reorientation, up to 50-70% of the total potential barrier. The same method has been used to analyse the molecular structures of $Cl_3P{=}NR$, with R = CMe_3, CEt_3, $C(CF_3)_3$, $CCl(CF_3)_2$, CCl_3, CCl_2CCl_3 and $CCl(CCl_3)_2$; the results were compared with the experimental ^{35}Cl NQR frequencies.[36] A correlation was suggested between the experimental ^{35}Cl NQR frequencies at 77 K and the theoretically calculated charges on the chlorine atoms in the PCl_3 group. The activation parameters contributing to the T-dependence of the SLR for ^{35}Cl nuclei in groups with thermally activated mobility have been analysed.[37] A linear correlation was found between the log of the pre-exponential factor b and the activation energy, similar to the Arrhenius equation but for solids.

^{35}Cl NQR spectra at 77 K have been recorded for three different phases of $HGeCl_3$.[38] One phase gave a single line, indicating that the $HGeCl_3$ molecule was situated on a three-fold axis coinciding with the Ge-H bond direction. Phase II gave two signals in a 2:1 intensity ratio, signifying that the molecule was situated on a symmetry plane passing through H-Ge-Cl bonds. Three equally intense lines were observed from phase III. Values of T_1 were also measured, and an equation relating the ^{35}Cl NQR frequency to the inverse spin-lattice relaxation time was found. This equation was used to distinguish the short- and long-range order contributions to the NQR frequency, and changes in the strength of the local electric field were ascertained.

The salts $[C_7H_6N_3O_2][HgX_3]$. H_2O (X = Cl **1**, Br **2** or I **3**) have been synthesised and characterised by various techniques, including ^{35}Cl (**1**), ^{81}Br (**2**) and ^{127}I (**3**) NQR spectroscopy.[39] Compounds **1** and **2** were isostructural, but **3** differed. Both **1** and **2** contained trigonal planar HgX_3^- ions, H-bonded with H_2O and cations into 2D sheets parallel to the *ac* plane. These sheets were connected by double chains of Hg----X interionic bonds along the *b* direction, resulting in a tbp about Hg with two long Hg-----X axial bonds. Compound **3** contained a dimeric $Hg_2I_6^{2-}$ anion, with tetrahedral coordination of Hg. In accord with the crystal structures, three NQR signals for ^{81}Br and ^{127}I ($1/2 \leftrightarrow 3/2$) were observed at RT for **2** and **3**, while two ^{35}Cl NQR lines assignable to terminal atoms were detected for **1**. A phase transition was found for the bromo-compound around 160 K, with the two higher frequency lines splitting. The ^{35}Cl resonances disappeared below *ca.* 110 K, possibly indicating a phase transition. All three ^{127}I signals were visible from 77–300 K.

The crystal and molecular structures of $(4\text{-}BrC_5H_4NH)_2\ SbBr_5$ have been determined at 193 K by single-crystal X-ray diffraction.[40] The anion dimer $Sb_2Br_{10}{}^{4-}$ was present, edge-sharing between two $SbBr_6$ octahedra. The cation was bound to apical Br atoms through NH----Br H-bonding. Seven ^{81}Br resonances, including those for two Br atoms in non-equivalent cations, were seen in the *T*-range 77–360 K, in accordance with the crystal structure, and there was no evidence of a phase transition. The signals were assigned from the crystal structure. Two of the resonances showed an unusual positive *T*-dependence. Possible *trans influence* and electronic redistribution in the 3c–4e (Br-Sb-Br) bonds with change of *T* were discussed. The quantum antiferromagnet $Cu_2Te_2O_5Br_2$ has been investigated by bromine NQR at 15 K, and by ^{125}Te and bromine NMR, in the paramagnetic phase.[41] The results led to determination of the Br hyperfine coupling constant and efg tensor, and of the spin polarisation of the Br *p* orbitals. The crucial role of bromine in the interaction paths between Cu spins was demonstrated.

In the crystal and molecular structures of $(CH_6N_3)_2\ [HgI_4]$ at 298 K, the anions and cations were interconnected by N-H----I H-bonds, with the Hg atom tetrahedrally coordinated.[42] The Hg-I bond lengths were consistent with the observation of four different ^{127}I NQR frequencies (for each transition) at R*T*. ^{127}I NQR spectra at 77 K have been recorded for the semiconducting layered crystals $(BI_3)_{(1\text{-}n)}(PbI_2)_n$ $(0 \leq n \leq 0.50)$.[43] For $0.05 \leq n \leq 0.10$, the crystals exhibited the properties of a doped BI_3 crystal containing intralayer PbI_2 clusters. In the range $0.10 \leq n \leq 0.20$, the material showed the properties of a $BI_3.PbI_2$ solid solution, and ^{127}I NQR could not be observed. At $n \sim 0.20$, the mixed crystal underwent a concentration phase transition, with the disappearance of the previous signal and the appearance of a new one at lower frequency. The impurity crystal changed into a mixed crystal $(BI_3)_{(1\text{-}n)}(PbI_2)_n$ for $0.20 \leq n \leq 0.50$, where the translational symmetry of the crystal could be retained.

3 Transition metals

3.1 Cobalt-59

Quadrupole coupling constants and η values for ^{59}Co nuclei at 77 K have been correlated for a series of complexes $(\pi\text{-}C_5H_4X)Co(\pi\text{-}(3)\text{-}1,2\text{-}B_9H_8(8\text{-}Z)C_2HY)$.[44] Three non-equivalent positions for Co atoms were found for symmetrical $(\pi\text{-}C_5H_5)Co(B_9H_9C_2H_2)$, rarely seen in Co systems, whereas all the other complexes gave one signal only for each transition. All three positions were deduced to correspond with dynamic variations of the efg as a result of the cooperative Opik – Price effect. Four different stable phases and an intermediate one have been recognised for Na_xCoO_2 samples $(0.67 \leq x \leq 0.75)$ from ^{59}Co NQR data at 4.2 K.[45] By this means, pure phase samples could be easily distinguished from multi-phase samples. Keeping samples at R*T* in contact with humid air was found to lead to loss of phase purity and reduction in Na content. The structure of one of the phases was determined, and confirmed by X-ray methods. The results in combination with more diffraction data should help to clarify the structures of the other phases. Powder samples of the bilayered hydrated superconducting Na_xCoO_2. y H_2O with different T_C values have been synthesised by changing the duration in a high humidity atmosphere after the intercalation of H_2O molecules.[46] From ^{59}Co NQR, the resonance frequency was found to decrease monotonically with increasing time after filtration. An electronic phase diagram was drawn, with two superconducting phases and an in-between magnetic phase. The lattice parameter along the *c*-axis was also found to decrease monotonically with increasing *T*. The correlation between the NQR frequency and the lattice parameter along the *c*-axis was preserved, irrespective of the synthetic method used to prepare the samples. ^{59}Co NQR spectra have been recorded at various temperatures below 10 K for $Na_{0.35}CoO_2$. y H_2O (y ~ 1.3).[47] The nuclear quadrupolar coupling constants and η values were estimated from the results. There was no evidence of charge ordering or charge density wave ordering around the magnetic transition temperature. The duration dependence of superconductivity in a 43 % humidity atmosphere for polycrystalline samples of bilayer hydrated Na_xCoO_2. y H_2O has been followed by ^{59}Co NQR spectroscopy.[48] A reentrant behaviour of superconductivity with respect to time was found. The SLR indicated a magnetic critical slowing down at 5 K in the magnetic phase, excluding a charge density wave ordering, and an enhancement at and just above T_C in a new superconducting phase, suggesting the possible coexistence of magnetic ordering and superconductivity.

^{59}Co NQR at various *T* has been recorded for a single-crystal sample of UCoGe, a ferromagnetic superconductor with T_{Curie} ~ 2.5 K.[49] The development [illegible] a local magnetic moment around T_{Curie} was investigated, and the ferrom[illegible]etic transition was suggested to possess weakly first-order character. [illegible]ysteresis was observed in the *T* variation of the spectral intensitie[illegible]R (and NMR) measurements have been used to probe stacking fau[illegible]d the metallic properties for the triangular lattice system CoO_2, wh[illegible]as three oxygen layers, with a prismatic oxygen environment be[illegible] the layers.[50] A significant distribution of ^{59}Co NQR frequencies w[illegible]und, due to stacking faults and short atomic coherence. The spin dynamics could be understood in terms of the relaxation mechanism for a weakly coordinated metal system. Li impurities contained in the specimens arose from minor phases of Li_xCoO_2 with x ≈ 0.92, or Li-intercalated spinel-type Co_3O_4. Possible applications to rechargeable Li batteries were considered.

3.2 Copper-63 and -65

A mini cubic anvil apparatus for high pressure NQR (and NMR) measurements has been developed, and used to record the ^{63}Cu NQR signal of Cu_2O at R*T* up to 7.2 GPa with high sensitivity.[51] Attempts to reach higher *P* using a ceramic MgO gasket had been unsuccessful so far, although calculations suggested that this modification should extend the range up to 10 GPa. A space-efficient opposed-anvil high pressure cell has been designed, and similarly used to record ^{63}Cu NQR at 300 K from Cu_2O at

P up to 8.8 GPa.[52] The Knight shifts of copper and platinum, and the in-plane shift of metallic tin, were also measured. The ^{63}Cu NQR frequency was considered to have excellent resolution in determining the pressure, while the Knight shifts of platinum and tin were suitable manometers when the addition of Cu_2O to the sample was to be avoided.

Pressure-induced superconductivity in $CeCu_2Si_2$ has been studied *via* ^{63}Cu NQR up to 4.8 GPa.[53] A linear increase in the NQR frequency with P was observed below 3 GPa, but deviation from linearity became apparent above 3.5 GPa. The downward deviation was associated with a change of valence, consistent with the results of band calculations for the structural parameters under P. The AFM spin fluctuations were drastically suppressed with increasing P, whereas T_C increased from 0.7 K at ambient P to 1.64 K at 4.2 GPa. The bulk superconductivity suddenly disappeared at 4.8 GPa, accompanied by a rapid decrease in the density of states at the Fermi level. A T^3 law without a coherence peak just below T_C was observed around 4 GPa. The incidence of doped holes in the CuO_2 plane of slightly doped $YBa_2(Cu_{1-z}Li_z)_3O_{6+x}$ ($x<0.1$) has been probed by NQR of the Cu(1) sites, including relaxation time measurements as a function of T.[54] An inhomogeneous distribution of doped holes from heterovalent substitution was detected. At lower T, the motion of the holes slowed down, and they were thought to locate in restricted regions ($\sim$3 lattice constants) in the Coulomb potential of the Li^+ ions. The results were compared with those for slightly doped $Y_{1-y}Ca_yBa_2Cu_3O_6$. The in-plane Li^+ ions in the lithium compounds were thought to bind the holes more strongly than the out-of-plane Ca^{2+} ions in the latter compounds. Detailed $^{63,65}Cu$ NQR (together with ^{17}O and ^{139}La NMR) studies have been carried out for a stripe ordered $La_{1.875}Ba_{0.125}CuO_4$ single crystal, and for oriented powder samples of $La_{1.8-x}Eu_{0.2}Sr_xCuO_4$.[55] A complete wipeout of Cu intensity was seen at low T where the spin stripe order set in, together with a partial wipeout of the ^{17}O NMR intensity, and a simultaneous drop in the ^{17}O efg. The SLR for ^{17}O demonstrated a peak similar to that for ^{139}La, which was of magnetic origin. The results indicated slow fluctuations of the Cu electronic spins at the highest temperatures close to optimal doping.

Various physical methods including $^{63,65}Cu$ NQR have been used to investigate the electronic structure and magnetic properties of the spin-gap compound $Cu_2(PO_3)_2CH_2$, together with theoretical calculations.[56] The results gave evidence of a spin gap of about 25 K. The compound could be described to a good approximation as coupled alternating AF Heisenberg chains. The leading AF exchange of about 75 K was assigned from the calculations to a coupling between structural Cu_2O_6 dimers, whereas coupling within these dimers was strongly reduced, due to sizeable ferromagnetic contributions. The evaluated Cu quadrupole frequencies were unusually high compared with typical values for other cuprates. The influence of cerium substitution on the valence phase transition in the $Yb_{1-x}Ce_xInCu_4$ system ($x=0$, 0.04, 0.08, 0.12 or 0.16) has been investigated by ^{63}Cu NQR.[57] A discontinuous change in the resonance frequency around the valence transition T was detected for $x=0$, 0.04 and 0.08. In the vicinity of $x=0.08$, a crossover was observed from a first order valence phase transition to a gradual change of electronic and magnetic properties for a more highly doped system. ^{63}Cu NQR (and NMR) have been monitored as a function of T for $Cu_2Sc_2Ge_4O_{13}$ and $Cu_2Fe_2Ge_4O_{13}$.[58] The iron compound has weakly coupled Cu dimers and Fe chains. In the AF state below 39 K, the Cu resonances were observed under the internal magnetic field at the Cu site. From the results, the 3*d* hole was deduced to be located mainly in the $d_{x^2-y^2}$ orbital. In the scandium compound, with Cu dimers only, these appeared to be magnetically well isolated. The SLR in this compound showed thermally-activated behaviour. A remarkable oscillation of the Cu NQR spin-echo intensity was ascribed to nuclear spin coupling within the dimer. Other results for Cu nuclei have been described in the sub-section on Group 15.[26]

3.3 Ruthenium-101

A nonmagnetic anomaly had been discovered previously around 17.5 K in URu_2Si_2, but not characterised in the low T phase, and therefore called "the hidden order".[59] No anomaly was apparent in ^{101}Ru NQR measurements around this T, with the results indicating that the four-fold axis of the Ru site survived even at lower T. Space group analysis has shown that one type of second-order phase transition did not require any kind of lattice distortion in the system, and allowed the Ru frequency to remain unchanged, thus providing a possible explanation. The characteristics of the hidden order were discussed, based on a local $5f^2$ electron picture.

3.4 Palladium-105

^{105}Pd NQR experiments at 6 K have been carried out for $NpPd_5Al_2$ and $CePd_5Al_2$.[60] Four resonances were observed in both systems, showing two inequivalent Pd sites in each case. The resonance frequency, η value and efg for each Pd site were evaluated.

3.5 Rhenium-185 and 187

^{185,187}Re NQR, (together with ac susceptibility measurements), has been performed for $Cd_2Re_2O_7$ ($T_C = 1$ K), to study the correlation between superconductivity and structural transitions under high P.[61] (At ambient P $T_{s1} = 200$ K and $T_{s2} = 120$ K.) T_C was almost P-independent below 1 GPa, but increased monotonously above 1 GPa to reach 2.3 K at 2.5 GPa. An increase of P caused broadening and a shift to lower frequency of the ^{185}Re and ^{187}Re NQR signals. This broadening increased suddenly around 2 GPa, which could be related to a distributed efg, and elimination of the structural transition at T_{s2}.

References

1 J. Yarwood, R. Douthwaite and S. B. Duckett (eds), *Spectroscopic Properties of Inorganic and Organometallic Compounds*, The Royal Society of Chemistry, Cambridge, UK, 2011, vol. 42.
2 K. Ishida, T. Hattori, Y. Ihara, Y. Nakai, N. K. Sato, K. Deguchi, N. Tamura and I. Satoh, Kotai Butsuri, 2010, **45**, 319; CA Selects Plus: Carbon & Heteroatom NMR, 2010, (19), 1082475.
3 M. Giahi and M. Mirzaei, *Z. Naturforsch.*, 2009, **64a**, 251.
4 K. Hanzawa, *J. Phys. Soc. Jpn.*, 2010, **79**, 084704.

5 Y. Kawamura, S. Edamoto, T. Takesaka, T. Nishioka, H. Kato, M. Matsumura, Y. Tokunaga, S. Kambe and H. Yasuoka, *J. Phys. Soc. Jpn.*, 2010, **79**, 103701.
6 A. Zorko, F. Bert, P. Mendels, K. Marty and P. Bordet, *Phys. Rev. Lett.*, 2010, **104**, 057202.
7 M. Yashima, S. Taniguchi, H. Miyazake, H. Mukuda, Y. Kitaoka, H. Shishido, R. Settai and Y. Ōnuki, *J. Phys.: Conf. Series*, 2010, **200**, 012238.
8 M. Yashima, S. Taniguchi, N. Tagami, Y. Mugino, H. Mukuda, Y. Kitaoka, Y. Ōta, H. Shishido, R. Settai and Y. Ōnuki, *Physica C*, 2010, **470**, S558.
9 M. Yashima, S. Taniguchi, N. Tagami, H. Mukuda, Y. Kitaoka, Y. Ōta, H. Shishido, R. Settai and Y. Ōnuki, *J. Phys.: Conf. Series*, 2010, **200**, 012087.
10 Y.-f. Yang, R. Urbano, N. J. Curro and D. Pines, arXiv.org, e-Print Archive, *Cond. Mat.*, 2009, 1–5, arXiv:0909.1341v1 [cond-mat. str-el].
11 N. apRoberts-Warren, A. P. Dioguardi, A. C. Shockley, C. H. Lin, J. Crocker, P. Klavins and N. J. Curro, *Phys. Rev. B*, 2010, **81**, 180403(R).
12 F. Kanetake, H. Mukuda, Y. Kitaoka, K.-i. Magishi, H. Sugawara, K. M. Itoh and E. E. Haller, *J. Phys. Soc. Jpn.*, 2010, **79**, 063702.
13 F. Kanetake, H. Mukuda, Y. Kitaoka, H. Sugawara, K. Magishi, K. M. Itoh and E. E. Haller, *Physica C*, 2010, **470**, S703.
14 D. Ya. Osokin, R. R. Khusnutdinov, N. Dogan and B. Z. Rameer, *Tech. Phys.*, 2010, **55**, 225.
15 K. Zhu, T. Su, X. He and J. Nui, *Proc. Eng.*, 2010, **7**, 57.
16 G. Yao, X. Fang and Y. Mao, *Chinese J. Magn. Reson.*, 2009, **26**, 504.
17 Z. Li, Y. Ooe, X.-C. Wang, Q.-Q. Liu, C.-Q. Jin, M. Ichioka and G.-q. Zheng, *J. Phys. Soc. Jpn.*, 2010, **79**, 083702.
18 I. Morozov, A. Boltalin, O. Volkova, A. Vasiliev, O. Kataeva, U. Stockert, M. Abdel-Hafiez, D. Bombor, A. Bachmann, L. Harnagea, M. Fuchs, H.-J. Grafe, G. Behr, R. Klingeler, S. Borisenko, C. Hess, S. Wurmehl and B. Bűchner, *Crystal Growth & Design*, 2010, **10**, 4428.
19 S. Kawasaki, T. Tabuchi, X. F. Wang, X. H. Chen and G.-q. Zheng, *Supercond. Sci. Tech.*, 2010, **23**, 054004.
20 H. Mukuda, M. Nitta, M. Yashima, Y. Kitaoka, P. M. Shirage, H. Eisaki and A Iyo, *J. Phys. Soc. Jpn.*, 2010, **79**, 113701.
21 S. Kitagawa, Y. Nakai, T. Iye, K. Ishida, Y. Kamihara, M. Hirano and H. Hosono, *Physica C*, 2010, **470**, S282.
22 T. Tabuchi, Z. Li, T. Oka, G. F. Chen, S. Kawasaki, J. L. Luo, N. L. Wang and G.-q. Zheng, *Phys. Rev. B*, 2010, **81**, 140509(R).
23 G. Lang, H.-J. Grafe, D. Paar, F. Hammerath, K. Manthey, G. Behr, J. Werner and B. Bűchner, *Phys. Rev. Lett.*, 2010, **104**, 097001.
24 Y. Kobayashi, E. Satomi, S. C. Lee and M. Sato, *J. Phys. Soc. Jpn.*, 2010, **79**, 093709.
25 E. Mammadov, P. C. Taylor, A. Reyes, S. Mehdiyeva and P. Kuhns, *Phys. Status Solidi A*, 2010, **207**, 635.
26 A. J. Rossini, H. Hamaed and R. W. Schurko, *J. Magn. Reson.*, 2010, **206**, 32.
27 D. V. Pushkin, V. N. Serezhkin, M. O. Karasev and E. A. Kravchenko, *Russ. J. Inorg. Chem.*, 2010, **55**, 523.
28 V. G. Orlov, G. S. Sergeev, T. Asaji, E. A. Kravchenko and Yu. F. Kargin, *J. Exptl. Theo. Phys.*, 2010, **110**, 296.
29 J. Seliger and V. Žagar, *J. Magn. Reson.*, 2010, **203**, 220.
30 K. Miyake, *J. Phys. Soc. Jpn.*, 2010, **79**, 024714.
31 V. P. Feshin and E. V. Feshina, *Russ. J. Gen. Chem.*, 2009, **79**, 2116.
32 V. P. Feshin and E. V. Feshina, *Russ. J. Gen. Chem.*, 2009, **79**, 2121.
33 V. P. Feshin and E. V. Feshina, *Russ. J. Gen. Chem.*, 2009, **79**, 2126.

34 V. P. Feshin and E. V. Feshina, *Russ. J. Gen. Chem.*, 2010, **80**, 1609.
35 S. N. Shurov, A. N. Vasyanin and G. B. Soifer, *Russ. J. Gen. Chem.*, 2009, **79**, 2293.
36 G. B. Soifer and S. N. Shurov, *Russ. J. Inorg. Chem.*, 2009, **54**, 1440.
37 G. B. Soifer, *Russ. J. Phys. Chem. A*, 2010, **84**, 960.
38 G. K. Semin, E. V. Bryukhova and V. M. Burbelo, *Russ. J. Phys. Chem. A*, 2010, **84**, 141.
39 R. P. Sharma, A. Singh, P. Venugopalan, H. Ishihara, M. Nakashima, H. Terao and K. Horiuchi, *J. Mol. Struct.*, 2010, **973**, 27.
40 H. Terao, S. Ninomiya, M. Hashimoto and K. Eda, *J. Mol. Struct.*, 2010, **965**, 68.
41 A. Comment, H. Mayaffre, V. Mitrović, M. Horvatić, C. Berthier, B. Grenier and P. Millet, *Phys. Rev. B*, 2010, **82**, 214416.
42 H. Terao, T. M. Gesing, H. Ishihara, Y. Furukawa and B. Thimme Gowda, *Acta Cryst. E*, 2009, **65**, m503.
43 Yu. P. Gnatenko, A. I. Barabash, I. G. Vertegel, E. D. Chesnokov, A. I. Ovcharenko and S. V. Pogrebnyak, *Functional Materials*, 2010, **17**, 24.
44 G. K. Semin and E. V. Bryukhova, *Russ. J. Phys. Chem. A*, 2010, **84**, 235.
45 T. A. Platova, I. R. Mukhamedshin, A. V. Dooglav and H. Alloul, *JETP Lett.*, 2010, **91**, 421.
46 H. Ohta, C. Michioka, Y. Itoh and K. Yoshimura, *Physica B*, 2009, **404**, 3227.
47 H. Ohta, M. Månsson, Y. Ikedo, J. Sugiyama, C. Michioka, K. Yoshimura, J. H. Brewer, E. J. Ansaldo, S. L. Stubbs, K. H. Chow and J. S. Lord, *Physica C*, 2010, **470**, S755.
48 H. Ohta, C. Michioka, Y. Itoh and K. Yoshimura, arXiv.org, e-Print Archive, *Cond. Mat.*, 2010, 1–4, arXiv:1007. 4873v1 [cond-mat. supr-con].
49 T. Hattori, K. Ishida, Y. Nakai, T. Ohta, K. Deguchi, N. K. Sato and I Satoh, *Physica C*, 2010, **470**, S561.
50 M. Onoda and A. Sugawara, *J. Phys.: Condens. Matt.*, 2010, **22**, 035602.
51 R. Kobayashi, Y. Uchida, K. Hirayama, T. Yamazaki, H. Fukazawa, Y. Kohori and N. Takeshita, *J. Phys.: Conf. Series*, 2010, **215**, 012183.
52 K. Kitagawa, H. Gotou, T. Yagi, A. Yamada, T. Matsumoto, Y. Uwatoko and M. Takigawa, *J. Phys. Soc. Jpn.*, 2010, **79**, 024001.
53 K. Fujiwara, Y. Okazaki, Y. Hata, K. Miyoshi, J. Takeuchi, T. C. Kobayashi, C. Geibel and F. Steglich, *Phys. Status Solidi B*, 2010, **247**, 608.
54 A. V. Savinkov, A. V. Dooglav, H. Alloul, P. Mendels, J. Bobroff, G. Collin and N. Blanchard, *JETP Lett.*, 2010, **91**, 83.
55 H.-J. Grafe, N. J. Curro, B. L. Young, A. Vyalikh, J. Vavilova, G. D. Gu, M. Hűcker and B. Bűchner, arXiv.org, e-Print Archive, *Cond. Mat.*, 2010, 1–4, arXiv:1007.0339v1 [cond-mat. supr-con].
56 M. Schmitt, A. A. Gippius, K. S. Okhotnikov, W. Schnelle, K. Koch, O. Janson, W. Liu, Y.-H. Huang, Y. Skourski, F. Weickert, M. Baenitz and H. Rosner, *Phys. Rev. B*, 2010, **81**, 104416.
57 A. A. Gippius, K. S. Okhotnikov, P. A. Anferova, M. Baenitz, N. V. Mushnikov and A. N. Vasiliev, *Solid State Phenomena*, 2009, **152–153**, 419.
58 J. Kikuchi, S. Nagura, H. Nakanishi and T. Masuda, *J. Phys.: Conf. Series*, 2010, **200**, 022024.
59 H. Harima, K. Miyaki and J. Flouquet, *J. Phys. Soc. Jpn.*, 2010, **79**, 033705.
60 H. Chudo, H. Sakai, Y. Tokunaga, S. Kambe, D. Aoki, Y. Haga, T. D. Matsuda, Y. Homma, Y. Nakano, F. Honda, R. Settai, Y. Ōnuki and H. Yasuoka, *J. Phys.: Conf. Series*, 2010, **200**, 012020.
61 K. Fujiwara, K. Morimoto, E. Maeda and S. Nishigori, *J. Phys.: Conf. Series*, 2010, **200**, 012043.

Simulation of spectroscopic properties of inorganic compounds

Trevor J. Dines
DOI: 10.1039/9781849734899-00229

In this review we look at applications of modern quantum chemical methods to the interpretation of the spectroscopic properties of inorganic species. The timespan covered in this review is from mid-2008 to mid-2011. Although in a few cases more rigorous theoretical methods have been employed density functional theory (DFT) has been the most widely used for inorganic molecules. DFT methods provide the dual benefits of accuracy of prediction of structures and spectra, with comparatively modest computational demands.

1 Introduction

The last decade has seen an explosion in the number of studies in which quantum chemical methods have been applied to the computation of molecular spectroscopic properties, fuelled by significant advances in both hardware and software. A brief explanation of the methods available was provided in my previous review in this series[1] and we focus here on the applications that have been reported in the last three years. These number several hundred and for this reason it has been necessary to be selective in the examples given here, which have been chosen to illustrate the very wide range of inorganic systems that have been investigated, and also the range of spectroscopic techniques that have been employed. It should also be pointed out that the same theoretical methods have been extensively employed for the investigation of structures and reactivity of inorganic species, but without any prediction of spectroscopic properties – these studies therefore lie outside the scope of this review.

Applications of quantum mechanical (QM) methods to the simulation of spectra of main group, transition metal, and lanthanide and actinide compounds are reviewed in sections 2, 3 and 4, respectively. Each of these sections contains extensive tables giving brief details of the species investigated, the spectroscopic properties that have been simulated, and the theoretical methods used in the simulation. Further elaboration is provided in the text for a small number of these studies, which have been chosen for their relevance to topics of current interest.

There are a multitude of software packages (both commercial and non-commercial) available for modern quantum chemical calculations on molecular systems, of which the most commonly used are ADF[2], GAMESS (US version),[3,4] GAMESS-UK,[5] *Gaussian09*,[6] Jaguar,[7] NWChem,[8] HyperChem,[9] Orca[10] and Turbomole.[11] Some of these are also capable of periodic calculations on solids and surfaces (especially ADF and NWChem) but there are other software packages specifically tailored for modelling of the solid state, in particular Castep[12] and Crystal.[13] Modelling of species in

Division of Electronic Engineering & Physics, University of Dundee, Dundee, DD1 4HN, UK; E-mail: t.j.dines@dundee.ac.uk

the solid state employs periodic boundary conditions to replicate the unit cell in three dimensions. Such calculations generally use plane-wave, rather than Gaussian basis functions for valence electrons and pseudopotentials for core electrons, rather than employing all-electron methods (although these are possible with some programs). For modelling of solids and surfaces it is the unit cell geometry that is optimised, prior to computation of spectroscopic properties, and generally pure DFT methods are used in preference to the hybrid SCF-DFT methods used for molecules (which are very time-consuming for solids). A recent review on the modelling of solid state species, employing the Castep program, has been published by Milman *et al.*[14] A full list of available software for quantum chemistry and solid state physics modelling can be found at Wikipedia.[15] All of these have some capability for the computation of molecular spectroscopic properties but in many cases additional software may be needed to obtain simulated spectra, or to carry out a detailed analysis of the computed properties. For example, normal coordinate analysis software may be used, in conjunction with computed force constants, to analyse computed vibrational spectra and obtain band assignments as potential energy distributions. There are a number of normal coordinate programs available, most of which are derived from those of Schachtschneider.[16]

Examination of the recent literature reveals that the hybrid B3LYP method is still the most popular DFT method among the spectroscopic community, despite having some shortcomings (especially for transition metals). Other functionals have been favoured by some groups and the Truhlar group have introduced the so-called Minnesota M05- and M06-class functionals.[17] Although these have not yet been widely used it is expected that they will become more popular within the next few years.

Algorithms are now available for DFT analytic polarizability derivatives, leading to Raman intensities calculated under both static (zero excitation frequency) and dynamic (frequency-dependent) conditions. This has dramatically reduced the time taken to compute Raman intensities using pure and hybrid DFT methods and resulted in better agreement with observed spectra. Basis sets for computation of Raman spectra must contain diffuse functions on all atoms because the largest polarizability changes are likely to occur in regions of a molecule furthest from the atomic nuclei. These methods are not applicable under resonance conditions; resonance Raman (RR) intensities are controlled by either the geometric changes which occur in the resonant electronic excited state (A-term RR effect) or vibronic coupling of the resonant state to another excited state (B-, C- and D-term RR effect).[18] For the simulation of RR spectra it is necessary to compute the geometries of one or more excited states (*vide infra*).

Although a wide range of molecular spectroscopic properties can be simulated most research until a few years ago focused upon IR and Raman spectroscopy because it is relatively straightforward to compute quadratic force constants by analytic differentiation methods, from which harmonic vibrational frequencies can be obtained. Of course these often differ markedly from the anharmonic fundamentals obtained from spectra and two alternative strategies have been employed for the comparison of calculated and experimental spectra. The more rigorous approach involves computation of cubic and quartic force constants, leading to anharmonicity constants and hence anharmonic frequencies, although only numerical differentiation methods are available and this is very time-consuming. This would normally be prohibitive for large molecules although it is possible, with some loss of accuracy, to compute cubic and quartic force constants using a smaller basis set than that used for computation of the quadratic force constants. Alternatively computed spectra can be adjusted for anharmonicity by scaling of force constants. Historically, scaling was widely used for the correction of force constants and vibrational spectra obtained by the Hartree-Fock (HF) method, which systematically overestimates vibrational frequencies by around 10–11%. DFT methods do not generate systematic errors in frequencies (any errors are in fact random) and it may be argued that DFT frequencies should not be scaled. Nevertheless, many workers do employ scaling for DFT vibrational spectra to "correct" the results for anharmonicity. In particular it is common to use different scaling for modes involving motion of hydrogen atoms because these generally exhibit a greater degree of anharmonicity.

In the last few years there has been an upsurge in the modelling of vibrational spectra obtained using other methods, such as inelastic neutron scattering (INS) and nuclear resonance vibrational spectroscopy (NRVS). The latter is a synchrotron-based vibrational technique, related to the Mössbauer effect and can be employed for molecules containing a Mössbauer active isotope such as ^{57}Fe. It provides the complete set of vibrations corresponding to modes that involve motion of the active atom, thereby providing selectivity but with the significant advantage that it is not subject to optical selection rules.

Other than vibrational spectroscopy the most widely simulated methods are NMR and EPR, and electronic absorption and emission (UV-Visible) spectroscopy. Many modern software packages contain algorithms for the computation of NMR chemical shifts and spin-spin coupling constants. In terms of the choice of basis sets for NMR calculations it is important to be aware that requirements differ from those for computation of vibrational spectra. Quadratic (and also cubic and quartic) force constants are closely related to the electron density in the regions between atomic nuclei (*i.e.* what chemists regard as bonds), and these are usually well predicted even by modest basis sets containing some polarization functions. On the other hand, NMR chemical shifts are a consequence of the influence of chemical environment on a nuclear property, thus it is important to choose a basis set which predicts accurate electron density in the regions close to the nuclei of interest. It is generally evident that, at the present time, the prediction of NMR spectra is not as accurate as the prediction of vibrational spectra.

The simulation of UV-visible absorption and emission (photoluminescence) spectra requires calculation of excited state properties, as also is the case for simulation of RR spectra and chiroptical methods such as CD and MCD. Time-dependent DFT (TD-DFT)[19] is now the most widely used method in inorganic systems for the computation of vertical excitation energies (*i.e.* obtained at the ground state geometry) and oscillator strengths, from which electronic absorption spectra are simulated. The simulation of photoluminescence spectra requires optimization of the molecular geometry in each electronic excited state, for which there are now analytic procedures available. This yields vertical excitation energies at the excited state geometry (which are observed in emission spectra). Comparison of the optimised geometries of the ground and excited states can then be used to simulate RR spectra by computational of the displacements of normal coordinates in the resonant excited state. Inorganic applications of TD-DFT to RR and other electronic spectroscopy methods, as well as NMR and EPR, have recently been reviewed by Neese.[20]

A substantial number of investigations have used solvation models to simulate vibrational and electronic spectra. Modern software packages also provide methods for incorporating the effect of a solvent using, for example, polarizable continuum models (PCM), where additional input is required for the solvent dielectric constant and to define the solute cavity. Continuum solvation models have recently been reviewed by Tomasi *et al.*[21] Two of the most widely used methods are the integral equation formalism variant (IEF-PCM[21]), and the conductor-like screening model (referred to as either COSMO or CPCM[22]). Application of solvation methods for analysis of vibrational spectra of species in solution is especially important where there are significant changes in structure between gas-phase and solution species. It has also generally been regarded as important to use solvation methods for determination of statistical thermodynamic properties from computed spectroscopic data, although in a recent study Ribeiro *et al.*[23] have shown that this is unnecessary if there are no significant differences between gas-phase and solution structures.

2 Main groups

A summary of computational studies of the spectra of compounds of Groups 1 to 3 is provided in Table 1. Studies on Group 2 compounds have included solid-state NMR and IR spectra interpreted using periodic DFT methods, in conjunction with plane-wave basis sets for valence electrons and core electrons represented by pseudopotentials. Cahill *et al.*[27] reported ^{25}Mg solid-state MAS NMR spectra of a range of organic and inorganic magnesium–oxyanion compounds containing natural abundance ^{25}Mg. Their DFT calculations demonstrated that ^{25}Mg NMR parameters are very sensitive to the local environment of Mg^{2+} and that the observed spectra were significantly influenced by the overall hydration state. The results addressed key problems in inorganic materials and the environment of metal ions in biological molecules. A comparison of the calculated and experimental NMR spectra of inorganic magnesium–oxyanion systems is shown in Fig. 1.

In another study the solid-state ^{43}Ca MAS NMR analysis of calcium benzoate trihydrate has been facilitated by plane-wave-pseudopotential DFT calculations.[28] This system provides an effective model for Ca^{2+} sites in biological materials due to the similarity of its calcium coordination to that in many calcium proteins and relevance to natural phenomena such as biomineralisation.

The localization and coordination of Mg^{2+} cations in the zeolite Mg-ferrierite was investigated using IR spectroscopy of CO probe molecule in

Table 1 Groups 1–3.

Species	Spectroscopic method	Theoretical method (method/basis set)	Ref.
(CsBr)n, (CsI)n in Kr and Xe matrices	IR	B3P86/Stuttgart-Cologne-pTZV	24
Alkali metal crown ether complexes	IR-UV double resonance	M05-2X/6-31 + G(d) wB97XD/6-311 + + G(d,p) ECP for Rb and Cs	25
Hydroxylated Mg powders	photoluminescence	B3LYP/6-31 + G(d)	26
Mg oxyanion compounds	^{25}Mg NMR	DFT, plane-wave/pseudopotential	27
$Ca(C_6H_5COO)_2 \cdot 3H_2O$	^{43}Ca NMR	DFT, plane-wave/pseudopotential	28
Mg^{2+} in ferrierite	IR	DFT, plane-wave/pseudopotential	29
$Be(acac)_2$	IR, Raman	HF, MP2,B3LYP/ 6-311 + + G(3df,2p)	30
Group II metal nitrate clusters	IRMPD	Several DFT functional 6-311 + G(3d) and other basis sets; SDD on Sr	31
$BeCl_2(Ph_2PCH_2PPh_2)_2$	IR	RI-BP86/TZ2P	32
$Mg(acac)_2$	IR, Raman	HF, MP2,B3LYP/6-311 + + G**	33
ScC_3 in Ar matrix	IR	B3LYP/6-311G(3df,3pd)	34
$LiSc(BH_4)_4$	IR, Raman	PBE/TZP	35
Y(Pc)(Por) and $[Y(Pc)(Por)]^-$ double-decker complexes	IR, Raman	B3LYP/LanL2DZ	36

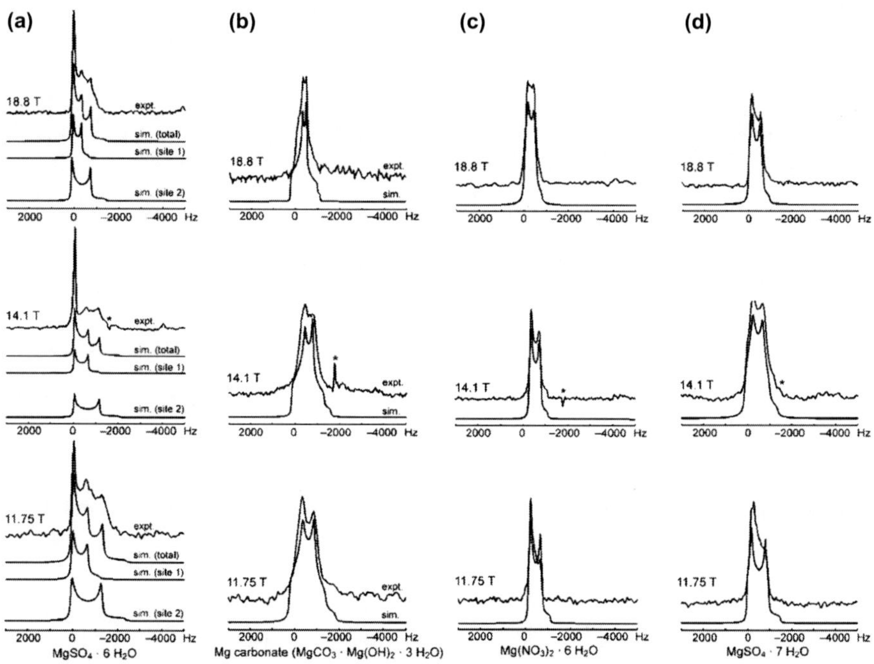

Fig. 1 Experimental and simulated ^{25}Mg MAS NMR spectra with increasing magnetic field for a) $MgSO_4 \cdot 6H_2O$, b) Mg carbonate ($MgCO_3 \cdot Mg(OH)_2 \cdot 3H_2O$), c) $Mg(NO_3)_2 \cdot 6H_2O$ and d) $MgSO_4 \cdot 7H_2O$. (Reproduced by permission from *Chem.-Eur. J.*, 2009, **15**, 9785).

conjunction with periodic DFT calculations.[29] In this study the vibrational frequency of CO adsorbed on monovalent extra-framework metal cations was calculated using the ω/r correlation for a reference set of molecules. Good agreement between calculated and experimental frequencies was achieved, enabling the experimental IR spectra to be interpreted in terms of mono-, di- and tricarbonyl complexes formed on particular Mg^{2+} sites.

A summary of computational studies of the spectra of compounds of the *p*-block elements (Groups 13 to 18) is provided in Table 2. Liakos and Simandiras[38] have used a molecular DFT method with a Gaussian basis set to simulate the structure and vibrational spectra of boron oxide glass, employing a $(B_2O_3)_{20}$ model cluster. Geometry optimisation of the cluster resulted in three minimum energy structures with B:O coordination numbers of 2.22, 2.20 and 2.14 in good agreement with an optimal coordination number of 2.40 for a covalent network. It was shown that all three models reproduced with good accuracy the IR and Raman spectra and also the radial distribution functions obtained from X-ray, neutron and electron diffraction. Rada *et al.*[43] modelled $(1-x)TeO_2 \cdot xB_2O_3$ glasses ($x = 0.3$ and

Table 2 Groups 13–18.

Species	Spectroscopic method	Theoretical method (method/basis set)	Ref.
$Al(acac)_3$	^{27}Al, ^{17}O and ^{13}C NMR	B3LYP, B3PW91,mPW1PW91/ 6-31++G**, D95++**	37
B_2O_3 glass	IR, Raman	HCTH/6-31G*	38
AlC_3 and AlC_3Al linear chains, matrix isolation	IR	B3LYP/6-311+G(3df)	39
GaN cluster models	UV-Vis	B3LYP/ECP	40
Salen-aluminum complexes	UV-Vis	B3LYP/6-31G(d)	41
Alq_3 and difluorinated derivatives	UV-Vis	B3LYP/6-31G(d,p)	42
$TeO_2 \cdot B_2O_3$ glasses	IR	B3PW91/CEP-4G	43
MBH_4 and MBF_4 (M = Li, Na and K)	IR, Raman	PBE/various basis sets	44
Al_2O_2 and Al_2O_3	IR, Raman	B3LYP/AUG-cc-pVDZ	45
GeC_3Ge, matrix isolation	IR	B3LYP/6-311+G(3df)	46
Bis(1,10-phenanthroline) silicon complexes	^1H NMR, UV-Vis	B3LYP/LanL2DZ	47
Silicon cluster cations in the gas phase	IR-MPD	B3LYP, BPW91/6-311+G(d)	48
Diorganotin(IV) complexes	Mössbauer	B3LYP/3-21G*, DZVP	49
difluoro(germylthio)-phosphine and difluoro(germylseleno)-phosphine	IR, Raman, NMR	MP2, B3LYP/6-311G(d,p)	50
M_4O_6, M_4O_{10} (M = P, As, Sb and Bi)	IR, Raman	B3LYP/LanL2DZ	51
Methylarsine and methylstibine	IR	HF, MP2, B3LYP/6-311G*	52
Phosphorus-containing dendrimers	IR, Raman	PBE/TZ2P	53–56
$H_4I_2O_{10}^{2-}$ cation	IR, Raman	B3LYP, BP86/6-31G*	57
Iodine clusters	RR	B3LYP/TZVPP, ECP46MWB	58
		B3PW91/MCPtzp+	59
Alkali metal bromates, matrix isolation	IR	DFT and MP2 methods	60
$[F_5SN(H)Xe][AsF_6]$	Raman	MP2, SVWN/SDB-cc-pVTZ	61

0.4) by geometry optimisation of clusters of unspecified size, where dangling Te-O bonds were terminated by hydrogen atoms. Their DFT calculations provided a good fit to experimental IR data.

Dendrimers are repetitively branched molecules that are typically symmetric around the core, and often adopting a spherical three-dimensional morphology. They are important as perspective materials for coordination chemistry, agrochemistry, catalysis and also the storage and delivery of drugs. In a series of papers Furer *et al.*[53–56] have investigated the IR and Raman spectra of phosphorus-containing dendrimers, supported by DFT calculations. In the first of these studies[53] they synthesised a zero generation dendrimer (P_0') built from an octasubstituted metal-free phthalocyanine core with oxybenzaldehyde terminal groups, and also a fourth generation dendrimer (P_4'). Such systems have possible relevance in photodynamic therapy, where aggregation causes a reduction in the photosensitising efficiency of the phthalocyanine core. Their results showed that the P_0' molecule exists in a stable conformation with a planar phthalocyanine core and the calculated geometrical parameters and harmonic vibrational frequencies were in good agreement with the experimental data. They also investigated second generation dendrimers built from the trifunctional thiophosphoryl core with the same repeated units and 4-oxibenzaldehyde terminal groups,[54] and dendrimers built from the hexafunctional cyclotriphosphazene core.[55,56]

3 Transition metals

A summary of computational studies of the spectra of compounds of the *d* block elements is provided in Tables 3 to 11. Most of these have been focused upon metal complexes, with rather less attention given to small covalent molecules of the transition metals.

3.1 Matrix isolation studies

DFT modelling of species formed in matrix isolation experiments are of particular importance because there are no X-ray structural data available

Table 3 Group 4.

Species	Spectroscopic method	Theoretical method (method/basis set)	Ref.
HCCMH and M-η^2-(C_2H_2), matrix isolation (M = Ti, Zr and Hf)	IR	B3LYP/6-311++G(3df,3pd) SDD for Zr and Hf	62
$(C_5R_5)_2MCl_2$ (R = H, Me; M = Ti, Zr, Hf) metallocenes	Cl K-edge XAS	B3LYP/6-31G* SDD for Ti, Zr and Hf	63
CH_2F_2 adsorbed on TiO_2	IR	B3LYP/cc-pVTZ (CH_2F_2), 86-51G* (O), 8-411G (Ti)	64
$Ti(OMe)_4$ oligomers	IR, Raman	B3LYP/6-31+G*	65
Ti(IV) complexes of porphyrazines and phthalocyanine	UV-Vis	LDA, BP86/TZ2P PCM	66
Hafnocenes	UVPES	B3LYP/SDD	67
$(ZrO_2)_n$ clusters	IR	B3LYP/6-31G*	68

for comparison with the spectroscopic results. Matrix isolation studies of species formed by transition metal atoms co-condensed with unsaturated organic ligands are important with respect to organometallic chemistry and in many cases have relevance to industrial catalytic processes. Cho *et al.*[62] investigated reactions of laser-ablated Ti, Zr and Hf atoms with acetylene in excess argon, and identified the products by IR spectroscopy, with assignments facilitated by DFT calculations of various isotopomers. It was found that all three metals formed ethynyl metal hydride complexes (HC≡C-MH) *via* oxidative C-H insertion during codeposition and subsequent photolysis. There was also evidence for the formation of π complexes whose low CC stretching frequencies implied that back-donation was much more extensive than in previously studied π complexes. In another study[79] they investigated reactions of Cr, Mo and W atoms with ethylene and observed different behaviour for Cr than that of Mo and W, and assigned IR spectra in terms of insertion and metallacyclopropane complex ($M–C_2H_4$) formation. On the other hand insertion and dihydrido products ($MH–CHCH_2$ and $MH_2–C_2H_2$) were identified for reactions of W and Mo with ethylene. The IR data and DFT results suggested a general trend of higher oxidation-state complexes becoming more stable on descending Group 6.

Oxidation of Fe is an important subject in material corrosion and biochemical processes. As part of a study of iron oxides Gong and Zhou[106] obtained the matrix IR spectra of two FeO_3 isomers produced from laser evaporation of a bulk Fe_2O_3 target and reaction with O_2 in Ar. With the aid of $^{16}O/^{18}O$ isotopic substitution and DFT calculations they identified these species as $(\mu^2\text{-}O_2)FeO$ (C_{2v} symmetry, $^5B_1$ ground state) and planar FeO_3 (D_{3h} symmetry, closed shell singlet ground state). The hydrides of Fe, Ru and Os are of interest with respect to catalytic processes such as hydrogenation and alkane activation. Wang and Andrews[122] looked at the matrix IR spectra of Group 8 atoms with H_2 in Ar and Ne, and also with pure H_2. Comparison of the IR results with DFT calculations of a wide variety of possible hydride complexes led to the conclusions that Fe, Ru and Os atoms

Table 4 Group 5.

Species	Spectroscopic method	Theoretical method (method/basis set)	Ref.
$(gu)_2[Nb(O_2)_3(quin\text{-}2\text{-}c)]\cdot H_2O$	UV-Vis	B3LYP/LanL2DZ	69
V_2O_5	Raman	LDA, plane-wave/pseudopotential	70
Vanadium haloperoxidase models	^{51}V NMR	B3LYP/several basis sets	71
$K_3[VO(O_2)_2(C_2O_4)]\cdot H_2O$	^{13}C, ^{51}V NMR	DFT PCM	72
SiO_2-supported vanadium oxide catalysts	IR, Raman	PBE, plane-wave/pseudopotential	73
$VO(NO_3)_3$	IR, Raman	B3LYP/6-311+G*	74
SiO_2-supported vanadium oxide catalysts	NEXAFS	BP86/TZVP	75
$Cs_4[H_2V_{10}O_{28}]\cdot 4H_2O$	^{51}V NMR	PW91, plane-wave/pseudopotential	76
Nb(V) phthalocyanine complex	EPR	B3LYP/TZVP, LanL2DZ PCM	77
$Nb(NO_3)_3$	IR, Raman	B3LYP, BP86/several basis sets	78

react with H_2 to produce MH molecules, $MH(H_2)_x$ complexes, FeH_2, and $MH_2(H_2)_x$ complexes (M = Fe or Ru).

Souvi *et al.*[150] studied the formation of Pd_2N_2 from the co-condensation of Pd and N_2 in Ne and Ar matrices and observed IR bands in the range

Table 5 Group 6.

Species	Spectroscopic method	Theoretical method (method/basis set)	Ref.
Group 6 metallacyclopropanes, matrix isolation	IR	B3LYP/6-311++G(3df,3pd) SDD for Mo and W	79
MoS_2 hydrodesulfurization catalyst	IR	PW91/DNP (all electron)	80
$[Mo_7O_{24}]^{6-}$ and $[W_7O_{24}]^{6-}$	IR, Raman	VWN, BP86/TZ COSMO	81
Mixed-ligand Cr(III) complex	UV-Vis	B3LYP/6-311G*, LanL2DZ	82
Heteroleptic tungsten dithiolene complexes	IR, Raman	B3LYP, B3PW91/LanL2DZ	83
$CrO_2(SO_3F)_2$	IR, Raman	B3LYP, B3P86, B3PW91/several basis sets	84
Molybdenyl complexes	MCD	BP86/TZ	85
Binuclear Cr and Mo complexes	RR	B3LYP/6-31G* 6-311G* for Cr, LanL08 for Mo	86
$[Cp_2Mo(dmit)]$ charge transfer salts	Raman	B3LYP/3-21G* MP2/LanL2DZ	87
$[W_6O_{19}]^{2-}$, $[W_6O_{19}]^{3-}$ and $[W_6O_{19}]^{4-}$	UV-Vis	Several DFT methods/ DZ and TZ basis sets COSMO	88

Table 6 Group 7.

Species	Spectroscopic method	Theoretical method (method/basis set)	Ref.
Tricarbonyl Re(I) complexes with a bridging ligand	UV-Vis	PBE0/6-31G(d,p) SDD for Re CPCM	89
Re(I) pyridinybenzoimidazole complexes	UV-Vis	B3LYP/LanL2DZ PCM	90
Binuclear mixed valence Mn complexes	EPR	B3LYP	91
Re(III) and Re(V) di-2-pyridylketone complexes	UV-Vis	B3LYP/LanL2DZ PCM	92
μ-O-bridged dinuclear Re complexes	UV-Vis	MP2, B3LYP/cc-pVDZ ECP for Re	93
$[Re(R_2bpy)(CO)_3X]$ complexes	UV-Vis	B3LYP/6-31G** ECP for Re PCM	94
$[Mn(NH_3)_6](NO_3)_2$	IR, Raman	B3LYP/LanL2TZ+, 6-311+G(d,p)	95
$Tc_2Cl_4(PMe_3)_4$ and $Tc_2Br_4(PMe_3)_4$	UV-Vis	DFT, CASSCF, CASPT2/ several basis sets	96
$[Tc_6Q_8(CN)_6]^{4-}$ (Q = S, Se)	IR, UV-Vis	B3LYP/LanL2DZ	97
Re(I) bpy complexes with cyclotriphosphazenes	RR, UV-Vis	B3LYP/LanL2DZ PCM	98
Binuclear Re(I) complexes with tetrathiafulvalenes	IR, UV-Vis	B3LYP/6-31G* LanL2DZ for Re	99
Hexanuclear chalcohalide Rhenium(III) clusters	UV-Vis	B3LYP, PBE1PBE, B2PLYP/ LanL2DZ CPCM	100

1800–2200 cm^{-1}. Selective irradiation in the near-IR and visible spectrum of the species formed in Ar caused transformation between three structures whose identification was supported by DFT calculations: a bridged T-shaped species (ν(NN) = 1990 cm^{-1}), side-on species (ν(NN) = 2178 cm^{-1}) and parallel species (ν(NN) = 1823 cm^{-1}).

3.2 Heterogeneous catalysts

Several transition metals are of industrial importance for their role in heterogeneous catalysis, which have for many years been subjected to spectroscopic investigation, especially by IR, in an endeavour to identify adsorbed species and the nature of active sites. DFT modelling of solid surfaces can be tackled by two entirely different strategies. In the first of these a cluster model is taken to represent the active site and adsorbed species, with dangling bonds terminated by hydrogen atoms, treating this cluster as a molecule and using Gaussian basis functions. It is generally found that good results can be obtained if the model is sufficiently large and truly representative of the surface species. Alternatively, a periodic DFT approach may be employed by representing the active site/adsorbed species by a supercell slab and in this case the accuracy is largely dependent on the depth of the slab, *i.e.* the number of atoms below the surface, although the size of the calculation increases markedly with increasing slab depth.

Scaranto and Giorgianni[64] investigated the IR spectrum of CH_2F_2 adsorbed on the anatase form of TiO_2 and carried out periodic DFT calculations using Gaussian basis functions. They modelled a slab with 12 atomic layers, with 2×2 periodicity, corresponding to a surface coverage of 0.25, chosen to limit interaction between neighbouring adsorbed molecules. The results established that adsorption occurred by an interaction between a F atom and a surface Ti^{4+} Lewis acid site and also a hydrogen bond between the CH_2 group and a surface Lewis basic site (O^{2-} or OH^-).

Supported vanadium oxide catalysts are widely used in selective oxidation reactions of, for example, ethane and propane and the selective reduction of NO_x emissions with NH_3. These catalysts consist of an active vanadium oxide phase grafted on the surface of a high-surface area oxide support such as SiO_2 and other oxides. An understanding of the catalytic activity requires spectroscopic characterisation of the vanadium oxide species under reaction conditions and how it is anchored to the surface of the support material. Islam *et al.*[73] carried out a systematic theoretical study of SiO_2-supported vanadium oxide and assigned the IR spectrum. The active site was modelled by a $VO(OH)_3$ unit attached to q $12.77 \times 17.64 \times 25.17$ Å^3 SiO_2 supercell corresponding to a vanadium coverage of 0.44 atoms nm^{-2}, close to experimental conditions. Calculation of the vibrational frequencies for the geometry-optimised structures established that the vibration at *ca.* 1064 cm^{-1} is due to the V=O stretch, whereas the 800–1000 cm^{-1} region is dominated by V-O-Si stretching modes. These exhibit dispersion due to the different nature of the supported vanadium oxide units. Cavalleri *et al.*[75] investigated the same system using a combination of NEXAFS measurements and DFT calculations involving cluster models. The simplest SiO_2

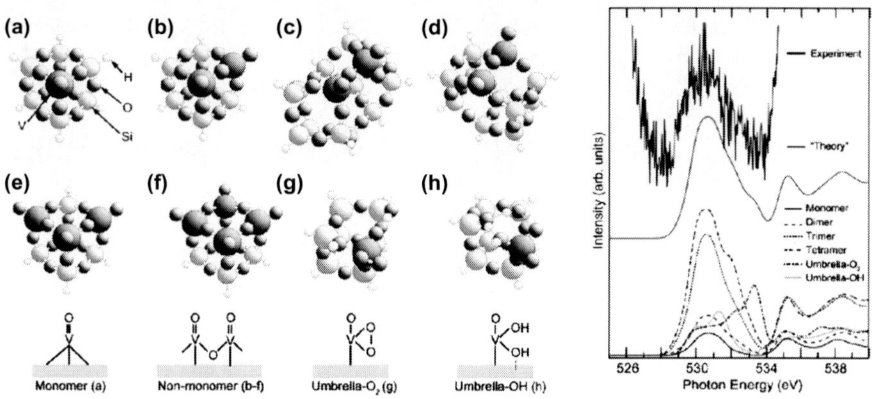

Fig. 2 Geometric structures of the clusters used to model SiO_2-supported vanadium oxide units and the experimental NEXAFS spectrum compared with theoretical O 1s core excitation spectra of the different model clusters. (Reproduced by permission from *J. Catal.*, 2009, **262**, 215).

models with ring structure are derived from a $Si_8O_{12}H_8$ cube with Si occupying the corners and O near the edge midpoints, and a monomeric pyramidal vanadium-oxygen, binding to three Si atoms was modelled by replacing one SiH moiety by a V=O group. A dimeric VOV unit was modelled by replacing a Si_2H_2O edge of the cube by a V_2O_3 group, and a trimeric species was modelled by a cluster in which a $Si_3H_3O_2$ edge was replaced by V_3O_5. Results of the modelling are shown in Fig. 2.

Molybdenum sulfide catalysts have been widely used in hydrodesulfurization (HDS) and hydrodenitrogenation (HDN) processes, for methanation and for the Fischer–Tropsch synthesis of light hydrocarbons by CO hydrogenation. The latter has been studied extensively because of the tolerance of these catalysts to sulfur poisoning. Shi *et al.*[80] used surface models employing periodic DFT methods to examine intermediates of the CO hydrogenation reaction steps and to compute vibrational spectra for the interpretation of experimental IR measurements. The DFT calculations used periodic slabs of MoS_2 where the molecular adsorbates and reactants were added to one side of the slab, and identified vibrations for the adsorbates CHOH, COH, CHO and CH_3CO in the range 1286–1336 cm^{-1}. However, the results could not confirm the existence of formate (HCOO) intermediates which have been identified in the experimental spectra.

NO_x emissions from engines have also been treated using Ag/Al_2O_3 catalysts and considerable effort has been devoted to the study of active catalysts and mechanisms for the selective catalytic reduction of NO_x with hydrocarbons in excess oxygen. Zhang *et al.*[164] have investigated temperature-programmed desorption and DRIFTS spectra of bridged, bidentate and monodentate surface nitrate species on Ag/Al_2O_3 catalysts. In this study four different surface species were modelled by DFT calculations of small clusters and, notwithstanding the simplicity of these models, remarkably good agreement was found with the observed IR spectra.

3.3 Biological molecules and model systems

DFT calculation of an entire metallo-protein or enzyme would be computationally very demanding, especially in view of the conformational flexibility of such molecules and most studies have employed model complexes in order to simulate the spectroscopic properties of the active site. A strategy which has been increasingly deployed in recent years has been to combine QM methods (usually DFT) with molecular mechanics (MM) and molecular dynamics (MD) procedures. For example, different parts of a biological macromolecule can be modelled using different levels of theory. This multi-layer approach has been likened to the layers of an onion and referred to as the "ONIOM" method. In this approach the active site is modelled at a high level of theory, typically a DFT method with a large basis set and the remainder of the protein treated by a molecular mechanics method – a two-layer ONIOM calculation. In a three-layer ONIOM calculation the region of the protein surrounding the active site may be treated quantum-mechanically at a lower level of theory (*e.g.* semi-empirical) with the remainder of the protein treated by a molecular mechanics method. This same approach has also been applied to cluster models of adsorption on heterogeneous catalysts. However, although there have been a number of such studies of structures and reactivity of catalysts there have not yet been any simulations of spectroscopic data.

Schweitzer *et al.*[71] have modelled hydrogen bonding interactions in vanadium haloperoxidases with *cis*-dioxovanadium(V) complexes based on N-salicylidenehydrazide ligands. Solid-state ^{51}V MAS NMR spectra of three model complexes were simulated by DFT calculations, employing several different basis sets and structural were compared with the experimental data to determine the best calculation strategy. It was shown that the calculations were capable of reliably determining the Euler angles describing the relative orientations of chemical shifts and quadrupolar tensor independently of basis set, providing a promising tool for study of the structure-function relationships of these enzyme models.

Guo *et al.*[115] used ^{57}Fe NRVS to study the Fe site in an iron-sulfur cluster-free hydrogenase. The spectra were interpreted by comparison with DFT calculations of model compound, $Fe(S_2C_2H_4)(CO)_2(PMe_3)_2$, and also by normal mode simulations of plausible active site structures using a modified form of the Vibratz program.[116] The simulations are shown in Fig. 3. The same group[129] also investigated the dynamics of the [4Fe-4S] cluster in *Pyrococcus furiosus* D14C ferredoxin using NRVS and RR spectroscopy in conjunction with DFT calculations. The observed vibrational frequencies enabled refinement of a Urey-Bradley force field for the oxidized site which satisfactorily reproduced both the high-frequency stretching modes and the low-frequency bending and torsional modes, and was used for a MM simulation. A simplified $[Fe_4S_4(SCH_2CH_3)_4]^{2-}$ oxidized form model was used for DFT simulations of the spectra, using an initial geometry taken from the high resolution crystal structure of the protein.

Soldatova *et al.*[126] studied $Fe^{III}(NO)$ adducts of myoglobin variants with altered polarity in the distal heme pockets by RR spectroscopy using both UV and visible excitation. The observed vibrational frequencies (but not RR band intensities) were interpreted from DFT calculations of model $Fe^{III}(NO)$ porphyrin complexes with and without distal imidazole ligands, assuming a singlet ground electronic state. They showed that the ν(FeN)

Table 7 Group 8.

Species	Spectroscopic method	Theoretical method (method/basis set)	Ref.
Ruthenium oligopyridine complexes	UV-Vis	B3LYP/LanL2DZ PCM	101
		B3LYP/DZ, ECP for Ru	102
Osmium (II) diimine complexes	UV-Vis	B3LYP/LanL2DZ PCM	103
Creutz-Taube complex	IR	PBE/DNP-ECP PCM	104
$Ru(acac)_2$(R-o-benzoquinone-diimine) complexes	IR, Raman, UV-Vis	B3LYP/LanL2DZ PCM	105
FeO_3 isomers, matrix isolation	IR	BP86/6-311+G(d)	106
Osmium(II) terpyridine complexes	UV-Vis	CIS, B3LYP/LanL2DZ	107
$(NH_3)_5)Ru^{III}$(2,4-dihydroxy-benzoate) and $(NH_3)_5)Ru^{III}$ (xanthine) chromophores	UV-Vis	Modified B3LYP	108
Iron porphine halides	FCS, RR	B3LYP/Ahlrichs VTZ, 6-31G(d)	109
Fe(II) spin crossover complexes	IR	DFT	110
Fe(III) monophosphate complexes	IR	B3LYP/SDD	111
DNA light-switch behavior of Ru(II) intercalators	UV-Vis	B3LYP/6-31G* SDD for Ru PCM	112
Iron(II)-based spin crossover compounds	IR, UV-Vis	Several DFT methods and basis sets	113
Diruthenium tetracarboxylates	UV-Vis	B3LYP/LanL2DZ PCM	114
Fe site in iron-sulfur cluster-free hydrogenase	IR, NRVS, RR	PWPW91, B3LYP/ LACV3P++	115
Ru(VI) and Os(VI) nitride complexes	UV-Vis	B3LYP	117
$[Fe(btpa)]^{2+}$ spin-crossover	TR^3	B3LYP/CEP-31G	118
Fe(III) complexes with P,N donor hemilabile ligands	IR, EPR, UV-Vis	B3LYP/6-31G* LanL2DZ for Fe	119
1-Ferrocenylethynyl-anthraquinone	UV-Vis	B3LYP/LanL2DZ	120
Fe(II) 2-pyrazinecarboxylate complexes	Moessbauer	B3LYP/6-311++G(d,p)	121
Fe, Ru, Os metal hydrides and dihydrogen complexes	IR	MP2, B3LYP/ 6-311+G(3df,3pd) SDD for Ru, Os	122
Fe trinuclear spin crossover complex	IR	DFT	123
Ferrocenes and their pentaphospholyl analogues	IR	OPBE/6-31G*, 6-311+G*, cc-pVTZ ECP on Fe	124
Ru(II) polypyridyl DNA intercalators	UV-Vis	QM/MM B3LYP/6-31G* ECP for Ru PCM	125
NO bonding in Fe(III) heme proteins	RR	B3LYP/6-31G* Ahlrichs VTZ for Fe	126
Potential anti-cancer drug, Ru-NAMI-A	IR. VCD	B3PW91/several basis sets	127
Fe(II) spin crossover complexes	XANES	BP86/several basis sets	128
[4Fe-4S] cluster	RR, NRVS	PW91/LACV3P++	129
Fe(II), Ru(II) and Os(II) tris (bipyridyl) complexes	CD	BP86/TZPTZ2P for Fe, Ru and Os	130

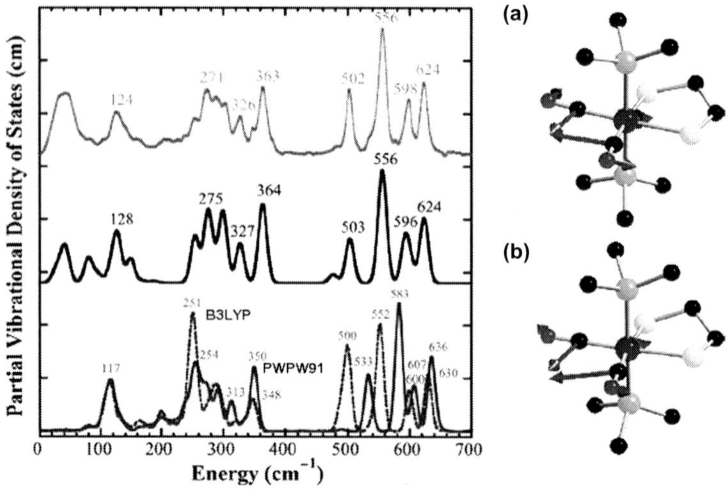

Fig. 3 Left: (top to bottom) NRVS PVDOS, Vibratz simulation, and DFT simulations using LACV3P**+ basis and PWPW91 and B3LYP functional for $Fe(S_2C_2H_4)(CO)_2(PMe_3)_2$. Right: atomic motion in normal modes at (a) 624 cm^{-1} in Vibratz simulation, (b) 636 cm^{-1} in DFT simulation using PWPW91 functional and LACV3P**+ basis. (Reproduced by permission from *Inorg. Chem.*, 2008, **47**, 3969).

Table 8 Group 9.

Species	Spectroscopic method	Theoretical method (method/basis set)	Ref.
Co(III) Schiff base complex	IR	B3LYP/6-311++G(d,p) LanL2DZ for Co	131
(ppy)(2)Ir(acac) derivatives	UV-Vis	B3LYP/LanL2DZ	132
Self-assembly Co(II) complex	IR, Raman, UV-Vis	B3LYP/several basis sets	133
Heteroleptic phosphorescent Ir(III) complexes	UV-Vis	B3LYP/LanL2DZ	134
Co(II) Schiff base complex	IR	B3LYP/6-31G(d)	135
Co(III) polypyridyl complexes with DNA	UV-Vis	B3LYP/LanL2DZ	136
Reactions of $Rh_2(CO)_4Cl_2$ with 1,5-cyclooctadiene and tetramethylallene	IR, Raman	PBEPBE/DGDZVP	137
Blue-emitting Ir(III) complexes	UV-Vis	B3LYP/6-31G(d) LanL2DZ for Ir	138
Cobalt(III) Schiff-base complexes	CD	B3LYP/6-31G(d) for C,H; 6-311+G(2d,p) for Co,N,O	139
Ir(III) complexes with phenylpyridine-based ligands	UV-Vis	PBE0/6-31G* LanL2DZ for Ir	140
Vitamin B_{12} and methylcobalamin	CD, MCD	CAM = B3LYP, BP86/ 6-31G(d), AUG-cc-pVDZ	141
Chiral rhodium complexes	CD, VCD	B3LYP/6-311++G(d,p) LanL2DZ for Rh	142

and ν(NO) band positions are reporters of heme pocket polarity and the accessibility of Lewis base donors.

The design of molecules for sequence-specific DNA recognition is of importance with respect to applications in sensitive diagnostics and

Table 9 Group 10.

Species	Spectroscopic method	Theoretical method (method/basis set)	Ref.
Platinum acetylide complexes	UV-Vis	B3LYP/6-31G* ECP for Pt	143
	Ps-TRIR	B3LYP/6-31G* SDD for Pt	144
PtN_2, PtP_2 and $PtAs_2$	IR, Raman	Several DFT methods and basis sets, periodic calc.	145
Pt(II) complexes with modified tetracycline	^{13}C NMR, UV-Vis	B3LYP/6-31 + G(2d,p) LanL2DZ for Pt PCM	146
Binuclear palladium complexes	UV-Vis	MP2 LSD X-α	147
Pd(II) and Pt(II) complexes in acidic solutions	IR, Raman	B3LYP/LanL2DZ	148
$PtBr_2$ derivatives	IR, Raman	B3LYP/6-31G(d) LanL2TZ for Pt	149
Pd_2N_2 in Ne and Ar matrices	IR	DFT using GGA functionals	150
Ni(II) glycinate-guanidoacetate complex	IR	B3LYP/6-31G, 6-311G	151
Phosphorescent Pt(II) complexes	UV-Vis	B3LYP/LanL2DZ	152
Pd(II) salen-type complexes	IR, UV-Vis	B3LYP/6-31G(d,p) ECP for Pt	153
cis-$[Pt(CH_3CN)_2Cl_2]$	IR, Raman	mPW1PW91, PBE1PBE1/ SDD	154
Ni(II) O,O-diethyl-dithiophosphate complex	IR, Raman	B3LYP/6-311 + + G(df,p)	155
N-C-N-coordinating Pt(II) complexes	UV-Vis	B3LYP/6-31G(d) LanL2DZ for Pt	156
Pt(II) and Pt(IV) anionic complexes in aqueous solution	^{195}Pt NMR	PBE, plane-wave/pseudopotential	157
Ni(II) 5,10-diphenylporphyrin	IR, Raman	B3LYP/6-31G(d)	158
Pyramidalized olefin Pt-complexes	IR, UV-Vis	B3LYP/LanL2DZ	159
Pt(II) complexes displaying	INS	DFT	160
anticancer properties	IR	PBE1PBE1/SDD	161
Ni, Pd, and Pt phthalocyanines	IR, RR	B3LYP/LanL2DZ	162

chemotherapeutics. It has been recognised for several years that ruthenium polypyridyl complexes may be used as DNA probes due to their intense ^{3}MLCT luminescence and excited-state redox properties, coupled with their ability to bind to DNA. Unusually, the complex $[Ru(bpy)_2(dppz)]^{2+}$ (dppz = dipyrido[3,2-*a*:2′,3′-*c*]phenazine) is non-emissive in aqueous solution but displays bright luminescence in organic solvents and upon binding to DNA, acting as a molecular "light-switch". Sun *et al.*[112] examined the emission spectra of a range of $[Ru(bpy)_2L)]^{2+}$ complexes, including L = dppz, and investigated the light-switch behaviour by TD-DFT methods in conjunction with the PCM solvation model. They found that differences in the emission behaviour of these complexes could be explained in terms of changes in the energy gap between the bright, highly emissive ^{3}MLCT excited state and the lowest-energy ^{3}MLCT with either mixed- or nonemissive character. For the light-switch complexes, the lowest-energy MLCT

Table 10 Group 11.

Species	Spectroscopic method	Theoretical method (method/basis set)	Ref.
Dicopper patellamides	IR, Raman	B3LYP/6-31G LanL2DZ for Cu	163
NO_3^- species on the Ag/Al_2O_3 catalyst surface	IR	B3P86/LanL2DZ	164
Formate species adsorbed on $Cu\text{-}Al_2O_3$ catalysts	IR	mPW1PW91/SDD	165
Cu complexes with cytosine	EPR	B3PW91/6-31G(d)	166
Cu(I) complexes of substituted 1,10-phenanthrolines	TR^3	B3LYP/6-31G(d)	167
Hypoxia selectivity of copper radiopharmaceuticals	UV-Vis	B3LYP/6-31++G(d,p)	168
Binuclear copper (II) complexes	UV-Vis	B3LYP/6-31G*	169
Dimethylgold(III) halides and carboxylates	IR, Raman	B3LYP/LanL2DZ	170
Copper-histamine complexes	IR	B3LYP/6-31+G(d,p)	171
Cu(II) and Ag(I) coumarin complexes	IR	B3LYP/6-31G(d) SDD for Cu, Ag	172
Cyclometalated gold(III) complexes	UV-Vis	B3LYP/6-31G(d) LanL2DZ for Au	173
cis-diammine(orotato)copper(II)	IR, Raman	MP2, B3LYP, M05-2X/ Several basis sets	174
Tetragonal Cu(II) model complexes with oxygen ligands	EPR	B3LYP/TZP	175
Silver and gold nanorod clusters	UV-Vis	BP86/DZP(frozen core)	176
Ag_{10}-pyridine complex	SERS	B3P86/TZP	177
Copper phthalocyanine	RR	B3LYP/6-311G(d,p)	178
Polypyridyl Cu(I)-based complexes	UV-Vis	B3LYP/6-31G(d). DZP and 3-21G mixed CPCM	179

state lies just below the bright state, such that the latter can be accessed thermally at room temperature. In contrast, the lowest-energy MLCT state of non-emissive complexes has much lower energy than the bright state, making the latter more difficult to populate. Ambrosek *et al.*[125] looked at the emission spectra of the similar complexes $[Ru(phen)_2(dppz)]^{2+}$ and $[Ru(tap)_2(dppz)]^{2+}$ (tap = 1,4,5,8 tetraazaphenanthrene) intercalated in two guanine–cytosine base pairs. Hybrid QM/MM calculations were used to optimize the ground state structures of these complexes intercalated in both the major and minor grooves of DNA using the two-base pair model. Excited state energies were computed from TD-DFT calculation with the PCM solvation model, providing simulated absorption spectra for different chemical environments. It was found that the Δ-enantiomers of both complexes intercalated either in the minor or major groove were more stable with respect to the Λ-enantiomers.

3.4 Spin-crossover complexes

Spin-crossover is a process in which a transition metal complex goes from a low- to high-spin configuration and has been widely studied for its potential applications in optical switching, data storage and optical display devices. Such a transformation may be induced by changes in temperature, pressure,

Table 11 Group 12.

Species	Spectroscopic method	Theoretical method (method/basis set)	Ref.
Cd(II)-cysteinate glycinate complex	IR, Raman	B3LYP/3-21G	180
tris(ethylenediamine) zinc(II) chloride	IR, VCD	B3LYP/6-31G**, 6-311G BPW91/6-31G**	181
Zn(II) bis-terpyridine complexes	Raman	B3P86/TZVP	182
Zn(II) 2,2′-biquinoline complexes	IR, Raman	B3LYP/6-31G(d,p)	183
Hg(II) mono- and bis-porphyrins	UV-Vis	B3LYP, M05/LanL2DZ	184
Zincocenes	IR, Raman	B3LYP/6-311+G**	185
Cd(II) 8-aminoquinoline complexes	UV-Vis	B3LYP/LanL2DZ	186
Diradical bis(α-iminopyridinato) zinc complex	EPR	B3LYP/Ahlrichs TZV	187
Mercury halides	IR, Raman	DFT	188
Zn(II) O,O-diethyl-dithiophosphate complex	IR, Raman	B3LYP/6-311++G(df,p)	189
Zn(II) bis(N-methylthio-semi-carbazonato) complex	UV-Vis	B3LYP/6-31G, 6-31G(d,p)	190
Zinc meso-phenylporphyrins	IR, RR	B3LYP/6-31G(d)	191
Zn-cysteine complex	IR, Raman	B3LYP/6-311++G(3df,2pd)	192
CdSe nanocluster-adenine complexes	UV-Vis	B3LYP/6-311++G**, LanL2DZ	193
$[Zn(dmit)_2]^{2-}$-cyclodextrin inclusion complexes	CD	B3LYP/6-31+G(d,p)	194
β-ethynylphenylene substituted zinc porphyrins	RR, UV-Vis	B3LYP/6-31G(d)	195

or the absorption of light. The most widely studied systems are octahedral Fe(II) complexes, where there is interconversion between a $^1A_1$ low-spin and $^5T_2$ high-spin state on a subpicosecond time scale. The dynamics of the process following photoexcitation have been probed using UV-Visible, IR and Raman, and X-ray techniques. Wolf *et al.*[110] used sub-picosecond transient IR spectroscopy to investigate the photoinduced low-spin to high-spin transition of $[Fe(btpa)](PF_6)_2$ and $[Fe(b(bdpa))](PF_6)_2$ in solution at room temperature. They observed negative and positive IR difference bands in the 1000–1065 cm^{-1} region. DFT calculations of the structures and IR spectra of both the low-spin and high-spin states were in good agreement with experimental data. In a subsequent study the same group[118] reported further DFT calculations, showing that the $[Fe(btpa)]^{2+}$ complex exhibits two high-spin isomers, one with a C_1 quasi hepta-coordinated structure (long-lived isomer) and the other with a C_2 hexa-coordinated structure. Both of these structures had been proposed previously based on time-resolved Raman and flash photolysis experiments.

Respondek *et al.*[113] reported IR and UV-Visible spectra of spin-crossover complexes of the type $[FeL\text{-}(pyN_4)]X_2$, where pyN4 is the penta-dentate ligand 2,6-bis(1′,3′-diamino-2′-methyl-prop-2′-yl)pyridine. In the IR spectrum clear changes were observed in the high-spin state in accordance with the results of DFT calculations employing various methods. All DFT methods predicted a significant softening of the Fe–N_6 symmetric stretch upon spin excitation, in agreement with experimental observations, and explained by the occupation of Fe–N antibonding orbitals in the high-spin case. TD-DFT calculations enabled comparison of simulated UV-Visible spectra for various high-spin/low-spin ratios with experimental, temperature-dependent spectra. They found that the computed energy difference between high-spin and low-spin states depends markedly on the DFT functional, in particular the amount of exact HF-like exchange and obtained better results with a modified form of the B3LYP functional with only 10% of exact exchange.

Van Kuiken and Khalil[128] employed TD-DFT to simulate the picosecond time-resolved iron K-edge X-ray absorption spectrum of the spin crossover complex, $[Fe(tren(py)_3)]$ $(PF_6)_2$ where tren = tris(2-aminoethyl)amine. They simulated the X-ray absorption spectrum in the low-spin ground state ($^1A_1$) and the high-spin excited state ($^5T_2$) and compared the results with those obtained for the static X-ray spectrum of a high-spin analogue complex, $[Fe(tren(6\text{-}Me\text{-}py)_3)](PF_6)_2$. The results identified some important aspects of the photoinduced spin-crossover process in $[Fe(tren(py)_3)]^{2+}$, in particular that it is accompanied by a large asymmetric lengthening of the Fe-N bonds. The geometric change is coupled to changes in the electronic structure that are manifest in the pre- and near-edge regions of the spectrum. The change in spin state is evident in the pre-edge region of the X-ray absorption where a single feature observed for the low-spin state is replaced by two peaks on excitation to the high spin state. The DFT calculations reveal that there is less π-backbonding in the high spin state.

3.5 Electroluminescent materials

Complexes of the third row transition metals, especially those of Ir(III) and Pt(II) with π-conjugated ligands, such as polypyridines and porphyrins, are of considerable interest as electroluminescent materials for optoelectronic devices *e.g.* organic light-emitting diodes (OLEDs). These complexes have the potential for use of both the singlet and triplet excitons upon electron–hole annihilation, resulting in theoretical 100% internal quantum efficiency. Several homoleptic and heteroleptic cyclometallated Ir(III) complexes have been examined, mostly based on the cyclometallating 2-phenylpyridine ligand. Tuning of the enission wavelength throughout the visible region is possible by substituting various electron-donating or electron-withdrawing groups at different positions of the coordinated ligands.

Zhang *et al.*[134] investigated the absorption and emission spectra of some heteroleptic 2-phenylpyridine Ir(III) complexes containing picolinate ligand with dimethylamino, trifluoromethyl, and methoxy substituents at the 4-position. They modelled the ground states of the complexes using B3LYP calculations and simulated the absorption spectra using TD-DFT methods. The geometries of the lowest energy triplet excited state were estimated using the single-excitation CIS method and these geometries were then used to determine the emission wavelengths from TD-DFT calculations, assuming gas-phase environment and acetonitrile solution (PCM method). It was found that bond distances and torsional angle were slightly different in the excited state, in particular most of the metal-ligand distances were increased upon excitation. The computed emission spectra of these complexes were found to display a blue shift in acetonitrile relative to the gas. It

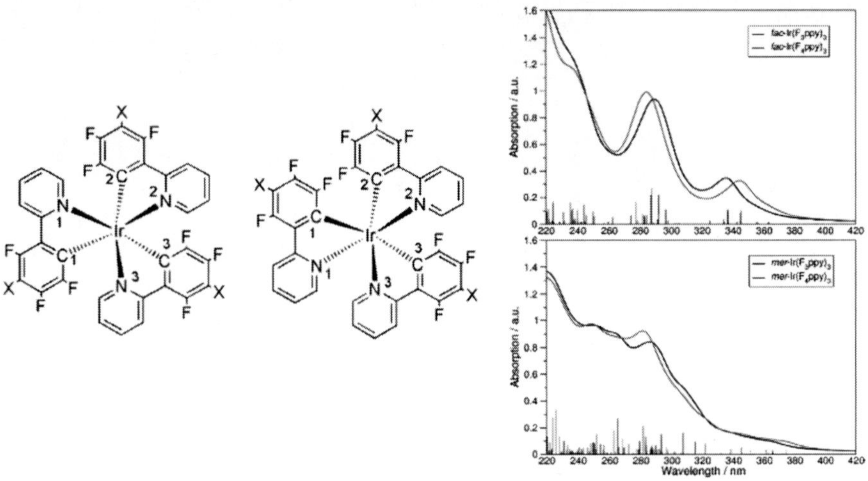

Fig. 4 Structures of *fac* and *mer* fluorinated 2-phenylpyridine Ir(III) complexes and their computed UV-Visible spectra from TD-DFT calculations. (Reproduced by permission from *Eur. J. Inorg. Chem.*, 2011, 2517).

was shown that the major transitions are of HOMO to LUMO character, indicating that electron-donating or electron-withdrawing substituents at the 4-position of the picolinate ligand do not change the order of the frontier orbitals but do alter their energies leading to a wavelength shift of the emitted radiation. This group used the same approach to investigate Pt(II) complexes containing fluorinated 2-phenylpyridine ligands[152] and found the same influence of the nature of the substituent on the phenylpyridine ligand on the distributions of HOMO and LUMO and their energies.

Shi *et al.*[138] used DFT methods to investigate the electronic structures and optoelectronic properties of several blue-emitting Ir(III) complexes containing fluorinated ligands. The injection abilities of holes and electrons were estimated by evaluating the ionization potentials and electron affinities and it was shown that the nature of the ligands had considerable influence on photophysical properties. Li *et al.*[140] investigated Ir(III) complexes with fluorinated 2-phenylpyridine ligands and simulated UV-Visible spectra (shown in Fig. 4) from TD-DFT calculations, and also accurately estimated phosphorescence rate constants and lifetimes. The calculations indicated that for the *fac*-isomers, the LUMO, LUMO + 1 and LUMO + 2 orbitals are localized on two or three ppy ligands, whereas in the *mer* isomers, these orbitals are localized on a single ppy ligand. It was found that the introduction of fluorine atoms in the ppy ligand resulted in a larger HOMO–LUMO energy gap and consequent blue-shift of the emission wavelength.

4 Lanthanides and actinides

A summary of computational studies of the spectra of lanthanide and actinide compounds is provided in Table 12. The development of theoretical methods suitable for compounds of the *f* block elements remains challenging due to the large numbers of electrons, relativistic effects and electron

Table 12 Lanthanides and Actinides.

Species	Spectroscopic method	Theoretical method (method/basis set)	Ref.
Tetraaquatris(succinate) diholmium(III) hexahydrate	IR, Raman	B3LYP/6-311G*, CEP-121G	196
Lanthanide complexes of a hexaaza macrocyclic ligand	NMR	B3LYP	197
U(VI) compounds	NMR	Comparison of several DFT methods and basis sets	198
SUO_2, matrix isolation	IR	B3LYP, BPW91/ 6-311 + G(3df) SDD for U	199
UO_2Cl and $UOCl_2$ monomers and dimers	IR	B3LYP/cc-pVDZ LanL2DZ for U	200 201
U(VI) nitrate monoamide complexes	IR	B3LYP/6-31G(d,p) RECP-60MWB for U	202
Uranyl(VI) complexation with aromatic acids	UV-Vis	B3LYP/AUG-cc-pVDZ ECP for U	203
Actinide oxides	IR	DFT, CASPT2/ANO-RCC	204
UN_2, NUO^+ and UO_2^{2+}	UV-Vis	CASPT2, RASPT2, CCSD(T) methods	205

correlation. However, reasonable results have been obtained using DFT methods with judicious choice of functionals. Core electrons are always represented by pseudopotentials which need to take account of relativistic effects. This is necessary because core electrons in atoms of high atomic number have large angular momenta, leading to velocities close to the velocity of light.

Iché-Tarrat and Marsden[198] carried out an in-depth study of U(VI) compounds testing several different DFT functionals for their suitability, in conjunction with two Stuttgart-Dresden pseudopotentials having different-sized cores and the larger-core LanL2DZ pseudopotential. They modelled experimental results including vibrational wavenumbers, thermochemical data and ^{19}F NMR chemical shifts for molecules of the type $UF_{6-n}Cl_n$. Predicted vibrational wavenumbers were satisfactory for pure GGA functionals as well as hybrid GGA with the large-core LanL2DZ pseudopotential. However, for thermochemistry, it was shown that hybrid DFT methods were more successful. By contrast, no combination of DFT method and basis functions provided a satisfactory prediction of trends in the ^{19}F chemical shifts of $UF_{6-n}Cl_n$ species. They also carried out CCSD(T) calculations and found these to be far superior to any of the DFT methods, but CCSD(T) is a very computationally demanding method and not suitable for general use with large atoms.

Umreiko *et al.*[200,201] investigated the IR spectra of UO_2Cl and $UOCl_2$ in the solid state in combination with B3LYP calculations of monomer and dimer models using the cc-pVDZ basis set for O and Cl atoms and the LANL2DZ pseudopotential for U. They established that dimer models better reproduced the spectral features of both molecules, indicating polymeric structures of these species in the solid state. Wiebke *et al.*[203] used TD-DFT calculations with the COSMO solvation model to investigate the UV-Visible spectra of uranyl (UO_2^{2+}) complexes with salicylhydroxamate, benzohydroxamate, and benzoate ligands in aqueous solution. The results indicated that UO_2^{2+} undergoes η^2-chelation *via* the hydroxamate O atoms in agreement with experimental data and that estimates of complex stabilities show that coordination to hydroxamates is preferred with respect to benzoate coordination.

5 Summary and future directions

The wide range of publications cited in this review clearly indicate the importance of QM methods, especially DFT, to the interpretation of the spectra of inorganic compounds. In order to achieve conciseness of this review I have deliberately been selective and the cited papers represent only about a quarter of those published in the last three years. Many of the studies reported here involve computation of the structures and spectra of isolated molecules in the gas phase, but increasing use is being made of solvation models, especially PCM. Nevertheless such implicit solvation models have their own drawbacks in that specific solute-solvent interactions such as hydrogen-bonding are not included, and require the addition of explicit solvent molecules in the model. The modelling of a huge solute-solvent cluster would normally represent a huge computational challenge but can be made more tractable by using a lower level of theory for solvent molecules outside the first coordination sphere, *e.g.* using QM/MM and QM/MD methods. I predict that such methods will be used increasingly in the next few years. The application of periodic DFT methods for modelling of solid-state structures and spectra has become much more widespread in the last few and, as predicted in my previous review, are now being applied in mainstream inorganic spectroscopic studies.

References

1 T. J. Dines, *Spectrosc. Prop. Inorg. Organomet. Compds.*, 2009, **40**, 171.

2 ADF2010, SCM, Theoretical Chemistry, Vrije Universiteit, Amsterdam, The Netherlands, http://www.scm.com.

3 M. W. Schmidt, K. K. Baldridge, J. A. Boatz, S. T. Elbert, M. S. Gordon, J. H Jensen, S. Koseki, N. Matsunaga, K. A. Nguyen, S. Su, T. L. Windus, M. Dupuis and J. A. Montgomery, *J. Comput. Chem.*, 1993, **14**, 1347.

4 M. S. Gordon and M. W. Schmidt, in *Theory and Applications of Computational Chemistry: the First Forty Years*, C. E. Dykstra, G. Frenking, K. S. Kim and G. E. Scuseria (eds), Elsevier, Amsterdam, 2005. pp. 1167–1189.

5 M. F. Guest, I. J. Bush, H. J. J. van Dam, P. Sherwood, J. M. H. Thomas, J. H. van Lenthe, R. W. A Havenith and J. Kendrick, *Mol. Phys.*, 2005, **103**, 719.

6 M. J. Frisch, G. W. Trucks, H. B. Schlegel, G. E. Scuseria, M. A. Robb, J. R. Cheeseman, G. Scalmani, V. Barone, B. Mennucci, G. A. Petersson, H. Nakatsuji, M. Caricato, X. Li, H. P. Hratchian, A. F. Izmaylov, J. Bloino, G. Zheng, J. L. Sonnenberg, M. Hada, M. Ehara, K. Toyota, R. Fukuda, J. Hasegawa, M. Ishida, T. Nakajima, Y. Honda, O. Kitao, H. Nakai, T. Vreven, J. A. Montgomery, Jr., J. E. Peralta, F. Ogliaro, M. Bearpark, J. J. Heyd, E. Brothers, K. N. Kudin, V. N. Staroverov, R. Kobayashi, J. Normand, K. Raghavachari, A. Rendell, J. C. Burant, S. S. Iyengar, J. Tomasi, M. Cossi, N. Rega, J. M. Millam, M. Klene, J. E. Knox, J. B. Cross, V. Bakken, C. Adamo, J. Jaramillo, R. Gomperts, R. E. Stratmann, O. Yazyev, A. J. Austin, R.

Cammi, C. Pomelli, J. W. Ochterski, R. L. Martin, K. Morokuma, V. G. Zakrzewski, G. A. Voth, P. Salvador, J. J. Dannenberg, S. Dapprich, A. D. Daniels, O. Farkas, J. B. Foresman, J. V. Ortiz, J. Cioslowski, and D. J. Fox, *Gaussian 09, Revision* ***C.01***, Gaussian, Inc., Wallingford CT, 2009.

7 http://www.schrodinger.com/products/14/7/.

8 M. Valiev, E. J. Bylaska, N. Govind, K. Kowalski, T. P. Straatsma, H. J. J. van Dam, D. Wang, J. Nieplocha, E. Apra, T. L. Windus and W. A. de Jong, *Comput. Phys. Commun.*, 2010, **181**, 1477.

9 http://www.hyper.com/Products/HyperChemProfessional/tabid/360/Default.aspx.

10 http://www.thch.uni-bonn.de/tc/orca/.

11 http://www.turbomole.com/.

12 S. J. Clark, M. D. Segall, C. J. Pickard, P. J. Hasnip, M. J. Probert, K. Refson and M. C. Payne, *Z. Kristall.*, 2005, **220**, 567.

13 R. Dovesi, R. Orlando, B. Civalleri, C. Roetti, V. R. Saunders and C. M. Zicovich-Wilson, *Z. Kristallogr.*, 2005, **220**, 571.

14 V. Milman, K. Refson, S. J. Clark, C. J. Pickard, J. R. Yates, S. P. Gao, P. J. Hasnip, M. I. J. Probert, A. Perlov and M. D. Segall, *J. Mol. Struct.*, 2010, **954**, 22.

15 http://en.wikipedia.org/wiki/List_of_quantum_chemistry_and_solid_state_physics_software.

16 J. A. Schachtschneider, *Vibrational Analysis of Polyatomic Molecules, Parts V and VI*, Technical Report Nos. 231 and 57, Shell Development Co., Houston TX, 1964 and 1965.

17 Y. Zhao and D. G. Truhlar, *Acc. Chem. Res.*, 2008, **41**, 157.

18 R. J. H. Clark and T. J. Dines, *Angew. Chem. Int. Ed. Engl.*, 1986, **25**, 131.

19 M. E. Casida, C. Jamorski, K. C. Casida and D. R. Salahub, *J. Chem. Phys.*, 1998, **108**, 4439.

20 F. Neese, *Coord. Chem. Rev.*, 2009, **253**, 526.

21 J. Tomasi, B. Mennucci and R. Cammi, *Chem. Rev.*, 2005, **105**, 2999.

22 M. Cossi, N. Rega, G. Scalmani and V. Barone, *J. Comp. Chem.*, 2003, **24**, 669.

23 R. F. Ribeiro, A. V. Marenich, C. J. Cramer and D. G. Truhlar, *J. Phys. Chem. B*, 2011, **115**, 14556.

24 C. P. Groen and A. Kovacs, *Vib. Spectrosc.*, 2010, **54**, 30.

25 Y. Inokuchi, O. V. Boyarkin, R. Kusaka, T. Haino, T. Ebata and T. R. Rizzo, *J. Am. Chem. Soc.*, 2011, **133**, 12256.

26 C. Chizallet, G. Costentin, H. Lauron-Pernot, J. M. Krafft, M. Che, F. Delbecq and P. Sautet, *J. Phys. Chem. C*, 2008, **112**, 19710.

27 L. S. Cahill, J. V. Hanna, A. Wong, J. C. C. Freitas, J. R. Yates, R. K. Harris and M. E. Smith, *Chem.-Eur. J.*, 2009, **15**, 9785.

28 D. Laurencin, C. Gervais, A. Wong, C. Coelho, F. Mauri, D. Massiot, M. E. Smith and C. Bonhomme, *J. Am. Chem. Soc.*, 2009, **131**, 13430.

29 R. Bulanek, I. Voleska, E. Ivanova, K. Hadjiivanov and P. Nachtigall, *J. Phys. Chem. C*, 2009, **113**, 11066.

30 S. F. Tayyari, T. Bakhshi, M. Ebrahimi and R. E. Sammelson, *Spectrochim. Acta Part A*, 2009, **73**, 342.

31 C. M. Leavitt, J. Oomens, R. P. Dain, J. Steill, G. S. Groenewold and M. J. Van Stipdonk, *J. Am. Soc. Mass Spectrom.*, 2009, **20**, 772.

32 G. Frenking, N. Holzmann, B. Neumueller and K. Dehnicke, *Z. Anorg. Allg. Chem.*, 2010, **636**, 1772.

33 S. F. Tayyari, T. Bakhshi, S. J. Mahdizadeh, S. Mehrani and R. E. Sammelson, *J. Mol. Struct.*, 2009, **938**, 76.

34 R. E. Kinzer, C. M. L. Rittby and W. R. M. Graham, *J. Chem. Phys.*, 2008, **129**, 184309.

35 H. Hagemann, M. Longhini, J. W. Kaminski, T. A. Wesolowski, R. Cerny, N. Penin, M. H. Sorby, B. C. Hauback, G. Severa and C. M. Jensen, *J. Phys. Chem. A*, 2008, **112**, 7551.

36 Y. X. Zhang, D. D. Qi, X. Cai and J. Z. Jiang, *Vib. Spectrosc.*, 2009, **51**, 184.

37 S. K. Amini and M. Tafazzoli, *Magn. Reson. Chem.*, 2008, **46**, 1045.

38 D. G. Liakos and E. A. Simandiras, *J. Phys. Chem. A*, 2008, **112**, 7881.

39 S. A. Bates, C. M. L. Rittby and W. R. M. Graham, *J. Chem. Phys.*, 2008, **128**, 234301.

40 X. L. Hu, J. Q. Li, Y. F. Zhang, H. H. Li and Y. Li, *Theor. Chem. Acc.*, 2009, **123**, 521.

41 K. Y. Hwang, H. Kim, Y. S. Lee, M. H. Lee and Y. Do, *Chem.-Eur. J.*, 2009, **15**, 6478.

42 S. H. Dong, W. L. Wang, S. W. Yin, C. Y. Li and J. Lu, *Synth. Met.*, 2009, **159**, 385.

43 S. Rada, M. Culea and E. Culea, *J. Non-Cryst. Solids*, 2009, **354**, 5491.

44 O. Zavorotynska, M. Corno, A. Damin, G. Spoto, P. Ugliengo and M. Baricco, *J. Phys. Chem. C*, 2011, **115**, 18890.

45 A. V. Mitin, *Struct. Chem.*, 2011, **22**, 411.

46 E. Gonzalez, C. M. L. Rittby and W. R. M. Graham, *J. Phys. Chem. A*, 2008, **112**, 10831.

47 I. Irgibaeva, A. Aldongarov, N. Barashkov and T. Schmedake, *Int. J. Quantum Chem.*, 2008, **108**, 2641.

48 J. T. Lyon, P. Gruene, A. Fielicke, G. Meijer, E. Janssens, P. Claes and P. Lievens, *J. Am. Chem. Soc.*, 2009, **131**, 1115.

49 G. F. de Sousa, E. Garcia, C. C. Gatto, I. S. Resck, V. M. Deflon and J. D. Ardisson, *J. Mol. Struct.*, 2010, **981**, 46.

50 L. Rhyman, H. H. Abdallah and P. Ramasami, *Polyhedron*, 2010, **29**, 220.

51 A. Chrissanthopoulos and C. Pouchan, *Vib. Spectrosc.*, 2008, **48**, 135 P. Carbonniere and C. Pouchan, *Chem. Phys. Lett.*, 2008, **462**, 169.

52 H. W. Kim, M. K. Patel and D. Zeroka, *Spectrochim. Acta Part A*, 2009, **73**, 730.

53 V. L. Furer, I. I. Vandyukova, A. E. Vandyukov, J. P. Majoral, A. M. Caminade and V. I. Kovalenko, *Chem. Phys.*, 2009, **358**, 177.

54 V. L. Furer, A. E. Vandyukov, J. P. Majoral, A. M. Caminade and V. I. Kovalenko, *J. Mol. Struct.*, 2009, **919**, 366.

55 V. L. Furer, I. I. Vandyukova, A. E. Vandyukov, S. Fuchs, J. P. Majoral, A. M. Caminade and V. I. Kovalenko, *Spectrochim. Acta Part A*, 2011, **79**, 462.

56 V. L. Furer, I. I. Vandyukova, A. E. Vandyukov, S. Fuchs, J. P. Majoral, A. M. Caminade and V. I. Kovalenko, *J. Mol. Struct.*, 2011, **987**, 144.

57 R. Jaquet and H. Haeuseler, *J. Raman Spectrosc.*, 2008, **39**, 599.

58 E. Hulkko, T. Kiijunen, T. Kiviniemi and M. Pettersson, *J. Am. Chem. Soc.*, 2009, **131**, 1050.

59 M. Otsuka, H. Mori, H. Kikuchi and K. Takano, *Comput. Theor. Chem.*, 2011, **973**, 69.

60 J. S. Ogden, J. T. Graham, J. T. Joy and F. Ferrante, *Phys. Chem. Chem. Phys.*, 2009, **11**, 650.

61 G. L. Smith, H. P. A. Mercier and G. J. Schrobilgen, *Inorg. Chem.*, 2008, **47**, 4173.

62 H. G. Cho, G. P. Kushto, L. Andrews and C. W. Bauschlicher, *J. Phys. Chem. A.*, 2008, **112**, 6295.

63 S. A. Kozimor, P. Yang, E. R. Batista, K. S. Boland, C. J. Burns, C. N. Christensen, D. L. Clark, S. D. Conradson, P. J. Hay, J. S. Lezama, R. L.

Martin, D. E. Schwarz, M. P. Wilkerson and L. E. Wolfsberg, *Inorg. Chem.*, 2008, **47**, 5365.
64 J. Scaranto and S. Giorgianni, *Spectrochim. Acta Part A*, 2009, **74**, 1072.
65 I. S. Ignatyev, M. Montejo and J. J. L. Gonzalez, *Vib. Spectrosc.*, 2009, **51**, 218.
66 X. Zarate, E. Schott and R. Arratia-Perez, *Int. J. Quantum Chem.*, 2011, **111**, 4168.
67 I. Novak and B. Kovac, *Chem. Phys. Lett.*, 2011, **512**, 195.
68 R. Jin, Y. H. Zhang, S. P. Huang, P. Wang and P. H. Tian, *Chin. J. Chem.*, 2011, **29**, 13.
69 A. Maniatakou, C. Makedonas, C. A. Mitsopoulou, C. Raptopoulou, I. Rizopoulou, A. Terzis and A. Karaliota, *Polyhedron*, 2008, **27**, 3398.
70 B. Zhou and D. Y. He, *J. Raman Spectrosc.*, 2008, **39**, 1475.
71 A. Schweitzer, T. Gutmann, M. Waechtler, H. Breitzke, A. Buchholz, W. Plass and G. Buntkowsky, *Solid State Nucl. Magn. Reson.*, 2008, **34**, 52.
72 B. R. Zeng, J. Zhang, R. Q. Fu, S. H. Cai and Z. Chen, *Inorg. Chem. Commun.*, 2009, **12**, 1259.
73 M. M. Islam, D. Costa, M. Calatayud and F. Tielens, *J. Phys. Chem. C*, 2009, **113**, 10740.
74 S. A. Brandan, C. Socolsky and A. Ben Altabef, *Z. Anorg. Allg. Chem.*, 2009, **635**, 582.
75 M. Cavalleri, K. Hermann, A. Knop-Gericke, M. Haevecker, R. Herbert, C. Hess, A. Oestereich, J. Doebler and R. Schloegl, *J. Catal.*, 2009, **262**, 215.
76 L. A. Truflandier, F. Boucher, C. Payen, R. Hajja, Y. Millot, C. Bonhomme and N. Steunou, *J. Am. Chem. Soc.*, 2010, **132**, 4653.
77 E. W. Y. Wong, C. J. Walsby, T. Storr and D. B. Leznoff, *Inorg. Chem.*, 2010, **49**, 3343.
78 M. V. Castillo, E. Romano, H. E. Lanus, S. B. Diaz, A. Ben Altabef and S. A. Brandan, *J. Mol. Struct.*, 2011, **994**, 202.
79 H. G. Cho and L. Andrews, *J. Phys. Chem. A.*, 2008, **112**, 12071.
80 X. R. Shi, H. J. Jiao, K. Hermann and J. G. Wang, *J. Mol. Catal. A-Chem.*, 2009, **312**, 7.
81 B. Courcot and A. J. Bridgeman, *J. Phys. Chem. A.*, 2009, **113**, 10540.
82 Y. P. Tong and Y. W. Lin, *Inorg. Chim. Acta*, 2009, **362**, 2167.
83 E. W. Reinheimer, I. Olejniczak, A. Lapinski, R. Swietlik, O. Jeannin and M. Fourmigue, *Inorg. Chem.*, 2010, **49**, 9777.
84 A. Ben Altabef and S. A. Brandan, *J. Mol. Struct.*, 2010, **981**, 146.
85 E. Hernandez-Marin, M. Seth and T. Ziegler, *Inorg. Chem.*, 2010, **49**, 6066.
86 R. E. Da Re, J. L. Eglin, C. N. Carlson, K. D. John, D. E. Morris, W. H. Woodruff, J. A. Bailey, E. Batista, R. L. Martin, F. A. Cotton, E. A. Hillard, C. A. Murillo, A. P. Sattelberger and R. J. Donohoe, *J. Am. Chem. Soc.*, 2010, **132**, 1839.
87 R. Swietlik, A. Lapinski, M. Fourmigue and K. Yakushi, *J. Raman Spectrosc.*, 2009, **40**, 2092.
88 D. Ravelli, D. Dondi, M. Fagnoni, A. Albini and A. Bagno, *J. Comput. Chem.*, 2011, **32**, 2983.
89 M. Mauro, M. Panigati, D. Donghi, P. Mercandelli, P. Mussini, A. Sironi and G. D'Alfonso, *Inorg. Chem.*, 2008, **47**, 11154.
90 X. N. Li, X. J. Liu, Z. J. Wu and H. J. Zhang, *J. Phys. Chem. A.*, 2008, **112**, 11190.
91 S. Schinzel and M. Kaupp, *Can. J. Chem.-Rev. Can. Chim.*, 2009, **87**, 1521.
92 B. Machura, J. Mrozinski, R. Kruszynski and J. Kusz, *Polyhedron*, 2009, **28**, 2821.

93 Y. Nakao, K. Saito and S. Sakaki, *Int. J. Quantum Chem.*, 2009, **109**, 2319.
94 L. Rodriguez, M. Ferrer, O. Rossell, F. J. S. Duarte, A. G. Santos and J. C. Lima, *J. Photochem. Photoobiol. A-Chem.*, 2009, **204**, 174.
95 K. Druzbicki and E. Mikuli, *Spectrochim. Acta Part A*, 2010, **77**, 402.
96 F. Poineau, P. M. Forster, T. K. Todorova, L. Gagliardi, A. P. Sattelberger and K. R. Czerwinski, *Inorg. Chem.*, 2010, **49**, 6646.
97 T. Yoshimura, T. Ikai, T. Takayama, T. Sekine, Y. Kino and A. Shinohara, *Inorg. Chem.*, 2010, **49**, 5876.
98 R. Horvath, C. A. Otter, K. C. Gordon, A. M. Brodie and E. W. Ainscough, *Inorg. Chem.*, 2010, **49**, 4073.
99 G. N. Li, T. Jin, L. Sun, J. Qin, D. Wen, J. L. Zuo and X. Z. You, *J. Organomet. Chem.*, 2011, **696**, 3076.
100 J. S. Gancheff and P. A. Denis, *J. Phys. Chem. A.*, 2011, **115**, 211.
101 K. Christopoulos, K. Karidi, A. Tsipis and A. Garoufis, *Inorg. Chem. Commun.*, 2008, **11**, 1341.
102 J. Bossert and C. Daniel, *Coord. Chem. Rev.*, 2008, **252**, 2493.
103 T. Liu, B. H. Xia, X. Zhou, Q. C. Zheng, Q. J. Pan and H. X. Zhang, *J. Mol. Struct.-Theochem.*, 2008, **869**, 11.
104 T. Todorova and B. Delley, *Inorg. Chem.*, 2008, **47**, 11269.
105 D. Kalinina, C. Dares, H. Kaluarachchi, P. G. Potvin and A. B. P. Lever, *Inorg. Chem.*, 2008, **47**, 10110.
106 Y. Gong and M. F. Zhou, *J. Phys. Chem. A*, 2008, **112**, 10838.
107 J. P. Zhang, X. Zhou, T. Liu, F. Q. Bai, H. X. Zhang and A. C. Tang, *Theor. Chem. Acc.*, 2008, **121**, 123.
108 K. Krogh-Jespersen, R. T. Stibrany, E. John, J. D. Westbrook, T. J. Emge, M. J. Clarke, J. A. Potenza and H. J. Schugar, *Inorg. Chem.*, 2008, **47**, 9813.
109 M. Kubo, F. Gruia, A. Benabbas, A. Barabanschikov, W. R. Montfort, E. M. Maes and P. M. Champion, *J. Am. Chem. Soc.*, 2008, **130**, 9800.
110 M. M. N. Wolf, R. Gross, C. Schumann, J. A. Wolny, V. Schuenemann, A. Dossing, H. Paulsen, J. J. McGarvey and R. Diller, *Phys. Chem. Chem. Phys.*, 2008, **10**, 4264.
111 A. Dhouib, C. Minot and M. Abderraba, *J. Mol. Struct.-Theochem.*, 2008, **860**, 161.
112 Y. Sun, D. A. Lutterman and C. Turro, *Inorg. Chem.*, 2008, **47**, 6427.
113 I. Respondek, L. Bressel, P. Saalfrank, H. Kaempf and A. Grohmann, *Chem. Phys.*, 2008, **347**, 514.
114 M. A. Castro, A. E. Roitberg and F. D. Cukiernik, *Inorg. Chem.*, 2008, **47**, 4682.
115 Y. S. Guo, H. X. Wang, Y. M. Xiao, S. Vogt, R. K. Thauer, S. Shima, P. I. Volkers, T. B. Rauchfuss, V. Pelmenschikov, D. A. Case, E. E. Alp, W. Sturhahn, Y. Yoda and S. P. Cramer, *Inorg. Chem.*, 2008, **47**, 3969.
116 http://www.shapesoftware.com/#anchor_vibratz.
117 M. X. Li, X. Zhou, H. X. Zhang, H. G. Fu and C. C. Sun, *Chem. J. Chin. Univ.-Chin.*, 2009, **30**, 2284.
118 J. A. Wolny, H. Paulsen, J. J. McGarvey, R. Diller, V. Schuenemann and H. Toftlund, *Phys. Chem. Chem. Phys.*, 2009, **11**, 7562.
119 P. Das, P. P. Sarmah, M. Borah and A. K. Phukan, *Inorg. Chim. Acta*, 2009, **362**, 5001.
120 M. Kondo, M. Uchikawa, K. Namiki, W. W. Zhang, S. Kume, E. Nishibori, H. Suwa, S. Aoyagi, M. Sakata, M. Murata, Y. Kobayashi and H. Nishihara, *J. Am. Chem. Soc.*, 2009, **131**, 12112.
121 G. J. Long, S. Tanase, F. Remacle, G. Periyasamy and F. Grandjean, *Inorg. Chem.*, 2009, **48**, 8173.
122 X. F. Wang and L. Andrews, *J. Phys. Chem. A*, 2009, **113**, 551.

123 J. A. Wolny, S. Rackwitz, K. Achterhold, Y. Garcia, K. Muffler, A. D. Naik and V. Schuenemann, *Phys. Chem. Chem. Phys.*, 2010, **12**, 14782.
124 T. P. Gryaznova, S. A. Katsyuba, V. A. Milyukov and O. G. Sinyashin, *J. Organomet. Chem.*, 2010, **695**, 2586.
125 D. Ambrosek, P. F. Loos, X. Assfeld and C Daniel, *J. Inorg. Biochem.*, 2010, **104**, 893.
126 A. V. Soldatova, M. Ibrahim, J. S. Olson, R. S. Czernuszewicz and T. G. Spiro, *J. Am. Chem. Soc.*, 2010, **132**, 4614.
127 M. Knapp-Mohammady and N. H. March, *Theor. Chem. Acc.*, 2010, **125**, 293.
128 B. E. Van Kuiken and M. Khalil, *J. Phys. Chem. A*, 2011, **115**, 10749.
129 D. Mitra, V. Pelmenschikov, Y. S. Guo, D. A. Case, H. X. Wang, W. B. Dong, M. L. Tan, T. Ichiye, F. E. Jenney, M. W. W. Adams, Y. Yoda, J. Y. Zhao and S. P. Cramer, *Biochemistry*, 2011, **50**, 5220.
130 J. Fan, J. Autschbach and T. Ziegler, *Inorg. Chem.*, 2010, **49**, 1355.
131 H. Eshtiagh-Hosseini, M. R. Housaindokht, S. A. Beyramabadi, S. Beheshti, A. A. Esmaeili, M. J. Khoshkholgh and A. Morsali, *Spectrochim. Acta Part A*, 2008, **71**, 1341.
132 T. Liu, B. H. Xia, X. Zhou, Q. C. Zheng, Q. J. Pan and H. X. Zhang, *Theor. Chem. Acc.*, 2008, **121**, 155.
133 W. Li, F. Yang, Z. D. Wang, J. M. Hu and J. S. Ma, *J. Phys. Chem. A*, 2009, **113**, 3375.
134 M. Zhang, Z. S. Li, Y. Li, J. Liu and J. Z. Sun, *Int. J. Quantum Chem.*, 2009, **109**, 1167.
135 E. A. Orabi, *J. Coord. Chem.*, 2010, **63**, 4017.
136 T. F. Miao, J. Li, S. Y. Liao, J. A. Mei and K. C. Zheng, *J. Mol. Struct.-Theochem*, 2010, **957**, 108.
137 F. Gao, C. Z. Li, E. Widjaja, C. Jacob and M. Garland, *J. Organomet. Chem.*, 2010, **695**, 2394.
138 L. L. Shi, B. Hong, W. Guan, Z. J. Wu and Z. M. Su, *J. Phys. Chem. A*, 2010, **114**, 6559.
139 X. L. Su, Y. K. Wang, Y. Wang, J. Jia and X. L. Gao, *Acta Phys.-Chim. Sin.*, 2011, **27**, 1633.
140 X. Li, B. Minaev, H. Agren and H. Tian, *Eur. J. Inorg. Chem.*, 2011, 2517.
141 H. Solheim, K. Kornobis, K. Ruud and P. M. Kozlowski, *J. Phys. Chem. B*, 2011, **115**, 737.
142 G. Szilvagyi, Z. Majer, E. Vass and M. Hollosi, *Chirality*, 2011, **23**, 294.
143 Z. D. Yang, J. K. Feng and A. M. Ren, *Inorg. Chem.*, 2008, **47**, 10841.
144 C. J. Adams, N. Fey, Z. A. Harrison, I. V. Sazanovich, M. Towrie and J. A. Weinstein, *Inorg. Chem.*, 2008, **47**, 8242.
145 M. Meier and R. Weihrich, *Chem. Phys. Lett.*, 2008, **461**, 38.
146 B. L. Marcial, L. A. S. Costa, W. B. De Almeida and H. F. Dos Santos, *J. Mol. Struct.-Theochem*, 2009, **916**, 94.
147 Y. X. Qiu, J. Li and S. G. Wang, *Acta Chim. Sin.*, 2009, **67**, 1585.
148 L. Kocsis, J. Mink, F. Jalilehvand, L. J. Laffin, O. Berkesi and L. Hajba, *J. Raman Spectrosc.*, 2009, **40**, 481.
149 P. A. Dub, O. A. Filippov, N. V. Belkova, M. Rodriguez-Zubiri and R. Poli, *J. Phys. Chem. A*, 2009, **113**, 6348.
150 S. M. Souvi, B. Tremblay, J. P. Perchard and M. E. Alikhani, *J. Chem. Phys.*, 2009, **130**, 074304.
151 J. M. Ramos, O. Versiane, J. Felcman and C. A. Tellez, *Spectrochim. Acta Part A*, 2009, **72**, 182.

152 M. Zhang, Y. Li, Z. S. Li and J. Z. Sun, *Int. J. Quantum Chem.*, 2009, **109**, 826.
153 J. Fonseca, J. Martinez, L. Cunha-Silva, A. L. Magalhaes, M. T. Duarte and C. Freire, *Inorg. Chim. Acta*, 2010, **363**, 4096.
154 H. W. Gao, X. J. Wei, X. T. Liu and C. J. Huang, *Spectrochim. Acta Part A*, 2010, **77**, 461.
155 F. Billes, A. Holmgren and H. Mikosch, *Vib. Spectrosc.*, 2010, **53**, 296.
156 B. Z. Yang, X. Zhou, T. Liu, F. Q. Bai and H. X. Zhang, *Int. J. Quantum Chem.*, 2010, **110**, 1605.
157 L. A. Truflandier and J. Autschbach, *J. Am. Chem. Soc.*, 2010, **132**, 3472.
158 Y. H. Zhang, W. Zhao, J. Wang and P. Jiang, *Spectrochim. Acta Part A*, 2010, **75**, 499.
159 C. I. Oprea, F. Moscalu, S. Ioannou, A. Nicolaides and M. A. Girtu, *J. Optoelectron. Adv. Mater.*, 2009, **11**, 1773.
160 L. A. E. B. de Carvalho, M. P. M. Marques, C. Martin, S. F. Parker and J. Tomkinson, *ChemPhysChem*, 2011, **12**, 1334.
161 H. W. Gao, F. Y. Xia, C. J. Huang and K. F. Lin, Spectrochim, *Acta Part A*, 2011, **78**, 1234.
162 Z. Q. Liu, Z. X. Chen, B. B. Jin and X. X. Zhang, *Vib. Spectrosc.*, 2011, **56**, 210.
163 R. Latifi, M. Bagherzadeh, B. F. Milne, M. Jaspars and S. P. de Visser, *J. Inorg. Biochem.*, 2008, **102**, 2171.
164 X. L. Zhang, H. He, H. W. Gao and Y. B. Yu, *Spectrochim. Acta Part A*, 2008, **71**, 1446.
165 H. W. Gao and T. X. Yan, *J. Mol. Struct.*, 2008, **889**, 191; H. W. Gao, T. X. Yan, C. B. Zhang and H. He, *J. Mol. Struct.-Theochem*, 2008, **857**, 38.
166 D. Krilov, V. Gomzi and J. N. Herak, *Spectrochim. Acta Part A*, 2008, **70**, 1238.
167 S. L. Howell and K. C. Gordon, *J. Raman Spectrosc.*, 2008, **39**, 813.
168 J. P. Holland, P. J. Barnard, D. Collison, J. R. Dilworth, R. Edge, J. C. Green and E. J. L. McInnes, *Chem.-Eur. J.*, 2008, **14**, 5890.
169 T. F. Miao, S. Li and J. H. Cai, *J. Mol. Struct.-Theochem*, 2008, **855**, 45.
170 A. A. Bessonov, T. V. Basova, V. G. Kiselev, L. A. Sheludyakova, N. B. Morozova and I. K. Igumenov, *Vib. Spectrosc.*, 2009, **51**, 283.
171 B. Xerri, J. P. Flament, H. Petitjean, C. Berthomieu and D. Berthomieu, *J. Phys. Chem. B.*, 2009, **113**, 15119.
172 I. Georgieva, N. Trendafilova, B. S. Creaven, M. Walsh, A. Noble and M. McCann, *Chem. Phys.*, 2009, **365**, 69.
173 B. Z. Yang, X. Zhou, T. Liu, F. Q. Bai and H. X. Zhang, *J. Phys. Chem. A.*, 2009, **113**, 9396.
174 K. Helios, R. Wysokinski, W. Zierkiewicz, L. M. Proniewicz and D. Michalska, *J. Phys. Chem. B.*, 2009, **113**, 8158.
175 W. M. Ames and S. C. Larsen, *J. Phys. Chem. A.*, 2009, **113**, 4305.
176 M. S. Liao, P. Bonifassi, J. Leszczynski, P. C. Ray, M. J. Huang and J. D. Watts, *J. Phys. Chem. A.*, 2010, **114**, 12701.
177 R. L. Birke, V. Znamenskiy and J. R. Lombardi, *J. Chem. Phys.*, 2010, **132**, 214707.
178 T. V. Basova, V. G. Kiselev, B. E. Schuster, H. Peisert and T. Chasse, *J. Raman Spectrosc.*, 2009, **40**, 2080.
179 X. Q. Lu, S. X. Wei, C. M. L. Wu, S. R. Li and W. Y. Guo, *J. Phys. Chem. C.*, 2011, **115**, 3753.
180 J. M. Ramos, F. O. Grisset, J. Felcman and A. T. S. Claudio, *Spectrochim. Acta Part A*, 2008, **71**, 1364.

181 N. Norani, H. Rahemi, S. F. Tayyari and M. J. Riley, *J. Mol. Model*, 2009, **15**, 25.
182 M. Presselt, B. Dietzek, M. Schmitt, J. Popp, A. Winter, M. Chiper, C. Friebe and U. S. Schubert, *J. Phys. Chem. C*, 2008, **112**, 18651.
183 A. E. Ozel, S. Kecel and S. Akyuz, *Vib. Spectrosc.*, 2008, **48**, 238.
184 Z. Valicsek, G. Lendvay and O. Horvath, *J. Phys. Chem. B*, 2008, **112**, 14509.
185 D. del Rio, I. Resa, A. Rodriguez, L. Sanchez, R. Koeppe, A. J. Downs, C. Y. Tang and E. Carmona, *J. Phys. Chem. A*, 2008, **112**, 10516.
186 H. Xu, L. F. Huang, L. M. Guo, Y. G. Zhang, X. M. Ren, Y. Song and J. L. Xie, *J. Luminesc.*, 2008, **128**, 1665.
187 M. van Gastel, C. C. Lu, K. Wieghardt and W. Lubitz, *Inorg. Chem.*, 2009, **48**, 2626.
188 J. Kim, H. Ihee and Y. S. Lee, *J. Chem. Phys.*, 2010, **133**, 144309.
189 F. Billes, A. Holmgren and H. Mikosch, *Vib. Spectrosc.*, 2010, **53**, 296.
190 M. Christlieb, J. P. Holland and J. R. Dilworth, *Inorg. Chim. Acta*, 2010, **363**, 1133.
191 Y. H. Zhang, W. Zhao, P. Jiang, L. J. Zhang, T. Zhang and J. Wang, *Spectrochim. Acta Part A*, 2010, **75**, 880.
192 M. Kieninger and O. N. Ventura, *Theor. Chem. Acc.*, 2010, **125**, 279.
193 H. S. Kim, S. W. Jang, S. Y. Chung, S. Lee, Y. Lee, B. Kim, C. Liu and D. Neuhauser, *J. Phys. Chem. B*, 2010, **114**, 471.
194 Q. Wang, R. He, X. A. Cheng and C. S. Lu, *J. Incl. Phenom. Macrocycl. Chem.*, 2011, **69**, 231.
195 J. C. Earles, K. C. Gordon, A. W. I. Stephenson, A. C. Partridge and D. L. Officer, *Phys. Chem. Chem. Phys.*, 2011, **13**, 1597.
196 M. C. Bernini, J. C. Garro, E. V. Brusau, G. E. Narda and E. L. Varetti, *J. Mol. Struct.*, 2008, **888**, 113.
197 C. Nunez, R. Bastida, A. Macias, M. Mato-Iglesias, C. Platas-Iglesias and L. Valencia, *Dalton Trans.*, 2008, 3841.
198 N. Iche-Tarrat and C. J. Marsden, *J. Phys. Chem. A*, 2008, **112**, 7632.
199 X. F. Wang, L. Andrews and C. J. Marsden, *Inorg. Chem.*, 2009, **48**, 6888.
200 D. S. Umreiko, M. B. Shundalau, A. P. Zazhogin and A. I. Komyak, *J. Appl. Spectrosc.*, 2010, **77**, 510.
201 D. S. Umreiko, M. B. Shundalau and O. V. Trubina, *J. Appl. Spectrosc.*, 2010, **77**, 626.
202 A. Prestianni, L. Joubert, A. Chagnes, G. Cote, M. N. Ohnet, C. Rabbe, M. C. Charbonnel and C. Adamo, *J. Phys. Chem. A*, 2010, **114**, 10878.
203 J. Wiebke, A. Weigand, D. Weissmann, M. Glorius, H. Moll, G. Bernhard and M. Dolg, *Inorg. Chem.*, 2010, **49**, 6428.
204 A. Kovacs and R. J. M. Konings, *J. Phys. Chem. A*, 2011, **115**, 6646.
205 F. Wei, G. S. Wu, W. H. E. Schwarz and J. Li, *Theor. Chem. Acc.*, 2011, **129**, 467.

Combined time-resolved X-ray scattering and spectroscopy methods

Wim Bras[a] and Andrew M. Beale[b]
DOI: 10.1039/9781849734899-00257

Introduction

X-ray Absorption Spectroscopy (XAS) is a tool that can render structural information hardly accessible with other methods at the length scale of the immediate local surrounding of probe atoms and up to several surrounding shells. In this it provides information not readily available with conventional X-ray diffraction (XRD) methods which rely on a longer range atomic order, typically several crystallographic unit cells large, before measurable information can be obtained. Low resolution techniques like Small Angle X-ray Scattering (SAXS) can fill some of the gaps between the size range probed by XAS in its different forms and X-ray diffraction since it is sensitive to spatially averaged electron density differences which allows one to observe 'clumps' of atoms with a diameter of 1 nm or larger but without providing information about the internal structure of the 'clumps'.

The complementarities with respect to the different length scales that can be accessed with these different X-ray based tools imply that a combination of the results of these techniques to follow the structural development of materials is a logical approach. Often it will not be required to do this in a single experiment and it will suffice to perform the experiments independently and to compare the data sets. This is the optimal situation since one can use sample environments and sample thicknesses which are optimised for the different experiments. The combination of independently obtained XAS and XRD results was utilised as early as 1987[1] and it has already been recognised for a long time that the complementary information from XRD and XAS can be instrumental in elucidating finer structural details where the XRD data in itself is not sufficient. However, in the cases that one follows structural evolution in time there are several reasons which can render it useful to obtain both datasets in a single experiment.[2] For instance in the case that the required time resolution is so high that it will be difficult to compare independent results or in the case that one deals with less homogeneous samples.

A research area that lends itself very well to the application of several techniques is catalysis since both the chemical state, as well as the structure, play an important role in the proper functioning of a catalyst.[3–5] However, catalysis is by no means the only field where one applies technique combinations including XAS. The range of applications is very wide. Specific examples of this include the study of corrosion processes of metallic cultural heritage materials,[6,7] the formation of reversed organic micelles,[8] materials

[a]*Netherlands Organization for Scientific Research (NWO), DUBBLE@ESRF, BP 220, F38043 Grenoble Cedex; E-mail: Wim.Bras@esrf.eu*
[b]*Inorganic Chemistry and Catalysis, Debye Institute for Nanomaterials Science, Universiteitsweg 99, Utrecht University 3584 CG, Utrecht, The Netherlands; E-mail: A.M.Beale@uu.nl*

for the storage of nuclear waste[9] and chemical geology in environmental studies.[10]

For XAS one requires a wider wavelength spectrum and higher X-ray intensity than can be usefully obtained with conventional X-ray tubes. Therefore one has to rely on synchrotron radiation sources which are also required for time resolved X-ray scattering experiments. Therefore it should not be surprising that efforts to combine X-ray spectroscopy and scattering on synchrotron radiation beamlines have been made through the years. The pioneering work, using energy dispersive methods, was performed around 1991 at the SRS in Daresbury[11–14] which was later followed by quick scanning EXAFS at the Hamburg synchrotron.[15] The success of the approach to combine spectroscopy and diffraction in a single time-resolved experiment triggered several similar efforts from which the combination of SAXS and Wide Angle X-ray Scattering (WAXS) was probably the most successful[16,17] but where also for instance Raman scattering, FTIR and UV-Vis spectroscopy were combined with X-ray scattering.[18]

In recent years several experimental facilities have been constructed that allow the simultaneous collection of XRD and XAS data.[19–22] The combination with SAXS is more difficult but is bound to gain more interest as well.[5,19]

In this text we have focussed mainly on the combinations from X-ray spectroscopy with X-ray diffraction and scattering in a time regime of 0.1 seconds/data frame and slower. This is a timing regime which is relevant in many materials science-related research fields. Faster XAS experiments are feasible but tend to focus on more fundamental problems in which structure formation is maybe the consequence but not concurrent in time and therefore combinations with X-ray scattering is less obvious.

Synchrotron radiation

For XAS a high intensity continuous X-ray spectrum is required which is most conveniently generated on a synchrotron radiation source. There are several excellent text books, at both introductory as well as advanced level, available on the theory and general application of synchrotron radiation[23,24] and therefore we will only treat the most relevant details required for the combinations of spectroscopy and scattering techniques, which is the subject of this manuscript, here.

Synchrotron radiation is generated when the path of charged particles moving with relativistic speeds is altered by sending them through a magnetic field. For the production of synchrotron radiation in practice only electrons and positrons are relevant. The magnetic structures can be a simple static field, in which the electrons are forced to follow a circular trajectory (bending magnet), or more complicated structures in which the particles are forced to follow sinusoidal or helical paths (undulators and wigglers). Different magnetic structures generate different X-ray spectra. The most simple spectrum is the bending magnet and wiggler radiation. Both these structures generate a continuous spectrum which is very useful when one wants to perform XAS. See Fig. 1 for a flux *vs.* photon energy spectrum of a typical bending magnet spectrum.

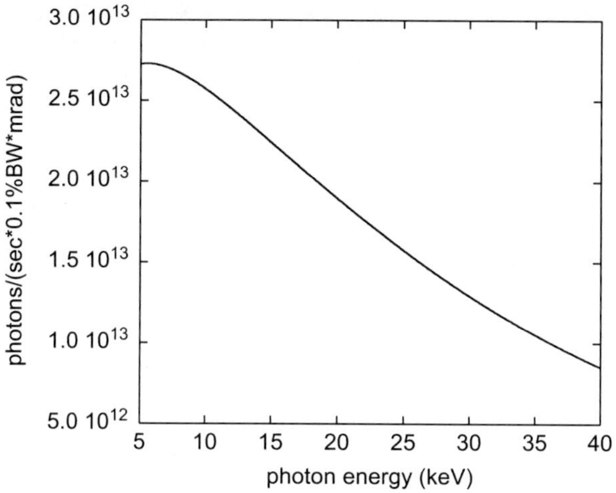

Fig. 1 A typical ESRF bending magnet spectrum from a 0.8 Tesla dipole magnet. For different synchrotron laboratories this curve can be shifted in both the vertical (intensity) as well as the horizontal (energy) direction. The important feature is that the spectrum is monotonous.

For comparison we mention here that a conventional X-ray generator is capable of producing about 10^9–10^{11} photons/sec but with a larger divergence compared to synchrotron sources which reduces the number of photons which can be delivered onto the sample with an increasing source-sample distance. Therefore a synchrotron beamline is capable of delivering several orders of magnitude more photons on a sample than a conventional X-ray source.

The undulator spectrum is better characterised by a series of very high intensity peaks on a continuous background. See Fig. 2. This spectrum can be tuned in energy by changing the gap of the undulator.

For scattering and diffraction experiments which are essentially carried out at fixed wavelengths/photon energies, the use of an undulator or bending magnet radiation source does not make much difference as long as the requirements regarding available flux and collimation are met. This is slightly different for spectroscopy methods where the response of the sample as a function of photon energy is measured. In the ideal case one would like to utilise a radiation source with a constant intensity for each wavelength. Since this does not exist the next best thing is a bending magnet source in which the intensity variation as function of energy does not change dramatically over the energy range typically used for an EXAFS energy scan ($\approx 600\,\mathrm{eV}$).

However, it is not the case that undulator beamlines cannot be used. In some cases where a very high flux or a very high collimation is required one resorts to the use of undulators as radiation sources but one often has to change the distance between the magnetic structures (gap) during an EXAFS scan. This adds an extra complication to the experiment but is not impossible. Undulators are also used to create high intensity 'pink beams' which cover a somewhat wider energy range sufficient for XANES experiments when using an energy dispersive configuration.

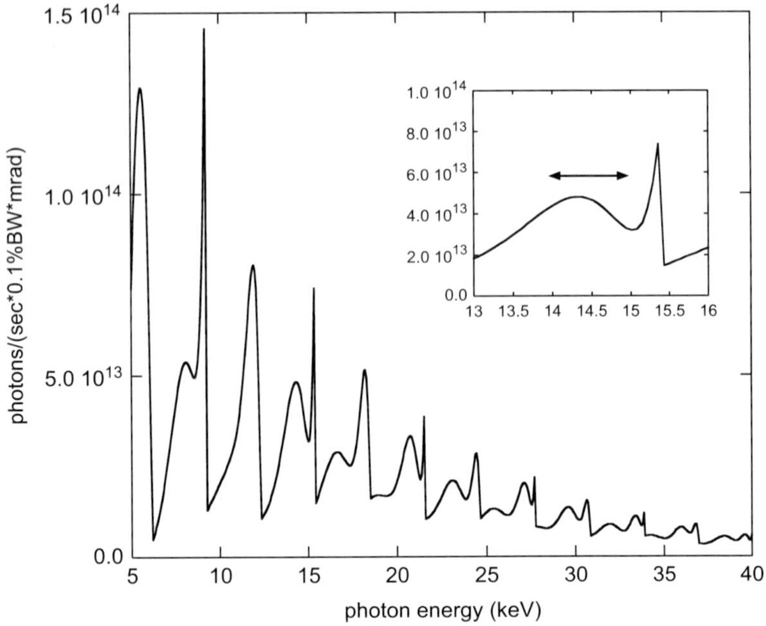

Fig. 2 An ESRF undulator spectrum. The inset shows a magnified section and an indication of the extent of an average energy scan used in an EXAFS experiment. An XANES scan is about 1/10th of the length of an EXAFS scan. Compared to the bending magnet spectrum the variation in intensity is larger and less smoothly varying.

X-ray scattering

Synchrotron radiation based X-ray diffraction and scattering methods have found many uses in the study of structural developments of materials when exposed to external perturbations like pressure, chemical or mechanical changes. The well known method of powder X-ray diffraction (PXRD) or also Wide Angle X-ray Scattering (WAXS) hardly requires any introduction since it is a standard tool in many laboratories and we refer here to the many existing text books that describe the possibilities and limitations of this technique.[†]

Small Angle X-ray Scattering is a less well known method and there is unfortunately a shortage of accessible text books that go beyond specialist applications.[25,26] A diagram of the experimental lay-out is given in Fig. 3. The sample-detector distance for the SAXS experiment is in general several meters in order to allow the scattered radiation to diverge out of the parasitic scatter cone around the direct beam.

In a combined time-resolved experiment one normally utilises a position sensitive detector but in the cases that a higher resolution WAXS pattern is required one can utilise a diffractometer set-up in which the spatial resolution is determined by the slit settings in front of the point detector. Due to

[†]In this text, just as with the majority of practitioners of combined techniques, the terminology XRD and WAXS will be used interchangeably. For the purists this might be unpleasant but the rationale is that many samples change during an experiment from amorphous to crystalline and thus from scattering to diffraction.

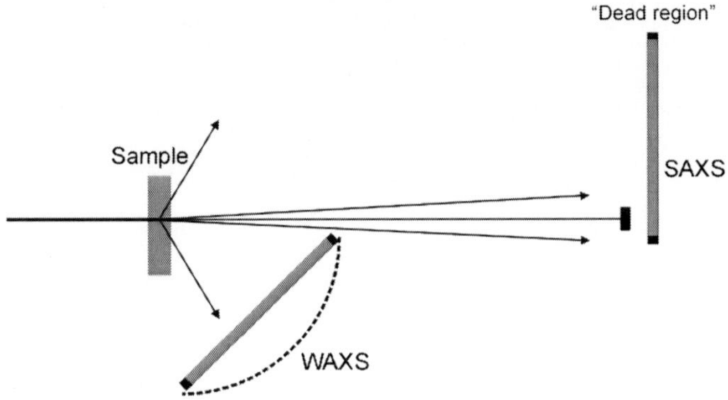

Fig. 3 Schematic lay-out of a combined SAXS-WAXS experiment. The figure is not to scale. The sample-detector distance for the SAXS detector is in general several meters. For the WAXS detector this is dependent on the detector size and the required spatial resolution and the constraints imposed by the sample environment but in general between 0.2–0.5 meters. For time resolved experiments one generally utilises a position sensitive WAXS detector but in some cases it is possible to utilise a fast scanning diffractometer set-up. The trajectory of such a diffractometer based detector is indicated by the dotted line.

the finite time that is required for radially scanning this detector the experimental time-resolution will be reduced.

The SAXS scattering contrast derives from electron density variations over length scales where the contributions of the individual atoms are indistinguishable and one only observes the spatially averaged electron density differences. Therefore it is intrinsically a low resolution technique. In this text we will assume that we are dealing with samples that scatter isotropically, which is the case when the scattering entities in the sample are randomly oriented.

The scattered intensity is given by

$$I(q) = \int \rho(r) e^{-iqr}\, dr = S(q)\, |F(q)|^2 \tag{1}$$

and is the Fourier transform of the electron density distribution $\rho(r)$, whilst the scattering vector q is defined by

$$q = \frac{4\pi \sin\theta}{\lambda} \tag{2}$$

This intensity can often be factored out into two components, one of which describes the spatial distribution of the scattering entities in space, the structure factor $S(q)$, and the other one describing the shape of the scattering entities. The latter is called the form factor $F(q)$. However, this is only true in the case that the particles are identical in shape and size. Until recently this was a rare situation and only encountered for biological samples. Recently SAXS has been recognized as providing supplementary information to protein crystallography especially where the proteins or protein complexes have to be studied in a near native environment.[27] Although with improved synthesis methods it has become more common in materials science to encounter monodisperse entities in a matrix or solvent[28] However, it is more likely that the particles will be polydisperse. This can be tackled in several ways[29–31] but in general one assumes a certain size distribution and fits the scattered intensity by using the width of the size

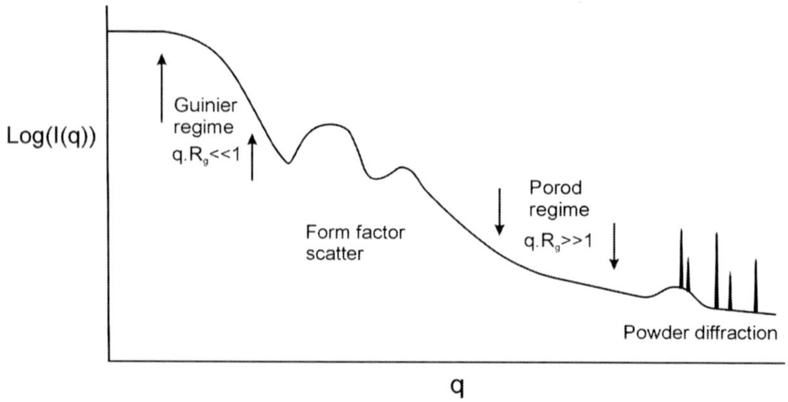

Fig. 4 A stylised diffraction curve obtained in a combined SAXS-WAXS experiment. The different scattering vector regimes and the information that can be obtained from these are further discussed in the text. The requirements on the scattering vector are indicated in the figure and are dependent on the size of the scattering entities. An indication on the dimensions of the scattering objects is the radius of gyration (R_g).

distribution as an adjustable parameter. Regarding the structure factor term, $S(q)$, one can consider this to be similar to the structure factor that is used in diffraction studies of liquid and amorphous materials but with a scattering vector scale which is shifted to lower angles.

A stylised scattering curve, in which we assume the interparticle scattering to be negligible, *i.e.* a diluted sample, is shown in Fig. 4.

In the scattering curve we can loosely define some regions that contain information on different aspects of the scattering entities. Close to the direct beam one can obtain information about the particle size by the so called Guinier approximation

$$I(q) \sim e^{-q^2 R_g/3} \tag{3}$$

which is valid in the case that the polydispersity is low and that one satisfies the condition that

$$qR_g \ll 1 \tag{4}$$

In this the parameter R_g is the radius of gyration which can be interpreted as an average particle radius. However, one should still exercise some caution when applying this formula since polydispersity and asymmetric shapes of the particles can give erroneous results.

At larger scattering angles the Porod formalism describes the scattered intensity. A general approximation is given by

$$\underset{q\to\infty}{I(q)} = K_1 + \frac{K_2 e^{-\sigma^2 q^2}}{q^4} \tag{5}$$

In the case of a smooth sharp surface one finds $\sigma = 0$ otherwise this parameter is related to the extent of the electron density gradient around the scattering particle. When the surface shows roughness, the exponent in the denominator can be affected and become smaller than 4. The importance of the Porod approximation is that it contains information on the interface of the particle with the matrix.

For most commonly occurring particle shapes the scattered intensity due to this shape is available in an analytical expression for the form factor.[32] Quite often the fitting of the data with respect to a form factor function can generate more accurate particle parameters than by using for instance the Guinier approximation.

The internal structure of a particle can sometimes be determined *via* XRD. However, since the particles detectable by SAXS are small, one should keep in mind that the diffraction peaks can be broadened or even unobservable. In such cases the information obtained by XAS can be invaluable since the long range order that is required to generate diffraction intensity is not required to obtain useful information regarding the local environment of probe atoms in small crystallites or amorphous particles.

Intensity in the SAXS regime is not only generated by well defined particles but can also result from non-particulate scattering. In this case it is not possible to retrieve shape information. However, if we take for instance the case where phase separation occurs in a sample then the scattering in the Porod regime, which is caused by the interface between the two phases, can still be approached *via* the Porod formalism. A rather interesting parameter is the total amount of scattering, unfortunately named the invariant Q. This name is unfortunate since in an evolving system Q is a parameter that can change. In a two phase system the invariant is given by

$$Q = \int_0^\infty I(q) q^2 dq = \langle n_e \rangle^2 \phi_1 \phi_2 \tag{6}$$

in which ϕ_i represents the volume fraction of phase i and n_e is the electron density difference between the phases. It is easy to see that in a two phase system where $\phi_1 = 1 - \phi_2$ this renders a quadratic relation. Generalisations to more phase systems can also be made[33] but are in practice less useful for the interpretation of data. For an evolving system the invariant is a parameter that can be very accurately determined due to the fact that the whole detector surface is used for its determination and which has a link with the changes in volume fractions.

When the data quality is sufficiently high one can sometimes obtain information about the total internal surface to volume ratio *via* the relation

$$\frac{K_2}{Q} = \frac{1}{\pi} \frac{S}{V} \tag{7}$$

However, this is not always feasible since the parameter K_2 is derived from the wider angle scattering regime where the scattered intensity falls off rapidly as $I(q) \sim q^{-4}$ and therefore it is difficult to obtain the statistical accuracy to derive the S/V ratio without a rather larger error bar.

Another type of morphology that can be studied with SAXS is the case when a system exhibits fractal properties. In this case one finds over a certain range a power law scattering behavior $I(q) \sim q^{-x}$. The parameter x is correlated with the fractal dimension.[34,35] In some cases the fractal dimension can be linked to certain types of growth processes. However, care should be taken before analyzing data in a fractal context since there are conditions which the scattering data has to satisfy and there are several other morphologies that also produce power law scattering in a certain scattering vector range. Such morphologies can for instance be lamellar, cylinders or even critical scattering when one approaches a phase transition. In order to be allowed to assume a fractal structure the power law scattering behaviour has to extend over several decades in both q-range as well as scattered intensity.

The first dedicated set-up for combined SAXS and WAXS was constructed in the late 1990's at the SRS, Daresbury Laboratory. This example has since then been followed at many synchrotron radiation sources worldewide where, especially at third generation sources, the X-ray beam collimation was so far improved that the application of both techniques has become nearly trivial.

Possibilities and limitations of scattering

In X-ray scattering experiments the low angle resolution depends on which SAXS beam line one utilises. However, even with the best collimated beam lines the useable real space size range is limited to about 200 nm in the transmission geometry.[36–38] At present the achievable time resolution is hardly dictated by the photon flux but more by the sample environments which are required to perturb the samples controlled and homogeneously over the dimensions of the X-ray beam size. X-ray experiments down to a millisecond are possible but the sample environments generally impose limits of around 0.1 sec/frame. Specifically in combination with XAS experiments the time-resolution is dominated by the amount of time it takes for the spectroscopy data to be collected. The fastest energy scans can be done in 5–10 seconds.

For very fast diffraction experiments one could choose a scheme in which one uses a 'white beam' and instead of measuring angle dispersive one measures the diffraction energy dispersive (EDXD).[39] In principle this could easily be combined with energy dispersive XAS methods as well. However, this would be at the expense of exposing the samples to very high instantaneous radiation doses which, especially in non-crystalline materials, might have less desirable effects.[40–42]

A limitation with respect to the application of combined experiments in scattering and spectroscopy is that practically the scattering data has to be obtained at photon energies between 5–15 keV. If the energy is too high the scattering patterns are spatially compressed and the low angle part of the scattering pattern will disappear into the parasitic scattering cone which is intrinsic to every X-ray beamline and the angular extent of which is not energy dependent. In principle one can counteract this to some degree by increasing the sample-detector distance but there are obviously limitations on what is practically feasible.[43] With too low an energy the experimental difficulties increase due to sample and air absorption. This can lead to very low signal rates and data contaminated by multiple scattering.

The sample detector distance and the photon energy ultimately determine the low angle resolution which is achievable. In many cases it is most beneficial to be able to observe as large a q-range as possible instead of optimising the low angle resolution. If both the Guinier as well as the Porod range can be observed in a single experiment it is possible to follow for instance particle growth processes in great detail.[44,45]

When considering XRD data collection the photon energy also has to be taken into consideration. With higher energies the diffraction pattern is compressed and depending on the combination of the spatial resolution of the detector and the intrinsic beam size one could lose diffraction resolution.

X-ray spectroscopy

X-ray absorption spectroscopy (XAS) is a powerful technique for providing element specific insight into the local structure and electronic state of an absorber. As such it can be used to probe crystalline or amorphous materials where the element of interest can be as low as 0.5 weight %. The information regarding the local state of the absorber manifests itself in the fine structure of the X-ray absorption spectrum and hence the technique is often referred to as X-ray absorption fine structure (XAFS). The XAFS spectra can be divided into two parts: the X-ray absorption near edge structure (XANES), which concentrates on the region around an absorption edge (*ca.* 5–100 eV above the Edge energy) and extended X-ray fine structure (EXAFS), which is concerned with the analysis of data recorded some 100–1000 eV above. Both regions are sensitive to changes in a specific element's oxidation/coordination state and in some cases also the absorbing element spin state.

$$I(t) = I_0 e^{-\mu(E)x} \tag{8}$$

The X-ray absorption of a sample is calculated according to equation 8. The absorption is determined by measuring the incoming (I_0) and transmitted beam intensities (I_t) and is a function of the sample thickness x and the linear absorption coefficient $\mu(E)$ as a function of the photon energy. When the energy of the incoming photons excites an electron from a 'core level' to a vacant excited state or to the continuum, a sharp rise in the absorption intensity appears. The fundamental process in XAFS spectroscopy therefore involves a core 1s (K-edge), 2s, 2p (L-edges) electron excitation. In XANES this excitation occurs to an empty state and, therefore the technique probes the low-lying unoccupied/partially occupied states of the electronic structure of the system. When the electron possesses significant kinetic energy it can escape into the continuum so as to probe the local geometry (~ 6 Å) around the absorber as EXAFS. Due to the excited electron possessing a high mean free path at comparatively low energies the analysis of XANES data is complicated and is more often used as a fingerprint technique. However detailed analysis of the XANES data can be performed using a multiple scattering formalism or else use electronic structure models such as density-functional theory to calculate the unoccupied density of states (DOS) and/or cluster simulations.[46] A stylised XAFS curve is shown in Fig. 5.

EXAFS on the other hand is better established at providing more detailed insight into the local structure around an absorber atom. EXAFS occurs when the photoelectron with a kinetic energy, viewed as a spherical wave, scatters off neighboring atoms (*e.g.*, electron density), producing a

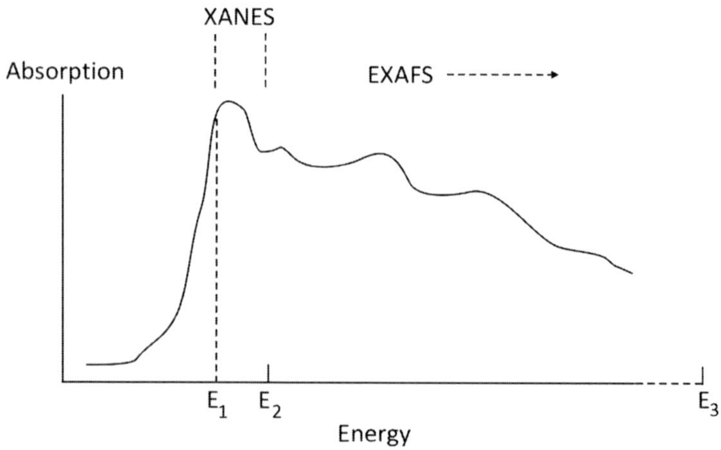

Fig. 5 A stylised XAFS curve. The approximate extension of the XANES as well as the EXAFS region is indicated. E_1 is the absorption edge energy, E_2 is the approximate extent of the XANES region ($E_2 = E_1 + \approx 100\,\text{eV}$) and E3 the maximum relevant extension of the EXAFS range ($E_2 = E_1 + \approx 700$ eV).

backscattered wave which interferes with the original outgoing wave leading to an interference pattern as a result of total construction (both waves are in phase) or total destruction (both waves are out of phase). Since the path lengths of the waves defines the final phase, the distance between the absorber and scatterer determines the interference pattern. This interference pattern thus modulates the absorption coefficient but can be distinguished from the elemental X-ray absorption coefficient. The term $\chi(E)$ is normally referred to as the EXAFS function, from which local geometric data are extracted. Using the simplest single scattering approximations, the contributions of all interference patterns (sine waves) scattered from all neighboring atoms can be approximated by equation 9:

$$\chi = \sum_{shells} \frac{n_{\bar{\chi}} S_0^2 f_{\bar{\chi}}(k) e^{-2k^2\sigma^2}}{kr^2} \sin(2kr + \alpha_{M\bar{\chi}}(k)) \tag{9}$$

The amplitude component contains information on the coordination number (n), the static and thermal disorder and since $f_x(k)$ represents the backscattering amplitude, it is possible to infer the nature of the back-scattering neighbours, all other things being considered. The S_0^2 component is known as the amplitude reduction factor caused by absorber shake-off due to relaxation processes after photoionisation. The components to the right hand side of the expression ($2kr + \alpha_{MX}(k)$) represent the sine components of the interference pattern and comprise the absorber-scatterer distance as well as a phase factor which takes into account changes in the kinetic energy of the photoelectron as it approaches/leaves the nucleus of the atomic species from whence it scatters. Analysis of the data is normally performed using least squares fitting of a proposed model to weighted isolated $\chi(E)$ function or else the Fourier Transformed $\chi(E)$ data to extract information on the local structure in terms of bond distance, coordination number, disorder, and Fermi (Edge) energy.[47]

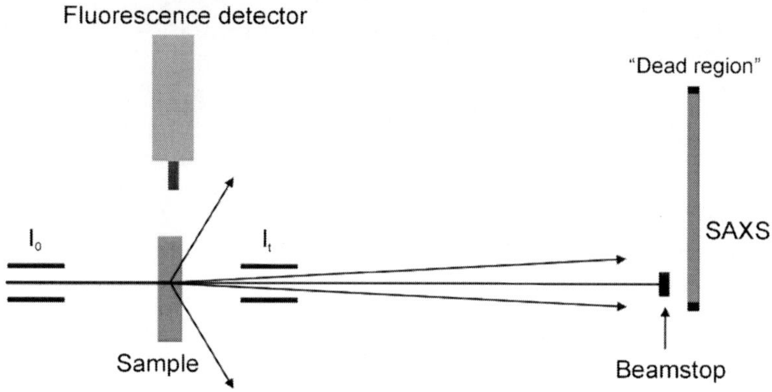

Fig. 6 Schematic aerial lay-out of a combined scattering-XAFS experiment in the energy scanning mode. For clarity the WAXS detector has been omitted. The primary beam intensity is in general measured with a conventional ion-chamber (I_0). There are several ways in which the XAFS spectrum can be determined. Either one utilises a second ion-chamber for the transmitted beam (I_t) or one measures the fluorescent intensity (indicated). One can also obtain XAFS data by the electron yield technique although this is normally reserved for measuring at low gas pressures due to the short mean free path of the electron in ambient pressures. An alternative approach, which experimentally is slightly more complicated, is to measure the transmitted intensity with a photo diode mounted in the beamstop which is required for the SAXS experiments.

X-ray absorption spectroscopy is normally measured in two ways, *i.e.*, transmission, using a combination of pre and post sample ion chambers and fluorescence/emission using a combination of pre ion chamber and multi-element fluorescence detector (Fig. 6). Spectra can normally be obtained with minute to second time resolution using standard scanning or quick-scanning (QEXAFS) modes, which scan across and above an absorption edge by moving a monochromator incrementally and fast.

Alternatively, by illuminating the sample using a range of X-ray wavelengths in what is known as energy dispersive mode (EDE) an entire XAS spectra acquisition can be obtained in ms or faster. Spectra are then acquired using detectors possessing high linearity and an ability to discriminate the signal spatially (*i.e.* such as a diode array detector or a CCD camera). To date EDE has been successfully employed to study highly dynamic processes occurring on supported catalysts (*i.e.* the behaviour of three-way catalyst components under component gas atmospheres, propane dehydrogenation *etc.*) normally in conjunction with non-SR based spectroscopic techniques (IR, UV-Vis, Raman, *etc.*)[48,49]

Recent developments in X-ray absorption spectroscopy

There are many X-ray absorption beamlines operational at present so as to render the technique almost routine albeit for a limited number of research groups. However, there are a number of limitations of the technique that often make it unsuitable for measuring certain samples. For example if two or more types of the same absorber atom but with a different valence are present in a sample, the resultant X-ray absorption spectrum shape will comprise a linear combination of the individual sites. Furthermore for heavy elements a shortening of the core hole lifetime leads to an energy broadening that dampens the features in the XANES spectra and the EXAFS oscillations. In both cases detailed interrogation of the local structure becomes more complicated.

A new development, which has become feasible with the use of newly developed high energy resolution detectors, (Fig. 7) is that it is nowadays possible to measure the X-ray absorption spectrum focusing upon a particular X-ray emission decay channel or line so as to be able to circumvent the mixed site/broadening problem.[46] For example, conventional fluorescence measurements for which very large core whole widths can be found ($\sim$5.3 eV for Pt K-edge) measure the $K_{\alpha 1}$ signal(s) originating from the decay of a 2p electron to fill the 1s core hole. By focussing a high resolution detector on the $2p_{3/2}$-3d emission channel and measuring the $L_{\alpha 1}$ signal (or for that matter any of the L emission lines), the resultant lifetime broadenings of 2 eV or less can be realised, resulting in high energy resolution fluorescence detected (HERFD) XAFS spectra.[50]

For multiple sites the local site for an element in a particular oxidation state can be examined by measuring the 1s-3p fluorescence peaks as a function of the incoming X-ray energy since these exhibit an energy dependency on oxidation state ($\sim\Delta$2–3 eV/charge).[51] Likewise it is possible to perform spin-selective X-ray absorption measurements so as to be able to characterise systems that carry a local moment on the transition-metal ion.[46] Alternatively the type of near neighbours can be probed by non resonant X-ray emission spectroscopy by measuring the valence-to-core XES, *i.e.*, the $K\beta''$ peak (often from a ligand 2s orbital) although since the transition is forbidden the signal is normally very weak so as to render detailed information on the local structure difficult to obtain.

Alternatively instead of collecting the emission signal, the inelastic scatter or X-ray Raman signal can be collected either at resonant or under non-resonant conditions. Again this inelastic signal is discriminated *via* the use of a high resolution energy detector. Crucially this makes it possible to measure soft-X-ray absorption edges (such as the C *K*-edge at 290 eV, or the O *K*-edge at 535 eV) with hard X-rays *i.e.* this approach circumvents the need for vacuum conditions to probe the local structure of soft edge

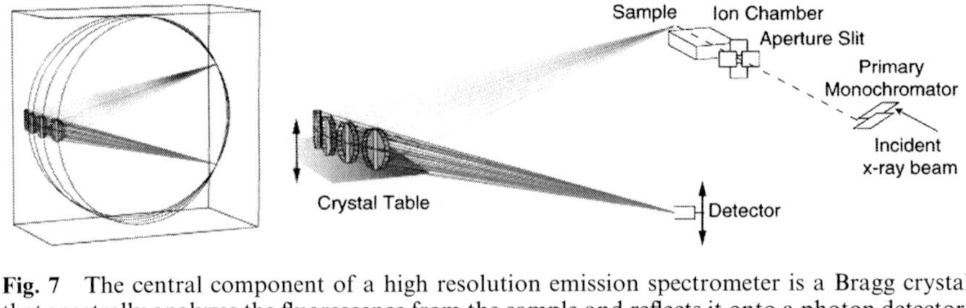

Fig. 7 The central component of a high resolution emission spectrometer is a Bragg crystal that spectrally analyzes the fluorescence from the sample and reflects it onto a photon detector. This particular example utilises a spherically bent Johann type crystal in a one-to-one focusing Rowland geometry in connection with a solid state detector. The solid angle of collection can be increased by increasing the number of analyzer crystals, all aligned to intersect at the two focal points.[52]

elements.[46] Clearly these techniques then offer great potential for characterising materials under *in situ* reaction conditions.

However, the number of beamlines offering this rather specialised infrastructure is still limited even though in the future undoubtedly new experimental stations will be constructed alongside the existing more conventional XAFS beamlines.

Possibilities and limitations of spectroscopy

Notwithstanding the general limitations of obtaining X-ray absorption data from synchrotrons in general (*i.e.* the requirement of a high spectral brilliance X-ray source and the possibility that X-rays could induce sample damage), perhaps the biggest challenges for the technique are energy range availability, time resolution and data handling.

Energy range availability. Many synchrotron radiation sources worldwide work with an optimum brightness around the 8–15 keV mark. Furthermore, the photon availability with the standard Si(111) double crystal monochromator arrangement limits the usable X-ray range to somewhere between 4–30 keV, restricting the elements that can be studied; at low energies photon availability is further compromised by strong absorption by the beamline components whilst at the other extreme, the low tilt angle of the monochromator reduces energy resolution. The switch, where possible, to Si(220) or Si(311) monochromators alleviates the angle problem although the resolution is then increasingly compromised by lifetime broadening. As such there is little real possibility of extending the energy range on standard XAFS beamlines.

Time resolution. Current XAFS time resolutions are limited to the second and in extreme cases sub-second for scanning experiments. This is, however, perfectly sufficient for measuring solid state transformation purposes. For measuring changes in molecular/atomic bonding the energy dispersive mode is required to collect data in the ms or even μs time domain – the main reason for using EDE mode for collecting so quickly is not due to photon statistics, but that scanning requires mechanical movement that introduces additional noise, affecting data quality.[53] Time resolutions (ns) can be obtained using laser pump/X-ray probe experiments although these require the use of specialist beamlines and apparatus.

In many cases time resolution is improved by virtue of brute force in that the brightness at the sample is increased so as to improve the signal:noise statistics. However, ultimately there is a limit to how hard a sample can be interrogated since a large number of photons can have serious consequences on the sample stability leading to both predictable and unexpected changes in composition.[40–42] This is however manageable through careful control of the measurement protocol.

Data handling. The processing of XAFS data is far from trivial and remains a bottleneck in the attempt to extract useful information from a series of data. Chief among the reasons for this problem has been software availability and good data treatment protocols. In recent years a number of freely available, intuitive software packages have been designed and developed for such purposes; these have also been accompanied by detailed literature on how to use the programmes and to treat the data in a step-by-step manner. This has gone a long way towards lowering the activation barrier of data analysis and in particular has reduced the propensity for erroneous analysis. Treatment of multiple data sets still lags behind that which has been achieved for XRD – *i.e.* parametric refinements and screening of the data sets by cluster analysis. However programmes do exist for performing batch data processing and extraction on a large number of data sets.[54,55]

Technique combinations experimental issues

The synergy from technique combinations

When discussing technique combinations one should distinguish between the combination of independently obtained data sets and the situation when several techniques are applied simultaneously on a single sample. From a data quality point of view the former situation is indisputably the best and we will discuss some examples of this later on in the text. Experimental conditions in separate experiments can be optimised for the different techniques in order to generate optimum quality data sets. Therefore there is no good reason to perform techniques simultaneously unless one is dealing with systems that evolve in time or when spatially scanning the sample with microfocus beams. In the first case the loss of data quality, which is inevitable when combing experiments, might be compensated by the synergy of having an accurate time correlation between the data sets. In the second case one has the advantage that one knows that the data is taken from the same sample area and that one eliminates artefacts that could be introduced by sample inhomogeneity or environment.

When discussing time-resolved experiments we will not take into consideration the experimental methods which utilise energy dispersive methods to collect the XAS data. This method is in principle suitable for very fast data collection but the radiation load on the sample is such that for experiments in which the temporal evolution as determined by the physical/chemical process under observation is on the order of seconds or longer, the sample integrity or the chemical process under study can be compromised due to radiation damage/interference.[40,41,56,57] Therefore we will limit ourselves to the quick, energy scanning methods.

The highest time resolution that technically can be achieved at present is obtained by using a cam-driven monochromator. Thus an EXAFS scan can be repeated with a framing rate of 40 Hz.[58] The overhead required to obtain the diffraction/scattering data, which has to be done at fixed photon energy, is on the same order of magnitude. Therefore the fastest overall repetition rate that can be achieved will be around 20 Hz. However, with such fast rates one is unlikely to obtain a high data quality and maybe the range of the energy scan will be limited to the XANES region. It is therefore up to the experimenter to decide if the statistical data quality is sufficient to be able to extract from the data the parameters in which they are interested.

Obviously one should only carry out such fast experiments if the time evolution of the sample is on the same scale. If one deals with slower systems it is better to reduce the data collection speed.

The photon energy for the XAS experiments is determined by the elemental absorption edges. As pointed out before, for diffraction data it is beneficial if the photon energy is not too high. Therefore the degree of difficulty increases when the XAS experiments require high photon energies to be used and it might be possible that realistically the experiment is not feasible due to the mechanical limitations imposed by the beam line monochromator.

Experimental lay-out

The earliest combinations from XAS with XRD were performed in an energy dispersive mode. In the earliest version a time-resolution of 30 seconds/data cycle was achievable. At present the energy dispersed scheme is used mainly in very fast (*i.e.* subsecond) experiments.[59]

A schematic lay-out for a combined scattering-diffraction experiment using the QEXAFS mode, where the photon energy is rapidly scanned, is shown in Fig. 8. There are several ways in which combined experiments can be implemented. The primary beam intensity is in general measured with an ion-chamber even though the parasitic scatter that this device intrinsically will generate is somewhat detrimental for SAXS measurements. In a XAS/XRD experiment the transmitted intensity can be measured by either a second ion chamber or a fluorescence detector and the XRD detector, or diffractometer can have free line of sight to the sample. This situation is different in SAXS experiments where the second ion chamber would make the SAXS signal unusable. Therefore this ion chamber has to be translated

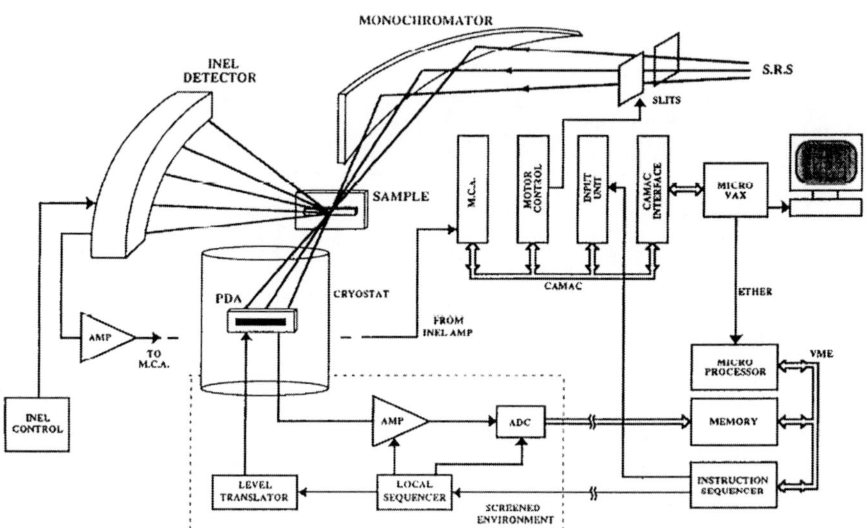

Fig. 8 Experimental lay-out for the earliest version of combined WAXS/XRD experiments implemented at the SRS Daresbury using non-monochromatic radiation. The white beam is energy dispersed by using a bent Si monochromator. After passing the sample the dispersed beam (range 700 eV) was collected by a linear position sensitive detector. The position on the detector is related to the photon energy. Monochromatic radiation (bandwidth 30 eV) required for the XRD experiments was obtained by reducing the slit opening before the monochromator. (Reproduced by permission).

out of the way or one can choose to implement a photo diode in the beam stop from the SAXS camera.

Obviously for the lower photon energies there could be appreciable air absorption if the distance between the sample and the WAXS detector is too large. In that case one can consider constructing a vacuum- or helium-chamber between the sample and the detector. When using SAXS this is in all cases required.

Data collection strategy

QEXAFS. An experiment in which the photon energy is scanned is not ideal for the collection of X-ray scattering data. Therefore the XAS and scattering experiments are carried out quasi-simultaneously. The scattering data is collected below the absorption edge before the XAS scan is triggered – see Fig. 9. Once the XAS scan is finished the monochromator is moved back to its starting position and kept there for several seconds in order to allow the next frame of scattering data to be collected. The duration of the scan depends on if one wants to collect only XANES or also EXAFS, but also on the required data quality.

In theory there is the possibility to not only collect X-ray scattering data below the absorption edge but also at some discrete points during the energy scan. The scattering contrast changes in this energy range considerably and this effect could be used to determine the positions of the metal atoms in the sample.[60,61] However, at present even in static situations it is still difficult to obtain quantitative results with these experiments and it can be predicted

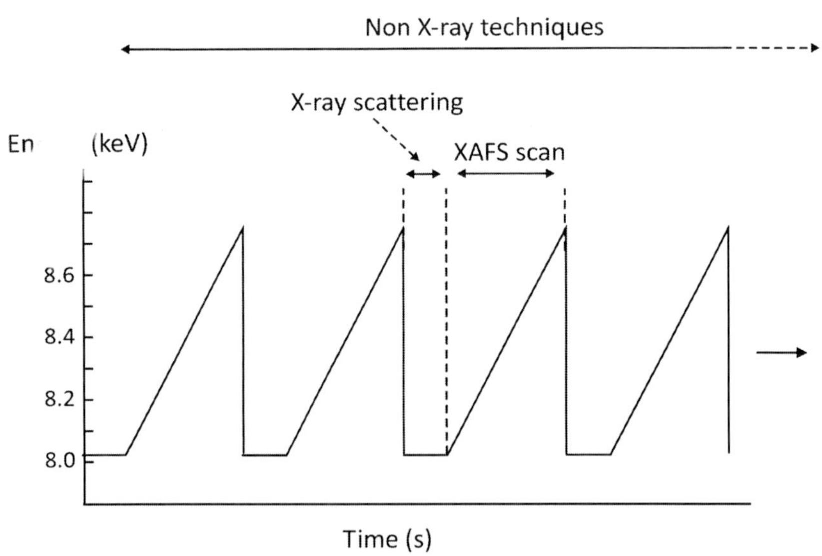

Fig. 9 A schematic diagram of time *vs.* photon energy in a combined QEXAFS-Scattering experiment. The vertical scale is dependent upon the element under investigation. On specialised beamlines it is feasible to carry out the XAFS scan in under a second. More usual is a scan time of up to 1 minute. The X-ray scattering data can be obtained in a few seconds. Non-X-ray based techniques can be carried out independent of the X-ray energy to which the sample is exposed.

that it will be several years of further development before this very interesting method can be used in a time-resolved way.

In practice the data acquisition system of the beamline will be the 'master' system. All other techniques that one would like to interface to be used simultaneously will have to do this in a 'master-slave' configuration in which the digital systems used for instance Raman or Drifts will be synchronized *via* pulses from the master system.

Sample environments and auxiliary techniques

Sample environments. A major component of the experimental chain is the environment in which the sample is kept and with which it is able to perturb the sample in a controlled and uniform way. In static structure determination using XAS it is quite often desirable to perform the experiments at cryogenic temperatures in order to reduce radiation damage and minimise the detrimental effect of temperature factor on the disorder of the system. For time-resolved experiments this is obviously not an option since the time scale of the experiments would become somewhat long.

The requirements which are important for cells which have to be used in the more complicated technique combinations are:

- Access for the incoming beam
- Exit windows for the XAFS detection
- Exit window for the diffracted/scattered X-rays
- Access for other probe beams
- Thermal and pressure control

The requirements on the exit windows for the XAFS detection are simple in the case of transmission XAFS. The materials and dimensions can be the same as the entrance window and the lay-out is straightforward. For fluorescence detection the exit window has to be placed at right angles with respect to the direct beam and also has to be larger in order to allow a reasonable solid angle of fluorescent photons to be detected. With electron yield EXAFS one has to place the sample in the same environment as the sample.

For SAXS experiments the window requirements are very similar to those of transmission XAFS but this situation is radically different for WAXS where a large angle opening is required. This can become a problem when high temperatures or high pressures are required.

The simplest parameter to control is the temperature and in the more sophisticated applications one does not only control this but one is also able to obtain thermodynamic data at the same time by using Differential Scanning Calorimetry (DSC).[62]

More complicated are sample environments suitable for catalysis[63] and electro-chemistry.[6] For such combined experiments the requirements on the sample environment can at time be rather complicated. If for instance one would like to perform catalysis experiments under realistic conditions, *i.e.* under flow and pressures in excess of 20 Bar, with a Ti functionalised catalyst one is faced with the problem that the windows should be both thin as well as be able to withstand the pressure. In transmission XAS this is still feasible, by choosing small window diameters, but once one incorporates

XRD the engineering problems become larger due to the requirement to have a large angle opening for the diffracted X-rays and thus a large X-ray transparent window. Sometimes one has to accept that some experiments are not feasible. However, with some ingenuity many complicated experiments are possible. An example of this is a cell in which it is possible to perform depth resolved electron yield EXAFS experiments in which the reaction gas also has the function of detection medium.[64] By applying a voltage between the sample and anode above the sample it is possible to obtain sufficient amplification to be able to detect the photo electrons energy resolved. The energy of the photons contains the depth information.

Auxiliary techniques. Although we mainly discuss the combination of X-ray based techniques in this manuscript there are strong tendencies at present to incorporate other experimental probes in the experiments as well. Therefore we briefly mention some of the other techniques which can be combined with time-resolved XAS measurements.

In catalysis it has become accepted practice that when studying the performance of a catalyst this has to be carried out in conditions that mimic the real operating conditions closely, *i.e.* operando conditions. In many university laboratories systems in operando conditions are studied with a variety of techniques before they are brought to the synchrotron laboratories to be studied with X-ray techniques. Among these techniques one can count Raman, FTIR spectroscopy and UV-Vis correlation of the results from these techniques with the X-ray data has led to a deeper understanding of the changes a material undergoes. A whole body of work exists for example detailing the benefits of the synchronous combination of Diffuse Reflectance Infra Red Fourier Transform Spectroscopy (DRIFTS) with an X-ray probe (EDE and high energy X-ray diffraction) for studying supported catalysts under oscillatory gas compositions (1–2 Hz); the setup has for example been able to reveal the dynamism of catalyst particle oxidation state, shape and alloying extent as a result of interaction/reaction of the gas atmospheres with the particles at the surface.[48,65–67] Combinations with Raman and X-ray techniques are more numerous having been successfully employed to study self-assembly and crystallisation of porous metal substituted aluminophosphates and non-porous bismuth molybdate materials, phase evolution and both gas-phase and liquid phase catalysis. On a number of occasions UV-Vis has been employed together with Raman spectroscopy mainly because it utilises fibre optic technology allowing for data collection in diffuse reflectance mode and is therefore easily integrated into the measurement setup. Although technically more challenging, all studies demonstrated clearly that the obtaining of congruent data with such multiple technique setups leads to a more definitive understanding of what is happening in a dynamic process. This has been used for example to reveal a darkening of a catalyst surface due to coke build up on supported Mo catalysts, and that the anaerobic oxidation of methanol over MoO_3 where simultaneous changes in Raman and XRD suggested that one of the three oxygens of the MoO_3 lattice was more labile than the others.[49,68] Interestingly a few studies abound in which XAFS and UV-Vis techniques have been employed to study reactions in the liquid phase. These include for example, the study under stopped flow conditions of the two step oxidative addition reaction of iodobenzene to $[(PPh_3)_2Pd(dba)]$); the two steps involving a loss of the dba ligand (seen by UV-Vis), and concomitant addition of phenyl iodide (seen by XANES and EXAFS).[69] Additionally, combined QEXAFS/Raman and UV-Vis were combined to study Ce^{4+} reduction during ethanol decomposition.[70]

A comprehensive review of the benefits of all these technique combinations is beyond the scope of this manuscript and in the following text we will concentrate on the combinations of XAFS with X-ray scattering.

Technique combinations examples

For the historical perspective we mention here that most of the early work in combining XAS with XRD was driven by catalysis and ceramics related research. Already in 1993 the experimental possibilities had developed to such a degree that the different steps in the thermal induced conversion process of a ZnO-zeolite B mixed composite to cordierite could be followed.[71] These experiments were carried out with sample temperatures of around a 1000 °C and with a time-resolution of 30 seconds/frame. The XRD data was collected with a curved INEL detector which had a somewhat limited angular resolution and count rate capacity but with it still was feasible to discern the different phases that were formed during the isothermal heat treatment. It was found that the zeolite initially amorphised before forming an intermediate phase of 'stuffed quartz'. Ultimately the material transformed into crystalline cordierite. Similar 2 stage behaviour has been observed for the crystallisation of the ordered spinel phases $CoAl_2O_4$ and $ZnAl_2O_4$ from Co^{2+} and Zn^{2+} exchanged zeolites A, B and X (Fig. 10).[72–75]

Similar methods have been used in the study of the self assembly processes by which both dense phase and microporous materials form.[76] Often these studies have revealed detailed and sometimes surprising insight into

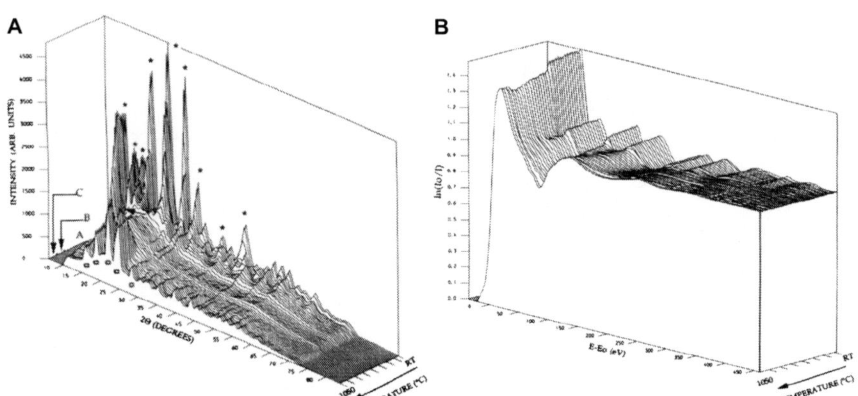

Fig. 10 Time-resolved XRD (panel A) and QEXAFS (panel B) data showing the thermally induced formation of a crystalline cordierite phase from Mg-exchanged zeolite B. This transformation takes place *via* an intermediate amorphous phase as is clearly visible in the XRD data.

the mechanisms by which materials crystallise from solution. The advantage of combining time resolved XAFS and XRD is perhaps best realised when the concentration of the species probed by XAFS is small (*i.e.* at substitution levels). In that way one can consider the behaviour of the 'XAFS atom' in some way to reflect the behaviour of all crystallographically similar atoms (*i.e.* Co^{2+} ions easily replace a small percentage of Al^{3+} ions in crystalline microporous aluminophosphates) during self-assembly, thereby affording atomic insight into self-assembly and crystallisation. A number of studies for example have revealed that during CoAlPO-5 formation, the local structure of Co^{2+} species changes dramatically even before crystalline material is detected. XAFS data showed that the initial reactant gel contains free octahedral $Co[(H_2O)_6]^{2+}$ species that eventually attach to aluminophosphate species before beginning to change coordination state from octahedral to tetrahedral as crystallisation begins.[76–78] Ideally armed with this information, it should have been possible to exact a more rational design process for microporous material synthesis. That this has not resulted may be due in part due to the differences in behaviour of the different substituents and the resultant crystalline microporous structure.[3,79,80]

A mini review of the development of the experimental possibilities and the scientific questions that were driving this catalysis oriented research at the different synchrotron laboratories in the first decade since the introduction of combined XAS-XRD can be found in a 1999 manuscript By G. Sankar and J.M. Thomas.[81] Further, and more recent reviews, oriented towards catalysis research have also been published.[4,82]

The ideas upon which these developments were initiated in this period are still relevant today. Obviously there has been progress due to improved instrumentation possibilities. More accurate monochromators and more efficient detectors have become available, and allow for faster experiments, and the modern generation of position sensitive detectors is better capable of taking advantage of the high angular resolution XRD experiments which are feasible with a synchrotron beam line. Also the control of environmental parameters like pressure and temperature in the sample environments, have improved considerably.

Example static EXAFS/WAXS combination

The combination of EXAFS with diffraction data can play a role in refining structural information when the information of neutron or X-ray diffraction/scattering on its own is limited. This can be the case when the amount of material, exhibiting long range order is small, like in very small crystallites, or if the information content of the scattering data is low like it is in the case of amorphous of liquid materials. An example is the case of amorphous $Se_{0.9}S_{0.1}$ alloys which were created by mechanical alloying.[83] In this case the essential parameters like average atomic distances can be obtained by X-ray scattering with which the radial distribution functions can be determined. However, the accuracy can be increased by using EXAFS data and corefining these together with the scattering data by using Reverse Monte Carlo (RMC) methods[84] and for instance the presence of small clusters can also be established. In the past this method was not very widespread since it required access to scarce neutron scattering beamlines[85] but over the last decade much progress has been made using this methodology since the more abundant synchrotron radiation sources have made the photon energy range >40 keV more accessible.[86] This high X-ray energy is required to access scattering vectors over a very wide range in order to be able to obtain an accurate Fourier Transform of the experimental data. An advantage of the total scattering method is that it is also possible to perform these experiments on lighter elements where XAS data is more difficult to obtain.[87]

Another reason to apply several techniques on the same sample might be for samples that are small, scarce and inhomogeneous. A nice example of this can be found in corrosion studies on ancient steel support beams from Amiens cathedral.[7] It will be obvious that understanding the deterioration mechanism as well as limiting the amount of available sample is crucial for the long term structural integrity of this structure which dates from 1947. See Fig. 11. In corrosion studies it is relevant to determine which different crystalline phases do exist in the samples and also how they are spatially distributed. The different crystalline and amorphous phases are only occurring in small amounts and therefore microprobes are required. However, due to the small size and texture of different inclusions diffraction data in general does not provide sufficient information and different spectroscopy techniques, among which is XANES, are required to obtain a full understanding of the materials.

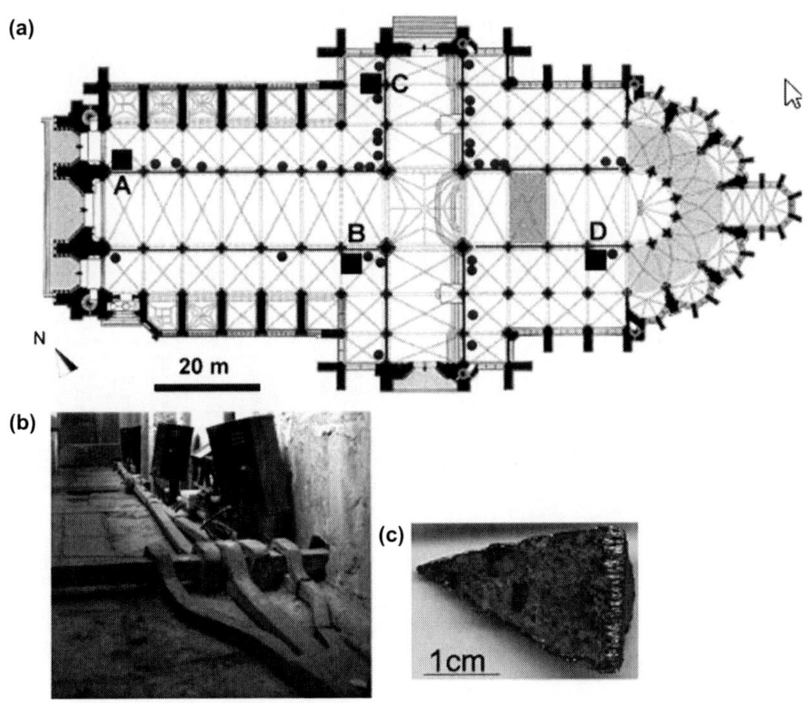

Fig. 11 An example of samples where the application of combined techniques could be interesting even if there is no aspect of time-resolution involved. Panel (a) shows the area in Amiens cathedral from where the samples were obtained. Panel (b) the larger structure of one of these collection places and panel (c) the actual sample. The scarcity of the sample and the probability that it is not homogeneous make it paramount that the data is collected from the same sample position.

Example time resolved EXAFS/XRD

Combined XAFS/XRD has already been used to understand a number of dynamic processes. In most cases one of the techniques proves more important than the other when trying to answer a specific question about a process under study. This will of course depend both upon the question to be answered and the material. On the odd occasion however the congruency of the information obtained from the technique combination allows for highly detailed mechanistic insight into self assembly as has been obtained during low temperature hydrothermal crystallisation of γ-Bi_2MoO_6; a bimetal oxide polymorph known to possess catalytic oxygen/nitrogen insertion capacity.[88] The structure can be considered to comprise of $[Bi_2O_2]^{2+}$ layers alternated with layers of corner-sharing MoO_6 octahedra (Aurivillius-type structure). Initial studies by Beale *et al.* demonstrated that the pre-cursor gel contained tetrahedral Mo^{6+} species, ascribable to $[MoO_4]^{2-}$ species known to be stable under the pH conditions of the reactant gel (pH ~ 7).[89] Further heat treatment leads to the appearance of reflections corresponding to the γ-Bi_2MoO_6 phase albeit with a larger unit cell than that typically found for the final crystalline phase. Further heating of the gel saw a concomitant increase in Mo^{6+} coordination and reduction in the unit cell leading the authors to conclude the initial large cell γ-Bi_2MoO_6 phase most likely comprised $Bi_2O_2^{2+}$ layers with $[MoO_4]^{2-}$ species intercalated in between. As heating progressed, they argued that the simultaneous shift of the entire diffraction pattern towards higher 2θ values with the increase in Mo^{6+} coordination, was caused by $Bi_2O_2^{2+}$ layers condensing together and forcing the Mo^{6+} to increase in coordination number as Bi-O-Mo links form. A subsequent follow up study by Kongmark *et al.* (Fig. 12) using a combination of XANES, high resolution XRD and Raman observed a

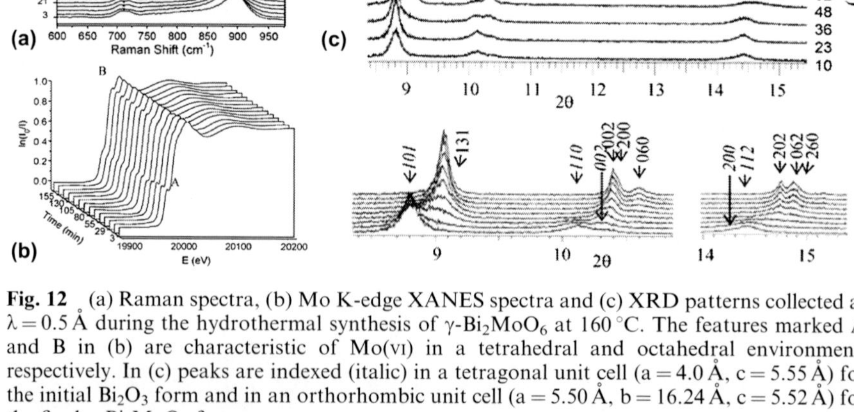

Fig. 12 (a) Raman spectra, (b) Mo K-edge XANES spectra and (c) XRD patterns collected at $\lambda = 0.5$ Å during the hydrothermal synthesis of γ-Bi_2MoO_6 at 160 °C. The features marked A and B in (b) are characteristic of Mo(VI) in a tetrahedral and octahedral environment, respectively. In (c) peaks are indexed (italic) in a tetragonal unit cell (a = 4.0 Å, c = 5.55 Å) for the initial Bi_2O_3 form and in an orthorhombic unit cell (a = 5.50 Å, b = 16.24 Å, c = 5.52 Å) for the final γ-Bi_2MoO_6 form.

similar two-step process.[90] The main differences between these studies was the use of Raman spectroscopy demonstrating a gradual Mo–O bond lengthening during formation of the γ-Bi_2MoO_6 phase, but perhaps more importantly this study also demonstrated that the high resolution XRD data allowed them to propose an initial bismuth oxide phase (β-Bi_2O_3-like) which they proposed acted like a 'skeleton' for the growth of the γ-Bi_2MoO_6 phase. Diffraction being a global technique could not however determine whether the β-Bi_2O_3 phase contained within its midst, substituted tetrahedral MoO_4 species. One could imagine however that the large mismatch in average bond distance for tetrahedral Bi^{3+} species (~ 2.20 Å) with that of Mo^{6+} species (~ 1.80 Å) would alter the unit cell if all or many of the Mo^{6+} species were incorporated into the β-Bi_2O_3 pre-phase and that this could be detectable. Since this was not seen, this suggests that Mo^{6+} species were not incorporated in the β-Bi_2O_3 phase or at least not to any large extent. Thus crystallisation of γ-Bi_2MoO_6 appears to involve a reaction between the solid β-Bi_2O_3 species and solution based MoO_4^{2-} species. Since the structure of the γ-Bi_2MoO_6 phase can be rationalised in terms of $Bi_2O_2^{2+}$ layers, linked by MoO_6 species and considering the rapidity of the formation process, a solid-state condensation process seems the most logical conclusion for the formation of the γ-Bi_2MoO_6 phase. Further experiments are required however in order to verify this.

Example EXAFS/SAXS

In the manufacturing process of many metal containing nano composite materials from a solvent there is a stage in which the nano particles are created and stabilised with the aid of surfactants. These particles can then at a later stage be used to form a scaffold for the formation of nanostructured materials. Due to the small size of the nanoparticles they will not be able to generate sufficient diffraction intensity which could be used to determine the internal structure and moreover would not give information about the metallic coordination state. At a larger scale it is relevant to determine how the surfactant molecules interact with the nano-particles and the solvent. The particle-surfactant size range is suitable to be studied with SAXS and therefore it makes sense the results of these two techniques.[8] This method has successfully been used to follow the synthesis of 'necklaces' of Co nanoparticles which were formed and stabilised by AOT reversed cylindrical micelles in a heptane solution. With the SAXS data it was shown that the cylinders were arranged in a hexagonal packing. See Fig. 13.

A similar approach was followed in the manufacture of Co-Ni nanoparticles which were chemically reduced in a surfactant solution.[91] The spherical particles formed spontaneously and had an average radius of 1.68 ± 0.15 nm. Further growth was stopped by the attachment of surfactant molecules to the formed particles. The EXAFS data showed that the coordination number of the metallic atoms in the first shell was lower than in bulk metals. This confirms that a disordered but mainly metallic entity is formed with the atoms on the outside having a higher oxidation degree than those on the inside. It should be remarked here that the 'error bar' on the radius derived from the SAXS data is not necessarily a real error but more likely represents the spread in particle diameter.

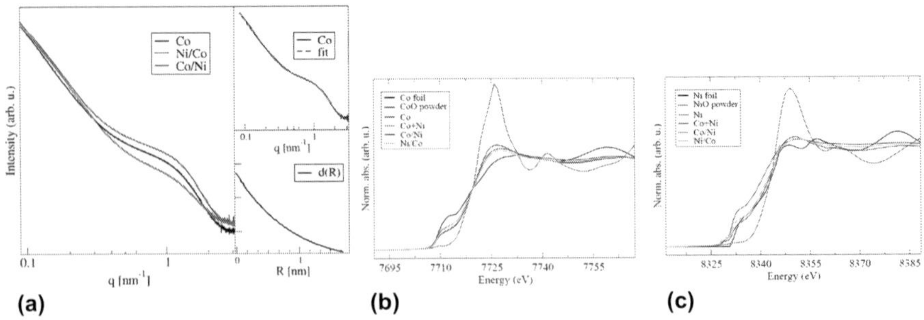

Fig. 13 SAXS (a) and XAFS (b,c) patterns of composite Co-Ni nanoparticles. The EXAFS data is obtained on both the Co (b) as well as the Ni-edge (c).

These examples show that the combination of low resolution techniques (SAXS) with a method to determine the internal structure of nanoparticles can give important insights in the production process of these nanostructured materials. With conventional XRD one would not be able to determine the internal structure of the particles due to either their small size and the resulting Scherrer broadening or due to the amorphous/poorly ordered state in which most particles at the early stages of growth will exist.

Example EXAFS/SAXS/WAXS

The combination of SAXS and WAXS in a single experiment became common place about 20 years ago when the first dedicated SAXS/WAXS station was constructed.[17] This development was feasible because of the good collimation and low beam divergence that synchrotron beamlines allow. This is required since in order to cover a reasonable wide scattering vector range one requires one detector positioned close to the sample and a second detector at several meters distance from the sample so that the scattered X-rays can diverge out of the parasitic scatter cone. Since beamline optics can only produce a single focal plane such a set-up is only feasible when the beam divergence is small enough so that the depth of field around the focal spot is long enough.

It is somewhat surprising that even though XRD/XAS and SAXS/WAXS experiments were carried out over 20 years ago that the logical next step, the combination of SAXS/WAXS/XAFS had to wait so long before it was implemented.[19] A schematic diagram of the first experimental set-up with which SAXS/WAXS/XAFS data could be collected is shown in Fig. 14. The first improvised experiments with this equipment were performed around 2006. Essentially the setup was constructed by adapting an existing XAFS/XRD setup by adding a SAXS camera ($\sim$ 1.8 m) and was used to study the formation of the microporous aluminophosphate, because the setup can deliver atomic-to-nano size and amorphous-to-crystalline phase information.

The equipment as sketched in Fig. 14 has since been upgraded with respect to WAXS and SAXS detection but the principle remains essentially the same. One of the first experiments carried out with this set-up concerned the crystallisation process of a Zn substituted aluminophosphate.[5] The thermally induced crystallisation reaction of this material was followed on-line

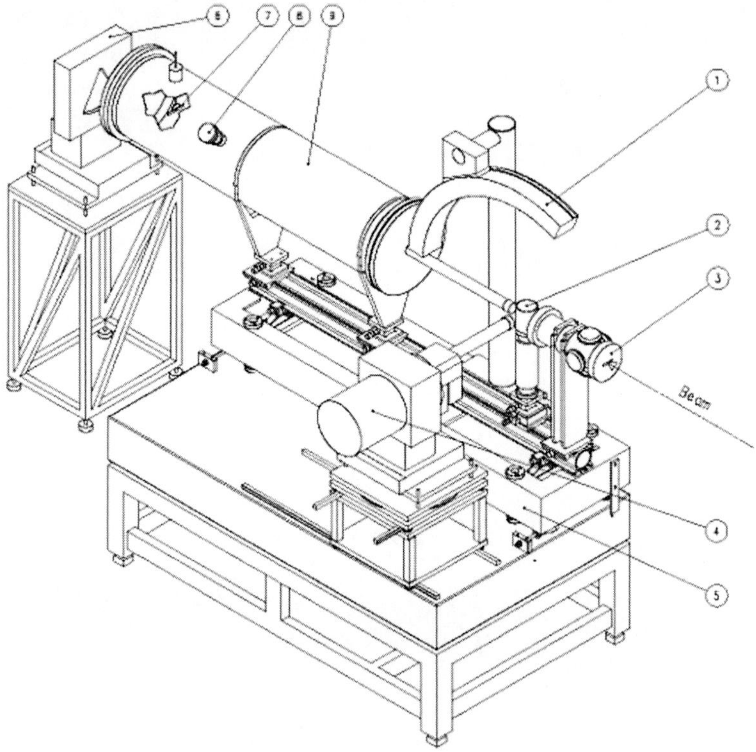

Fig. 14 The first experimental set-up with which it was possible to collect quasi-simultaneously SAXS/WAXS/XAFS data. 1-position sensitive WAXS detector, 2-sample position and sample environment, 3-ion chamber for incoming beam intensity monitoring, 4-fluoresence detector, 5-optical bench, 6-SAXS detector, 7-beam stop in which a photo diode for transmitted beam intensity monitoring is integrated, 9-evacuated SAXS flight tube.

with a time-resolution of around 1 minute/frame. The XAFS data showed that the initial gel mainly contained tetrahedral coordinated Zn^{2+} and WAXS data showed essentially an absence of any long range crystalline order. During the reaction the XANES data showed that the coordination of the Zn^{2+} changed to a local environment which was compatible with an incorporation of the Zn^{2+} ions into a Zn-AlPO-34 structure. The reaction/crystallisation kinetics could be followed by using the integrated intensity of the diffraction peaks in the WAXS pattern. By using the Avrami formalism it was made plausible that the growth mechanism was reaction limited and that the reaction products were three dimensional and not layered two dimensional features as was previously expected. See Fig. 15.

The SAXS data here plays a crucial role in the elucidation of the full process. At first sight the data does not appear to be very interesting. There is only a small change in domain size observable and both the SAXS data, *via* a determination of the R_g, as well as the WAXS data, *via* a size determination using the Scherrer peak broadening, render a domain size of around 12.3 nm. However, if one determines the invariant, Q, one notices that this is within the experimental error also a constant value. See Fig. 16.

If we now remind ourselves of the fact that the invariant correlates the volume fraction of two phases and the electron density difference between

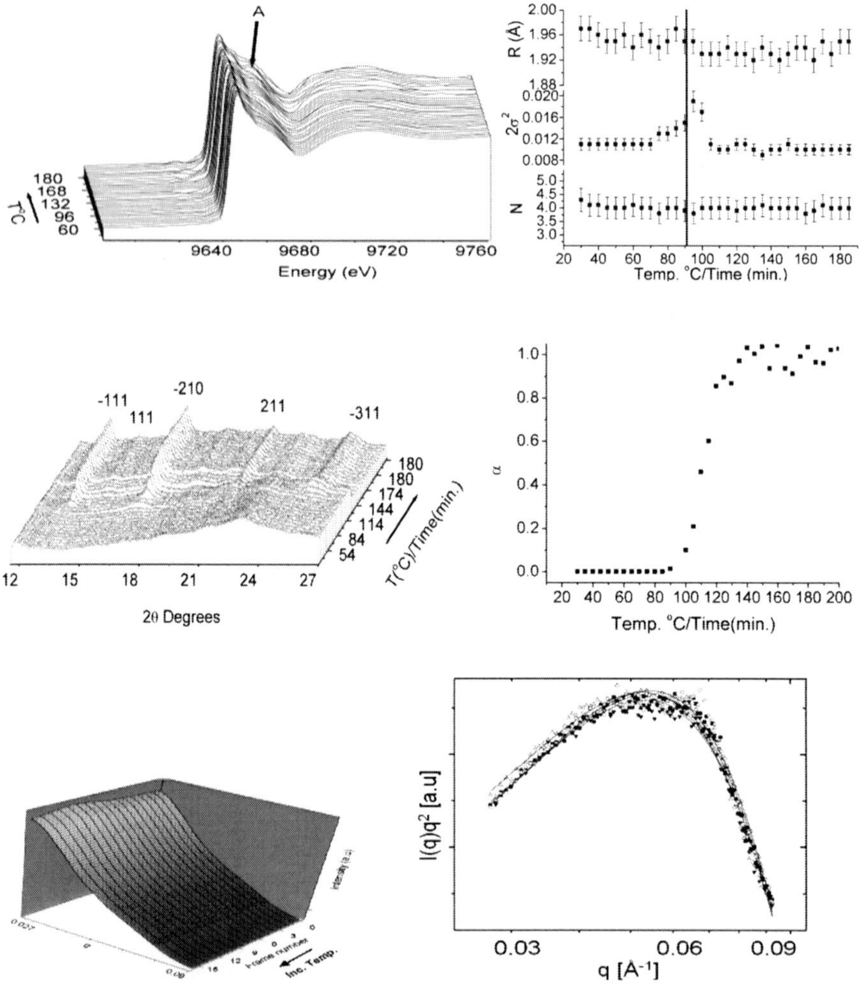

Fig. 15 Results of a combined SAXS/WAXS/XAFS experiment on Zn-AlPO-34 structure formation. The top row shows the time resolved EXAFS data and the derived EXAFS parameters. The middle row contains the XRD data and the degree of conversion derived from the XRD data. The bottom row shows the SAXS data and the derived SAXS data upon which one can base the estimate for the domains size. The left hand panels show the experimental data whilst the right hand panels show the parameters derived from these data sets. For a more complete description one is referred to the text.

these two phases (see equation [6]) then we see a remarkable result. The WAXS and XAFS data show that a crystallisation process takes place and thus that there are structural changes appearing within the sample. With a constant invariant this means that there are two scenarios feasible. In the first the electron density difference between the two phases is exactly compensated by the difference in the crystalline volume fraction as one phase grows at the expense of the other one. The second scenario is that in the very early stages a phase separation has taken place and only one of the phases crystallises whilst the other remains amorphous. The average electron density of the two different phases remains the same. When comparing these

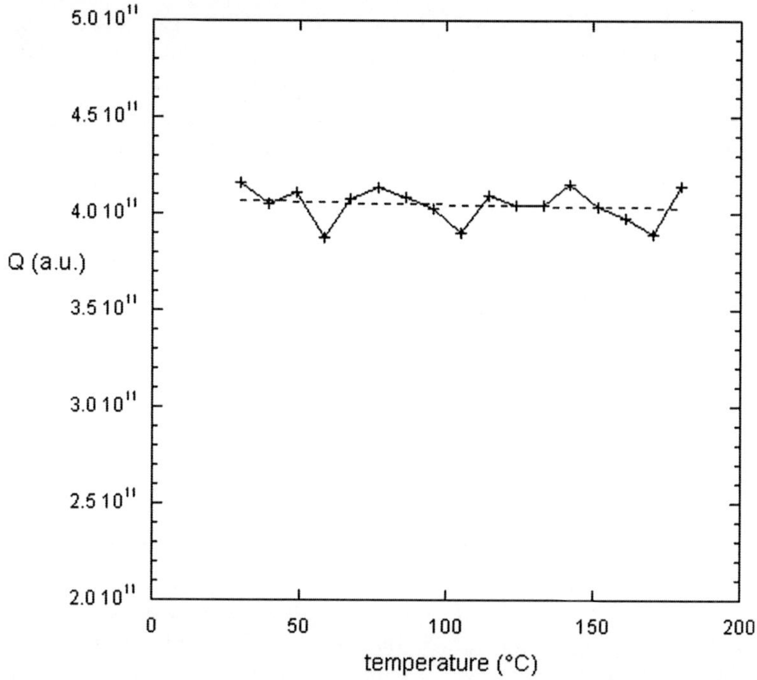

Fig. 16 The development of the SAXS invariant $Q=\int_0^\infty I(q)q^2dq=\langle n_e\rangle^2\phi_1\phi_2$ as function of temperature. The absence of change means that either the change in electron density difference between the two phases is exactly compensated by the change in relative volume fractions of the two different domains or that the electron density difference between the two phases is constant and only a single phase crystallises.

two scenarios and a judicious application of Ockham's razor one can only conclude that the latter scenario is the most plausible.

The above example is one where the power of combining X-ray spectroscopy and scattering is rather evident. One obtains the information about the changes in the direct environment of the Zn atoms whilst in the same experiment one obtains information on structural changes, and the kinetics of these changes, in the size range from inter-atomic distances to potentially several hundreds of nanometers, *i.e.* 2–3 decades in size range.

Perhaps a major drawback though of the coupled XAFS techniques is that it is not possible to study the 'soft' elements that comprise the majority of the porous material framework structure *i.e.* Al^{3+} and Si^{4+}. This is mainly due to their possessing of an absorption edge energy which is strongly attenuated by air and therefore this requires that such edges are measured in vacuum. Thus in the main, XAFS studies have been restricted to examining 'probe' elements (*e.g.* Zn or Co) substituted in relatively small quantities into the framework. The assumption here is that the substituent accurately mimics the major components and to an extent this is true, although it has also been shown that Co^{2+} ions in CoAPO-5 retard the crystallization process in comparison to unsubstituted AlPO-5.

Obviously not all experiments will render a positive result. The experiment for which the first improvised SAXS/WAXS/XAFS set-up was constructed was the thermally induced crystallisation of a Cr doped glass.[44,45]

This experiment rendered a negative result due to the too low concentration of the probe atomic species on which the XAFS could be performed and the XAFS data could only be obtained on quenched samples in a non-time-resolved mode. However, even in this case the combination of the time-resolved scattering data sets with the static XAFS data provided a clearer picture of the crystallisation process than would have been possible without using the complementarities of the data sets.

Conclusions

The combination of different experimental X-ray based techniques in a single time-resolved experiment became possible due to the evolution in X-ray beam quality which has been based upon the improvements in synchrotron radiation sources and the optical elements required harnessing these X-ray beams with respect to energy resolution, beam size, intensity and focal depth. The introduction of non-X-ray based techniques in the same experiment has the advantage of bridging the information that can be obtained in a home laboratory with the results only obtainable in a synchrotron radiation laboratory. At present a wealth of opportunities exists which regretfully has not been utilised as much as is possible.

However, as in all technique combinations one should carefully weigh the advantages of combining experiments against the (slight) loss in data quality that is intrinsic to these technique combinations. In general the synergy of simultaneously obtained data sets is only useful in the case of faster time-resolved experiments or in special cases where there is only a limited amount of sample material available.

The lack of a sufficient number of beamlines suitable to combine these experiments with respect to the X-ray hardware as well as the experimental infrastructure required for the sample handling is one of the main limiting factors for further development.

Notwithstanding this limitation a considerable number of research areas hav[illegible]atly benefited from a combination of XAFS and X-ray scattering. In [illegible]ial science one has seen beautiful results in for instance the field of ca[illegible]s, where one can now carry out experiments in operando conditions, and [illegible] study of the structural development of nano-technological materials. Also the usefulness has been proven in the case where the sample material is inhomogeneous and scarce like for instance in archaeological artefacts.

The opportunity to elucidate structural developments from the smallest length scale possible, where crystallographic order is still absent, to the nanometer length scales relevant for the structural integrity or proper functioning of for instance catalysts, in one single experiment provides the researcher with a very powerful method which should not be lacking from the experimental tool box.

Acknowledgements

We would like to acknowledge the input from our colleagues Gophinatan Sankar, Andy Dent and Claudio Ferrero in this work. A special mention is deserved for Neville Greaves who, while working in SRS Daresbury

Laboratory in the early 1990's, was the tireless champion of the application of technique combinations. WB would like to thank his colleagues at ALS Lawrence Berkeley Laboratory for enabling the sabbatical period in which this review was conceived.

References

1 H. J. Robota, R. W. Broach, J. W. A. Sachtler and S. A. Bradley, *Ultramicroscopy*, 1987, **22**(1–4), 149–157.
2 W. Bras, S. Nikitenko, G. Portale, A. Beale, A. van der Eerden and D. Detollenaere, in: *Xiv International Conference on Small-Angle Scattering*, G. Ungar, (Ed.) 2010; Vol. 247.
3 A. M. Beale, M. G. O'Brien, M. Kasunic, A. Golobic, M. Sanchez-Sanchez, A. J. W. Lobo, D. W. Lewis, D. S. Wragg, S. Nikitenko, W. Bras and B. M. Weckhuysen, *Journal of Physical Chemistry C*, 2011, **115**(14), 6331–6340.
4 M. A. Newton and W. van Beek, *Chemical Society Reviews*, 2010, **39**(12), 4845–4863.
5 A. M. Beale, A. M. J. van der Eerden, S. D. M. Jacques, O. Leynaud, M. G. O'Brien, F. Meneau, S. Nikitenko, W. Bras and B. M. Weckhuysen, *Journal of The American Chemical Society*, 2006, **128**(38), 12386–12387.
6 M. G. Dowsett and A. Adriaens, *Analytical Chemistry*, 2006, **78**(10), 3360–3365.
7 J. Monnier, D. Neff, S. Reguer, P. Dillmann, L. Bellot-Gurlet, E. Leroy, E. Foy, L. Legrand and I. Guillot, *Corrosion Science*, 2010, **52**(3), 695–710.
8 A. Longo, G. Portale, W. Bras, F. Giannici, A. M. Ruggirello and V. T. Liveril, *Langmuir*, 2007, **23**, 11482–11487.
9 E. Curti, R. Dahn, F. Farges and M. Vespa, *Geochimica Et Cosmochimica Acta*, 2009, **73**(8), 2283–2298.
10 K. S. Savage, D. K. Bird and P. A. O'Day, *Chemical Geology*, 2005, **215**(1–4), 473–498.
11 J. W. Couves, J. M. Thomas, D. Waller, R. H. Jones, A. J. Dent, G. E. Derbyshire and G. N. Greaves, *Nature*, 1991, **354**(6353), 465–468.
12 A. J. Dent, M. P. Wells, R. C. Farrow, C. A. Ramsdale, G. E. Derbyshire, G. N. Greaves, J. W. Couves and J. M. Thomas, *Review of Scientific Instruments*, 1992, **63**(1), 903–906.
13 E. Dooryhee, C. R. A. Catlow, J. W. Couves, P. J. Maddox, J. M. Thomas, G. N. Greaves, A. T. Steel and R. P. Townsend, *Journal of Physical Chemistry*, 1991, **95**(11), 4514–4521.
14 G. Sankar, P. A. Wright, S. Natarajan, J. M. Thomas, G. N. Greaves, A. J. Dent, B. R. Dobson, C. A. Ramsdale and R. H. Jones, *Journal of Physical Chemistry*, 1993, **97**(38), 9550–9554.
15 R. Frahm, *Nuclear Instruments & Methods In Physics Research Section A-Accelerators Spectrometers Detectors And Associated Equipment*, 1988, **270**(2–3), 578–581.
16 M. Bark, H. G. Zachmann, R. Alamo and L. Mandelkern, *Makromolekulare Chemie-Macromolecular Chem. And Phys.*, 1992, **193**, 2363.
17 W. Bras, G. E. Derbyshire, A. J. Ryan, G. R. Mant, A. Felton, R. A. Lewis, C. J. Hall and G. N. Greaves, *Nuclear Instruments & Methods in Physics Research Section a-Accelerators Spectrometers Detectors and Associated Equipment*, 1993, **326**(3), 587–591.
18 W. Bras and A. J. Ryan, *Advances in Colloid and Interface Science*, 1998, **75**(1), 1–43.

19 S. Nikitenko, A. M. Beale, A. M. J. van der Eerden, S. D. M. Jacques, O. Leynaud, M. G. O'Brien, D. Detollenaere, R. Kaptein, B. M. Weckhuysen and W. Bras, *Journal of Synchrotron Radiation*, 2008, **15**, 632–640.
20 M. W. Tew, H. Emerich and J. A. van Bokhoven, *Journal of Physical Chemistry C*, 2011, **115**(17), 8457–8465.
21 F. Coppari, A. Di Cicco, E. Principi, A. Trapananti, N. Pinto, A. Polian, S. Chagnot and A. Congeduti, *High Pressure Research*, 2010, **30**(1), 28–34.
22 A. I. Frenkel, Q. Wang, N. Marinkovic, J. G. Chen, L. Barrio, R. Si, A. L. Camara, A. M. Estrella, J. A. Rodriguez and J. C. Hanson, *Journal of Physical Chemistry C*, 2011, **115**(36), 17884–17890.
23 G. Margaritondo, *Introduction to Synchrotron Radiation*. Oxford University Press, 1988.
24 P. J. Duke, *Synchrotron Radiation Production and properties*. Oxford Science Publications: 2000.
25 O. Glatter and O. Kratky, *Small Angle X-ray Scattering*. Academic Press: 1982.
26 A. Guinier and G. Fournet, *Small Angle Scattering of X-rays*. John Wiley & Sons: 1955.
27 M. H. J. Koch and W. Bras, *Annual Reports Section "C" (Physical Chemistry)*, 2008, **104**, 35–80.
28 M. Megens, C. M. vanKats, P. Bosecke and W. L. Vos, *Langmuir*, 1997, **13**(23), 6120–6129.
29 G. Beaucage, H. K. Kammler and S. E. Pratsinis, *Journal of Applied Crystallography*, 2004, **37**, 523–535.
30 R. Mittelbach and O. Glatter, *Journal of Applied Crystallography*, 1998, **31**, 600–608.
31 J. Pencer, S. Krueger, C. P. Adams and J. Katsaras, *Journal of Applied Crystallography*, 2006, **39**, 293–303.
32 J. S. Pedersen, *Advances in Colloid and Interface Science*, 1997, **70**, 171–210.
33 J. Goodisman and H. Brumberger, *Journal of Applied Crystallography* 1971, **4**, (Oct1), 347–351.
34 G. Beaucage, *Journal of Applied Crystallography*, 1995, **28**, 717–728.
35 J. Teixeira, *Journal of Applied Crystallography*, 1988, **21**, 781–785.
36 W. Bras, I. P. Dolbnya, D. Detollenaere, R. van Tol, M. Malfois, G. N. Greaves, A. J. Ryan and E. Heeley, *Journal of Applied Crystallography*, 2003, **36**, 791–794.
37 T. Narayanan, O. Diat and P. Bosecke, *Nuclear Instruments & Methods In Physics Research Section A-Accelerators Spectrometers Detectors And Associated Equipment*, 2001, **467**, 1005–1009.
38 H. Tsuruta and T. Irving, *Current Opinion in Structural Biology*, 2008, **18**(5), 601–608.
39 R. Caminiti and V. R. Albertini, *International Reviews in Physical Chemistry*, 1999, **18**(2), 263–299.
40 J. G. Mesu, A. M. Beale, F. M. F. de Groot and B. M. Weckhuysen, *Journal of Physical Chemistry B*, 2006, **110**(35), 17671–17677.
41 V. Martis, S. Nikitenko, S. Sen, G. Sankar, W. van Beek, Y. Filinchuk, I. Snigireva and W. Bras, *Crystal Growth & Design*, 2011, **11**(7), 2858–2865.
42 M. Eichelbaum, K. Rademann, A. Hoell, D. M. Tatchev, W. Weigel, R. Stosser and G. Pacchioni, *Nanotechnology*, 2008, **19**, 13.
43 X. Pan, X. L. Wu, X. Chen, K. Mo, J. Almer, D. R. Haeffner and J. F. Stubbins, *Journal of Nuclear Materials*, 2010, **398**(1–3), 220–226.
44 W. Bras, S. M. Clark, G. N. Greaves, M. Kunz, W. van Beek and V. Radmilovic, *Crystal Growth & Design*, 2009, **9**(3), 1297–1305.

45 W. Bras, G. N. Greaves, M. Oversluizen, S. M. Clark and G. Eeckhaut, *Journal of Non-Crystalline Solids*, 2005, **351**(27–29), 2178–2193.
46 F. de Groot, *Chemical Reviews*, 2001, **101**(6), 1779–1808.
47 D. C. Koningsberger, B. L. Mojet, G. E. van Dorssen and D. E. Ramaker, *Topics in Catalysis*, 2000, **10**(3–4), 143–155.
48 M. A. Newton, D. G. Burnaby, A. J. Dent, S. Diaz-Moreno, J. Evans, S. G. Fiddy, T. Neisius and S. Turin, *Journal of Physical Chemistry B*, 2002, **106**(16), 4214–4222.
49 A. M. Beale, A. M. J. van der Eerden, K. Kervinen, M. A. Newton and B. M. Weckhuysen, *Chemical Communications*, 2005, **24**, 3015–3017.
50 O. V. Safonova, M. Tromp, J. A. van Bokhoven, F. M. F. de Groot, J. Evans and P. Glatzel, *Journal of Physical Chemistry B*, 2006, **110**(33), 16162–16164.
51 P. Glatzel, L. Jacquamet, U. Bergmann, F. M. F. de Groot and S. P. Cramer, *Inorganic Chemistry*, 2002, **41**(12), 3121–3127.
52 P. Glatzel and U. Bergmann, *Coordination Chemistry Reviews*, 2005, **249**(1–2), 65–95.
53 A. J. Dent, *Topics in Catalysis*, 2002, **18**(1–2), 27–35.
54 T. Ressler, *Journal of Synchrotron Radiation*, 1998, **5**, 118–122.
55 B. Ravel and M. Newville, *Journal of Synchrotron Radiation*, 2005, **12**, 537–541.
56 H. D. Barth, E. A. Zimmermann, E. Schaible, S. Y. Tang, T. Alliston and R. O. Ritchie, *Biomaterials*, 2011, **32**(34), 8892–8904.
57 B. M. Weon, J. S. Lee, J. H. Je and K. Fezzaa, *Physical Review E*, 2011, **84**, 3.
58 R. Frahm, M Richwin and D. Lutzenkirchen-Hecht, *Physica Scripta*, 2005, **T115**, 974.
59 O. Mathon, T. Mairs and S. Pascarelli, in: *Sri 2009: The 10th International Conference on Synchrotron Radiation Instrumentation*, R. Garrett, I. Gentle, K. Nugent, S. Wilkins, (Eds.) 2010; Vol. 1234, pp. 117–120.
60 H. Brumberger, D. Hagrman, J. Goodisman and K. D. Finkelstein, *Journal of Applied Crystallography*, 2005, **38**, 324–332.
61 D. Tatchev, *Philosophical Magazine*, 2008, **88**(12), 1751–1772.
62 W. Bras, G. E. Derbyshire, A. Devine, S. M. Clark, J. Cooke, B. E. Komanschek and A. J. Ryan, *Journal of Applied Crystallography*, 1995, **28**, 26–32.
63 J. S. Girardon, A. Y. Khodakov, M. Capron, S. Cristol, C. Dujardin, F. Dhainaut, S. Nikitenko, F. Meneau, W. Bras and E. Payen, *Journal of Synchrotron Radiation*, 2005, **12**, 680–684.
64 A. Vollmer, J. D. Lipp, J. R. I. Lee, G. E. Derbyshire and T. Rayment, *Analytical Chemistry*, 2003, **75**(23), 6571–6575.
65 M. A. Newton, *Chemical Society Reviews*, 2008, **37**(12), 2644–2657.
66 A. J. Dent, J. Evans, S. G. Fiddy, B. Jyoti, M. A. Newton and M. Tromp, *Faraday Discussions*, 2008, **138**, 287–300.
67 M. A. Newton, M. Di Michiel, A. Kubacka and M. Fernandez-Garcia, *Journal of the American Chemical Society,* **132**(13), 4540.
68 M. G. O'Brien, A. M. Beale, S. D. M. Jacques, T. Buslaps, V. Honkimaki and B. M. Weckhuysen, *Journal of Physical Chemistry C*, 2009, **113**(12), 4890–4897.
69 G. Guilera, M. A. Newton, C. Polli, S. Pascarelli, M. Guino and K. K. Hii, *Chemical Communications*, 2006, **41**, 4306–4308.
70 V. Briois, D. Lutzenkirchen-Hecht, F. Villain, E. Fonda, S. Belin, B. Griesebock and R. Frahm, *Journal of Physical Chemistry A*, 2005, **109**(2), 320–329.
71 G. Sankar, P. A. Wright, S. Natarajan, J. M. Thomas, G. N. Greaves, A. J. Dent, B. R. Dobson, C. A. Ramsdale and R. H. Jones, *J. Phys. Chem.*, 1993, **97**(38), 9550.
72 L. M. Colyer, G. N. Greaves, A. J. Dent, S. W. Carr, K. K. Fox and R. H. Jones, *Tracing the Production of Spinel Based Ceramics from the Heat-induced*

Transformations of Zinc and Cobalt Exchanged Zeolite-A Using Combined XRD/XAFS Techniques, 1994, Vol. 84, 387–394.
73 A. M. Beale and G. Sankar, *Chemistry of Materials*, 2006, **18**(2), 263–272.
74 A. M. Beale, S. D. M. Jacques, O. Leynaud and G. Sankar, *Microporous and Mesoporous Materials*, 2007, **106**(1–3), 201–211.
75 W. Schmidt and C. Weidenthaler, *Chemistry of Materials*, 2001, **13**(2), 607–612.
76 G. Sankar, J. M. Thomas, F. Rey and G. N. Greaves, *Journal of the Chemical Society-Chemical Communications*, 1995, **24**, 2549–2550.
77 D. Grandjean, A. M. Beale, A. V. Petukhov and B. M. Weckhuysen, *Journal of the American Chemical Society*, 2005, **127**(41), 14454–14465.
78 K. Simmance, G. Sankar, R. G. Bell, C. Prestipino and W. van Beek, *Physical Chemistry Chemical Physics*, **12**(3), 559–562.
79 A. M. Beale, A. M. J. van der Eerden, D. Grandjean, A. V. Petukhov, A. D. Smith and B. M. Weckhuysen, *Chemical Communications*, 2006, **42**, 4410–4412.
80 M. G. O'Brien, A. M. Beale and B. M. Weckhuysen, *Chemical Society Reviews*, **39**(12), 4767–4782.
81 G. Sankar and J. M. Thomas, *Topics in Catalysis*, 1999, **8**(1–2), 1–21.
82 J. D. Grunwaldt and B. S. Clausen, *Topics in Catalysis*, 2002, **18**(1–2), 37–43.
83 K. D. Machado, D. F. Sanchez and S. F. Brunatto, *Journal of Non-Crystalline Solids*, 2010, **356**(50–51), 2865–2868.
84 R. L. McGreevy, *Journal of Physics-Condensed Matter*, 2001, **13**(46), R877–R913.
85 W. Bras, R. Xu, J. D. Wicks, F. Vanderhorst, M. Oversluizen, R. L. McGreevy and W. Vanderlugt, *Nuclear Instruments & Methods in Physics Research Section a-Accelerators Spectrometers Detectors and Associated Equipment*, 1994, **346**(1–2), 394–398.
86 P. J. Chupas, X. Y. Qiu, J. C. Hanson, P. L. Lee, C. P. Grey and S. J. L. Billinge, *Journal of Applied Crystallography*, 2003, **36**, 1342–1347.
87 S. J. L. Billinge and M. G. Kanatzidis, *Chemical Communications*, 2004, **7**, 749–760.
88 R. K. Grasselli and J. D. Burrington, *Advances in Catalysis*, 1981, **30**, 133–163.
89 A. M. Beale, L. M. Reilly and G. Sankar, *Applied Catalysis a-General*, 2007, **325**(2), 290–295.
90 C. Kongmark, V. Martis, A. Rubbens, C. Pirovano, A. Lofberg, G. Sankar, E. Bordes-Richard, R. N. Vannier and W. Van Beek, *Chemical Communications*, 2009, **32**, 4850–4852.
91 L. Sciortino, F. Giannici, A. Martorana, A. M. Ruggirello, V. T. Liveri, G. Portale, M. P. Casaletto and A. Longo, *Journal of Physical Chemistry C*, 2011, **115**(14), 6360–6366.

Solid state NMR of immobilized catalysts and nanocatalysts

Anna Grünberg, Hergen Breitzke and Gerd Buntkowsky*
DOI: 10.1039/9781849734899-00289

In the last decade solid state NMR techniques have evolved into a major tool for the characterization of immobilized homogeneous catalysts, tethered to a solid support material. Of particular interest are supports consisting of porous materials, which are periodically structured on the mesoscopic length scale, and have very large specific surfaces and pore volumes. Combining one- and two-dimensional ^{13}C-, ^{29}Si- and ^{31}P-solid state MAS-NMR techniques, it is now possible to elucidate details of the binding of the catalyst molecule to the surface on the molecular level. These solid state NMR techniques do not require crystalline samples or special clean well defined surfaces like the conventional surface science techniques, but can work with typical ill-defined real-world systems. This review discusses, after a short introduction, the salient features of these materials and the applied NMR experiments to give the reader a basic knowledge of the systems and the experiments. The rest of the review focuses on recent examples, which demonstrate the power of the method. First the structural properties of the Wilkinson's catalyst, bound to the surface of SBA-3 are analyzed in detail. It is shown that two of the three phosphine ligands of the catalyst are replaced by a linker. Then a short overview about solid state NMR characterization of surface reactions and surface species of metallic nanocatalysts is given. The review is closed with an outlook about very recent developments in solid state NMR methodology. It is reported that fast-MAS, indirect detection and hyperpolarization solid state NMR techniques lead to astonishing sensitivity improvements, which bring the structural characterization of technical catalysts with low specific surface areas into reach.

1 Introduction

Most organic reactions are conducted in isotropic organic solvents and billions of pounds of organic solvents are used each year in the fine chemical and pharmaceutical industries for reaction and product isolation purposes. These huge amounts of solvents cause environmental problems because volatile organic compounds are a major source of industrial pollution. To maintain a clean world for future generations and sustain a long-term economic growth, chemists have developed the concept of green chemistry, which is one of the major challenges of chemistry in the 21st century. Its goal is the development of new, environmentally benign reaction schemes, which comprise effective organic transformations that are fully selective for the desired product, use a minimum of resources and energy and produce minimum amounts of pollutants. Since most organic transformations are controlled by catalysts, optimizing their properties is the major way to achieve this goal. Despite the impressive advances in homogeneous catalysis, heterogeneously catalysed reactions remain more productive.

Institute of Physical Chemistry, Technical University Darmstadt, Petersenstrasse 22, D-64287 Darmstadt, Germany, Fax: +49 6151 16-4347; Tel: +49 6151 16-2607; E-mail: gerd.buntkowsky@chemie.tu-darmstadt.de

Heterogeneous catalysts are used in most large-scale industrial production processes and tons of heavy metal catalytic species are employed every year. A rational design of these systems requires a detailed atomistic understanding of the local surface structures and reactive processes at the solid-liquid and solid-gas interfaces. Establishing this knowledge is still a major challenge of current research in catalysis, not the least because of a lack of appropriate methodology to acquire this information. Thus, the quest is out to develop appropriate methods that help to bridge this current gap.

Organometallic catalysts are widely used in organic synthesis, as for example in pharmaceutical and natural products chemistry, and the synthesis of fine chemicals. Homogenous catalysts offer high activity and good selectivity; however, difficulties in separation, recycling and regeneration of catalysts cause higher costs on an industrial scale and heavy metal pollution. In contrast, heterogeneous catalysts can be easily separated and reused in the continuous flow reactions, which plays important role in industrial processes and in "green chemistry". Consequently, owing to the favourable properties of heterogeneous catalysts, a lot of research has been carried out to develop new supported catalytically active analogues of homogenous catalysts.[1–20] In particular, continuous expansion and improvement of new organic-inorganic hybrid materials led to successful applications in various catalytic reactions.[20–26] In recent years immobilization of homogeneous catalysts on solid supports played a crucial role in combining the advantages of homogeneous and heterogeneous catalytic processes. The successful use of these materials necessitates better understanding of the catalysis involved, and innovations in the areas of materials science, surface science, metal-organic chemistry and physical chemistry.

Many attempts have been made to establish the characteristic properties of a perfect support material.[15] In general, it ought to be stable in multiple cycles and its chemical and physical properties should not negatively influence the catalytic activity. Moreover, it has to be easily separable from the reaction system and harmless to the environment. Among various studies on immobilization of catalytically active molecules, inorganic oxides and polymeric materials are the most common carriers, especially for immobilization of metal catalysts (see Fig. 1). In these so called immobilized or heterogenized catalysts, the organic or organometallic functional groups are tethered on a solid inorganic support by binding to the outer surface of nanoparticles or to the inner surfaces of micro- or mesoporous materials. Depending on the intended application, the functional group is immobilized on the surface by various techniques.[15,27,28]

A wide range of organic polymers have been used as a support for metal catalysts and a great number of investigations has been completed in this field.[29–31] In comparison to organic polymers, however, inorganic oxides like zeolites, silica nanoparticles or silica mesoporous materials provide numerous advantages. For example the structures of such frameworks are chemically and thermally stable under different reaction conditions, and not influenced by swelling processes. [32]

The use of inorganic oxides as support material has expanded continuously since the first report of the encapsulation of metal phtalocyanines inside the zeolite Na-Y was published in 1977.[33–35] Microporous

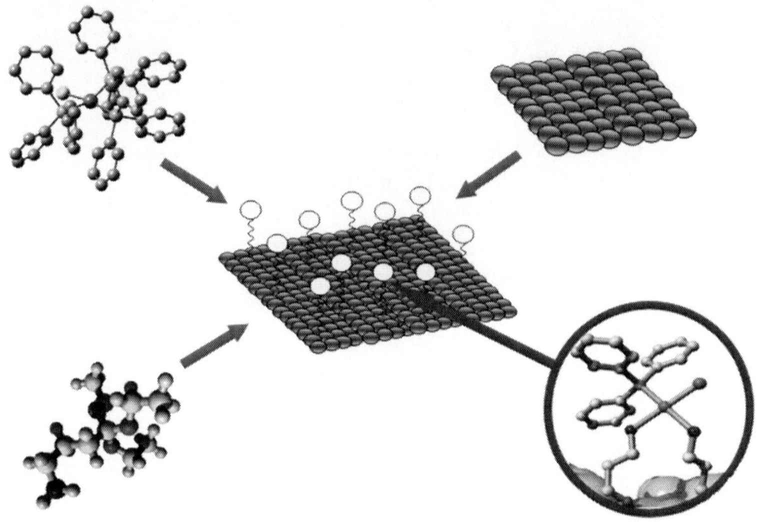

Fig. 1 Immobilization of a homogeneous catalyst on a solid support.

aluminosilicate materials, belonging to the family of molecular sieves, find wide applications due to the unique physical and chemical properties of their framework. These applications include catalytic chemistry, especially in the petrochemical industry, as well as sorption and separation. However, limitations caused by the microporous framework do not allow the employment of this material in the heterogeneous catalysis of big molecules.

The discovery of ordered mesoporous silica materials by scientists from Mobil Corporation in 1992 was a significant breakthrough in the field of porous materials.[36–38] This class of materials, whose most prominent representatives are MCM- and SBA-type materials, offers unique potential for the immobilization of catalysts.[36,37] Their unprecedented properties, such as large pore spaces, ordered tuneable pore sizes and large surface area open up new possibilities for the catalytic conversion of substrates with larger molecular size. The physical properties of their inner surfaces, such as surface acidity, can be chemically modified, affording additional flexibility in catalyst design.[39]

The immobilization of a metal complex on a solid support can be carried out following one of the two main strategies. In the first approach the so-called supported homogeneous catalyst is obtained by substitution of its ligands with functional groups of the organic linker attached to the surface of the solid carrier. Alternatively, the metal complex can be attached directly to the oxide support, which plays the role of a ligand. This approach, referred to as surface organometallic chemistry by Coperet and Basset, differs from the supported homogenous catalysts in that the metal complex loses its molecular character.[40] Several recent studies contributed to this important area (see refs. 41–43 and references therein).

The field of heterogeneous catalysis employing immobilized homogeneous catalysts continues to expand. Several transition metal complexes and their derivatives have been immobilized on the surface of silica materials showing high activity in different catalytic reactions.[7,12–16,20,22–26,44–49] However, researchers investigating supported metal complexes face limitations that do not pertain to such compounds in solution or in crystalline state.[50]

The solid support plays a crucial role in the immobilization of catalytic active species, thus proper structural characterization of its surface is of great importance for understanding the properties of attached molecules. Among various spectroscopic methods,[51] such as IR, X-ray or EXAFS, high resolution solid state NMR techniques, often utilizing magic angle spinning (MAS) and cross-polarization (CP), have evolved into a major spectroscopic tool for the investigation of intermolecular interactions at the interfaces of nano-particles, zeolites or mesoporous silicates. A major reason for the success of NMR techniques is that they do not require clean and well-defined surfaces, like those studied by conventional surface science techniques, but can work with the typical ill-defined real-world systems such as for example technical silica materials. Since NMR is a bulk method, it reflects the whole surface and not just a small area. NMR techniques allow also the study of modified silica materials, prepared for example under low surfactant concentration[52] or functionalized by organic groups.[53] By combining different NMR measurements it is possible to study the adsorption/desorption kinetics of the guests, reveal their translational and rotational dynamics on or near the surface and inside the pores, and probe the binding conformation.[54–67] The NMR methods applicable to the studies of dynamic processes inside the pores are discussed in detail in the textbook by Schmidt-Rohr and Spiess[68] and the recent reviews by Böhmer *et al.*,[69] Vogel[70] and Buntkowsky *et al.*[39]

High-resolution ^{29}Si NMR spectra of solids can be obtained using ^{1}H-^{29}Si CPMAS, as demonstrated by Lippmaa *et al.* in 1981.[71] ^{29}Si NMR spectroscopy, originally applied to study silicate solutions and silicon-organic compounds in the liquid phase, has since evolved into an important tool for the investigation of various classes of crystalline and amorphous solids.[72–76] Klinowski and co-workers provided remarkable new insights into the structure and catalytic activity of diverse aluminosilicates employing ^{1}H, ^{29}Si, ^{27}Al and ^{17}O MAS NMR, as well as an array of two dimensional solid state NMR methods.[77–86]

The ^{29}Si chemical shift is most sensitive to the immediate chemical and structural surrounding of the silicon atoms. Chemically and/or crystallographically inequivalent Si sites can thus be detected by distinct resonances in the spectra. Moreover empirical and theoretical relations have been established between the chemical shift and the type, number and structural arrangement of the nearest and second nearest neighbour atoms of the silicon atoms.[72]

An alternative to silica based immobilized catalysts are metal nanoclusters or nanoparticles (NPs).[87] They have found widespread application as catalysts in various important heterogeneous reactions (see for example refs. 88–90 and references therein). These particles typically assume more or less spherical shapes and contain several hundred atoms. To avoid aggregation, the metal particles can be embedded in organic matrices or decorated with organic ligands. Their interaction with hydrogen has been

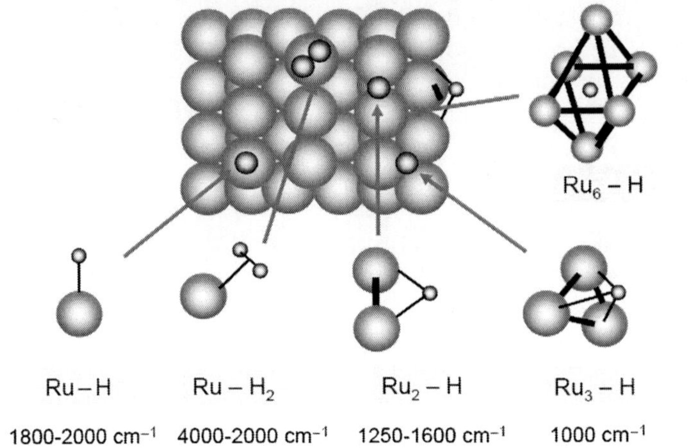

Fig. 2 Hydrogen species on metal surfaces. The frequencies in cm^{-1} refer to hydrogenic vibrations.

studied extensively owing to the technological importance of this substrate. Numerous studies of single-crystal surfaces have significantly contributed to the understanding of hydrogen-metal interactions, as well.[91] During the adsorption process, dihydrogen molecules are firstly physisorbed to the surfaces and in a second step chemisorbed. Generally, this process consists of a dissociation of the dihydrogen molecule. The atoms can then form bonds to one or more surface atoms, and/or diffuse into the metal lattice as illustrated schematically in Fig. 2. In the case of Ru it is now commonly accepted that hydrogen atoms occupy 3-fold coordinated fcc and hcp sites at all coverages.[92–96] However, hydrogen atoms may also be located in two-fold sites forming Ru_2H surface species, or reside "on top", *i.e.* are bound to a single metal atom.[97–99] In the case of palladium, evidence for non-dissociated chemisorbed dihydrogen bound to a single metal atom has been observed.[100]

As compared to model metal surfaces, characterization of surface hydrogen species on nanoparticles is difficult using conventional surface techniques including vibrational spectroscopy. However, we have recently shown that interesting information about surface hydrogen interacting with Ru-nanoparticles can be obtained by variable temperature solid-state ^2H NMR.[101,102] Using this method, one can measure the deuteron quadrupolar coupling constants Q_{cc} and the asymmetry parameters η which define the local electric field gradients (EFGs) at the ^2H sites. The quadrupolar spectra provide information about the local structure, including bond directions and charge distributions of electrons.[103–107] The quadrupolar and chemical shift tensors can be interpreted employing quantum chemical DFT calculations.[108–112] In addition, anisotropic or isotropic motions can be detected which lead to a partial or full averaging of the quadrupolar coupling tensors.

2 Solid state NMR spectroscopy

The characterization of the structure and dynamics of molecules in solid phases is a very important spectroscopic task in instrumental analytical chemistry. In cases where the material is crystalline, this task may be solved by X-ray or neutron diffraction. However, many solids of practical or basic scientific interest do not exhibit the long-range order necessary for X-ray diffraction. Moreover, there are also dynamic aspects of solid state structures, stemming from the mobility of individual molecules or motions of molecular groups or chemical exchange processes. Finally, the dynamic range accessible in the solid is much larger than the dynamic range accessible in liquids or liquid solutions, because systems can be studied at much lower temperatures, since there are no limitations imposed by the solubility of a compound or by the freezing of a solvent.

Nuclear Magnetic Resonance[113,114] (NMR) is a spectroscopic technique which covers a very large range of application areas, due to the multitude of different NMR techniques which are available today. Roughly three principal NMR domains can be distinguished, namely liquid state NMR spectroscopy,[115] spatially resolved NMR techniques[116–118] and solid state NMR spectroscopy.[68,119] While the main application of liquid state NMR spectroscopy refers to the area of chemical and biochemical analysis of liquid or soluble compounds, spatially resolved NMR techniques are primarily employed for medical and technical applications. Solid state NMR spectroscopy is devoted to the chemical analysis of insoluble compounds, to the study of electronic structures in conducting systems, and mainly to the characterization and investigation of structural and dynamic properties of solid systems. A recent overview about the application of solid-state NMR technique to the study of organic inorganic hybride materials is given by Geppi and co-workers.[120]

Proton enhancement,[121] magic angle spinning[122,123] (MAS) and CP-MAS NMR techniques[124,125] have provided the major route to obtain liquid-like spectra from powder samples, but the line widths obtained in these experiments are not as narrow as in high resolution NMR of liquids. Moreover, magnetic dipolar interactions, which are the main nuclear spin interactions between neighbouring atoms in the solid state, are destroyed in the course of the MAS experiment. These dipolar interactions carry the information about the distances of the nuclei. Therefore it is either necessary to recover these interaction by the application of suitable dipolar recoupling techniques[126] like rotational echo double resonance[127,128] (REDOR) or fast flips of the rotation axis,[129] or to do without sample spinning and study single crystals[130] or oriented systems.[131,132]

In the area of structural characterization, solid state NMR is in collaboration and competition with other spectroscopic techniques like X-ray and neutron diffraction. Roughly speaking, the application areas can be divided into the following classes: if structural properties of well-crystallizing systems are desired, X-ray diffraction or to some extent neutron diffraction is the technique of choice, in particular if one is interested in the positions of heavy atoms. If the system is only available as a polycrystalline or amorphous solid, or if one is interested in hydron positions (*i.e.* hydrogen isotopes), or if dynamical processes inside the system are to be analyzed, or if disordered structures in a solid phase are investigated, then a problem is encountered which is well suited for the application of solid state NMR.

As far as the dynamic properties of solid systems are concerned, NMR can cover a very large dynamical range: It is possible to study processes on a time scale of 10^{-12} sec (indirect detection of a reaction kinetics) to processes on a time scale of 10^{+7} sec (slow dynamic processes like conformational changes of a molecule or slow chemical reactions) and observe directly or indirectly molecular structures and their transformations on these time scales.

NMR in the solid state

This section gives a brief summary of the necessary background of solid state NMR experiments. Details about the interactions are given in the appendix. Only for few materials single crystals of suitable size for NMR studies are available. Most organic materials, however, exist only as polycrystalline or amorphous powder samples, where a statistical distribution of molecular and thus tensor orientations exists, causing a corresponding distribution of resonance frequencies. Accordingly, in most solid state NMR experiments non-oriented powder samples are studied, so called powder-samples. In these samples a statistical distribution of molecular orientations is present, which causes broad spectra, due to the orientation dependence of dipolar, quadrupolar and chemical shielding interactions. The NMR signal is the sum of the contributions of all different orientations present in the sample. If all molecular orientations have equal probability (non-oriented powder sample), this is simply the integral over all possible orientations ($t = t_1, t_2, \ldots, t_n$ denotes the number of time-dimensions of the experiment).

$$M_+(t) = \frac{1}{4\pi} \int_0^{2\pi} \int_0^{\pi} M_+(t, \phi, \vartheta) \sin\vartheta \, d\vartheta \, d\phi \tag{1}$$

The NMR spectrum is simply the Fourier transform of the time signal. For some special cases (pure chemical shielding, quadrupolar or heteronuclear dipolar interactions present) an analytic solution of this double integral in terms of elliptic integrals exists. However, in most cases a numerical evaluation of the integral has to be performed (for details see any standard text-book about solid-state NMR).

Magic-Angle-Spinning (MAS) NMR

Probably the most important technique in solid state NMR is the so called Magic Angle sample Spinning or MAS experiment. The basic idea of the MAS experiments is closely related to liquid state NMR. In liquid state NMR motional averaging of the orientation dependent NMR interactions, due to fast isotropic rotational diffusion of the molecule, removes the anisotropic part of the interaction tensors and leads to narrow lines. The same effect can be achieved utilizing the special orientation dependence of the anisotropic interactions. It is found that fast spinning of the sample around an axis, which is tilted $\vartheta_m = 54.7°$ (the magic angle) with respect to the direction of the external field, can remove this orientation dependence for spin 1/2 systems[122] or at least convert it into a manifold of narrow spinning side bands.

The necessary spinning frequency for spin-1/2 necessary for achieving a full line-narrowing of the side bands depends on the system under investigation. If the Hamiltonian of the system is self commuting, which is for example the case for a diluted system of X-nuclei, it is possible to fully remove the anisotropic part by spinning faster than the largest anisotropic interaction. However, already slower spinning frequencies are usually sufficient in this case, because the spectrum splits into narrow spinning side bands and the intensities of the non-central side bands decrease fast with increasing rotation frequency. If, however, the Hamiltonian is not self commuting, for example when both CSA and homonuclear dipolar interactions are present, only very fast sample spinning leads to a reasonable narrowing of the lines. This is particularly a problem for ^{1}H-MAS NMR, due to the strong homonuclear ^{1}H-^{1}H dipolar interaction. However, recently suitable MAS probes have become available to get well-resolved spectra of strongly dipolar coupled protons.

For quadrupolar nuclei ($I > 1/2$) only in the case of small quadrupolar interactions (first-order interactions) the anisotropic powder spectrum is completely converted into a manifold of narrow spinning side bands. For strong quadrupolar interactions, second order quadrupolar interactions cause a broadening of the spinning side-bands, which prohibits true high-resolution. The first successful techniques to overcome this resolution limit where the double rotation[133] (DOR) and dynamic angle spinning[134] (DAS) experiments, which were both introduced in 1988. However, since these techniques demand specialized hard-ware and are technically quite demanding they were not widely used. The situation changed with the introduction of the multiple-quantum MAS NMR (MQ-MAS) technique by Frydman and co-workers.[135,136] The MQ-MAS technique provided a robust experiment, which can be routinely applied on a standard MAS probe. As an alternative to the MQ-MAS the satellite-transition MAS (ST-MAS) experiment was introduced by Gan in 2000,[137] which can exhibit better sensitivity.[138]

Both the mathematical description and the simulation of MAS-experiments are fairly advanced and beyond the scope of this review. For this reason we refer the reader to the text-book of Schmidt-Rohr[68] and the review by Eden.[139] Fortunately, for the practical evaluation of MAS-experiments there are publicly available software packages like SIMPSON,[140] SPINEVOLUTION,[141] or others.

3 Applications to immobilized catalysts

The aim of this section is to provide an overview about the development and application of solid state NMR spectroscopic techniques for the study of molecular structures and dynamics on the molecular and intermolecular length scale (1 Å–10 Å). In particular, anisotropic magnetic nuclear spin interactions like chemical shielding anisotropy (CSA), magnetic dipolar interaction and quadrupolar interaction are used as probes for interatomic distances and orientations of molecular groups, *i.e.* structures, and changes of these interactions are monitored and used as a measure of dynamic processes inside the system.

Applications of standard MAS techniques to immobilized catalysts

Nowadays MAS measurements are the standard experiments which are routinely used in the majority of Solid State NMR. CP-MAS, the combination of MAS and CP experiments has been routinely utilized to obtain high-resolution NMR spectra of dilute spins such as ^{13}C, ^{15}N, and ^{31}P Nuclei in solid materials.[142] CP-MAS plays a key role in the characterization of homogeneous catalysts immobilized on solid supports, providing direct information about the structure of support and the metal complex. Accordingly a number of NMR studies were performed in order to investigate the morphology and chemical properties of several types of pure and functionalized silica materials (ref. 52 and references therein).[71,143–147]

Among the large number of catalytically active transition metal complexes, Wilkinson's catalyst (Fig. 3),[148] is one of the most important and probably the most widely studied of all known homogenous hydrogenation catalysts. For this reason, the immobilization of this complex and its derivates on solid supports and their application in heterogeneous reactions are intensely investigated. [22–24,46,149–151]

Silica as support material

Owing to its favourable chemistry, silica based materials of different morphologies are ideal support materials for the immobilization of a catalyst. Their surface can be easily functionalized by different organic linkers like for example APTES (3-aminopropyltriethoxysilane, Fig. 4),[152] which bind to the silanol groups and serve as anchoring points for the tethering of the catalyst.

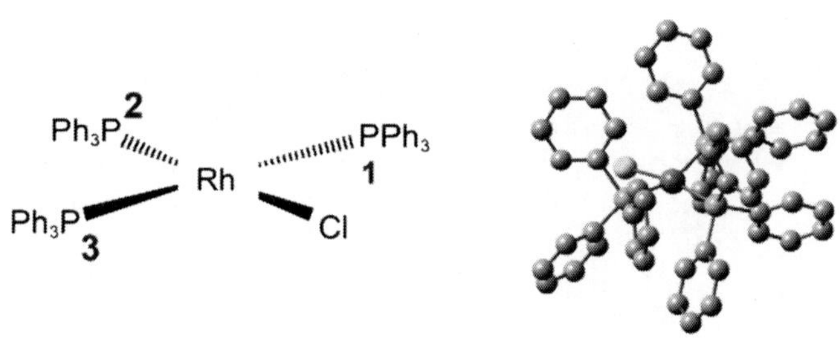

Fig. 3 Wilkinson's catalyst.

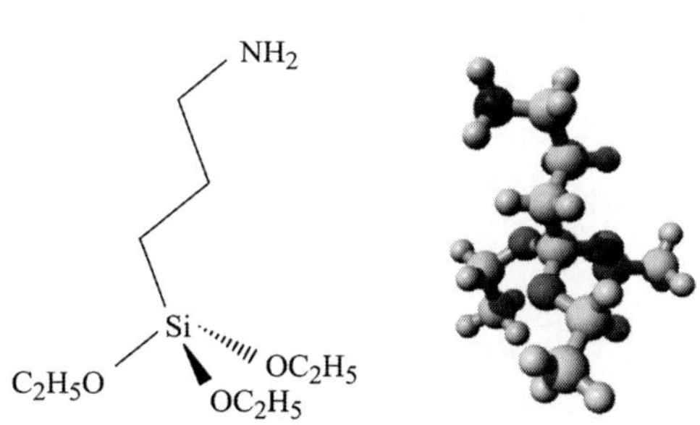

Fig. 4 3-aminopropyltriethoxysilane (APTES).

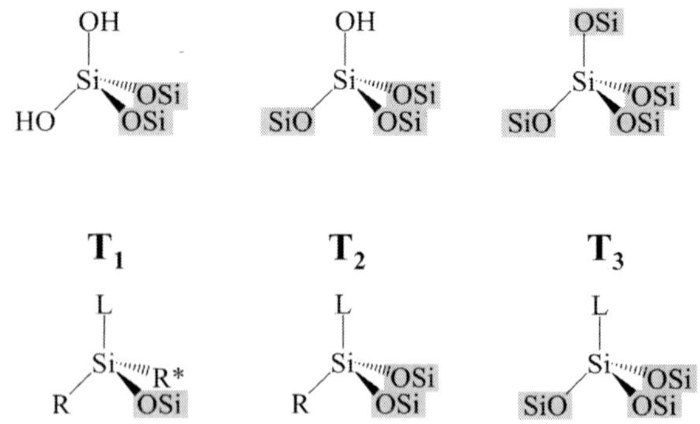

Fig. 5 Upper panel: The different types of silicon atoms in pure silica materials are denoted as Q_n (n = 2–4). Lower panel: the corresponding groups in functionalized silica are denoted as T_n (n = 1–3).

In the case of pure silica materials, four different types of silica sites are possible, which are characterized by the number n of –O-Si ligands as Q_n-sites (Fig. 5). In typical silica materials like MCM-41 or SBA-15 only sites with n = 2, 3, 4 (Q_2,Q_3 and Q_4) are found. The corresponding silicon atoms of the organic linker covalently attached to the surface silanol groups are denoted as T_n-moieties (see Fig. 5).

^{29}Si CP-MAS and MAS techniques are very widely employed for the characterization of these silica materials. As long as sufficiently large surface areas are available, ^{29}Si MAS and ^{29}Si CP-MAS NMR are possibly the fastest and most efficient spectroscopic techniques for the characterization of the success of functionalization. They allow a direct monitoring of the surface silanol groups, the bulk silica atoms, and, if standard linkers like APTES are employed, also of the linker molecules attached to the surface. In general ^{29}Si-CP-MAS techniques are faster, but they strongly favour surface near groups and are thus less quantitative than simple ^{29}Si-MAS techniques, without cross-polarization. If the recycle time of the experiment is suitably long, they allow a quantitative determination of the different silica species. ^{29}Si-CP-MAS experiments on the other hand monitor directly those silica groups, which are reacting with the catalyst or the linker. Since the line-widths of the different chemical species are typically larger than their spectral distance, both types of spectra exhibit substantial line overlap. In many cases the interpretation of the spectra is already feasible with some sensible assumptions on line positions and line shapes. In more complicated cases or for a quantitative analysis of the obtained signals, deconvolution and line fitting procedures are necessary.

A typical example of this type of experiments is shown in Fig. 6. Two broad sets of signals are observable which are attributed to the Q_n- and T_n- groups present at the surface of the silica material SBA-3. The presence of different T_n-groups confirms not only the successful surface modification but also characterizes the method of attachment of APTES to the silica surface. This example clearly demonstrates that ^{29}Si CP-MAS is the tool of choice for a detailed characterization of the modified surface of silica support materials.

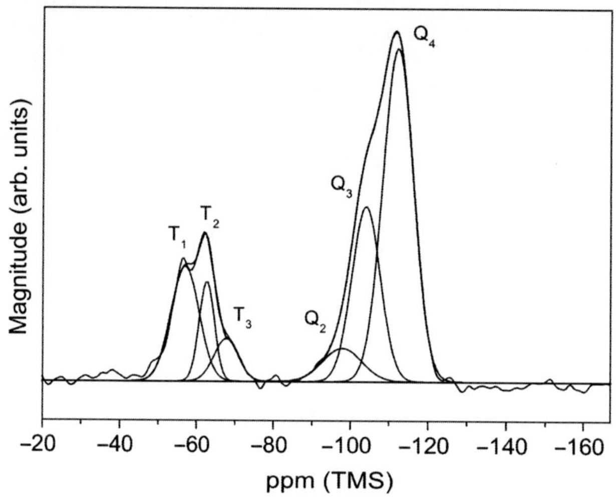

Fig. 6 Typical ^{29}Si-MAS-NMR spectrum of APTES functionalized silica SBA-3 (see text).

^{13}C MAS-NMR

While ^{29}Si MAS- and CP-MAS-NMR monitor mainly the changes of the surface of the support material due to the immobilization, ^{13}C MAS- and CP-MAS NMR monitor the changes in the structure of the organic linker or the immobilized molecule. Thus ^{29}Si and ^{13}C NMR are complementary methods. For example, ^{13}C CP-MAS-NMR techniques confirm the binding of alkoxysilane organic linkers on the silica surface. They also provide information about the coordination sphere of the metal centre. Therefore ^{13}C CP-MAS experiment are often employed as a characterization tool for supported metal catalysts.

Several studies, including research of our group, have probed the structure of the organic linker attached to the surface of silica materials employing ^{13}C NMR measurements.[153–158] A typical example of these experiments is shown in Fig. 7, which displays the ^{13}C CP-MAS NMR spectrum of the mesoporous silica material SBA-3 functionalized with APTES (A) and after tethering of the Wilkinson's catalyst (B). The spectrum exhibits characteristic resonance peaks which are attributed to carbon atoms of the APTES molecule. The signals at 8, 26 and 46 ppm correspond to the carbon atoms of the alkyl chain, C_1, C_2 and C_3, respectively. Two additional signals at 16 ppm and at 58 ppm are due to the presence of carbon atoms of the ethoxy group, CH_3 and CH_2, respectively. The set of the signals visible in the spectrum corroborates the results of ^{29}Si MAS NMR spectra, indicating successful modification of the silica surface. Moreover the presence of the aromatic carbons in the right panel at *ca.* 130 ppm is a clear indication of the phenyl-carbons of the Wilkinson's catalyst.

^{31}P MAS NMR experiments

Whereas ^{29}Si and ^{13}C CP-MAS experiments permit a characterization of the silica surface before and after the process of functionalization and reveal the characteristic binding of the linker to the surface, they are only of limited

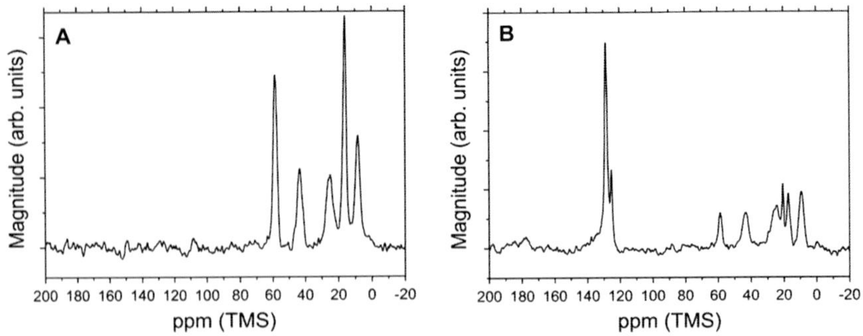

Fig. 7 ^{13}C CP-MAS NMR spectrum of the mesoporous silica material SBA-3 functionalized with APTES (left panel) and after tethering of Wilkinson's catalyst (right panel, adapted from ref. 152).

value for the characterization of the binding situation of the metal centre. The main reason is that the metal centre is usually coordinated to heteroatoms. So, we can confirm the presence of the ligand coordinated to the metal centre, but we are not able to obtain detailed information about the attachment of the catalyst to the linker molecules. Since the latter is of great importance in understanding of the chemical activity of supported molecules, it is necessary to apply measurements which give detailed information about the first coordination sphere of the homogenous catalysts, by directly monitoring the nuclei in the first coordination sphere.

Since phosphine ligands play a very important role in organometallic chemistry, they found also applications as efficient functional groups coordinating metal complexes to the surface of solid supports. Therefore a detection of the phosphorus atoms with ^{31}P solid state NMR methods plays a crucial role in the structural characterization of heterogenized complexes.

^{31}P NMR spectroscopy is an excellent technique for the study of a great variety of phosphorus containing molecules. In solution, many of the transition metal phosphine catalysts are kinetically labile. Thus, low temperature experiments are often required to properly characterize phosphine ligands in solution NMR. As an alternative, it is often advantageous to measure the ^{31}P NMR spectra in the solid state. The high sensitivity of the spin $\frac{1}{2}$ phosphorus nucleus ^{31}P with its high gyromagnetic ratio, and its natural abundance 100% gives a probe, which is sensitive to structural variations in the transition metal phosphine complexes. Furthermore, it removes many problems arising from the low complex concentrations in supported systems.[159] Since the pioneering ^{31}P NMR investigations of Fyfe and co-workers were published,[160,161] the combination of cross polarization (CP) and magic angle spinning (MAS) has made ^{31}P Solid State NMR spectroscopy a standard tool for studying phosphorus containing metal complexes. The studies of Fyfe and Wasylishen determined the morphology of Pd, Pt and Ni containing phosphine complexes in the crystalline state or attached to a solid support. They could show that both ^{31}P chemical shifts and coupling constants are useful spectroscopic observables for the characterization of metal complexes immobilized at a surface.

Following these seminal investigations, many studies contributed to the determination of the geometry of phosphine organic linkers and metal phosphine complexes immobilized on a solid support. Komoroski *et al.* provided investigations of tertiary phosphine-palladium complexes bound to a silica surface.[162] In these studies, the application of high resolution solid state ^{31}P NMR and ^{13}C NMR methods allowed the identification of the composition and stereochemistry of the palladium complexes. Moreover, similarly to work presented by Fyfe, ^{31}P NMR not only confirmed the immobilization of the metal complex but gives in addition information about reaction side products such as phosphine oxides present at the surface. In a series of publications, the Blümel group contributed to the study of immobilized species, especially different phosphine linkers and their morphology and interactions with the surface of the silica support.[32,163–168] Their ^{31}P CP-MAS studies provided unprecedented insights into the chemistry of phosphine ligands at the surface of silica support materials. In particular their results give detailed information about surface functionalization through attachment of different phosphine organic linkers. This allows a prediction of a successful process of catalyst immobilization and an avoiding of leaching during the immobilization and the catalytic reactions.

In addition to the standard solid state MAS-NMR methods, Blümel and co-workers employed high-resolution magic angle spinning NMR of suspensions, which they abbreviated as HRMAS experiments, for the investigation of the structure of phosphine linkers and homogenous catalyst immobilized on the silica surface.[16,32,169,170] The resulting HRMAS NMR spectra exhibited very narrow lines, compared to standard ^{31}P CP-MAS techniques. These results opened up new possibilities for a detailed structural characterization of immobilized species.

In our group we focus on the investigation of the morphology of ruthenium and rhodium complexes immobilized on the mesoporous silica SBA-3 functionalized with 3-aminopropyltriethoxysilane (APTES).[152,171] In order to characterize changes of the catalysts structure we employed ^{31}P CP-MAS experiments for immobilized species and comparing the spectra with the results of the neat catalyst, we successfully confirmed attachment of the catalyst to the supporting material.[171]

Fig. 8 compares the centre bands of the ^{31}P-CP-MAS spectrum of the bulk catalysts and the catalyst tethered to the surface of mesoporous SBA-3. The spectrum of the neat catalyst can be divided into three parts, with isotropic chemical shifts of 47.5, 30 and 21 ppm. The interpretation of this spectrum is in agreement with the results presented by Wu and Wasylishen[172] and the results of solution state NMR.[173] The low-field doublet at 47.5 ppm is attributed to the J-coupling between the phosphorus atom P2 and the rhodium atom, with coupling constant J(Rh-P) = 194 Hz. The multiplet present in the higher field corresponds to the two magnetically non-equivalent phosphorus atoms P1 and P3, with groups of line at 21 ppm and at 32 ppm, respectively. These groups contain doublets of equally spaced lines with spacing of 184 Hz, which is attributed to a J-coupling to the central Rh atom. Apparently, the groups are further splitted by a ^{31}P-^{31}P ^{2}J-coupling of 394 Hz, which is typical for the homonuclear J-coupling of trans phosphorus atoms, which is usually larger than cis J-coupling (0-50 Hz).[174]

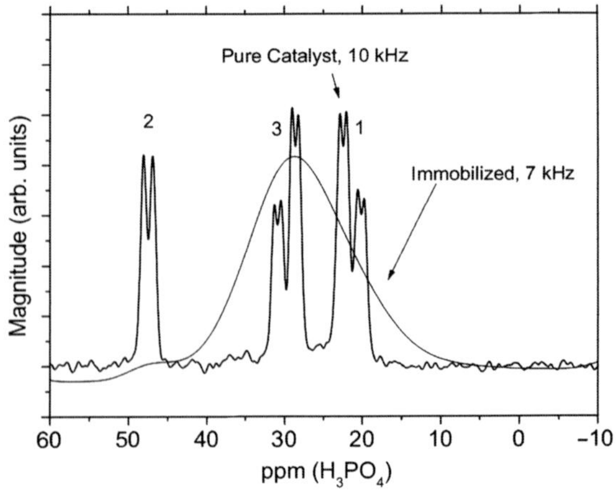

Fig. 8 Central region of the ^{31}P CP-MAS spectrum of the neat Wilkinson's catalyst and after its immobilization on mesoporous silica SBA-3 (adapted from ref. 171).

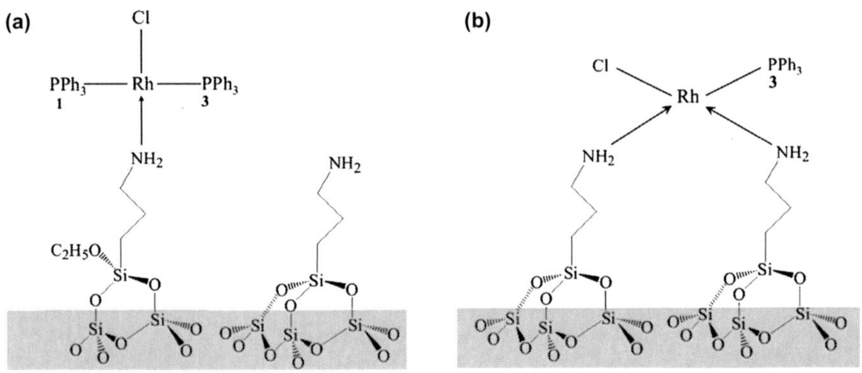

Fig. 9 Two possibilities of the immobilization of the Wilkinson's catalyst at the surface of amine modified silica material SBA-3 (monopodal or bipodal tethering).

The ^{31}P CP-MAS spectrum of the immobilized Wilkinson's catalyst differs significantly from that of the neat catalyst. It consists of only one broad asymmetric peak visible at 32 ppm, which covers the area of P1 and P3. Additionally, there is a broad weak signal at the position of the former P2 signal, which corresponds to less than 5% of the integral intensity and can be assigned most probably to a small amount of not immobilized Wilkinson's catalyst. These results show that P2 has evidently been replaced by an amine group of APTES linker and confirm the successful grafting on the surface.

However, these results are not yet sufficient for a detailed determination of the structure of immobilized rhodium catalyst. The reason is that the broadening of the signal at 32 ppm can result either from a single phosphorus ligand or from two coalescent ^{31}P chemical shifts related to the phosphine ligands with P3 and P1 phosphorus atoms. In other words, there are two possible ways of binding of the Wilkinson's catalyst to the silica surface, namely a monopodal and a bipodal scenario. (see Fig. 9). In the first case (monopodal tethering), only one phosphine ligand of the metal complex is substituted by NH_2 group of APTES molecule, and the heterogenized catalyst is attached to the surface only by a single coordination bond. The remaining *trans*- phosphine ligands, P1 and P3 are chemically equivalent, therefore only one broad signal is visible in the spectrum. The second possible situation (bipodal tethering) results from the substitution of two phosphine ligands P1 and P2 and the anchored catalyst is immobilized on the surface through two linkers.

By standard 1D ^{31}P MAS-NMR experiments it is impossible to distinguish between these two scenarios. For this reason, it is necessary to apply advanced 2D ^{31}P NMR techniques to characterize in detail the structure of the immobilized Wilkinson's catalyst.

Applications of advanced MAS techniques to immobilized catalysts

In the previous section it was shown that the combination of different classical one-dimensional MAS NMR methods gives the possibility of a detailed characterization of the surface of the silica support material, the geometry of the attached organic linkers and their interactions with the solid surface. In addition they provide information about the immobilization of the homogenous catalyst on the support. However, the exact geometry of the catalytically active metal centre of the tethered catalyst cannot be resolved only by means of standard MAS techniques. This is a particular problem if the attachment of the metal is performed directly to the silica surface without any tethering linkers. To solve this problem there are two possible strategies, which of course can be combined. On the one hand the solid state NMR techniques can be combined with other physicochemical characterization techniques like nitrogen sorption analysis (BET, BJH), FTIR or XRD and on the other hand more advanced multi-dimensional solids state NMR techniques can be applied.

The first strategy was employed for example by the Basset group in a series of papers,[48,175–181] where they combined standard solid state MAS measurements with other methods like X-ray or EXAFS. Their results showed that these combinations open up the possibility to directly investigate the behaviour and the structure of metal atoms present at the surface.

The necessity to extend the solid state NMR investigation to two or more dimensions stems from the recognition that the properties of a molecular system cannot be fully characterized by a conventional one-dimensional (1D) NMR spectrum. Two dimensional (2D) NMR spectroscopy is a general concept which makes it possible to acquire more detailed information about the system under investigation in physics, chemistry, biology materials sciences or medicine.[115] This technique has become an effective tool in the study of molecular structures and dynamics, in particular in solution. Moreover, in the last two decades a large number of 2D NMR methods were adapted to or developed for the investigation of solid systems. In solid systems they reveal information about anisotropic interactions, which are averaged in solution. Owing to these favourable properties they were already in the nineties applied by a series of papers of the Wasylishen group (see refs. 159, 172 and references therein) to the study of neat solid organometallic catalysts. In this review we will focus mainly on advanced

2D ^{31}P NMR techniques like correlation spectroscopy (HETCOR, COSY) and J-resolved experiments (Off-MAS, slow spinning MAS), which play a significant role in the detailed characterization of the structure of an immobilized catalysts.

The advantages of applying advanced 2D ^{31}P NMR techniques were first realized by the seminal work of Wu and Wasylishen.[159,172,182] Employing 2D J-resolved ^{31}P NMR experiments, they studied the structure of neat metal complexes with phosphine ligands in the solid state in detail. They showed that in favourable instances, the signals in solid-state NMR spectra can be made sufficiently narrow by MAS to resolve the homo- and heteronuclear J-couplings. These couplings can be exploited for correlation spectroscopy in much the same way as in the liquid-state NMR.[115] Moreover one has to note at this point that there is a crucial difference between homonuclear 2D J-resolved spectroscopy in the liquid state and the solid state NMR. In those cases where the isotropic chemical shifts and coupling patterns of the interacting nuclei are equal, *i.e.* if they are chemically and magnetically equivalent, they form an A_2 spin system and it is not possible to resolve the J-coupling in liquid state NMR. Due to the presence of anisotropic interactions in the solid, the magnetic equivalence is lifted in many solid systems and this limitation can be lifted by suitable solid state NMR techniques (see below).

In the previous section it was demonstrated that it is not possible to reveal the morphology of the immobilized $RhCl(PPh_3)_3$ employing standard 1D ^{31}P Solid State NMR method. Since the structure of the heterogenized Wilkinson's catalyst is not only of principal scientific interest but also of great importance for the further application of this material in catalysis, it is obligatory to apply more advanced ^{31}P solid-state NMR methods. For this reason we employed 2D J-resolved ^{31}P NMR experiments, which allow the determination of the number of phosphine ligands coordinating the rhodium in the investigated sample, by analysing the spin-spin interactions in the molecule.

This technique was recently applied to analyze the binding of the Wilkinson's catalyst to mesoporous SBA-3.[171] In the first step the 2D ^{31}P J-resolved NMR spectrum of the bulk $RhCl(PPh_3)_3$ was recorded (Fig. 10). It confirms the results of the 1D ^{31}P CP-MAS experiments and results presented by Wu and Wasylishen.[159] The well-resolved pattern of heteronuclear $^{1}J(Rh,P)$ coupling is clearly visible along the f_2 dimension, while the characteristic large homonuclear J-coupling $^{2}J(P1, P3)$ appears along the f_1 dimension.

Fast spinning MAS spectra, even in the case of J-resolved 2D experiments; do not allow to distinguish between the presences of one or two chemically equivalent nuclei since under fast spinning MAS conditions the situations is comparable to liquid state NMR *i.e.* chemical shift anisotropies are averaged out. However, under static conditions, the chemical shift tensors of two chemically equivalent nuclei do not coincide and both nuclei have in general different chemical shifts at certain orientations of the molecules with respect to the magnetic field axis, *i.e.* they are magnetically inequivalent. Thus under static conditions, the spin system may be described as an AB or AX system, instead of an A_2 spin system.

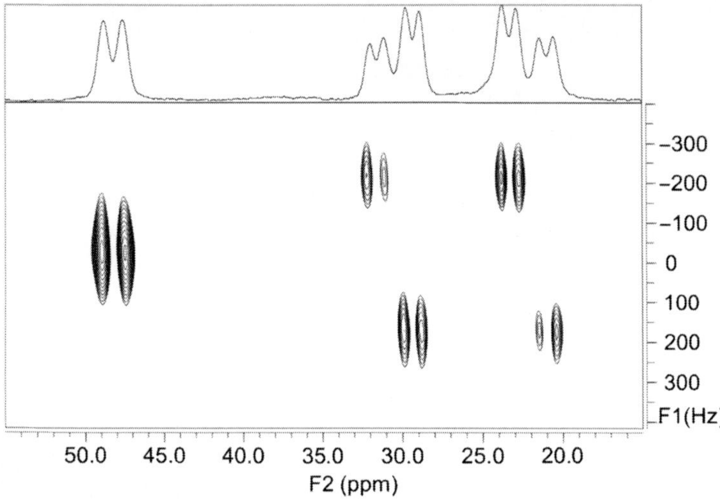

Fig. 10 ^{31}P 2-D J-resolved NMR spectrum of the pure $RhCl(PPh_3)_3$, spinning speed 10 kHz (adapted from ref. 171).

Slow spinning 2D-MAS

Slowly spinning such a molecule at the magic angle will give rise to a modulated spin system, which is no longer describable by first order averaged Hamiltonian theory. Instead, higher order correction terms have to be taken into account, and the transition probability between J-coupled levels becomes considerably larger than zero. This effectively reintroduces the J-coupling into the spectra. Thus at slow spinning speeds the J-coupling interactions between chemically equivalent nuclei may be observed in solid-state J-resolved MAS NMR and J-couplings may be employed to determine the number of binding ligands. A comprehensive theoretical description of the Hamiltonian under slow spinning conditions, validated by experimental investigations is given in ref. 182 and references therein.

As an example Fig. 11 displays the slow spinning ^{31}P 2D J-resolved experiment of the immobilized Wilkinson's catalyst. The slow spinning ^{31}P 2D J-resolved spectrum of $RhCl(PPh_3)_3$ after the process of immobilization clearly shows the absence of any ^{31}P-^{31}P J-couplings within the spectral resolution. This suggests the presence of only a single P3 phosphorus atom in the catalyst molecule attached to the silica surface. Owing to the partial overlap of the center band with the spinning side bands the broad base of the center band is not visible.

Fast spinning 2D-OFF-MAS

Since the slow spinning spectra depends on chemical shift anisotropy, direct dipolar- and J-coupling, which all are unknown in the case of an immobilized catalyst; the absence of resolved J-couplings at 2.5 kHz does not unequivocally prove the presence of only one phosphine ligand. In order to finally confirm the result obtained by the slow spinning experiment, the J-resolved measurement have to be repeated under off-magic angle spinning

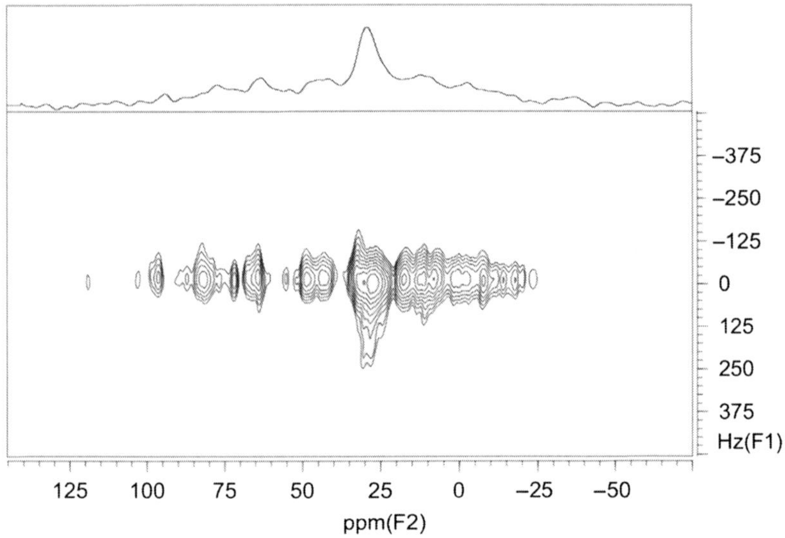

Fig. 11 The two dimensional ^{31}P J-resolved MAS slow spinning (2.5 kHz) spectrum of the Wilkinson's catalyst $RhCl(PPh_3)_3$, immobilized on mesoporous silica, reveals only a single ^{31}P ligand. This shows that two ligands are replaced by the immobilization.

conditions, which are more relaxed regarding the detection of J-couplings.[183] Furthermore, fast spinning under off-magic angle conditions offers higher spectral resolution, considerably increases the signal to noise ratios and drastically reduces experimental times, with respect to slow spinning conditions. Under fast spinning off-magic angle conditions the averaged Hamiltonian resembles a two spin AB system in liquid state NMR.[183] That is, the anisotropic parts of the chemical shift tensors of both nuclei are not completely averaged out under off-magic angle conditions. These anisotropic parts render the different nuclei magnetically inequivalent. As a result of this inequivalency the J-coupling between the nuclei becomes visible in the spectrum, similar to the transition from an A_2 to an AB or AX spin system.

As an example the ^{31}P 2-D J-resolved Off-MAS spectrum of the immobilized Wilkinson's catalyst is shown in Fig. 12. In the spectrum there is evidently no visible J-coupling. This confirms not only the results of the slow spinning measurements but also proves the presence of only a single remaining phosphorus atom, P3. The presence of one phosphine ligand in the molecule of heterogenized catalyst proves the assumption deduced from the 1D-spectrum that during the immobilization process two phosphine ligands are substituted by amine groups of organic linker. The phosphine with phosphorus atom labeled as P3 remains in the molecule (see Fig. 3).

The presented 2D ^{31}P J-resolved NMR experiments provide clear evidence that Wilkinson's catalyst immobilized on the amine functionalized silica SBA-3 material, possess only one triphenylphosphine ligand. The rhodium centre is coordinated to the silica surface through the two amine ligands. It reveals significant information about the structure of

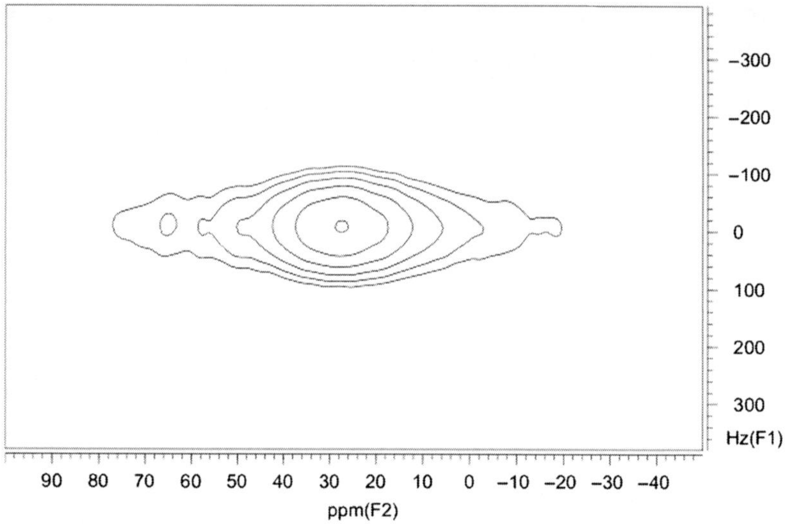

Fig. 12 2-D J-resolved Off-MAS spectrum of the immobilized Wilkinson's catalyst.

heterogenized spieces and gains the new insights into the immobilization of metal phosphine homogenous catalysts.

4 Characterization of reaction intermediates in heterogeneous catalysis by solid state NMR

Solid state NMR techniques not only allow a characterization of the immobilization of a homogeneous catalyst, but permit under favourable conditions also a monitoring of reaction intermediates and their dynamics. In particular for hydrogenation reactions the study of the mobility and bonding situation of hydrogen/deuterium ligands on the surface of the nanoparticles stabilized by mesoporous systems is essential.

In a set of papers we have shown that the combination of variable temperature ^{2}H solid-state NMR line-shape analysis, ^{2}H MAS NMR spectroscopy and quantum chemical calculations is a powerful approach for the characterization of the binding of hydrogen/deuterium to transition metal atoms and clusters by virtue of the chemical shift and quadrupolar interaction strength of the deuteron.[103,106,152,184–191] In our studies we focused on molecular complexes and well-defined magic-cluster nanoparticles as models to identify the characteristic NMR parameters of the deuteron. Then the characteristic values of quadrupolar coupling constants and asymmetry parameters can be attributed to particular deuterium-ruthenium bonding situations and deuterons exhibit a tremendous amount of dynamics adsorbed on metal nano-particles.[101,187] We also demonstrated that this technique can be applied to the study of deuterium interaction with metallic ruthenium immobilized on the mesoporous silica SBA-3. Using this method, also the reaction intermediates (methyl-groups bound to the surface of the Ru nanoparticles) created by the hydrogenation reaction of ethene could be detected (see Fig. 13).[102] Here ^{2}H-solid state NMR is clearly

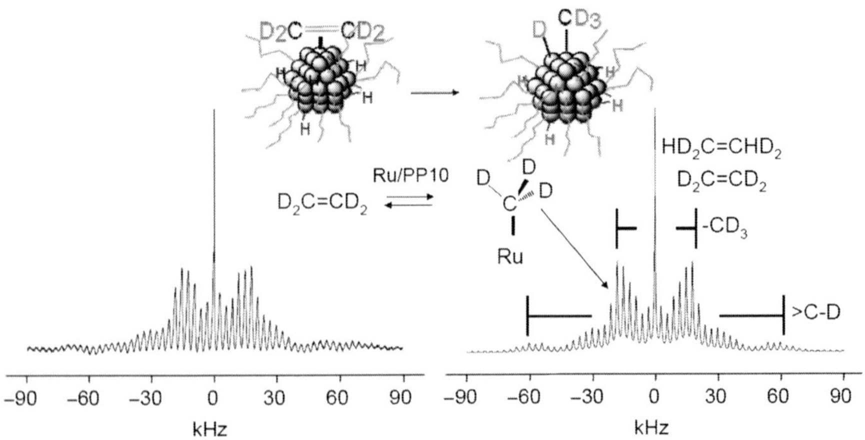

Fig. 13 The solid state 14 Tesla ^{2}H-NMR spectrum of ethylene-d_4 reveals the formation of immobile methyl groups, characterized by Qcc ≈ 50 kHz, immobile >CD groups and two liquid like Lorentzians. (adapted from ref. 102).

advantageous to ^{1}H-solid state NMR, since in the latter case the presence of the organic linkers necessary to stabilize the nanoparticles, causes very strong ^{1}H-background signals.

5 Outlook and perspectives

Despite this success the application of solid-state NMR techniques for the investigation of solid/liquid and solid/solid interfaces is still severely hampered by the inherent low sensitivity of NMR techniques, which limits the applicability of solid-state NMR to these systems. One example is the important case of carbon nuclei, where the NMR active ^{13}C isotope has only 1% natural abundance and a comparatively small gyromagnetic ratio, resulting in relatively small signal amplitudes. In turn, a very large specific surface area (typically 50 m^2/g and above) is required to have sufficient sensitivity for good spectra in a reasonable time. Moreover, the spin lattice relaxation in compounds without mobile groups is slow rendering signal acquisition slow and tedious. Furthermore, it is often difficult to separate surface near chemical building units from the more intense bulk signals. A second example is the acquisition of two-dimensional solid-state NMR spectra, which are often required to achieve the necessary spectral resolution.

Two different strategies were developed in the course of the last ten years, to solve the sensitivity problem, namely ultra-fast MAS NMR combined with indirect detection and nuclear hyperpolarization.

Fast MAS-NMR and indirect detection

MAS-NMR techniques have been developed for quite some time in solid state NMR. The main limiting factors in their applicability to small molecules were resolution and sensitivity. In particular the observation of protons, which are by far the most useful nuclei in liquid state NMR, is hampered by the strong homonuclear dipolar interactions in the solid.

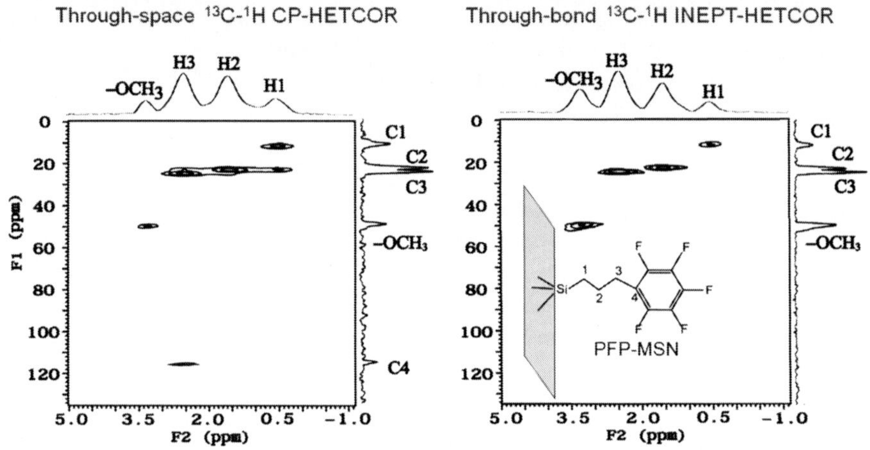

Fig. 14 Indirectly detected though space and through bond ^{13}C-^{1}H 2D-HETCOR-MAS NMR spectra of Mesoporous Silica Nanoparticles (MSN) functionalized with 3-(pentafluorophenyl)-^{n}prop-1-yl (-CH_2-CH_2-CH_2-C_6F_5, PFP) (adapted by permission from Pruski *et al.*[19]; Copyright (2009) Elsevier Limited).

To overcome this limitation, ingenious experiments were developed, which finally resulted in the combined rotation and multi-pulse spectroscopy (CRAMPS) experiment by Gerstein and co-workers.[192] In this experiment it is possible to obtain highly-resolved ^{1}H-MAS NMR spectra at relatively low spinning speeds. However, CRAMPS-type experiments are difficult to set up, demanding to perform, and not suitable as a routine technique in most NMR labs. The situation in high-resolution solid state MAS NMR of protons changed dramatically in the last decade, owing to tremendous advances in MAS probe technology, with the fastest achievable MAS-rotation speeds currently approaching 80 kHz.[193,194] These fast MAS-NMR probes combine excellent sensitivity per spin with efficient cross-polarization,[53,195] very high radiofrequency fields and the possibility of using low-power RF schemes during preparation, mixing and decoupling periods.[196] An additional advantage of fast-MAS NMR is the increased spectral width in the indirect dimension in rotor-synchronized experiments, which is determined by the rotation frequency.

The Pruski group demonstrated in a seminal study[53] that already at 40 kHz spinning frequency a CRAMPS-like resolution is achievable in mesoporous organic–inorganic hybrid materials, where the dipolar line-width was reduced by molecular dynamics. Further improvements in the sensitivity and resolution where obtained by indirect detection schemes, pioneered in solid state NMR by Ishii and Tycko.[197]

In these schemes the X-nucleus of interest is detected *via* the far more sensitive protons. Employing indirect detection (Fig. 14) the Pruski group succeeded to detect two-dimensional (2D) ^{1}H-^{13}C-HETCOR spectra of functionalized silica materials in natural abundance 15 times faster than with conventional detection.[198] Finally combining indirect detection with 2D and advanced homonuclear decoupling sequences it is even possible to detect through bond couplings,[19,199] similar to the COSY type experiments of liquid state NMR spectroscopy. In the last two years the Pruski group has employed these techniques for series of studies of functionalized silica materials with unprecedented resolution and sensitivity.[200–203]

Hyperpolarization

Hyperpolarization is a term which collects a set of different signal-enhancement techniques in NMR spectroscopy. They have in common that they create population differences (polarizations) of the nuclear spin, which are far larger than the thermal equilibrium value (Boltzmann polarization). Following a proposal by Overhauser[204] in 1953, which was experimentally verified by Carver and Slichter[205,206] Dynamic Nuclear Polarization (DNP) schemes[207,208] were developed. These techniques employ an auxiliary reservoir of fast and effectively polarizable spins to generate spin order which is subsequently transferred to the spin system of interest. In case of DNP experiments the strong polarization of electron spins is transferred to the nuclear spins by means of microwave irradiation. In the early days DNP experiments were restricted to relatively low fields (^{1}H-frequency below 60 MHz) due to the technical limitations of klystrons used as microwave sources.

Here the situation changed strongly with the availability of gyrotrons as a reliable high-frequency microwave sources with high output power (see review by Felch *et al.*[209]), which has triggered a fascinating new development in the field of high field DNP. In particular the pioneering work by the Griffin group in the last fifteen years has shown that tremendous gain in signal/noise ratio is also achievable at high magnetic fields.[210] An overview about the current state of the art is given in their recent review.[211] The polarizing agents in ssDNP studies used so far are mainly TEMPO and TOTAPOL.[211,212]

DNP techniques open up completely new application fields for surface NMR, for example in the field of the characterization of immobilized catalyst or reaction intermediates on heterogeneous catalysts. As an example the Emsley group[213,214] showed recently that the combination of DNP with magic angle spinning (MAS) NMR delivers roughly a 50 fold increase in the sensitivity of ^{13}C signals for phenol or an imidazolium derivative grafted on mesoporous silica (see Fig. 15). In their experiment they studied the grafting of phenol or an imidazolium derivative on mesoporous silica support. The dry samples were impregnated with a 30 mM solution of TEMPO or a 25 mM solution of TOTAPOL, as the DNP polarization source, respectively. The high polarization of the protons of the radical is then transferred to the molecules immobilized on the surface by spin-diffusion. In these experiments they found an enhancement of more than 56 for ^{13}C and 400 for ^{29}Si.

An alternative possibility to increase the sensitivity of solid-state NMR for the study of catalytic processes is the application of spin-polarized para-hydrogen.[215] The most prominent example of this technique is the so-called "Para Hydrogen Induced Polarization" or short "PHIP" experiment,[216,217] originally invented for liquid-state NMR studies of homogeneous catalysis. PHIP is not restricted to homogeneous hydration reactions and liquid state NMR, but is also applicable to studies of heterogeneous hydration

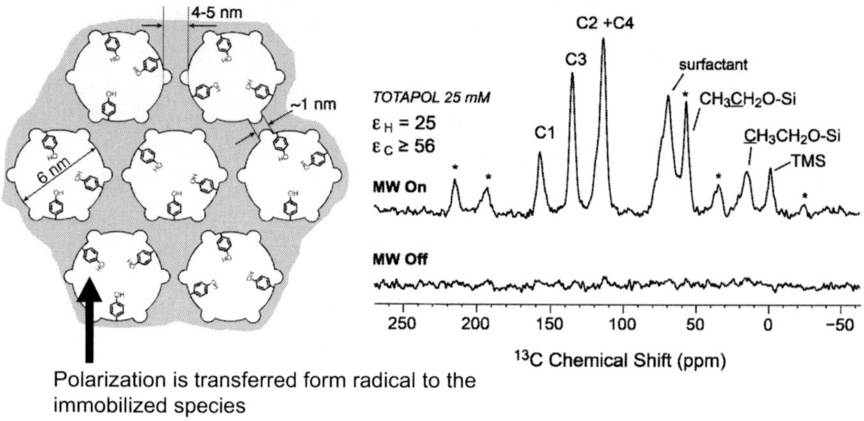

Fig. 15 Enhancing the sensitivity of ^{13}C-MAS-NMR of immobilized surface groups by DNP (adapted with permission from Lesage *et al.*[213]; Copyright (2010) American Chemical Society) on a 400 MHz solid-state NMR spectrometer with 263 GHz microwave irradiation gives an enhancement factor of up to 56 for ^{13}C.

reactions. An example being hydration reactions on zinc oxide surfaces, where adsorption of molecular hydrogen results in two hydrogen atoms chemisorbed in close proximity to each other,[218,219] which allows to study the binding of hydrogen to the ZnO surface by PHIP, as recently demonstrated by the Weitekamp group.[220] In these experiments short burst of para-H_2 were repeatedly given inside the magnet to the ZnO surface and the resulting spectra indicated the existence of a reversible hydrogen binding site. From the PHIP-spectrum the distance of the two hydrogen atoms in this chemisorbed state could be estimated as 2.6 Å. These experiments clearly demonstrated that the differential nature of the PHIP effect, which causes a partial cancellation of the positive and negative signal components in a non-oriented powder patterns, is no severe obstacle for the application of PHIP to the study of surfaces.[220] However, they are only applicable in the case of hydrogenation reactions with the additional constraint that the atoms of the para-hydrogen molecule remain in close proximity to avoid the loss of spin-coherence between the two hydrogen atoms.

Similar to DNP and para-hydrogen methods, hyperpolarized (HP) noble gases can significantly increase the nuclear polarization of noble gas atoms.[221,222] The transfer of spin polarization is more efficient for larger noble gas species (*e.g.*, xenon) as the spin exchange rate depends on the interaction time between the alkali metal and noble gas atoms, leading to enhancements of 10^4–10^5. The chemical shift of ^{129}Xe is highly sensitive to its local chemical environments, making it a particularly useful probe of internal and external surfaces of catalysts and porous materials.[223–225] Recent developments have utilized this spin polarization reservoir to elucidate NMR insensitive nuclei on solid surfaces *via* polarization transfer techniques. Spin polarization nuclear Overhauser effect (SPINOE) is perhaps the most commonly used transfer method.[226,227] It has been successfully applied to surface ^{1}H and ^{13}C signals of methoxylated silica, reaching maximum enhancement factors of 3.4.[228] Magic angle spinning (MAS) probes have been adapted for continuous flow of HP ^{129}Xe gas, allowing for increased spectral resolution and signal averaging. While surface signal enhancements as large as 1200% have been observed, results suggest the need for long T_1 relaxation times of surface nuclei as xenon-surface cross relaxation rates are remarkably slow.[228,229]

Hartmann-Hahn cross polarization (CP) offers an alternative approach. Recent ^{129}Xe-^{13}C CP experiments on methoxylated silica demonstrate the utility of this technique: not only is it more selective toward surface nuclei, it is applicable at temperatures near and above room temperature and applicable to materials with short T_1 relaxation times (<1 s) and small surface areas ($<5\,m^2/g$).[230,231] As such, selective enhancement of the magnetization of surface nuclei is possible provided a coupling exists between the adsorbed HP ^{129}Xe and the nuclei of interest.

Conclusions

Employing solid state NMR techniques it is nowadays easily feasible to study homogeneous catalysts immobilized on a solid support surface. From these studies it is possible to reveal the structure of the catalyst on the surface and its modifications, due to binding. Combined with the recent improvement in NMR methodology (hyperpolarization, indirect detection and FAST-MAS), the detailed study of surface processes of technical catalysts on a molecular level by solid-state NMR is now within reach.

Appendix

Interactions in solid state NMR

The basic principles of solid state NMR spectroscopy can be most easily understood by discussing the relevant NMR interactions. In contrast to most other types of spectroscopy NMR has the unique feature, that the full quantum mechanical interaction Hamilton operators (Hamiltonians) of the spin system are usually known. As usual all energies are measured in units of the angular velocity (rad/sec), *i.e.* all energies are divided by $\hbar$.

The Zeeman interaction

The dominating interaction in a conventional NMR-experiment is the Zeeman interaction between a nuclear spin and an external magnetic field $\vec{B}$. In most NMR experiments the magnetic field $\vec{B}$ consists of a constant homogeneous field $\vec{B}_0 = (0, 0, B_0)$ which by convention is chosen to determine the z-direction of the laboratory frame and a transversal time dependent field $\vec{B}_1 = (B_{1x}(t), B_{1y}(t), 0)$. The corresponding Hamiltonians of the Zeeman interaction are $\hat{H}_0$ respectively $\hat{H}_1$ (where γ is the gyromagnetic ratio of the spin):

$$\hat{H}_0 = -\gamma B_0 \hat{I}_z \tag{A1}$$

and

$$\hat{H}_1 = -\gamma (B_{1x}(t)\hat{I}_x + B_{1y}(t)\hat{I}_y) \tag{A2}$$

The local interactions

Apart of the global Zeeman interaction $\hat{H}_0$, which is merely an adjustable experimental parameter, there are the interactions caused by local fields. They are the interesting contributions to the spin Hamilton operator, because only they contain non--trivial information about the system. For organic solids essentially the following contributions are of importance: the dipolar interaction with other nuclear spins $\hat{H}_D$, the screening of the external magnetic field by the electron cloud, *i.e.* the chemical shift Hamiltonian $\hat{H}_{CS}$, the interaction of a nuclear quadrupolar moment with an electric field gradient, *i.e.* the quadrupolar interaction and finally the indirect spin-spin- interaction, *i.e.* the J-coupling $\hat{H}_J$. In addition to these interactions there are several others (for example Knight-Shift, Para-magnetic Shift), which behave similar to the chemical shift and are of lesser importance in the present context. All these local interactions add to the Zeeman interaction to form the total spin Hamiltonian of the system.

$$\hat{H} = \hat{H}_0 + \hat{H}_1 + \hat{H}_D + \hat{H}_{CS} + \hat{H}_Q + \hat{H}_X + \hat{H}_J \tag{A3}$$

Orientation dependence of local interactions

In solids the strength of most NMR-interactions depends on the relative orientation of the studied material to the external magnetic field. In Cartesian coordinates these orientation dependence is describable *via* interaction tensors (real symmetric 3 × 3 matrices)

$$\tilde{T} = \begin{pmatrix} T_{11} & T_{12} & T_{13} \\ T_{21} & T_{22} & T_{23} \\ T_{31} & T_{32} & T_{33} \end{pmatrix} \tag{A4}$$

If $\vec{b}_0$ is the unity vector in the direction of the external magnetic field in polar coordinates

$$\vec{b} = \frac{\vec{B}_0}{B_0} = \begin{pmatrix} \cos\alpha \sin\beta \\ \sin\alpha \sin\beta \\ \cos\beta \end{pmatrix} \tag{A5}$$

the corresponding strength of the interaction is given by

$$T_{zz} = \vec{b}_0 \tilde{T} \vec{b}_0 = \vec{b}_0 \begin{pmatrix} T_{11} & T_{12} & T_{13} \\ T_{21} & T_{22} & T_{23} \\ T_{31} & T_{32} & T_{33} \end{pmatrix} \vec{b}_0 \tag{A6}$$

If only a single interaction tensor is present, it is always possible to employ the principle axis (PAS) system of the tensor as the coordinate system, resulting in the following expression:

$$T_{zz}(\alpha, \beta) = T_{11} \cos^2\alpha \sin^2\beta + T_{22} \sin^2\alpha \sin^2\beta + T_{33} \cos^2\beta \tag{A7}$$

If more than one interaction tensor is present (for example two different CS-tensors or simultaneous presence of dipolar and CS-tensor) the relative angles of the PAS systems of these tensors have to be taken into account, employing the Euler angles between the different coordinate systems.

The Magnetic Dipolar Interaction is the major tool for the structure determination of solids by NMR. It is the result of the classical interaction of two magnetic dipoles ($\vec{r}_{jk}$ is the distance vector of the two interacting magnetic moments):

$$\hat{H}_D = -\frac{\mu_0}{4\pi}\hbar\sum_{j<k}\gamma_j\gamma_k\frac{3(\hat{\vec{I}}_j\vec{r}_{jk}/r_{jk})(\hat{\vec{I}}_k\vec{r}_{jk}/r_{jk}) - \hat{\vec{I}}_j\hat{\vec{I}}_k}{r_{jk}^3} = \sum_{j<k}\hat{\vec{I}}_j\tilde{D}_{jk}\hat{\vec{I}} \qquad \text{(A8)}$$

Here the dipolar interaction tensors $\tilde{D}_{jk}$ have been introduced as a convenient short hand notation of the geometry dependent expression inside the double sum.

The Chemical Shift is the next contribution to the local interactions. It is a result of the interaction of the external magnetic field with the electron cloud surrounding the nucleus. In a semiclassical description two different effects are observed: On one hand the external magnetic field causes a partial orientation, *i.e.* a polarization, of the magnetic moments of the electron (paramagnetic shift). On the other hand the external magnetic field induces ring currents in the electron cloud, which produce, according to the Lenz rule, a field opposite to the field (diamagnetic shift). Due to the fact that both effects have the same linear dependence on the external field, they are indistinguishable in the NMR and are therefore combined in a common chemical shift Hamiltonian with the CS-tensor. The corresponding Hamiltonian is:

$$\hat{H}_{CS} = \gamma\hat{\vec{I}}\tilde{\sigma}\,\overrightarrow{B_0} \qquad \text{(A9)}$$

For solid state NMR of immobilized catalysts and nanocatalysts the ^{13}C, ^{29}Si and ^{31}P-chemical shift values are of particular importance, because of the relatively wide range of shifts of these nuclei and the good resolution as compared to protons for example. In addition proton shifts can be of high diagnostic value if the necessary resolution is obtainable, for example due to high-mobility, fast-MAS or isotope-depletion (deuteration).

The nuclear Quadrupolar Interaction is a contribution which is visible in the spectra of nuclei with $I > 1/2$ (for example deuterons or aluminium). It describes the coupling of an electric nuclear quadrupolar moment, *i.e.* a non spherical charge distribution inside the nucleus, to an electric field gradient, employing the so called quadrupolar tensor. Due to the fact that the electric field and thus the field gradients are mainly determined by bond directions and charge distributions of electrons or ions, the quadrupolar tensor is a very sensitive tool for structural and dynamical studies. The Hamiltonian of the quadrupolar interaction is:

$$\hat{H}_Q = \frac{eQ}{2I(2I-1)\hbar}\hat{\vec{I}}\tilde{V}\hat{\vec{I}} \qquad \text{(A10)}$$

In this relation V denotes the electric field gradient tensor, *i.e.* the partial derivatives of an inhomogeneous electric field:

$$V_{kl} = \frac{\partial^2}{\partial x_k \partial x_l}\Phi, \quad k, l = x, y, z \qquad \text{(A11)}$$

The Indirect or Spin-Spin coupling is a collection of several interactions which all have the same structure. In particular there is the polarization of

Table 1 Secular parts of the solid state NMR Hamiltonians.

$\hat{H}_{CS} = \gamma B_0 \hat{I}_z$	$\hat{H}_D = \begin{cases} 2D_{12}\hat{I}_{1z}\hat{I}_{2z} \\ D_{12}\left(3\hat{I}_{1z}\hat{I}_{2z} - \hat{\vec{I}}_1\hat{\vec{I}}_2\right) \end{cases}$
$\hat{H}_Q = \frac{eQ}{2I(2I-1)\hbar} V_{zz}\left(\hat{I}_z^2 - \hat{\vec{I}}\hat{\vec{I}}\right)$	$\hat{H}_J = \begin{cases} J_{12}\hat{I}_{1z}\hat{I}_{2z} \\ J_{12}\hat{\vec{I}}_1\hat{\vec{I}}_2 \end{cases}$

the electron spins caused by a nuclear spin, which leads to a coupling of two nuclear spins and the exchange interaction of two spin 1/2 nuclei. The J-coupling is also described by a tensor, but in contrast to the dipolar interaction this tensor is not traceless, resulting in a non vanishing coupling also in isotropic media.

$$\hat{H}_J = \hat{\vec{I}}_k \tilde{J}_{kl} \hat{\vec{I}}_l \qquad \text{(A12)}$$

The High-Field Approximation: In most NMR experiments the nuclear Zeeman interaction with the static external magnetic field is much stronger than all other interactions of the nuclear spins. As a result of these differences in the size, it is usually possible to treat these interactions in first order perturbation theory, *i.e.* use only those terms which commute with the Zeeman Hamiltonian, the so called secular terms. This approximation is called the high field approximation. While the single particle interactions like CSA or quadrupolar interaction have a unique form, for bilinear interactions, one has to distinguish between a homonuclear and a heteronuclear case. The secular parts of Hamiltonians discussed in the previous section are collected in Table 1.

In the last two equations the upper line is for the hetero- and the lower line for the homonuclear case. There are two specialities for quadrupolar nuclei ($I > 1/2$): (i) for strong quadrupolar interactions there are additional non-diagonal terms in the quadrupolar Hamiltonian which influence the spectra; (ii) for half-integer quadrupolar nuclei, there exists a central transition ($-1/2$ <-> $+1/2$) which is only in second order influenced by the quadrupolar interaction and thus an important observable for the chemical characterization of these nuclei. Details of these second order broadening are beyond the scope of this review and can be found for example in the recent reviews by Jerschow[232] or Wimperis *et al.*[138]

Acknowledgements

Financial support by the Deutsche Forschungsgemeinschaft under contract Bu-911-12-1/2 is gratefully acknowledged.

References

1 H. Brunner, E. Bielmeier and J. Wiehl, *J. Organomet. Chem.*, 1990, **384**, 223.
2 B. E. Handy, I. Gorzkowska, J. Nickl, A. Baiker, M. Schramlmarth and A. Wokaun, *Ber. Bunsen. Phys.*, 1992, **96**, 1832.
3 H. Werner and U. Mohring, *J. Organomet. Chem.*, 1994, **475**.
4 J. Bluemel, *Ber. Bunsen. Chem.*, 1995, **99**.

5 J. U. Koehler and H. L. Krauss, *Journal of Molecular Catalysis a-Chemical*, 1997, **123**.
6 J. Evans, A. B. Zaki, M. Y. El-Sheikh and S. A. El-Safty, *J. Phys. Chem. B*, 2000, **104**.
7 A. J. Sandee, D. G. I. Petra, J. N. H. Reek, P. C. J. Kamer and P. W. N. M. van Leeuwen, *Chem-Eur J*, 2001, **7**, 1202–1208.
8 J. K. Lee, T. J. Yoon and Y. K. Chung, *Chem., Comm.* 2001.
9 W. H. Cheung, W. Y. Yu, W. P. Yip, N. Y. Zhu and C. M. Che, *J. Organomet. Chem.*, 2002, **67**.
10 R. Becker, H. Parala, F. Hipler, O. P. Tkachenko, K. V. Klementiev, W. Grunert, H. Wilmer, O. Hinrichsen, A. Birkner, M. Muhler, C. Woll, S. Schafer and R. A. Fisher, *Angew. Chem., Int. Ed.*, 2004, **43**, 2839–2842.
11 J. O. Krause, O. Nuyken, K. Wurst and M. R. Buchmeiser, *Chem-Eur J*, 2004, **10**, 777–784.
12 A. Crosman and W. E. Hoelderich, *J. Catal.*, 2005, **232**, 43–50.
13 A. M. J. Rost, H. Schneider, J. P. Zoller, W. A. Herrmann and F. E. Kuhn, *J. Organomet. Chem.*, 2005, **690**, 4712–4718.
14 M. Hartmann and C. Streb, *Journal of Porous Materials*, 2006, **13**, 347–352.
15 X. S. Zhao, X. Y. Bao, W. Guo and F. Y. Lee, *Materials Today*, 2006, **9**, 32.
16 T. Posset and J. Blumel, *J. Am. Chem. Soc.*, 2006, **128**, 8394–8395.
17 J. W. Wiench, C. Michon, A. Ellern, P. Hazendonk, A. Iuga, R. J. Angelici and M. Pruski, *J. Am. Chem. Soc.*, 2009, **131**, 11801–11810.
18 J. L. Rapp, Y. L. Huang, M. Natella, Y. Cai, V. S. Y. Lin and M. Pruski, *Solid State Nucl Mag*, 2009, **35**, 82–86.
19 K. Mao, J. W. Wiench, V. S. Y. Lin and M. Pruski, *J. Magn. Reson.*, 2009, **196**, 92–95.
20 T. Joseph, S. S. Deshpande, S. B. Halligudi, A. Vinu, S. Ernst and M. Hartmann, *Journal of Molecular Catalysis a-Chemical*. 2003, **206**, 13–21.
21 H. Z. Li, F., H. Yin, Y. Wang and Y. Lu, *Green Chem*, 2005, **9**.
22 L. Huang, J. C. Wu and S. Kawi, *React. Kinet. Catal. Lett.*, 2004, **82**, 65–71.
23 L. Huang and S. Kawi, *Catalysis Letters*, 2003, **90**, 165–169.
24 L. Huang and S. Kawi, *Catalysis Letters*, 2004, **92**, 57–62.
25 V. Dufaud, F. Beauchesne and L. Bonneviot, *Angew. Chem., Int. Ed.*, 2005, **44**, 3475–3477.
26 E. Duliere, M. Devillers and J. Marchand-Brynaert, *Organometallics*, 2003, **22**, 804–811.
27 S. Hermes, M. K. Schröter, R. Schmid, L. Khodeir, M. Muhler, A. Tissler, R. W. Fischer and R. A. Fischer, *Angewandte Chemie International Edition*, 2005, **44**, 6237–6241.
28 F. Hoffmann, M. Cornelius, J. Morell and M. Fröba, *Angewandte Chemie International Edition*, 2006, **45**, 3216–3251.
29 J. Lu and P. H. Toy, *Chem Rev*, 2009, **109**, 815–838.
30 B. M. L. Dioos, I. F. J. Vankelecom and P. A. Jacobs, *Adv Synth Catal*, 2006, **348**, 1413–1446.
31 M. Ohtaki, M. Komiyama, H. Hirai and N. Toshima, *Macromolecules*, 1991, **24**, 5567–5572.
32 J. Bluemel, *Coord.Chem.Rev.*, 2008, **252**, 2410–2423.
33 B. F. G. Johnson, S. A. Raynor, D. S. Shephard, T. Mashmeyer, J. M. Thomas, G. Sankar, S. Bromley, R. Oldroyd, L. Gladden and M. D. Mantle, *Chem. Comm.*, 1999, 1167–1168.
34 P. Sutra and D. Brunel, *Chem. Comm.*, 1996, 2485–2486.
35 D. Brunel, F. Fajula, J. B. Nagy, B. Deroide, M. J. Verhoef, L. Veum, J. A. Peters and H. van Bekkum, *Applied Catalysis a-General*, 2001, **213**, 73–82.

36 D. Y. Zhao, Q. S. Huo, J. L. Feng, B. F. Chmelka and G. D. Stucky, *J. Am. Chem. Soc.*, 1998, **120**, 6024–6036.
37 D. Y. Zhao, J. L. Feng, Q. S. Huo, N. Melosh, G. H. Fredrickson, B. F. Chmelka and G. D. Stucky, *Science*, 1998, **279**, 548–552.
38 J. S. Beck, J. C. Vartuli, W. J. Roth, M. E. Leonowicz, C. T. Kresge, K. D. Schmitt, C. T. W. Chu, D. H. Olson, E. W. Sheppard, S. B. Mccullen, J. B. Higgins and J. L. Schlenker, *J. Am. Chem. Soc.*, 1992, **114**, 10834–10843.
39 G. Buntkowsky, H. Breitzke, A. Adamezyk, F. Roelofs, T. Emmler, E. Gedat, B. Grunberg, Y. Xu, H. H. Limbach, J. Shenderovich, A. Vyalikh and G. Findenegg, *Phys Chem Chem Phys*, 2007, **9**, 4843–4853.
40 C. Coperet and J. M. Basset, *Adv Synth Catal*, 2007, **349**, 78–92.
41 C. Coperet, M. Chabanas, R. P. Saint-Arroman and J. M. Basset, *Angew. Chem., Int. Ed.*, 2003, **42**, 156–181.
42 F. Lefebvre and J. M. Basset, *Journal of Molecular Catalysis a-Chemical*, 1999, **146**, 3–12.
43 F. Lefebvre, P. Putaj, J. M. Basset, X. X. Wang and X. Z. Fu, *Science China-Chemistry*, 2010, **53**, 1862–1869.
44 S. W. Song, K. Hidajat and S. Kawi, *Langmuir*, 2005, **21**, 9568–9575.
45 H. X. Li, F. Zhang, H. Yin, Y. Wan and Y. F. Lu, *Green Chem*, 2007, **9**, 500–505.
46 J. Bluemel, R. Fetouaki, M. Bogza and C. Merckle, *Abstracts of Papers of the American Chemical Society*, 2004, **227**, U1443–U1444.
47 K. Bogar, P. Krumlinde, Z. Bacsik, N. Hedin and J. E. Backvall, *European Journal of Organic Chemistry*, 2011, 4409–4414.
48 M. Chabanas, C. Coperet and J. M. Basset, *Chem-Eur J*, 2003, **9**, 971–975.
49 A. Fihri, M. Bouhrara, B. Nekoueishahraki, J. M. Basset and V. Polshettiwar, *Chemical Society Reviews*, 2011, **40**, 5181–5203.
50 J. Guzman and B. C. Gates, *Dalton T*, 2003, 3303–3318.
51 J. W. Niemantsverdriet, *Spectroscopy in Catalysis*, Wiley-VCH, 2000.
52 J. Trebosc, J. W. Wiench, S. Huh, V. S. Y. Lin and M. Pruski, *J. Am. Chem. Soc.*, 2005, **127**, 3057–3068.
53 J. Trebosc, J. W. Wiench, S. Huh, V. S. Y. Lin and M. Pruski, *J. Am. Chem. Soc.*, 2005, **127**, 7587–7593.
54 F. Stallmach, A. Graser, J. Kärger, C. Krause, M. Jeschke, U. Oberhagemann and S. Spange, *Microporous Mesoporous Mater.*, 2001, **44**, 745–753.
55 E. Gedat, A. Schreiber, G. Findenegg, I. Shenderovich, H.-H. Limbach and G. Buntkowsky, *Magn.Res.Chem.*, 2001, **39**, 149.
56 L. Wasyluk, B. Peplinska, J. Klinowski and S. Jurga, *Phys Chem Chem Phys*, 2002, **4**, 2392–2397.
57 E. Gedat, A. Schreiber, J. Albrecht, I. Shenderovich, G. Findenegg, H.-H. Limbach and G. Buntkowsky, *J.Phys.Chem.B.*, 2002, **106**, 1977.
58 M. D. Jones, M. J. Duer, S. Hermans, Y. Z. Khimyak, B. F. G. Johnson and J. M. Thomas, *Angew. Chem., Int. Ed.*, 2002, **41**, 4726–4729.
59 I. Shenderovich, G. Buntkowsky, A. Schreiber, E. Gedat, S. Sharif, J. Albrecht, N. S. Golubev, G. H. Findenegg and H. H. Limbach, *J. Phys. Chem. B*, 2003, **107**, 11924–11939.
60 F. Babonneau, L. Yeung, N. Steunou, C. Gervais, A. Ramila and M. Vallet-Regi, *J Sol-Gel Sci. Techn.*, 2004, **31**, 219–223.
61 B. Grünberg, T. Emmler, E. Gedat, I. Shenderovich, G. H. Findenegg, H. H. Limbach and G. Buntkowsky, *Chemistry*, 2004, **10**, 5689–5696.
62 Y. Komori and S. Hayashi, *Microporous Mesoporous Mater.*, 2004, **68**, 111–118.
63 W. Masierak, T. Emmler, E. Gedat, A. Schreiber, G. H. Findenegg and G. Buntkowsky, *J.Phys.Chem. B*, 2004, 18890–18896.

64 S. Pizzanelli, S. Kababya, V. Frydman, M. Landau and S. Vega, *J. Phys. Chem. B*, 2005, **109**, 8029–8039.
65 T. Azais, C. Tourne-Peteilh, F. Aussenac, N. Baccile, C. Coelho, J. Devoisselle and F. Babonneau, *Chem. Mater.*, 2006, **18**, 6382–6390.
66 A. Vyalikh, T. Emmler, E. Gedat, I. Shenderovich, G. H. Findenegg, H.-H. Limbach and G. Buntkowsky, *Solid State NMR*, 2005, **28**, 117–124.
67 A. Vyalikh, T. Emmler, I. Shenderovich, Y. Zeng, G. H. Findenegg and G. Buntkowsky, *Phys. Chem. Chem. Phys.*, 2007, **9**, 2249–2257.
68 K. Schmidt-Rohr and H. W. Spiess, *Multidimensional Solid State NMR and Polymers*, Academic Press, London, 1994.
69 R. Bohmer, G. Diezemann, G. Hinze and E. Rossler, *Prog. Nucl. Magn. Reson. Spectrosc.*, 2001, **39**, 191–267.
70 M. Vogel, *Eur.Phys.J. Special Topics*, 2010, **189**, 47–64.
71 E. Lippmaa, M. Magi, A. Samoson, G. Engelhardt and A. R. Grimmer, *J. Am. Chem. Soc.*, 1980, **102**, 4889–4893.
72 G. Engelhardt and D. Michel, *"Hig-Resolution Solid-State NMR of Silcates and Zeolites"*, John Wiley and Sons Ltd., 1987.
73 B. Pfleiderer, K. Albert, E. Bayer, L. Vandeven, J. Dehaan and C. Cramers, *J. Phys. Chem.*, 1990, **94**, 4189–4194.
74 K. L. Walther, A. Wokaun and A. Baiker, *Mol. Phys.*, 1990, **71**, 769–780.
75 A. Tuel, H. Hommel, A. P. Legrand, Y. Chevallier and J. C. Morawski, *Colloids Surf.*, 1990, **45**, 413–426.
76 S. Leonardelli, L. Facchini, C. Fretigny, P. Tougne and A. P. Legrand, *J. Am. Chem. Soc.*, 1992, **114**, 6412–6418.
77 J. Klinowski, J. M. Thomas, C. A. Fyfe and J. S. Hartman, *J. Phys. Chem.*, 1981, **85**, 2590–2594.
78 J. Klinowski, S. Ramdas and H. Hamdan, *Abstracts of Papers of the American Chemical Society*, 1987, **194**, 309-INOR.
79 J. Klinowski, T. A. Carpenter and J. M. Thomas, *J Chem Soc Chem Comm*, 1986, 956–958.
80 J. Klinowski and M. W. Anderson, *Magn. Reson. Chem.*, 1990, **28**, S68–S81.
81 J. Klinowski, *Anal. Chim. Acta*, 1993, **283**, 929–965.
82 J. Klinowski, *Chem Rev*, 1991, **91**, 1459–1479.
83 J. Klinowski, *Colloids Surf.*, 1989, **36**, 133–154.
84 J. Klinowski, *Annual Review of Materials Science*, 1988, **18**, 189–218.
85 C. A. Fyfe, J. M. Thomas, J. Klinowski and G. C. Gobbi, *Angewandte Chemie-International Edition in English*, 1983, **22**, 259–275.
86 C. A. Fyfe, G. C. Gobbi, J. S. Hartman, J. Klinowski and J. M. Thomas, *J. Phys. Chem.*, 1982, **86**, 1247–1250.
87 J. M. Thomas and R. Raja, *Top Catal*, 2010, **53**, 848–858.
88 M. T. Reetz and E. Westermann, *Angew. Chem., Int. Ed.*, 2000, **39**, 165–168.
89 A. Roucoux, J. Schulz and H. Patin, *Chem Rev*, 2002, **102**, 3757–3778.
90 A. Roucoux and K. Philippot, in *Handbook of Homogenous Hydrogenations*, ed. G. d. Vries, WILEY-VCH, Weinheim, 2007, pp. 217–256.
91 K. Christmann, in *Hydrogen Transfer Reactions*, ed. J. T. K. Hynes, J. Limbach, H. H. Schowen, R. L., Wiley-VCH Germany Weinheim, 1997.
92 P. Feulner and D. Menzel, *Surf. Sci.*, 1985, **154**, 465–488.
93 M. Y. Chou and J. R. Chelikowsky, *Phys. Rev. Lett.*, 1987, **59**, 1737–1740.
94 M. Lindroos, H. Pfnur, P. Feulner and D. Menzel, *Surf. Sci.*, 1987, **180**, 237–251.
95 M. Sokolowski, T. Koch and H. Pfnur, *Surf. Sci.*, 1991, **243**, 261–272.
96 P. J. Feibelman, J. E. Houston, H. L. Davis and D. G. Oneill, *Surf. Sci.*, 1994, **302**, 81–92.

97 G. Lauth, E. Schwarz and K. Christmann, *J. Chem. Phys.*, 1989, **91**, 3729–3743.
98 R. Doell, L. Hammer, K. Heinz, K. Bedurftig, U. Muschiol, K. Christmann, A. P. Seitsonen, H. Bludau and H. Over, *J. Chem. Phys.*, 1998, **108**, 8671–8679.
99 L. A. Truflandier, I. Del Rosal, B. Chaudret, R. Poteau and I. C. Gerber, *Chemphyschem*, 2009, **10**, 2939–2942.
100 P. K. Schmidt, K. Christmann, G. Kresse, J. Hafner, M. Lischka and A. Gross, *Phys. Rev. Lett.*, 2001, **8709**.
101 T. Pery, K. Pelzer, G. Buntkowsky, K. Philippot, H. H. Limbach and B. Chaudret, *Chemphyschem*, 2005, **6**, 605–607.
102 J. Garcia-Anton, M. R. Axet, S. Jansat, K. Philippot, B. Chaudret, T. Pery, G. Buntkowsky and H. H. Limbach, *Angew. Chem., Int. Ed.*, 2008, **47**, 2074–2078.
103 I. del Rosal, T. Gutmann, L. Maron, F. Jolibois, B. Chaudret, B. Walaszek, H. H. Limbach, R. Poteau and G. Buntkowsky, *Phys Chem Chem Phys*, 2009, **11**, 5657–5663.
104 M. Mirzaei and N. L. Hadipour, *Chem. Phys. Lett.*, 2007, **438**, 304–307.
105 M. Mirzaei and N. L. Hadipour, *J. Comput. Chem.*, 2008, **29**, 832–838.
106 B. Walaszek, A. Adamczyk, T. Pery, Y. P. Xu, T. Gutmann, N. D. Amadeu, S. Ulrich, H. Breitzke, H. M. Vieth, S. Sabo-Etienne, B. Chaudret, H. H. Limbach and G. Buntkowsky, *J. Am. Chem. Soc.*, 2008, **130**, 17502–17508.
107 A. Wong, R. Ida, X. Mo, Z. H. Gan, J. Poh and G. Wu, *J. Phys. Chem. A*, 2006, **110**, 10084–10090.
108 J. Autschbach, in *Density Functional Theory in Inorganic Chemistry*, eds. N. Kaltsoyannis and J. E. McGrady, Springer Verlag, Berlin/Heidelberg, 2004, vol. 112, pp. 1–48.
109 H. Behzadi, N. L. Hadipour and M. Mirzaei, *Biophysical Chemistry*, 2007, **125**, 179–183.
110 I. del Rosal, L. Maron, R. Poteau and F. Jolibois, *Dalton T*, 2008, 3959–3970.
111 M. Pavanello, B. Mennucci and J. Tomasi, *Theoretical Chemistry Accounts*, 2006, **116**, 711–717.
112 L. Truflandier, M. Paris, C. Payen and F. Boucher, *J. Phys. Chem. B*, 2006, **110**, 21403–21407.
113 A. Abragam, *Principles of Nuclear Magnetism*, Clarendon Press, Oxford, 1961.
114 C. P. Slichter, *Principles of Magnetic Resonance 3rd Edition*, Springer Verlag Berlin, Heidelberg, New York, 1990.
115 R. Ernst, G. Bodenhausen and A. Wokaun, *Principles of NMR in One and Two Dimensions*, Clarendon Press, Oxford, 1987.
116 P. Mansfield and P. G. Morris, *NMR Imaging in Biomedicine*, Academic Press, New York, 1982.
117 P. T. Callaghan, *Principles of Nuclear Magnetic Resonance Microscopy*, Clarendon Press, Oxford, 1991.
118 R. Kimmich, *NMR Tomography Diffusometry Relaxometry*, Springer, Berlin, 1997.
119 M. Mehring, *High Resolution NMR Spectroscopy in Solids*, Springer Verlag, Berlin, Heidelberg, New York, 1983.
120 M. Geppi, S. Borsacchi, G. Mollica and C. A. Veracini, *Applied Spectroscopy Reviews*, 2008, **44**, 1–89.
121 A. Pines, M. G. Gibby and J. Waugh, *Chem. Phys. Lett.*, 1972, **15**, 373.
122 E. R. Andrew, A. Bradbury and R. G. Eades, *Nature*, 1958, **182**, 1659.
123 I. J. Lowe, *Phys. Rev. Lett.*, 1959, **2**, 285.
124 J. Schaefer and E. O. Stejskal, *J. Am. Chem. Soc.*, 1976, **98**, 1031.

125 J. Schaefer, E. O. Stejskal and R. Buchdahl, *Macromolecules*, 1975, **8**, 291.
126 A. E. Bennett, R. G. Griffin and S. Vega, *Springer Series in NMR*, 1994, **33**, 1.
127 T. Gullion and J. Schaefer, *J.Magn. Res.*, 1989, **81**, 196.
128 T. Gullion and J. Schaefer, *Adv. in Magn. and Opt. Res. Ed.W.S. Warren*, 1989, 13 , 57.
129 A. Bax, N. M. Szeverenyi and G. E. Maciel, *J.Magn.Res.*, 1983, **55**, 494.
130 U. Haeberlen, *High Resolution NMR in Solids. Selective Averaging*, Academic Press, New York, 1976.
131 B. Bechinger and S. J. Opella, *J.Magn. Res.*, 1991, **95**, 585.
132 P. McDonnel, K. Shon, Y. Kim and S. J. Opella, *J.Mol.Biol.*, 1993, **233**, 447.
133 A. Samoson, E. Lippmaa and A. Pines, *Mol. Phys.*, 1988, **65**, 1013–1018.
134 A. Llor and J. Virlet, *Chem. Phys. Lett.*, 1988, **152**, 248–253.
135 A. Medek, J. S. Harwood and L. Frydman, *J. Am. Chem. Soc.*, 1995, **117**, 12779–12787.
136 L. Frydman and J. S. Harwood, *J. Am. Chem. Soc.*, 1995, **117**, 5367–5368.
137 Z. Gan, *J. Am. Chem. Soc.*, 2000, **122**, 3242–3243.
138 S. E. Ashbrook and S. Wimperis, *Prog. Nucl. Magn. Reson. Spectrosc.*, 2004, **45**, 53–108.
139 M. Edén, *Concepts in Magnetic Resonance*, 2003, **17**, 117–154.
140 M. Bak, J. T. Rasmussen and N. C. Nielsen, *J. Magn. Res.*, 2000, **147**, 296.
141 M. Veshtort and R. G. Griffin, *J. Magn. Reson.*, 2006, **178**, 248–282.
142 G. Wu and R. E. Wasylishen, *Inorg. Chem.*, 1994, **33**, 2774–2778.
143 I. M. ElNahhal, J. J. Yang, I. S. Chuang and G. E. Maciel, *J. Non-Cryst. Solids*, 1996, **208**, 105–118.
144 G. E. Maciel and D. W. Sindorf, *J. Am. Chem. Soc.*, 1980, **102**, 7606–7607.
145 D. W. Sindorf and G. E. Maciel, *J. Phys. Chem.*, 1983, **87**, 5516–5521.
146 G. E. Maciel and T. Tao, *Abstracts of Papers of the American Chemical Society*, 2000, **219**, U321–U321.
147 E. Lindner, T. Schneller, F. Auer and H. A. Mayer, *Angew. Chem., Int. Ed.*, 1999, **38**, 2155–2174.
148 J. A. Osborn, F. H. Jardine, J. F. Young and G. J. Wilkinson, *J. Chem. Soc. A*, 1966, 1711–1732.
149 V. Dufaud, G. P. Niccolai, J. Thivollecazat and J. M. Basset, *J. Am. Chem. Soc.*, 1995, **117**, 4288–4294.
150 C. Merckle and J. Blumel, *Adv Synth Catal*, 2003, **345**, 584–588.
151 E. Lindner, B. Keppeler, H. A. Mayer, K. Gierling, R. Fawzi and M. Steimann, *J. Organomet. Chem.*, 1996, **526**, 175–183.
152 A. Adamczyk, Y. Xu, B. Walaszek, F. Roelofs, T. Pery, K. Pelzer, K. Philippot, B. Chaudret, H. H. Limbach, H. Breitzke and G. Buntkowsky, *Top Catal*, 2008, **48**, 75–83.
153 A. S. M. Chong, X. S. Zhao, A. T. Kustedjo and S. Z. Qiao, *Microporous Mesoporous Mater.*, 2004, **72**, 33–42.
154 A. S. M. Chong and X. S. Zhao, *J. Phys. Chem. B*, 2003, **107**, 12650–12657.
155 A. M. Liu, K. Hidajat, S. Kawi and D. Y. Zhao, *Chem. Comm.*, 2000, 1145–1146.
156 T. Tao and G. E. Maciel, *J. Am. Chem. Soc.*, 2000, **122**, 3118–3126.
157 J. J. Yang, I. M. ElNahhal, I. S. Chuang and G. E. Maciel, *J. Non-Cryst. Solids*, 1997, **209**, 19–39.
158 J. J. Yang, I. M. ElNahhal and G. E. Maciel, *J. Non-Cryst. Solids*, 1996, **204**, 105–117.
159 G. Wu and R. E. Wasylishen, *Organometallics*, 1992, **11**, 3242–3248.
160 L. Bemi, H. C. Clark, J. A. Davies, C. A. Fyfe and R. E. Wasylishen, *J. Am. Chem. Soc.*, 1982, **104**, 438–445.

161 L. Bemi, H. C. Clark, J. A. Davies, D. Drexler, C. A. Fyfe and R. Wasylishen, *J. Organomet. Chem.*, 1982, **224**, C5–C9.
162 R. A. Komoroski, A. J. Magistro and P. P. Nicholas, *Inorg. Chem.*, 1986, **25**, 3917–3925.
163 B. Beele, J. Guenther, M. Perera, M. Stach, T. Oeser and J. Blumel, *New J. Chem.*, 2010, **34**, 2729–2731.
164 R. Fetouaki, A. Seifert, M. Bogza, T. Oeser and J. Blumel, *Inorg. Chim. Acta*, 2006, **359**, 4865–4873.
165 T. Posset, F. Rominger and J. Blumel, *Chem. Mater.*, 2005, **17**, 586–595.
166 F. Piestert, R. Fetouaki, M. Bogza, T. Oeser and J. Blumel, *Chem. Comm.*, 2005, 1481–1483.
167 M. Bogza, T. Oeser and J. Blumel, *J. Organomet. Chem.*, 2005, **690**, 3383–3389.
168 G. Tsiavaliaris, S. Haubrich, C. Merckle and J. Blumel, *Synlett*, 2001, 391–393.
169 T. Posset, J. Guenther, J. Pope, T. Oeser and J. Blumel, *Chem. Comm.*, 2011, **47**, 2059–2061.
170 J. Guenther, J. Reibenspies and J. Blumel, *Adv Synth Catal*, 2011, **353**, 443–460.
171 A. Gruenberg, X. Yeping, H. Breitzke and G. Buntkowsky, *Chem-Eur J*, 2010, **16**, 6993.
172 G. Wu and R. E. Wasylishen, *Inorg. Chem.*, 1992, **31**, 145–148.
173 R. H. Crabtree, *The Organometallic Chemistry of the Transition Metals*, John Wiley & Sons, 2005.
174 J. W. Diesveld, E. M. Menger, H. T. Edzes and W. S. Veeman, *J. Am. Chem. Soc.*, 1980, **102**, 7935–7936.
175 F. Blanc, J. M. Basset, C. Coperet, A. Sinha, Z. J. Tonzetich, R. R. Schrock, X. Solans-Monfort, E. Clot, O. Eisenstein, A. Lesage and L. Emsley, *J. Am. Chem. Soc.*, 2008, **130**, 5886–5900.
176 S. R. Petroff, A. de Mallmann, F. Lefebvre and J. M. Basset, *Abstracts of Papers of the American Chemical Society*, 2002, **224**, U271–U272.
177 H. Provendier, C. C. Santini, J. M. Basset and L. Carmona, *European Journal of Inorganic Chemistry*, 2003, 2139–2144.
178 N. Rendon, R. Berthoud, F. Blanc, D. Gajan, T. Maishal, J. M. Basset, C. Coperet, A. Lesage, L. Emsley, S. C. Marinescu, R. Singh and R. R. Schrock, *Chem-Eur J*, 2009, **15**, 5083–5089.
179 G. Saggio, A. de Mallmann, B. Maunders, M. Taoufik, J. Thivolle-Cazat and J. M. Basset, *Organometallics*, 2002, **21**, 5167–5171.
180 X. X. Wang, L. Veyre, F. Lefebvre, J. Patarin and J. M. Basset, *Microporous Mesoporous Mater.*, 2003, **66**, 169–179.
181 X. X. Wang, H. B. Xu, X. Z. Fu, P. Liu, F. Lefebvre and J. M. Basset, *Journal of Molecular Catalysis a-Chemical*, 2005, **238**, 185–191.
182 G. Wu and R. E. Wasylishen, *J. Chem. Phys.*, 1993, **98**, 6138–6149.
183 G. Wu and R. E. Wasylishen, *J. Chem.Phys.*, 1994, **100**, 5546.
184 G. Buntkowsky, H. H. Limbach, F. Wehrmann, I. Sack, H. M. Vieth and R. H. Morris, *J. Phys. Chem. A*, 1997, **101**, 4679–4689.
185 F. Wehrmann, T. P. Fong, R. H. Morris, H. H. Limbach and G. Buntkowsky, *Phys Chem Chem Phys*, 1999, **1**, 4033–4041.
186 G. Buntkowsky, B. Walaszek, A. Adamczyk, Y. Xu, H. H. Limbach and B. Chaudret, *Phys Chem Chem Phys*, 2006, **8**, 1929–1935.
187 F. Schroeder, D. Esken, M. Cokoja, M. W. E. van den Berg, O. I. Lebedev, G. van Tendeloo, B. Walaszek, G. Buntkowsky, H. H. Limbach, B. Chaudret and R. A. Fischer, *J. Am. Chem. Soc.*, 2008, **130**, 6119–6130.

188 S. Macholl, J. Matthes, H. H. Limbach, S. Sabo-Etienne, B. Chaudret and G. Buntkowsky, *Solid State Nucl Mag*, 2009, **36**, 137–143.
189 B. Walaszek, Y. Xu, A Adamczyk, H. Breitzke, K Pelzer, H.-H. Limbach, J. Huang, H. Li and G. Buntkowsky, *Solid State NMR*, 2009, **35**, 164–171.
190 T. Gutmann, B. Walaszek, Y. Xu, M. Wächtler, I. d. Rosal, A. Grünberg, R. Poteau, R. Axet, G. Lavigne, B. Chaudret, H.-H. Limbach and G. Buntkowsky, *J. Am. Chem. Soc.*, 2010, **132**, 11759–11767.
191 I. del Rosal, T. Gutmann, B. Walaszek, I. Gerber, B. Chaudret, H.-H. Limbach, G. Buntkowsky and R. Poteau, *Phys Chem Chem Phys*, 2011, **13**, 20199–20207.
192 B. Gerstein, R. Pembleton, R. Wilson and L. Ryan, *The Journal of Chemical Physics*, 1977, **66**, 361.
193 L. S. Du, A. Samoson, T. Tuherm and P. Clare, *Chem. Mater.*, 2000, **12**, 3611–3616.
194 A. Samoson, ed., *Extended Magic-angle Spinning*, 2002.
195 M. Ernst, A. Detken, A. Böckmann and B. H. Meier, *J. Am. Chem. Soc.*, 2003, **125**, 15807–15810.
196 M. Ernst, M. A. Meier, T. Tuherm, A. Samoson and B. H. Meier, *J. Am. Chem. Soc.*, 2004, **126**, 4764–4765.
197 Y. Ishii and R. Tycko, *J. Magn. Reson.*, 2000, **142**, 199–204.
198 J. W. Wiench, C. E. Bronnimann, S. Y. L. Victor and M. Pruski, *J. Am. Chem. Soc.*, 2007, **129**, 12076–12077.
199 K. Mao and M. Pruski, *J. Magn. Reson.*, 2009, **201**, 165–174.
200 T. M. Hsin, S. Chen, E. Guo, C. H. Tsai, M. Pruski and V. S. Y. Lin, *Top Catal*, 2010, **53**, 746–754.
201 K. Mao, T. Kobayashi, J. W. Wiench, H. T. Chen, C. H. Tsai, V. S. Y. Lin and M. Pruski, *J. Am. Chem. Soc.*, 2010.
202 T. Kobayashi, K. Mao, S. G. Wang, V. S. Y. Lin and M. Pruski, *Solid State Nucl. Mag.*, 2011.
203 S. Nedd, T. Kobayashi, C. H. Tsai, I. I. Slowing, M. Pruski and M. S. Gordon, *J. Phys. Chem. C*, 2011.
204 A. W. Overhauser, *Phys. Rev.*, 1953, **92**, 411.
205 T. P. Carver and C. P. Slichter, *Phys. Rev.*, 1953, **92**, 212.
206 T. P. Carver and C. P. Slichter, *Phys. Rev.*, 1956, **102**, 975.
207 A. Abragam and W. G. Proctor, *C.R. Acad. Sci. Paris*, 1958, **246**, 2253.
208 K. H. Hausser and D. Stehlik, *Advances in Magnetic Resonance*, 1968, **3**, 79.
209 K. L. Felch, B. G. Danly, H. R. Jory, K. E. Kreischer, W. Lawson, B. Levush and R. J. Temkin, *Proc. IEEE*, 1999, **87**, 752.
210 L. R. Becerra, G. J. Gerfen, R. J. Temkin, D. J. Singel and R. G. Griffin, *Phys. Rev. Lett.*, 1993, **71**, 3561.
211 T. Maly, G. T. Debelouchina, V. S. Bajaj, K. Hu, C. Joo, M. L. Mak-Jurkauskas, J. R. Sirigiri, P. C. A. v. d. Wel, J. Herzfeld, R. J. Temkin and R. G. Griffin, *J. Chem. Phys.*, 2008, **128**, 052211.
212 C. Song, K. N. Hu, C. G. Joo, T. M. Swager and R. G. Griffin, *J. Am. Chem. Soc.*, 2006, **128**, 11385–11390.
213 A. Lesage, M. Lelli, D. Gajan, M. A. Caporini, V. Vitzthum, P. Mieville, J. Alauzun, A. Roussey, C. Thieuleux, A. Mehdi, G. Bodenhausen, C. Coperet and L. Emsley, *J. Am. Chem. Soc.*, 2010, **132**, 15459–15461.
214 M. Lelli, D. Gajan, A. Lesage, M. A. Caporini, V. Vitzthum, P. Mieville, F. Heroguel, F. Rascon, A. Roussey, C. Thieuleux, M. Boualleg, L. Veyre, G. Bodenhausen, C. Coperet and L. Emsley, *J. Am. Chem. Soc.*, 2011, 86–121.
215 A. Farkas, *Orthohydrogen, Parahydrogen and Heavy Hydrogen*, Cambridge Univ. Press, Cambridge, 1935.

216 C. R. Bowers and D. P. Weitekamp, *Phys. Rev. Lett.*, 1986, **57**, 2645.
217 T. C. Eisenschmidt, R. U. Kirss, P. P. Deutsch, S. I. Hommeltoft, R. Eisenberg and J. Bargon, *J. Am. Chem. Soc.*, 1987, **109**, 8089.
218 A. Rozovskii and G. Lin, *Kinetics and Catalysis*, 1999, **40**, 773–794.
219 J. Martins, C. Taft, S. Lie and E. Longo, *Journal of Molecular Structure-Theochem*, 2000, **528**, 161–170.
220 P. Carson, C. Bowers and D. Weitekamp, *J. Am. Chem. Soc.*, 2001, **123**, 11821–11822.
221 J. Brossel and A. C. R. Kastler, *Hebd. Seances Acad. Sci.*, 1949, **229**, 1213.
222 W. Happer, E. Miron, S. Schaefer, W. A. Schreiber, W. A. v. Wijngaarden and X. Zeng, *Phys. Rev. A*, 1984, **29**, 3092.
223 C. R. Bowers, T. Pietrass, E. Barash, A. Pines and R. K. Grubbs, *J. Phys. Chem.*, 1994, **98**, 9400.
224 c. J. Jameson, A. K. Jameson and S. M. Cohen, *J. Phys. Chem.*, 1973, **59**, 4540.
225 T. Pietrass, A. Bifone and A. Pines, *Surf. Sci.*, 1995, **334**, L730.
226 R. J. Fitzgerald, K. L. Sauer and W. Happer, *Chem. Phys. Lett.*, 1998, **284**, 87.
227 G. Navon, Y. Q. Song, T. Room, S. T. R. E. Appelt and A. Pines, *Science*, 1996, **271**, 1848.
228 L. J. Smith, J. Smith, E. MacNamara, K. Knagge and D. Raftery, *J. Phys. Chem. B*, 2001, **105**, 1412.
229 E. MacNamara, C. V. Rice, J. Smith, L. J. Smith and D. Raftery, *Chem. Phys. Lett.*, 2000, **317**, 165.
230 E. MacNamara, G. Fischer, J. Smith, C. V. Rice, S. J. Hwang and D. Raftery, *J. Phys. Chem. B*, 1999, **103**, 1158.
231 J. Smith, K. Knagge, E. MacNamara and D. Raftery, *J. Magn. Res.*, 2002, **159**, 111.
232 A. Jerschow, *Prog. Nucl. Magn. Reson. Spectrosc.*, 2005, **46**, 63–78.